DIGITAL DESIGN

Basic Concepts and Principles

Mohammad A. Karim
Xinghao Chen

CRC Press
Taylor & Francis Group
Boca Raton London New York

CRC Press is an imprint of the
Taylor & Francis Group, an **informa** business

CRC Press
Taylor & Francis Group
6000 Broken Sound Parkway NW, Suite 300
Boca Raton, FL 33487-2742

© 2008 by Taylor & Francis Group, LLC
CRC Press is an imprint of Taylor & Francis Group, an Informa business

10 9 8 7 6 5 4 3

International Standard Book Number-13: 978-1-4200-6131-4 (Hardcover)

Library of Congress Cataloging-in-Publication Data

Karim, Mohammad A.
 Digital design : basic concepts and principles / by Mohammad A. Karim and Xinghao Chen.
   p. cm.
 Includes bibliographical references and index.
 ISBN-13: 978-1-4200-6131-4
 ISBN-10: 1-4200-6131-3
 1. Digital electronics. 2. Logic design. I. Chen, Xinghao, 1957- II. Title.

TK7868.D5K365 2008
621.381--dc22   2007020768

Visit the Taylor & Francis Web site at
http://www.taylorandfrancis.com

and the CRC Press Web site at
http://www.crcpress.com

To the genius, place and time that mastered "zero" and a lot more, all who came thereafter to digitize the world, our students, and our families.

Contents

Preface

Much of modern digital design activities are realized using computer-aided design (CAD) applications, which provide a quick turn-around time with ready access to circuit synthesis and simulation applications. The benefit of using CAD is the increased productivity of design activities. One of the main features of this book is the integration of digital design principles with design practices using one of the industry's most popular design applications, the Xilinx *WebPACK*, which is provided to the public as a free-of-charge download at the Xilinx website. The book integrates many challenging issues critical to modern digital design practices, such as discussions on hazards and logic minimization in Chapters 3 and 11, finite-state-machine synthesis in Chapters 8 and 9, and testability issues in Chapter 12, along with classic theories of digital systems for which the book provides thorough treatment.

The contents of the book are suitable for a sophomore- or junior-level course on digital logic design. Advanced and exploratory topics are also included for advanced and interested students. There are plenty of illustrative examples on each subject, with many end-of-chapter problems, and more than 500 circuit drawings and 100 worked-out examples. Discussions on advanced topics in optical computing, neural networks, and sequential arithmetic processors are also sufficiently provided for advanced study.

Each chapter of this book starts with an introduction of the subject. Then related concepts and principles are discussed methodically in sufficient detail, accompanied by many demonstrative and illustrative examples and with the level of intended student population in mind. Each chapter also ends with a short summary and a list of exercise problems for students to solve and to help them master the subject.

Chapter 1 provides a comprehensive treatment of number systems, both conventional for current computing systems and unconventional for possible use in quantum and optical computing systems. It explains data representation, number system conversions, binary arithmetic and codes, as well as error detection concepts. As the use of CAD in digital design has become critical in practice, its importance is also described.

Chapter 2 discusses Boolean algebra and its algebraic operations. Boolean algebra has been used to script digital logic operations. It is a fundamental tool in analyzing digital designs as well. The process of extracting Boolean functions from truth tables is both explained and demonstrated.

Chapter 3 illustrates logic minimization techniques with the application of Karnaugh maps. Logic minimization is an important step in digital design to reduce design complexity; the Karnaugh map is a simple and straightforward technique exercised. Additionally, techniques for hazards analysis and single- and multioutput Q-M tabular reduction are also discussed.

Chapter 4 starts with simple implementation techniques using NAND and NOR, as well as XOR and XNOR, functions. It then discusses advanced implementation techniques, such as using multiplexers, ROM, and transmission gates, as these techniques are widely employed in modern digital implementation practices.

Chapter 5 introduces students to VHDL, a digital design language commonly used in the IC industry in conjunction with CAD applications. It explains some of the basic

VHDL components and statement syntax, with demonstrative examples in circuit synthesis, optimization, and simulation.

Chapters 6 and 7 discuss the design of various building blocks often used in digital systems. The building blocks include decoders and encoders, error-detection and error-correction circuits, arithmetic and logic units, code converters and comparators, as well as sequential elements, such as different types of latches and flip–flops.

Chapters 8, 9, and 10 are devoted to the design concepts and principles for finite-state-machine systems. Chapter 8 emphasizes models, state assignment techniques, and design algorithms while Chapter 9 emphasizes the design of various sequential building blocks, such as counters and registers. Chapter 10 introduces students to the design of sequential arithmetic circuits, which often result in simpler implementations.

Chapter 11 introduces students to asynchronous sequential circuits, which is a topic for advanced study. Although asynchronous digital systems are rather difficult to design, techniques for designing asynchronous digital components and modules have been known for several decades. This chapter discusses some of the concepts and techniques.

Chapter 12 introduces students to some of the basic concepts of testability with digital circuits and systems. Testability is a measure of the cost (or simplicity) of verifying digital designs and testing the implementations as designs go through development and manufacturing processes. A good digital design must be easy to verify and test with the lowest possible cost. Some of the basic concepts are discussed.

We are indebted to a large number of authors and their many books on this general subject, some of which are available in the market and many that are out of print. More books came and left the market during the last 40 years or so. Mohammad Karim was the coauthor (along with Everett L. Johnson then of Wichita State University) of such a PWS Publishers book, *Digital Design: A Pragmatic Approach*, published originally in 1987. Much of digital logic design is both well known and basic and those that are relevant and developed by him then for the earlier book have been retained. We attempted to incorporate in this book the many facets that changed dramatically since then in technology, technique, and the speed with which digital design is approached now. Also, possibilities in fields such as optical and quantum computing now require interested students to garner backgrounds in nonbinary digital systems.

As we started developing the current book, we originally thought of including extensive reference and reading lists for each chapter, not only to refer interested students to explore further but also to give credit to those who were instrumental in conceptualizing many of the design concepts and principles. Then we examined this from the intended level of students' perspectives and asked students for their input, and realized that most sophomore- and junior-level students do not use these lists but prefer to Google-search on the subjects if and when needed. We decided to include a handful of such references and further-reading lists, and acknowledge the contributions of the hundreds of researchers and practitioners in the field of digital design. We dedicate this book in honor of these pioneers, our many students past and present, and our families.

Acknowledgments

Contributions of a few individuals were of particular importance in bringing this project to a closure. Lutfi Karim of Raytheon Integrated Defense Systems provided a thorough review of VHDL and testability pertinent material. Shams Rashid, one of our former students who is a doctoral student at the State University of New York (SUNY) at Stony Brook prepared most of the figures. Lamya Karim, a doctoral student at Rensselaer Polytechnic Institute, went through much of the text and assisted in smoothing out language. Sheng (Alan) Yu, a doctoral student at City University of New York, reviewed and verified many of the VHDL design examples. Our publisher Nora Konopka, project coordinator Jessica Vakili, and project editor Richard Tressider, on behalf of CRC Press/Taylor and Francis Group, and project manager John Normansell, on behalf of Newgen Imaging Systems Ltd in India, had been our partners in this project. They were all wonderful in nudging us all forward as we went through various hoops.

Authors

Mohammad A. Karim is vice president for research at the Old Dominion University (ODU) in Norfolk, Virginia. He is the North American editor of *Optics and Laser Technology* and an associate editor of *IEEE Transactions on Education*, and has served as guest editor of 16 journal special issues. Before joining ODU, Karim served as dean of engineering at the City College of New York of the City University of New York, head of electrical and computer engineering at the University of Tennessee, and chair of electrical and computer engineering, and founding director of electro-optics at the University of Dayton in Ohio. Karim is author or coauthor of 8 text and reference books, over 325 research papers, 6 book chapters, 10 book reviews, 2 U.S. patents, and numerous technical reports. He is a fellow of the Optical Society of America, the Society of Photo Instrumentation Engineers, the Institute of Physics, the Institution of Engineering & Technology, and Bangladesh Academy of Sciences and a Distinguished Engineering Fellow of the University of Alabama.

Xinghao Chen is an associate professor in electrical and computer engineering at City College of the City University of New York (CCNY/CUNY). Before joining CCNY/CUNY, he was an advisory engineer/scientist at IBM Microelectronics Division (1995–2002) at its Endicott, New York, facility and as DARPA assistant research professor at the CAIP Research Center (1993–1995) at Rutgers University, Piscataway, New Jersey. He has authored and coauthored 2 books, 15 research papers, and numerous technical reports and disclosures. He holds two U.S. patents, with five more pending. He is a senior member of IEEE and a Golden Core Member of the IEEE Computer Society.

Authors

1

Data Type and Representations

1.1 Introduction

The vast majority of the devices and consumer instruments that we encounter in industries, work places, homes, or laboratories are electronic in nature. While some of these are analog in type many others are based on digital principles. Analog systems typically process information that is represented by continuous-time electrical signals. Digital systems, on the other hand, are characterized by discrete-time electrical signals. *Discrete* implies distinct or separate as opposed to *continuous*. The words digit and digital were coined to refer to counting discrete numbers. Digital systems are thus systems that process discrete information. Such systems receive input, process or control data manipulation, and output information in a discrete format. It is understood that analog devices and systems have played a significant role in the advancement of the electronics industry; however, most of the newer and more sophisticated electronic systems are turning out to be digital. This is in part because digital devices are inexpensive, reliable, and flexible. The ongoing development and continuous refinement of silicon technology, in particular, have opened up so many previously unthought-of application areas that we may hardly think of a man-made machine now that has not incorporated one or more digital components in it.

A *digital system* is often a combination of circuit components interconnected in a manner such that it is capable of performing a specific function when all its input variables are allowed only a finite number of discrete states. Two-valued discrete systems, in particular, are referred to as *binary systems*. Both the inputs and outputs in binary systems can assume either of two allowable discrete values. By combining such discrete states, one may represent numbers, characters, codes, and other pertinent information. There are several advantages that binary systems have over the corresponding *analog systems*. The electronic devices used in digital circuits are extremely reliable, inexpensive, and remarkably consistent in their performances so long as they are maintained in either of two logical states. Also, because binary circuits are maintained in either of two allowable states, they are much less susceptible to variations of environment and have tolerable accuracy.

Number systems provide the basis for quantifying information for operations in digital processing systems. The binary (two-valued) number system, in particular, serves as the most important basis for understanding digital systems since the electronic devices involved can assume only two output values. In this chapter, we shall study the binary number system, its relationship with other number systems, and then show how they can be represented in binary coded form. Many of these other number systems are of relevance to optical and quantum computing systems that are currently under development.

In Chapter 5, an introduction to computer-aided design (CAD) systems is provided so that it can be used by students. Before that, in Chapter 2, we shall describe interactions of various system variables in terms of logical operations. Logical operations describing the desired outputs of the to-be-designed system in terms of the inputs allow us to explore the various design strategies that can be employed for designing the system in question.

1.2   Positional Number Systems

The number system that is routinely used by us is the base-10 or decimal system. It uses *positional* number representation, which implies that the value of each digit in a multibit number is a function of its relative position in the number. The term *decimal* comes from the Latin word for "ten." Counting in decimal numbers and the numbers of fingers or toes are probably interrelated. We have become so used to this number system that we seldom consider the fact that even though we use only ten Arabic digits 0, 1, 2, 3, 4, 5, 6, 7, 8, and 9, we are able to represent numbers in excess of nine. To understand its relevance, consider expressing 1998, for example, in the roman number system. The roman equivalent of 1998 is given by MCMXCVIII requiring nine digits! The roman number XI, for example, represents decimal 11 as it is the sum of the values for the two component digits: X (symbol for 10) and I (the symbol for 1). The value for IX, however, is determined by subtracting the value of the digit on the left (i.e., 1) from that of the digit on the right (i.e., 10). Worse yet, the roman numbers just cannot represent a zero.

We can use positional number representation to also denote numbers that may include an integer part and a fractional part set apart by a decimal point. The value of the number is obtained by assigning weights to the positions of the respective digits. For each of the digits in a number, the digit and the corresponding weight are multiplied and then summed together to yield the value of the number. For instance, 2637.25 in the base-10 system can be written in polynomial notation as

$$(2637.25)_{10} = 2 \times 1000 + 6 \times 100 + 3 \times 10 + 7 \times 1 + 2 \times 0.1 + 5 \times 0.01$$

$$= (2 \times 10^3) + (6 \times 10^2) + (3 \times 10^1) + (7 \times 10^0) + (2 \times 10^{-1}) + (5 \times 10^{-2})$$

Each position in the decimal system, for example, is ten times more significant than the position on the right. This number is also referred to as a *fixed-point* decimal because the decimal point is in its proper place. Since operands and results in an arithmetic unit are often stored in registers of a fixed length, there are only a finite number of distinct values that can be represented with an arithmetic unit. Thus, the main difficulty of a fixed-point representation is that the range of numbers that could be represented within a finite size of digits is limited. Any arithmetic operation that attempts to produce a result that lies outside the range will always produce an incorrect result. In such cases, the arithmetic unit should typically indicate that the said result is an error. This is usually called an *overflow* condition.

An alternative representation of numbers is known as the *floating-point* format. It can overcome the range limitation problem otherwise associated with fixed-point decimals. In this representation, the radix point is not placed at a fixed place; instead, it floats around so that more digits can be assigned to the left or the right of the radix point. The number $(2637.25)_{10}$, for example, can be expressed as a floating number by changing the position of the decimal point and the value of the exponent. This number can be written also as $(2.63725 \times 10^3)_{10}$, or as $(26.3725 \times 10^2)_{10}$, or even as $(263725.0 \times 10^{-2})_{10}$. The correct location of the decimal point (as in a fixed-point number) is indicated by the location of the floating point and by the exponent of 10. If the exponent is $+m$, the floating point needs to be shifted to the right by m digits. On the other hand, a negative exponent would imply shifting the floating point to the left.

In general, a conventional number system is defined by the set of values that each digit can assume and by the mapping rule that defines the mapping between the sequences of the digits and their corresponding numerical values. The base (also referred to as *radix*) of a number system is the number of symbols that the system contains. As stated earlier, a sequence of digits in an arithmetic unit can represent a mixed number that has both a fractional part and an integer part. The number is partitioned into two sets: the integer part and the fractional part. In a fixed-point base-R number system, a number N is expressed as

$$(N)_R = (a_m a_{m-1} a_{m-2} a_{m-3} \ldots a_0 . a_{-1} a_{-2} \ldots a_{-n}) \qquad (1.1)$$

↑   ↑   ↑

integer part   radix   fraction part

point

where $0 \le a_i \le (R-1)_R$ such that there are $m+1$ integer digits and n fractional digits in the number. The integer digits are separated from the fractional digits by the *radix point*. Note that if there is no radix point, it is assumed to be to the right of the rightmost digit. The decimal equivalent number $(N)_{10}$ of a given base-R number system is obtained as follows:

$$(N)_{10} = a_m R^m + \cdots + a_0 R^0 + a_{-1} R^{-1} + \cdots + a_{-n} R^{-n} \qquad (1.2)$$

where the part consisting of the coefficients a_m through a_0 corresponds to the integer portion, and the part consisting of the coefficients a_{-1} through a_{-n} corresponds to the fractional part, and R^s is the weight assigned to the digit a_s in the s-th position. Except for possible leading and trailing zeros, the representation of a number in a positional number system is unique. Typically, the radix point is not stored in the arithmetic unit but it is understood to be in a fixed position between the n least significant digits and the $m+1$ most significant digits. The leftmost digit in such a number is known as the *most significant digit* (MSD) and the rightmost digit is referred to as the *least significant digit* (LSD). The conventional number systems are thus *positional*, *weighted*, and *nonredundant* number systems. In a nonredundant number system every number has a unique representation; in other words, no two different digit sequences will have identical numerical value.

In this section, we shall examine base-2 (binary) number system, in particular, and also study its interrelationships with base-8 (octal), -10 (decimal), and -16 (hexadecimal) number systems. R can take any integer value; however, we have restricted R to only four values (i.e., 2, 8, 10, and 16) since we intend to study electronic digital devices only. Table 1.1 shows corresponding numbers (1_{10} through 17_{10}) in these systems. It is obvious that if we were to use voltage response of a device for representing a number, it would be easiest to do so in the binary number system because then the system requires only two values. The corresponding binary number, however, may require a longer sequence of digits than those in other number systems. For example, 1998_{10}, which in roman number is represented by MCMXCVIII, can be expressed in binary system as 11111001110_2 but in hexadecimal number system only as $7CE_{16}$. Each binary digit is known as a *bit*. The large number of bits present in the binary representation presents no significant problem since in most digital systems most operations occur in parallel.

TABLE 1.1

Correspondence Table

$(N)_{10}$	$(N)_2$	$(N)_8$	$(N)_{16}$
0	0	0	0
1	1	1	1
2	10	2	2
3	11	3	3
4	100	4	4
5	101	5	5
6	110	6	6
7	111	7	7
8	1000	10	8
9	1001	11	9
10	1010	12	A
11	1011	13	B
12	1100	14	C
13	1101	15	D
14	1110	16	E
15	1111	17	F
16	10000	20	10
17	10001	21	11

The advantage of using the binary system is that there are many devices that manifest two stable states. Examples of such binary states include on–off, positive–negative, and high–low choices. One of the two stable states can be assigned the value *logic 1* while the other can be treated as *logic 0*. Bit values are often represented by voltage values. We define a certain range of voltages as logic 1 and another range of voltages as logic 0. A forbidden voltage range typically separates the two permissible ranges. The approximate nominal values used for many technologies are 5 V for a binary value of 1 and 0 V for a binary value of 0. Since there can be numbers represented in different systems, it can be confusing at times. It is thus prudent whenever we are working with a particular number system to use the radix as its subscript. Examples of a few binary numbers and their decimal equivalents are as follows:

$$10111_2 = 1 \times 16 + 1 \times 4 + 1 \times 2 + 1 \times 1 = 23_{10}$$

$$101.11111_2 = 1 \times 4 + 1 \times 1 + 1 \times 0.5 + 1 \times 0.25 + 1 \times 0.125 + 1 \times 0.0625 = 5.9375_{10}$$

$$1100.1001_2 = 1 \times 8 + 1 \times 4 + 1 \times 0.5 + 1 \times 0.0625 = 12.5625_{10}$$

Similar to what is known as a decimal point in decimal number system, the radix point in a binary number system is called a binary point.

Octal (radix 8) and hexadecimal (radix 16) numbers are used primarily as convenient shorthand representations for binary numbers because their radices are powers of 2. Converting a multibit number to either octal or hexadecimal (hex for short) equivalent thus reduces the need for lengthy indecipherable strings. The octal number system uses digits 0 through 7 of the decimal system. When a position digit exceeds the value represented by the radix, a carry into the next higher position is produced as shown in Table 1.1. The hexadecimal system, however, supplements the decimal digits 0 through 9 with letters A (equivalent to 10) through F (equivalent to 15). Both upper- and lowercase characters can be used for hex numerals, each representing a distinct value. Like all other positional number

TABLE 1.2

Bases and Characters

Radix	Base	Characters
2	Binary	0, 1
3	Ternary or trinary	0, 1, 2
...		...
8	Octal	0, 1, 2, 3, 4, 5, 6, 7
...	...	...
10	Decimal	0, 1, 2, 3, 4, 5, 6, 7, 8, 9
...	...	...
16	Hexadecimal	0, 1, 2, 3, 4, 5, 6, 7, 8, 9, A, B, C, D, E, F

systems, when the digit exceeds the radix value at any given digit position, a carry into the next higher-order digit position is produced. We could have used any other universally agreed upon six characters besides A through F. Here, characters A through F do not represent letters of the alphabet but hex numerals. Table 1.2 lists several more popular bases and the numerals used in their respective number systems.

Conversion between binary, octal, and hexadecimal numbers is straightforward because corresponding radix values are all powers of 2. One can extrapolate from Table 1.1 that the octal digits can be represented by a combination of three binary digits (000 through 111) while the hexadecimal digits can be represented by a combination of four binary digits (0000 through 1111). More about these and other conversions are explored in the next section.

1.3 Number System Conversions

In this section, we restrict our discussion to conversion of positive numbers only. The conversion of negative numbers is handled in a different way and will be discussed in a later section. Except when both radices are powers of the same number, one cannot convert a number representation in one radix into its equivalent representation in another radix by substituting digit(s) in one radix to their equivalents in this radix. In most cases, the conversion involves a complex process that requires arithmetic operations. Accordingly, we consider two different conversion processes: the conversion of a base-R number to its base-10 equivalent, and the conversion of a base-10 number to its base-R equivalent. Each of these conversion processes in turn includes two minor processes: conversion of the integer part and conversion of the fraction part.

1.3.1 Integer Conversion

The conversion of an m-digit, base-R integer to its equivalent base-10 integer is readily accomplished using Table 1.1 and Equation 1.2. Each of the base-R digits is first converted to its base-10 equivalent using Table 1.1 and then multiplied by the respective R^j value as in Equation 1.2, where j denotes the respective positional value of the integer between 0 and m. The products obtained thereby are then added to give the equivalent decimal integer.

The following two examples illustrate the mechanics of integer conversion:

Example 1.1

Convert 62523_8 to its decimal equivalent.

Solution

The positional values of 6, 2, 5, 2, and 3 are respectively 4, 3, 2, 1, and 0. Therefore, the equivalent decimal is represented as follows:

$$(6 \times 8^4) + (2 \times 8^3) + (5 \times 8^2) + (2 \times 8^1) + (3 \times 8^0)$$
$$= [(6 \times 4096) + (2 \times 512) + (5 \times 64) + (2 \times 8) + (3 \times 1)]_{10}$$
$$= (24576 + 01024 + 00320 + 00016 + 00003)_{10}$$
$$= (25939)_{10}$$

Example 1.2

Find base-10 equivalent of $(AC6B2)_{16}$.

Solution

From Table 1.1, we find that in base-10 system, A = 10, C = 12, and B = 11. Therefore,

$$(AC6B2)_{16} = [(10 \times 16^4) + (12 \times 16^3) + (6 \times 16^2) + (11 \times 16^1) + (2 \times 16^0)]_{10}$$
$$= (655360 + 049152 + 001536 + 000176 + 000002)_{10}$$
$$= (706226)_{10}$$

To convert a decimal integer to its equivalent base-R integer, each of the integer coefficients $(a_0, a_1, a_2, \ldots, a_m)$ of Equation 1.2 must first be determined. The integer part of Equation 1.2 represented by base-10 integer NI can be rewritten as

$$(NI)_{10} = a_m R^m + a_{m-1} R^{m-1} + \cdots + a_1 R^1 + a_0 R^0$$
$$= ((\cdots ((((a_m)R + a_{m-1})R + a_{m-2})R + a_{m-3}) \cdots)R + a_1)R + a_0 \qquad (1.3)$$

where $0 \le a_j \le R$. That is, we start with a sum of 0; beginning with the MSD, we multiply the sum by R and add the next less significant digit to the sum, repeating until all digits have been considered. For example, we can express

$$A23F1_{16} = ((((10)16 + 2)16 + 3)16 + 15)16 + 1$$

Equation 1.3 forms the basis for a very convenient method for converting a decimal number to its equivalent in radix R. It lends itself to an iterative division procedure that may be employed to convert a decimal integer into its base-R equivalent. When we divide NI by R, the parenthesized part of Equation 1.3 represents the quotient

$$Q = (\ldots((((a_m)R + a_{m-1})R + a_{m-2})R + a_{m-3})\ldots R + a_1)R \qquad (1.4)$$

while the least significant coefficient, a_0, is obtained as the reminder. If we next divide Q by R, a_1 will be obtained now as the remainder. Successive divisions of the quotients by R would yield the remaining coefficients, a_2, a_3, and so on, until all coefficients have been determined. This *successive division* conversion scheme is illustrated by the two examples that follow:

Example 1.3
Convert $(1623)_{10}$ to its binary equivalent.

Solution
Successive division by 2 is carried out to obtain the remainders as follows:

1623/2	=	811	remainder 1	(LSD)
811/2	=	405	remainder 1	
405/2	=	202	remainder 1	
202/2	=	101	remainder 0	
101/2	=	50	remainder 1	
50/2	=	25	remainder 0	
25/2	=	12	remainder 1	
12/2	=	6	remainder 0	
6/2	=	3	remainder 0	
3/2	=	1	remainder 1	
1/2	=	0	remainder 1	(MSD)

The last remainder is referred to as the MSD, while that obtained first is the LSD. The rest of the remainders fall between the MSD and the LSD and in the order of their appearances. Therefore,

$$(1623)_{10} = (11001010111)_2$$

We can check the result by examining $\sum_n a_n R^n$. And indeed,

$$(1 \times 2^0) + (1 \times 2^1) + (1 \times 2^2) + (1 \times 2^4) + (1 \times 2^6) + (1 \times 2^9) + (1 \times 2^{10})$$
$$= 1 + 2 + 4 + 16 + 64 + 512 + 1024 = (1623)_{10}$$

If a 1 occurs in a multibit base-2 number, the value associated with its position is calculated and added to the values for all other positions where a 1 occurs. Positions holding a 0 are just not taken into consideration in calculating the total.

Example 1.4

Find the octal equivalent of $(2865)_{10}$.

Solution

By proceeding as in the previous example,

$$
\begin{aligned}
2865/8 &= 358 \quad \text{remainder} \quad 1 \\
358/8 &= 44 \quad \text{remainder} \quad 6 \\
44/8 &= 5 \quad \text{remainder} \quad 4 \\
5/8 &= 0 \quad \text{remainder} \quad 5
\end{aligned}
$$

Therefore,

$$(2865)_{10} = (5461)_8$$

We can verify this result to be true since

$$(5 \times 8^3) + (4 \times 8^2) + (6 \times 8^1) + (1 \times 8^0)$$
$$= 2560 + 256 + 48 + 1$$
$$= (2865)_{10}$$

There is an additional scheme, known as the *add-the-weights* technique, for converting a decimal integer number to its binary equivalent. This technique involves identifying the positional weights of a binary number. Recall that 2^n is the positional weight of bit n. The positional weights are listed in order so that the largest positional weight is at the extreme left and the smallest positional weight 1 is at the extreme right. Given a decimal integer number, its binary equivalent is determined by placing a 1 under the largest positional weight either equal to or less than the given number. Starting with the next largest positional weight, a 1 is placed under only that positional weight, which when added to the other similarly identified positional weight(s), does not exceed the value of the given decimal integer. This process is continued until the sum of all such positional weights equals the value of the given decimal number. Finally, 0s are placed under each of the unused positional weights. The bit string formed of the combination of these 1s and 0s gives the binary number that is equivalent to the given decimal number. The next example will clarify this conversion scheme.

Example 1.5

Convert 183_{10} to its equivalent binary number using the add-the-weights scheme.

Solution

Since we are interested in converting 183_{10}, the positional weights may not have to exceed the following range:

$$\ldots \ 256 \ 128 \ 64 \ 32 \ 16 \ 8 \ 4 \ 2 \ 1$$

Since $183 = 128 + 32 + 16 + 4 + 2 + 1$, the 1s and the 0s are listed as follows:

Positional weights:	256	128	64	32	16	8	4	2	1
0s and 1s:		0	1	0	1	1	0	1	1	1

Therefore, $183_{10} = 10110111_2$

The bases 2, 8, and 16 can be obtained by raising 2 respectively to the powers of 1, 3, and 4. Accordingly, it is relatively easy to convert numbers back and forth between the binary, octal, and hexadecimal number systems. The traditional method of converting a number from octal, for example, to its binary equivalent is not that simple. The octal number is first converted to its decimal equivalent, and then this decimal number is converted to its binary equivalent. The equivalent numbers listed in Table 1.3 can be used to eliminate such a tedious process. Since $2^3 = 8$ and $2^4 = 16$, each octal number may be readily represented by a combination of three bits and each hexadecimal number may be represented by a combination of four bits. When a string of binary numbers is grouped into groups of four bits, starting from the binary point, the equivalent hexadecimal number may be obtained by replacing each group of four bits by its hexadecimal equivalent. Similarly, if the bits are grouped into groups of three bits and if each such group is directly substituted with its respective octal value, the equivalent octal number is obtained. Again, to convert from octal or hexadecimal to binary equivalent, we use Table 1.3 to simply write three bits for each octal or four bits for each hexadecimal digit and concatenate them all together. Note that to make proper groupings one may have to add zeros to the extreme left of the most significant integer digit.

TABLE 1.3

Short-Cut Conversion between Binary and Octal and Binary and Hex Numbers

Octal to Binary		Hexadecimal to Binary	
0	000	0	0000
1	001	1	0001
2	010	2	0010
3	011	3	0011
4	100	4	0100
5	101	5	0101
6	110	6	0110
7	111	7	0111
		8	1000
		9	1001
		A	1010
		B	1011
		C	1100
		D	1101
		E	1110
		F	1111

Example 1.6
Convert $(82171C)_{16}$ to its base-2 and base-8 equivalents.

Solution
Using Table 1.3 entries, we can express the number as

$$(82171C)_{16} = (1000\ 0010\ 0001\ 0111\ 0001\ 1100)_2$$
$$= (100000100001011100011100)_2$$

To find its octal equivalent, the base-2 equivalent of the hex number may be grouped into groups of three bits. Table 1.3 entries can be used to determine the corresponding octal equivalent. Accordingly, by regrouping and then assigning octal values, we find that,

$$(82171C)_{16} = (100\ 000\ 100\ 001\ 011\ 100\ 011\ 100)_2$$
$$= (40413434)_8$$

Therefore,
$$(271.A)_{16} = (1001110001.101)_2 = (1161.5)_8$$

Digital systems often have a set of switches as one means of input. If a set of 32 switches, for example, were to be used to input a 32-bit quantity, it is much easier to remember or reproduce an 8-digit hex number like $(7BC56D01)_{16}$ rather than $(01111011110001010110110100000001)_2$. In any event such a system will need to have a circuit that converts the eight hex key depressions to its corresponding 32 bits of binary code. In later chapters, we shall be acquiring skills to design such a circuit as well as more complicated digital systems.

1.3.2 Conversion of Fractions

An n-digit, base-R fraction can be converted to its equivalent base-10 fraction in a manner identical to that used in converting the base-R integers. Each of the base-R digits is first multiplied by its respective value of R^j as in Equation 1.2, where j denotes the positional value (between -1 and $-n$) of the fractional bits. The products for all of the digits are then added together to obtain the equivalent decimal fraction.

Example 1.7
Determine base-10 equivalent of $(0.11011)_2$.

Solution
Each of the bits is first multiplied by its corresponding positional multiplier, R^j and then added to obtain

$$(0.11011)_2 = (1 \times 2^{-1}) + (1 \times 2^{-2}) + (1 \times 2^{-4}) + (1 \times 2^{-5})_{10}$$
$$= (0.5 + 0.25 + 0.0625 + 0.03125)_{10}$$
$$= (0.84375)_{10}$$

The last conversion process to be considered in this section involves converting decimal fractions into their equivalent base-R fractions. This requires that we determine all the fraction coefficients $(a_{-1}, a_{-2}, \ldots, a_{-n})$ as identified in Equation 1.2. The base-10 fraction NF can be expressed as

$$
(NF) = \sum_{k=1}^{n} \frac{a_{-k}}{R^k}
$$
$$
= R^{-1}(a_{-1} + R^{-1}(a_{-2} + R^{-1}(a_{-3} + R^{-1}(a_{-4} + \cdots)))) \tag{1.5}
$$

Equation 1.5, in particular, also lends itself to an iterative process for converting a base-10 fraction to its equivalent base-R fraction. If a base-10 fraction NF is multiplied with R, we obtain a mixed number with a_{-1} as its integer part and $R^{-1}(a_{-2} + R^{-1}(a_{-3} + R^{-1}(a_{-4} + \cdots)))$ as its fractional part. The integer a_{-1} of the product becomes a numeral in the new base-R fractional number. If the resulting fractional part, on the other hand, is multiplied again with R, the integer part of the product will yield the next coefficient, a_{-2}. Continuing on with this successive multiplication process, we can obtain the remaining coefficients, a_{-3} through a_{-n}. The multiplication process is carried on successively until the fractional portion of the product reaches zero or until the number is carried out to sufficient digits and truncation occurs.

Unlike the integer conversions, conversions between decimal and binary fractions do not always lead to an exact solution. It may require an infinite string of digits in the new base for complete conversion. This occurs when during successive multiplication process, the fractional portion of the product does not become zero. Consider the binary equivalent of $(0.153)_{10}$ which is $(0.00100000110001001\ldots)_2$. Herein, the fractional portion of the product does not converge to zero. This may not always constitute a problem in practice, since the process can be terminated after p steps. However, besides this nonconvergence problem, there are many fractions that themselves are inexact. The decimal fraction $2/7$, for example, is $(0.285714285714285714\ldots)_{10}$—a recurring fraction. The following examples illustrate the application of successive multiplication technique. In comparison, the conversions between binary, octal, and hexadecimal systems are always accomplished exactly. One should employ the short-cut method of conversion between binary, octal, and hexadecimal fractional digits. By grouping binary digits in three or four starting from the radix point, we can readily convert a binary number respectively to its octal or hex equivalents.

Example 1.8
Determine octal equivalent of $(0.575195313)_{10}$.

Solution
Since,

$$
\begin{aligned}
0.575195313 \times 8 &= 4.6015625 \quad \text{integer 4} \quad \text{MSD} \\
0.6015625 \times 8 &= 4.8125 \quad \text{integer 4} \\
0.8125 \times 8 &= 6.5 \quad \text{integer 6} \\
0.5 \times 8 &= 4.0 \quad \text{integer 4} \quad \text{LSD}
\end{aligned}
$$

Therefore,

$$
(0.575195313)_{10} = (0.4464)_8
$$

We can verify this to be true since

$$(4 \times 8^{-1}) + (4 \times 8^{-2}) + (6 \times 8^{-3}) + (4 \times 8^{-4})$$
$$= 0.500000000 + 0.062500000 + 0.011718750 + 0.000976563$$
$$= (0.575195313)_{10}$$

Example 1.9
Obtain the hex equivalent of $(110.3271484375)_{10}$.

Solution
The integer part of the equivalent hex number is determined by successively taking divisions as follows:

$$110/16 = 6 \quad \text{remainder E} \quad \text{(LSD)}$$
$$6/16 \ = 0 \quad \text{remainder 6} \quad \text{(MSD)}$$

The fractional part of the hex equivalent is determined next by successively performing multiplication as follows:

$$0.3271484375 \times 16 = 5.2343750 \quad \text{coefficient 5} \quad \text{(MSD)}$$
$$0.2343750 \times 16 \ \ = 3.75 \qquad \quad \text{coefficient 3}$$
$$0.750 \times 16 \qquad = 12 \qquad \qquad \text{coefficient C} \quad \text{(LSD)}$$

The hex equivalent number is, therefore, $(6E.53C)_{16}$. We can verify this to be true since

$$(6 \times 16^{1}) + (14 \times 16^{0}) + (5 \times 16^{-1}) + (3 \times 16^{-2}) + (12 \times 16^{-3})$$
$$= 96 + 14 + 0.3125 + 0.01171875 + 0.0029296875$$
$$= (110.3271484375)_{10}$$

During conversion of a decimal fraction to its equivalent nondecimal fraction, or vice versa, the number of significant digits must be enough to maintain at least the same degree of accuracy. In case of nonconvergence, if any, when converting the decimal fraction 0.4356, for example, to its equivalent binary fraction, we shall have to determine the number of binary bits so as to distinguish between 0.4355, 0.4356, and 0.4357. By trial and error, one can determine that at least a 14-bit binary fraction will allow for an accuracy of 10^{-4}, which is just less than $1/2^{14}$. On the other hand, a 13-bit binary fraction is unable to provide an accuracy value of 10^{-4} or less.

The nonconvergence problem involving conversion of the fractions may now be rephrased to determine the smallest number of binary bits b that will represent a value of 10^{-d}. For a d-digit decimal fraction, the relationship between the number of bits, b, and d is as follows:

$$\left(\frac{1}{2}\right)^{b} \leq \left(\frac{1}{10}\right)^{d} \tag{1.6}$$

which, when solved, gives $b \geq 3.32d$. Since b has to be an integer just large enough so that it satisfies Equation 1.6, it is given by

$$b = \text{Int}(3.32d + 1) \tag{1.7}$$

where the function $\text{Int}(x)$ is the integer value of x obtained by discarding its fractional part. If instead, we are converting a decimal fraction into its binary equivalent, the inequality in Equation 1.6 will need to be reversed and then solved for d. The minimum number of decimal digits required to retain the accuracy of a b-bit binary number is then given by

$$d = \text{Int}(0.3b + 1)3 \tag{1.8}$$

Equations 1.7 and 1.8 primarily relate to conversions between binary and decimal fractions. In conversions involving octal and hexadecimal fractions, one needs to be cognizant of the fact that three bits are needed to represent an octal digit and four bits are needed to represent a hexadecimal digit. In real systems, the number of digits to represent a fraction may be limited by the digital system itself. That limitation is reflective of the inherent inaccuracies of the system.

Example 1.10

Convert $(A45.11)_{16}$ to its decimal equivalent.

Solution

In this number, there are two hex digits that form the fractional part. Thus, $b = 8$ since it takes eight binary digits to represent two hex digits. Using Equation 1.8, we obtain

$$d = \text{Int}[(0.3 \times 8) + 1] = 3.$$

Since only three decimal digits are needed to represent the fractional part, we determine that

$$
\begin{aligned}
(A45.11)_{16} &= [(10 \times 16^2) + (4 \times 16^1) + (5 \times 16^0) + (1 \times 16^{-1}) + (1 \times 16^{-2})]_{10} \\
&= (2560 + 64 + 5 + 0.0625 + 0.00390625)_{10} \\
&= (2629.06640625)_{10} \\
&= (2629.066)_{10}
\end{aligned}
$$

Example 1.11

Convert $(212.562)_{10}$ to its binary equivalent.

Solution

Using $d = 3$ in Equation 1.7, we obtain

$$b = \text{Int}[(3.32 \times 3) + 1] = 10.$$

Therefore,

$$(212.562)_{10} = (11010100.100011111101111\ldots)_2$$
$$= (11010100.1000111111)_2$$

1.4   Negative Numbers

In earlier sections, we limited our discussions to only positive numbers. However, as soon as we encounter subtraction or subtraction-related operations, we shall see that we cannot just be satisfied with positive numbers. Typically, the numbers that we use most often follow a *sign-and-magnitude* format. In this representation, the number consists of two parts: the magnitude and the sign. In our day-to-day use, when no sign appears before the number, we treat it as positive. However, such a convention requires the introduction of a separate symbol such as "–" sign for representing the negative values in our number system. Since the idea behind using binary system is to use as few discrete values as possible, it does not make sense to now add a third symbol to the binary symbols to accommodate negative numbers. This problem is avoided in sign-and-magnitude representation by introducing an additional bit, called the *sign bit*. The leftmost bit of the sign-and-magnitude number is reserved for the sign bit. According to the convention, the sign bit of a positive number is indicated by a 0, and that of the negative number is represented by $R-1$. The magnitude portion of the positive number is not different from that of the corresponding negative number. A number and its negative are only different in their sign bits. For example, a 10-digit sign-and-magnitude representation may be used as follows to denote:

$$+65_{10} = 0\ 001000001_2$$
$$-65_{10} = 1\ 001000001_2$$

However, if this same decimal number 65 were to be expressed say in hexadecimal form, the sign-and-magnitude representation will take on the values:

$$+65_{10} = 0\quad 041_{16}$$
$$-65_{10} = F\quad 041_{16}$$

A sign-and-magnitude binary integer with n bits covers the numbers ranging from $-(2^{n-1}-1)$ to $+(2^{n-1}-1)$. Since one digit space is reserved for the sign, only $2R^{n-1}$ out of the R^n possible sequences are utilized. Consider further that the $(n-1)$ digits representing the magnitude are partitioned further into p integral digits and q fractional digits. The range of positive numbers that can be covered is then $\{0, R^p - R^{-q}\}$ and that of negative numbers is $\{-(R^p - R^{-q}), -0\}$. Notice that we have two different representations for zero (+0 and −0).

A major disadvantage of the sign-and-magnitude representation is that the operation to be performed may often depend on the signs of the operands. The sign-and-magnitude representation requires us to frequently compare both the signs and the magnitudes of the operands. Consider for example, addition of two sign-and-magnitude numbers $D1 = \{S1, M1\}$ and $D2 = \{S2, M2\}$ to generate a sum digit $DS = \{SS, MS\}$. If the two signs $S1$ and $S2$ are the same, we simply add the magnitudes to obtain MS and the sign of

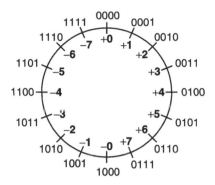

FIGURE 1.1
A counting circle representing 4-bit sign-and-magnitude numbers.

the sum SS takes the sign of the operands. But if $S1$ and $S2$ are different, we compare the magnitudes. If the magnitudes are the same, then MS becomes 0 but if one magnitude is larger than the other, MS takes on the value of the difference and SS takes the sign of the larger of the two magnitudes. It follows therefore that subtraction is equivalent to addition once the subtrahend has been complemented. These sign-related issues require that the to-be-designed logic circuit that may take in sign-and-magnitude numbers as input would have to make decisions such as "ifs," "adds," "compares," and "subtracts." Such requirements thus lead to logic circuit complexity and, therefore, to both excessive cost and processing time. It is obvious, therefore, that to add sign-and-magnitude numbers, an adding machine would have to have the capabilities for (a) identifying the sign, (b) comparing magnitudes, (c) adding numbers, and (d) subtracting numbers. The counting circle for the sign-and-magnitude representation of 4-bit numbers, for example, is shown in Figure 1.1. This counting circle covers the entire range of numbers from -7 to $+7$. Surprisingly, incrementation or decrementation of numbers is not directly realized by cycling around the counting circle. While incrementing, for example, we have to move around the counting circle clockwise for the positive numbers and counterclockwise for the negative numbers.

For an m-digit base-R number N, the signed radix complement of N is obtained by subtracting N from R^m as,

$$N_{2's\ compl} = R^m - N \tag{1.9}$$

Alternatively, we can obtain the signed radix complement without subtraction, by rewriting Equation 1.9 as follows:

$$N_{2's\ compl} = R^m - N = \{(R^m - 1) - N\} + 1 \tag{1.10}$$

The number $R^m - 1$ in Equation 1.10 consists of m digits whose value is $R - 1$. Defining the digit complement $a_{compl} = (R - 1) - a$, we get

$$(R^m - 1) - N = ((R - 1) - a_{m-1,compl})((R - 1) - a_{m-2,compl}) \dots ((R - 1) - a_{0,compl})$$

$$= N_{1's\ compl} \tag{1.11}$$

Consequently, Equation 1.9 can be rewritten as

$$N_{2's\ compl} = N_{1's\ compl} + 1 \tag{1.12}$$

In both radix complement and diminished-radix complement methods, a positive number is represented in exactly the same way as in the sign-and-magnitude method. A negative number $-A$ is represented by $(\mathfrak{R} - A)$ where $\mathfrak{R}$ is a to-be-determined constant. This representation satisfies the basic principle that $-(-A) = A$ since the complement of $(\mathfrak{R} - A)$ is $\mathfrak{R} - (\mathfrak{R} - A) = A$. Unlike sign-and-magnitude representation, in the radix complement methods regardless of the value of $\mathfrak{R}$ there are no decisions to be made during arithmetic operations. If a positive number A and a negative number B are to be added, the addition is described simply as $A + (\mathfrak{R} - B) = \mathfrak{R} - (B - A)$. When $B > A$, then the negative result $-(B - A)$ is already represented in the same complement representation, that is, as $\mathfrak{R} - (B - A)$. Accordingly, the operation requires no special decision pertaining to the sign. On the other hand, if $A > B$, the correct result is $(A - B)$ while $A + (\mathfrak{R} - B) = \mathfrak{R} + (A - B)$ is obtained as the result. The additional term $\mathfrak{R}$, therefore, must be discarded from the result. In fact, the value of $\mathfrak{R}$ may be selected such that this correction step can be eliminated completely.

Consider adding A and $-A$, each of size $(p + q)$ bits. Since the complement $a_{j,\text{compl}}$ of digit a_j is $(R - 1) - a_j$ where R is the radix, the computed sum and then corrected sum (obtained by adding a 1 at the least significant position) are obtained as,

$A =$	a_p	a_{p-1}	$\cdots$	a_{-q+1}	a_{-q}
$-A =$	$a_{p,\text{compl}}$	$a_{p-1,\text{compl}}$	$\cdots$	$a_{-q+1,\text{compl}}$	$a_{-q,\text{compl}}$
	$R - 1$	$R - 1$		$R - 1$	$R - 1$
1	0	0	$\cdots$	0	0

yielding $A + (-A) + R^{-q} = R^{p+1}$ where R^{-q} represents a 1 at the least significant bit. When the result is stored in a storage device of length $n = p + q + 1$, the MSD is automatically discarded and the result is simply a 0. Storing the correct result in a fixed n-bit storage device is equivalent to taking the remainder after dividing by R^{p+1}.

Understanding that $A + (-A) + R^{-q} = R^{p+1}$, if we were to choose $\mathfrak{R}$ as R^{p+1}, then $\mathfrak{R} - A = R^{p+1} - A = (-A) + R^{-q}$. The complement $(\mathfrak{R} - A)$ of A is thus independent of the value of p. We refer to this scheme of number representation as the *radix complement* representation. No correction is needed when $A + (\mathfrak{R} - B)$ is positive, that is, when $A > B$, since $\mathfrak{R} = R^{p+1}$ is discarded during the evaluation of $\mathfrak{R} + (A - B)$. On the other hand, when $\mathfrak{R}$ is set equal to $R^{p+1} - R^{-q}$, we refer to the corresponding number representation as the *diminished-radix complement* representation for which $\mathfrak{R} - A = (R^{p+1} - R^{-q}) - A = (-A)$. However, a correction is needed when computing $\mathfrak{R} + (A - B)$ if and only if $A > B$ as will be explained later in the section. This correction that involves addition of a 1 when a carry is generated out of the most significant bit is also referred to as an *end around carry*.

In most computing systems, the numbers are represented not using sign-and-magnitude representation but using (a) signed $(R - 1)$s or diminished radix or (b) signed Rs or radix representation. This choice allows us to overcome the limitation of the sign-and-magnitude arithmetic. In both of these representations, sign bit of a positive number is a 0 and that of the negative number is $R - 1$. This convention provides for an unambiguous representation of sign for all three representations. In *signed* $(R - 1)$s representation, the complement is obtained by subtracting each digit of the corresponding positive number from $R - 1$. On the other hand, a 1 is added to the least significant position of the $(R - 1)$'s complement to obtain the corresponding R's complement number. Consequently, the 8-bit binary positive number $(0 \; 0101010)_2$ when complemented yields $(1 \; 0101010)_2$ in sign-and-magnitude representation, $(1\ 1010101)_2$ in signed 1's complement representation, and $(1\ 1010110)_2$ in signed 2's complement representation.

When converting an n-bit complement number to an equivalent m-bit number (where $m > n$), one must be very careful. For $m > n$, we must append $m - n$ copies of the sign bit to the left of the given number. This is referred to as *sign extension* that amounts to padding a positive number with 0s and a negative number with 1s. On the other hand, if $m < n$, $n - m$ leftmost bits of the given number are discarded; however, the result is valid only if all of those discarded bits are the same as the sign bit of the result.

The binary complementation technique can be extended to complementing a base-R number. For example, a negative hex number may be represented using either F's complement or $(10)_{16}$'s complement. First, F's complement is obtained by subtracting every one of the digits of the positive hex number from F. Finally, $(10)_{16}$'s complement is found by adding a 1 to the F's complement number. For example, the positive 8-digit hex number $(0\ 0AB2B53)_{16}$ when complemented yields $(F\ 0AB2B53)_{16}$, in sign-and-magnitude representation. The corresponding signed F's complement number is given by $(F\ FFFFFFF)_{16} - (0\ 0AB2B53)_{16} = (F\ F54D4AC)_{16}$ while the corresponding $(10)_{16}$'s complement number is obtained as $(F\ F54D4AC) + (0\ 0000001)_{16} = (F\ F54D4AD)_{16}$.

Systems that use the 2's complement number system can represent integers in the range $\{-2^{n-1}, 2^{n-1} - 1\}$, where n is the number of bits available to represent the number. The range of 2's complement system is slightly asymmetric since there is one negative number more than the positive numbers. Accordingly, whenever we take a complement operation, an overflow may occur. The numerical value A of an n-bit 2's complement number $(a_{n-1}a_{n-2}\ldots a_1a_0)$ can be determined as follows.

If $a_{n-1} = 0$, then

$$A = \sum_{j=0}^{n-1} x_j 2^j \tag{1.13}$$

However, if $a_{n-1} = 1$, the given sequence of the 2's complement bits is first complemented and then Equation 1.13 is employed to determine the value of the given number. Alternatively, A can be determined as

$$A = -a_{n-1}2^{n-1} + \sum_{j=0}^{n-2} x_j 2^j \tag{1.14}$$

For example, 10100010 in 2's complement number system is equivalent to $-2^7 + 2^5 + 2^1 = -128 + 32 + 2 = -94$. When $a_{n-1} = 0$, Equation 1.14 reduces to Equation 1.13. However, when $a_{n-1} = 1$, the binary number is given by

$$-\{(-A) + 1\} = -\left[\sum_{j=0}^{n-2} a_{j,\text{compl}} 2^j + 1\right] \tag{1.15}$$

where 1 accounts for addition of a 1 that has been placed at the least significant bit position.

Example 1.12

Show that the numerical value A for a 2's complement representation is indeed given by Equation 1.14.

Solution

The numerical value of the number is obtained by adding the complement of the given number and a 1 placed at its least significant bit position. It is given by

$$
\{(-A) + 1\} = -\left[\sum_{j=0}^{n-2} a_{j,\text{compl}} 2^j + 1\right]
$$

$$
= -\left[\sum_{j=0}^{n-2}(1 - a_j)2^j + 1\right].
$$

$$
= -\left[\sum_{j=0}^{n-2} 2^j - \sum_{j=0}^{n-2} a_j 2^j + 1\right]
$$

$$
= -\left[\sum_{j=0}^{n-2}(2^{j-1} - 1) - \sum_{j=0}^{n-2} a_j 2^j + 1\right]
$$

$$
= -2^{n-1} + \sum_{j=0}^{n-2} a_j 2^j
$$

which is the same as Equation 1.14.

In the 1's complement representation, in comparison, the range of supportable numbers is symmetric. Consequently, finding the complement of a 1's complement number is very easy. The numerical value A for an all-integer 1's complement number is given by

$$
A = -a_{n-1}(2^{n-1} - 1) + \sum_{j=0}^{n-2} a_j 2^j \tag{1.16}
$$

Therefore, 10100010, for example, in 1's complement number system is equivalent to $-(2^7 - 1) + 2^5 + 2^1 = -127 + 32 + 2 = -93$.

In all three complement representations the positive numbers are treated exactly alike, but the negative numbers are expressed differently. For the binary number system, however, the rules for finding the complement are much more straightforward.

Rule for 1's Complement: Complement every bit including the sign bit by changing each 1 to a 0 and each 0 to a 1.

Rules for 2's Complement: (i) Add a 1 (by placing it at the least significant bit) and the corresponding 1's complement number while ignoring the carry-out of the sign bit, or (ii) while scanning the positive number from right to left, complement all bits (by changing each 1 to a 0 and each 0 to a 1) that are encountered after the first encountered 1.

For the binary number system, finding the $(R-1)$'s complement of a number is relatively simple. Rule for 1's complement follows from the fact that difference between $R - 1$ (i.e., 1)

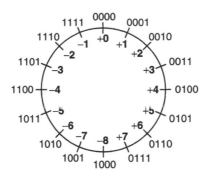

FIGURE 1.2
A modular counting circle representing 4-bit 2's complement numbers.

and each of the 0s and 1s of any given number respectively gives 1s and 0s. While Rule (i) for obtaining the 2's complement follows directly from the very definition of 2's complement, Rule (ii) for 2's complement is somewhat ad hoc but its validity can be verified by inspection.

An n-bit number allows for 2^n different bit combinations. If all these combinations were used to represent only unsigned positive values, numbers in the magnitude range of 0 through 2^{n-1} can be accounted for. However, when both negative and positive values are to be represented, some of these 2^n combinations will have to be used for representing negative numbers. Considering that the number of positive and negative numbers are approximately equal, an n-bit number can range from $-2^{n-1}+1$ to $2^{n-1}-1$.

Figure 1.2 shows the modular counting circle for the signed 4-bit 2's complement numbers, for example. When compared to that of Figure 1.1, this counting circle is more like a true counter. While incrementing, we have to move clockwise in the case of both positive and negative numbers. Conversely when decrementing, we have to move around counterclockwise for both positive and negative numbers. To add N to a given number, one needs to move N positions clockwise from the position of the given number. Again to subtract N, we have to cycle N positions counterclockwise starting from the position of the given number. These addition or subtraction operations provide the correct result as long as the result lies somewhere in between -8 and $+7$. The counting circle of Figure 1.2 can be used to also subtract N (or add $-N$) by moving $16 - N$ positions in clockwise direction. But since $16 - N = 2^4 - N = 1111_2 - N + 1_2$, $16 - N$ is 2's complement of N. Thus, the modular counting circle of Figure 1.2 is in agreement with the claim that addition of a negative number (in 2's complement form) may be realized using ordinary addition process.

Figure 1.3 shows the counting circle for 4-bit signed 1's complement number. Indeed, it has both incrementing and decrementing characteristics that are the same as those in Figure 1.2. But for both addition and subtraction results to be correct, we must not cross the discontinuity between not only -7 and $+7$ but also -0 and $+0$. This additional constraint associated with the signed 1's complement representation arises from the fact that there are now two different representations of zero (for $+0$ and -0). In fact, the 2's complement technique is the only means by which both $+0$ and -0 have the same representation. Consider Table 1.4 that lists and compares 4-bit fractional fixed-point numbers in all three representations. A significant difference between the 2's complement and the other two representations is their behavior under truncation. The advantages of using the signed 2's complement representation are thus obvious. Almost all digital computers thus use 2's complement representation for data processing. Not only does it have a single representation of zero, but it also allows for both correct addition and subtraction as long as the result has not exceeded the range: -2^{n-1} through $2^{n-1}-1$.

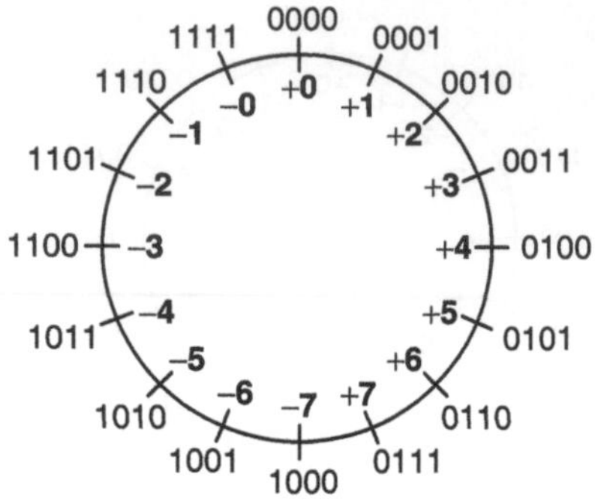

FIGURE 1.3
A counting circle representing 4-bit 1's complement numbers.

TABLE 1.4

Representation of 4-Bit Fractional Fixed-Point Numbers

Number	Sign-and-Magnitude	1's Complement	2's Complement
+7/8	0 111	0 111	0 111
+3/4	0 110	0 110	0 110
+1/2	0 100	0 100	0 100
+1/4	0 010	0 010	0 010
+0	0 000	0 000	0 000
−0	1 000	1 111	0 000
−1/4	1 010	1 101	1 110
−1/2	1 100	1 011	1 100
−3/4	1 110	1 001	1 010
−7/8	1 111	1 000	1 001

TABLE 1.5

Binary Addition

Augend	Addend	Carry-In	Carry-Out	Sum
0	0	0	0	0
0	0	1	0	1
0	1	0	0	1
0	1	1	1	0
1	0	0	0	1
1	0	1	1	0
1	1	0	1	0
1	1	1	1	1

1.5 Binary Arithmetic

There are many higher-level operations that require performing binary addition and taking either 1's complement or 2's complement. Most of the rules applicable to decimal numbers are equally applicable to binary numbers also. The rules governing binary addition are all summarized in Table 1.5. The addition process is not the same at all bit locations. When faced with the addition of multibit numbers, we need to understand that at the least significant position, binary addition involves two input operands: *addend* and *augend*. At other than the LSD, an additional input operand referred as the *carry-in* needs to be considered. It is the carry-out generated as an addition output at the next less significant digit position. At all bit positions, however, there are only two outputs: *sum* and *carry-out*. The sum output is a 1 when either only one input is a 1 ($1 + 0 + 0 = 1$ and no carry-out) or when all three inputs are 1 ($1 + 1 + 1 = 1$ and carry-out is a 1). The carry-out is a 1 if two or more inputs are equal to 1.

Example 1.13
Add the decimal numbers 11 and 14 using binary arithmetic.

Solution
Table 1.5 is used in determining the sum. For illustration, both carry-in and carry-out are also identified. Adding addend, augend, and carry-in arriving at the column under consideration generates each of the carry-outs. The carry-out at column j appears as the carry-in at column $j + 1$.

$$
\begin{array}{ll}
0\ 0001110 & \text{carry-out} \\
0\ 0011100 & \text{carry-in} \\
\end{array}
$$

$$
\begin{array}{ll}
(11)_{10} = (0\ 0001011)_2 & \text{augend} \\
(14)_{10} = (0\ 0001110)_2 & \text{addend} \\
\hline
(0\ 0011001)_2 & \text{sum} \\
\end{array}
$$

Therefore,

$$(11)_{10} + (14)_{10} = (0\ 0011001)_2 = (25)_{10}$$

Very much as in the case with addition, there are three input operands for subtraction: the *minuend*, the *subtrahend*, and the *borrow-in*, and the output variables are, respectively, the *borrow-out* and the *difference*. Table 1.6 lists the subtraction rules and Example 1.14 illustrates the implementation of these subtraction rules. However, as stated earlier, binary subtraction may also be accomplished using addition. According to this scheme, the subtrahend is first complemented and then added to the minuend.

TABLE 1.6
Binary Subtraction

Minuend	Subtrahend	Borrow-In	Borrow-Out	Difference
0	0	0	0	0
0	0	1	1	1
0	1	0	1	1
0	1	1	0	1
1	0	0	1	0
1	0	1	0	0
1	1	0	0	0
1	1	1	1	1

Example 1.14

Subtract $(31)_{10}$ from $(73)_{10}$ in binary.

Solution

Each of the borrow-outs is determined by the combination of the minuend, subtrahend, and borrow-in at the column under consideration. It is a 1 when the subtraction operation needs to borrow from the next more significant column. When the borrow-in is a 1 in a column, it indicates that a less significant column has already borrowed a 1 from the column under consideration. The borrow-out at a column j appears as the borrow-in at column $j + 1$.

$$
\begin{array}{lll}
& 0\ 0111110 & \text{borrow-out} \\
& 0\ 1111100 & \text{borrow-in} \\
\\
(73)_{10} = (0\ 1001001)_2 & & \text{minuend} \\
(31)_{10} = (0\ 0011111)_2 & & \text{subtrahend} \\
\hline
(0\ 0101010)_2 & & \text{difference}
\end{array}
$$

Therefore,

$$(73)_{10} - (31)_{10} = (0\ 0101010)_2 = (42)_{10}.$$

Many students find subtraction to be a more difficult operation to follow. Fortunately it can be avoided since we can show that the result of $A - B$ operation is also obtained by determining the sum of A and the complement of B. In fact, addition operation is the most fundamental arithmetic operation since it encompasses all other arithmetic operations. By repeatedly performing addition, we can implement the equivalent of a multiplication operation. When a multibit number is added to itself n times, the resulting sum is the same as the product of the given number and n. We may note that division, on the other hand, can be realized by repeatedly performing subtraction operations. But since subtraction itself can be realized via addition, we can indeed realize division also by means of addition. Understanding that negative numbers may be represented in either 1's complement or 2's complement form, the rules for the corresponding binary addition respectively are as follows:

Rule for 1's Complement Subtraction: All bits including the sign bit are added. The carry-out of the sign bit is then added to the sum to yield the difference. If the sign bit is a 0, the resultant sum (which is the difference for our purpose) is a sign-and-magnitude number. On the other hand, if the sign bit is a 1, the difference is negative and the number is in the 1's complement form.

Rule for 2's Complement Subtraction: All bits including the sign bits are added to yield the difference, and the carry-out of the sign bit is discarded. If the resultant sign bit is a 1, the sum (which is the difference for our purpose) is negative and it is in the 2's complement form. If the sign bit is a 0, the difference is positive and the number appears as a positive sign-and-magnitude number. These conditions are true as long as overflow is absent.

Recall that arithmetic operations involving numbers represented in 1's complement form do not work when the resultant number lies beyond the range of -2^{n-1} through $2^{n-1} - 1$. But more importantly, if the increment or decrement operation takes one either from -0 to $+0$

or from $+0$ to -0, the computed result turns out to be off by 1 from the correct result. Accordingly, the aforementioned Rule for 1's complement subtraction takes care of this discrepancy by adding the carry-out of the sign bit to the LSD of the computed result.

When two n-digit numbers are added and the sum occupies $n + 1$ bits, we say that an *overflow* has occurred. An overflow is indicated either by the presence of a 1 in the sign bit when the two to-be-added numbers are positive, or by the presence of a 0 in the sign bit when the two to-be-added numbers are negative. Overflow cannot occur when the two to-be-added numbers have opposite signs. The sum in that case has a magnitude smaller than at least one of the given numbers. Overflow may occur only when both of the numbers are either positive or negative. In the 2's complement representation, overflow occurs when the sum of n-bit numbers is either less than -2^{n-1} or larger than $2^{n-1} - 1$. Consider, for example, the following additions of 5-bit numbers:

$$(-12)_{10} + (-11)_{10} \rightarrow (10100)_2 + (10101)_2 = (01001)_2 = (9)_{10}$$
$$(+10)_{10} + (+8)_{10} \rightarrow (10110)_2 + (11000)_2 = (01110)_2 = (14)_{10}$$
$$(-10)_{10} + (-10)_{10} \rightarrow (10110)_2 + (10110)_2 = (01100)_2 = (12)_{10}$$

where the carry-out of the sign bit has been discarded. Unfortunately, the above additions result in 9_{10}, 14_{10}, and 12_{10} instead of -33_{10}, 18_{10}, and -20_{10}, respectively. In all these cases, the true result lies outside the range of -16 through $+15$. To adequately comprehend the scope of overflow condition, one may compare the carry propagation in addition both in the presence of overflow and in its absence. Consider the following two 5-bit 2's complement numbers:

Carry-out		1 0100		1 1100
Carry-in		0 1000		1 1000
Addend	$(-12)_{10}$	1 0100	$(-4)_{10}$	1 1100
Augend	$(-11)_{10}$	1 0101	$(-2)_{10}$	1 1110
Sum	$(+9)_{10}$	0 1001	$(-6)_{10}$	1 1010

In the example that results in an overflow, we see different carry-in and carry-out at the sign bit. However, when there is no overflow, the carry coming into and out of the sign bit are the same. Table 1.7 lists the various 2's complement addition and subtraction cases (corresponding to positive digits A and B) as well as overflow conditions.

Fortunately, there are two rules that are used for detecting overflow in addition. The first method involves examining the sign bit of the sum. The second method involves

TABLE 1.7

2's Complement Arithmetic

Case	Carry-Out	Sign Bit	Condition	Overflow
$A + B$	0	0	$(A + B) \leq 2^{n-1} - 1$	No
	0	1	$(A + B) > 2^{n-1} - 1$	Yes
$A - B$	1	0	$A \leq B$	No
	0	1	$A > B$	No
$-A - B$	1	1	$-(A + B) \geq -2^{n-1}$	No
	1	0	$-(A + B) < -2^{n-1}$	Yes

examination of both carry-in and carry-out at the sign bit. When the two carries are different, an overflow is said to have already occurred. This was demonstrated in the two examples considered above. An overflow is a rather serious problem in digital systems since the number of bits used to store each value is fixed. However, if the hardware can recognize the occurrence of an overflow, the user could eliminate the corresponding problem or correct the resulting discrepancy.

Example 1.15

Add $(-8)_{10}$ and $(12)_{10}$ using binary arithmetic.

Solution

Since $(8)_{10} = (0\ 001000)_2$,

$$(-8)_{10} = 1\ 110111_2 \quad \text{in signed 1's complement; and}$$
$$(-8)_{10} = 1\ 111000_2 \quad \text{in signed 2's complement}$$

Using the signed 1's complement representation,

$$(12)_{10} + (-8)_{10} = (0\ 001100)_2 + (1\ 110111)_2$$
$$= (10\ 000011)_2 \text{ with a carry-out of the sign bit}$$

The next step then involves performing an end-around carry. Accordingly, we obtain in the signed 1's complement representation

$$(12)_{10} + (-8)_{10} = (0\ 000011)_2 + (0\ 000001)_2 = (0\ 000100)_2 = (4)_{10}$$

On the other hand, in the signed 2's complement representation,

$$(12)_{10} + (-8)_{10} = (0\ 001100)_2 + (1\ 111000)_2$$
$$= (10\ 000100)_2 \text{ with a carry-out of the sign bit}$$

The resulting carry-out of the signbit is discarded to obtain the result given in the signed 2's complement representation as

$$(12)_{10} + (-8)_{10} = (0\ 000100)_2 = (0\ 000100)_2 = (4)_{10}$$

Example 1.16

Subtract $(13)_{10}$ from $(7)_{10}$ using binary arithmetic.

Solution

Since $(13)_{10} = (0\ 01101)_2$, $(-13)_{10}$ can be expressed as

$$(-13)_{10} = (1\ 10010)_2 \quad \text{in signed 1's complement; and}$$
$$(-13)_{10} = (1\ 10011)_2 \quad \text{in signed 2's complement.}$$

Using the signed 1's complement arithmetic, therefore, we obtain

$$(7)_{10} - (13)_{10} = (7)_{10} + (-13)_{10}$$
$$= (0\ 00111)_2 + (1\ 10010)_2$$
$$= (1\ 11001)_2 \quad \text{with no carry-out}$$

Since 1's complement of $(1\quad 11001)$ is $(0\quad 00110)$, therefore, $(7)_{10} + (-13)_{10} = -(0\ 00110)_2 = -(6)_{10}$. Now if we are to use instead the signed 2's complement arithmetic, we get

$$(7)_{10} - (13)_{10} = (7)_{10} + (-13)_{10}$$
$$= (0\ 00111)_2 + (1\ 10011)_2$$
$$= (1\ 11010)_2 \quad \text{with no carry-out}$$

But since the 2's complement of $(1\ 11010)$ is $(0\ 00110)$, $(7)_{10} + (-13)_{10} = -(0\ 00110)_2 = -(6)_{10}$. Indeed, therefore, both of the complement techniques result in the same correct answer.

Consider binary multiplication next. It follows the same train of thought as that employed for realizing decimal multiplication. In decimal multiplication, first-level products are first determined by multiplying the multiplicand with each digit of the multiplier. Each such first-level product is successively shifted to the left by one digit. The relatively shifted versions are all added together to yield the final product. For example,

$$
\begin{array}{r}
13 \\
\times 14 \\
\hline
52 \\
130 \\
\hline
182
\end{array}
$$

When we repeat the same process but in binary multiplication, we obtain

$$
\begin{array}{r}
1101 \\
\times 1110 \\
\hline
0000 \\
11010 \\
110100 \\
1101000 \\
\hline
10110110
\end{array}
$$

As we can see, finding the first-level products is rather trivial in binary multiplication, since the only possible values of the multiplier digits are now 0 and 1. The first-level product is thus either the multiplicand itself or 0. On the other hand, instead of listing all the shifted first-level products and then adding them, it is more convenient to add each shifted first-level product as it is generated to a cumulative *partial product*. When we have finished adding the first-level products to the cumulative partial product, the last cumulative product becomes equal to the final product. Correspondingly, this same binary multiplication can be carried out as follows:

Multiplicand	1101
Multiplier	×1110
Initial partial product	0000
Shifted multiplicand	0000
Partial product	00000
Shifted multiplicand	1101<
Partial product	11010
Shifted multiplicand	1101<<
Partial product	1001110
Shifted multiplicand	1101<<<<
Final product	10110110

For an n-bit by m-bit multiplication, the shift-and-add algorithm requires m first-level products and additions to obtain the final product. The first addition is trivial since the beginning partial product is just a string of 0s. After p such additions, the resulting partial product may have up to $n + p$ bits. The beauty of this scheme is that we can design simple logic circuits to determine the final product. As we shall show in later chapters, such a product determination will require an adder (to add), shift registers (to store the two numbers as well as partial products), and appropriate control logic (for making necessary decisions).

Multiplication of signed numbers can be also realized using a similar technique. Typically, we perform an unsigned multiplication of the magnitudes and make the product positive (or negative) if the operands have the same (or different) signs. This is rather convenient in sign-and-magnitude number system since the sign and magnitude are distinct. Multiplication of 2's complement numbers, however, may not be that simple. Obtaining the magnitude of a negative number and then negating the unsigned product are both nontrivial operations. Consequently, we seek an alternate multiplication algorithm for the complement number system.

Unsigned multiplication is equivalent to successively performed unsigned additions of the shifted multiplicands. In this process, the number of shifts equals the position of the corresponding multiplier bit under consideration. The bits in a 2's complement number have the same weights as in an unsigned number except for the most significant number that has a negative weight. Accordingly, 2's complement multiplication is realized by a sequence of 2's complement additions of shifted multiplicands. At the last step, however, the shifted multiplicand must be negated before it is added to the cumulative partial product. In this algorithm, one has to be rather careful when dealing with the most significant bit. Since we may gain an extra bit at each step, before adding each shifted multiplicand and the $n + p$-bit partial product, we transform them by sign extension to $n + p + 1$ significant bits. The resulting sum has also $n + p + 1$ bits and the carry-out of the most significant bit is discarded.

Example 1.17
Obtain the product of $(-13)_{10}$ and $(-14)_{10}$ using 2's complement binary arithmetic.

Solution
Since 2's complement of -13 (i.e., -1101) and -14 (i.e., -1110) are respectively 10011 and 10010, the multiplication is carried out as follows:

Multiplicand	10011
Multiplier	10010
Partial product	000000
Shifted multiplicand	000000
Partial product	0000000
Shifted multiplicand	110011<
Partial product	11100110
Shifted multiplicand	000000<<
Partial product	111100110
Shifted multiplicand	000000<<<
Partial product	1111100110
Shifted/negated multiplicand	001101<<<
Final product	0010110110

The binary division process is based also on a shift-and-subtract algorithm typically used in decimal division. According to this scheme, one compares the reduced dividend with multiples of the divisor to determine if the shifted divisor can be subtracted or not. This process is continued until the reduced dividend is no longer divisible. Fortunately, in binary number system, there are only two choices for the multiples of the divisor, 0 and the divisor itself. When the dividend and divisor are respectively of size $n + p$ bits and p bits, the corresponding quotient and remainder turn out to be of size n bits and p bits, respectively. For an overflow condition, the divisor is either 0 or the quotient occupies more than n bits. This division scheme for the unsigned numbers can be extended to handle signed numbers as well. Typically, one performs an unsigned division of the magnitudes and makes the quotient positive (or negative) if the two operands have the same (or different) signs. The remainder always has the same sign as the dividend. Accommodation of 2's complement numbers, for example, will require further adjustments similar to that already considered in the case of 2's complement multiplication

1.6 Unconventional Number Systems

We have already seen that the 2's complement representation of negative numbers lends itself to rather simple arithmetic operations. Yet, there are occasions when one may explore the use of negative-radix number system and redundant number system. Certain signal processing and optical computing algorithms, for example, make use of a *negabinary* number system where $R = -2$. Its application is still somewhat limited. In general, the value of the n-tuple $(x_{n-1}, \ldots, x_0)$ in *a negative-radix number system* is given by

$$A = \sum_{j=0}^{n-1} x_j(-\rho)^j \tag{1.17}$$

where ρ is a positive number. The weight of the negative-radix digit at position j is ρ^j if j is even and $-\rho^j$ if j is odd. For example, $(1299)_{-10} = 1(-1000) + 2(100) + 9(-10) + 9(1) = -1000+200-90+9 = -881_{10}$ and similarly $(0299)_{-10} = 0(-1000)+2(100)+9(-10)+9(1) = 0 + 200 - 90 + 9 = 119_{10}$. In fact, the largest possible 4-digit negadecimal number is $(0909)_{-10} = 909_{10}$ and the smallest possible 4-digit negadecimal number is $(9090)_{-10} = -9090_{10}$. This range of numbers from -9090 through 909 for the 4-digit numbers, for example, is asymmetric since it can accommodate about ten times more negative numbers than positive numbers. This is true since the total number of digits is even. The opposite is true when the number of digits is odd. For example, a 5-digit negadecimal number can represent numbers in the range from $(90909)_{-10} = 90909_{10}$ through $(09090)_{-10} = -9090_{10}$.

In general, in a negative-radix number system, there is no need for a separate sign digit and consequently, complementing schemes are not necessary. The sign of the number is typically determined by the position of the most significant nonzero digit. Since there is no distinction between the positive and negative numbers, however, arithmetic operations are somewhat indifferent to the sign of the operands. Accordingly, the arithmetic algorithms used to process negative-radix numbers are far more complex than those used to process the positive-radix number system. Consider, for example, the following negabinary addition:

Weights	−32	16	−8	4	−2	1
-5_{10} =	0	0	1	1	1	1
19_{10} =	0	1	0	1	1	1
	0	1	0	0	1	0

In the column for weight $= -2$, there is a carry-in that is equivalent to 2 and the sum of augend and addend is equivalent to -4. Thus the overall sum of augend, addend, and carry-in is -2. Accordingly, we assign a 1 for that sum bit and 0 for the corresponding carry-out. Similarly, for example, the carry-in at the column corresponding to weight -8 is equivalent to 8. The sum of augend and addend there equals 1, that is, -8_{10}. With a carry-in, therefore, the overall sum becomes 0 with no carry-out. We continue on with columnwise addition. The resulting sum is found to be $(010010)_{-10} = 14_{10}$.

Example 1.18
Add the numbers -3_{10} and -21_{10} using the negabinary scheme.

Solution

Weights	−32	16	−8	4	−2	1
-3_{10} =	0	0	1	1	0	1
-21_{10} =	1	1	1	1	1	1
	1	1	1	0	0	0

The resulting sum is $(111000)_{10}$. It is correct since $(111000)_{-10} = -24_{10}$.

TABLE 1.8

Signed-Binary Addition Rules

Bits at column j	$00/1\bar{1}$	$01/10$	$01/10$	$0\bar{1}/\bar{1}0$	$0\bar{1}/\bar{1}0$	11	$\bar{1}\bar{1}/1\bar{1}$
Bits at column $j-1$	–	Both positive	Otherwise	Both negative	Otherwise	–	–
σ_j	0	$\bar{1}$	1	1	$\bar{1}$	0	0
χ_j	0	1	0	$\bar{1}$	0	1	$\bar{1}$

The digit set in fixed-radix number systems that we have considered so far is given by $\{0, \ldots, R-1\}$. A *signed-digit number system*, on the other hand, allows the digit set to be $\{\bar{R}, \ldots, \bar{1}, 0, 1, \ldots, R\}$ where $\bar{R}$ is defined as $-R$. As in the negative-radix number system, the signed-digit number is either positive or negative and does not require a separate sign bit. Unlike the negative-radix number system, however, the signed-digit numbers are redundant numbers. Consider a 4-digit signed-binary number, for example. It can accommodate the numbers in the range $\bar{1}\bar{1}\bar{1}\bar{1}$ (i.e., -15_{10}) through 1111 (15_{10}) which amounts to 31 different numbers. However, since each digit can take any one of three values {i.e., $\bar{1}, 0, 1$}, a 4-digit signed-binary number can represent up to $3^4 = 81$ numbers. This confirms that 50 ($=81-31$) of the representations (61.7%) are redundant numbers. For example, the signed-digit numbers $0\bar{1}10$, $00\bar{1}0$ and $\bar{1}110$ all represent -2_{10}. As we shall show here, it is sometimes advantageous in certain arithmetic operations to have a built-in redundancy. One of its key attractive features is that we may be able to eliminate significantly the need to account for carry propagation while adding numbers. However, one must be cognizant of the fact that excessive redundancy may sometimes become an expensive endeavor.

Two n-bit signed-binary numbers can be added in two steps. First, an interim sum σ_j and interim carry-out χ_j are determined for each of the j-th bit (where j ranges from 0 through $n-1$) based on the values of the operands at j and $j-1$ bit positions. The exact values of the two interim quantities for all combinations of the bits are listed in Table 1.8. Note that the pairs of bits (a) 11, 11, and 00, (b) 01 and 10, and (c) 01 and 10 all contribute to the same set of interim values. In its final step, the intermediate sum σ_j and intermediate carry-in χ_{j-1} are added columnwise to obtain the sum. This process of determining the sum for all bit positions in parallel is very significant since it allows us to realize fast algorithms for addition and, therefore, also for multiplication and division.

Example 1.19
Add the numbers 5_{10} and 2_{10} using 5-bit signed-binary number system.

Solution
Consider, for example, $1\bar{1}0\bar{1}\bar{1}$ and $01\bar{1}\bar{1}0$ as representing respectively the two numbers. The addition is then carried out as follows:

$$
\begin{array}{rccccccc}
5_{10} & = & & 1 & \bar{1} & 0 & \bar{1} & \bar{1} \\
2_{10} & = & & 0 & 1 & \bar{1} & \bar{1} & 0 \\
\sigma & & & 1 & 0 & 1 & 0 & \bar{1} \\
\chi & & 0 & 0 & \bar{1} & \bar{1} & 0 & \\
\hline
S & & 0 & 1 & \bar{1} & 0 & 0 & \bar{1}
\end{array}
$$

The resulting sum is correct since $(0\,1\,\bar{1}\,0\,0\,\bar{1}) = 7_{10}$.

1.7 Binary Codes

When interacting with computers, for example, we enter pieces of information such as an alphabetic or numeric character or a control character. These key inputs are then converted to an equivalent binary form before the computer can process them. This is usually accomplished by assigning a specific pattern of bits so that it is available somewhere in the computer's memory to correspond to each keyboard entry. Many other reasons exist for coding information; among them are error detection and correction and encryption. In this section, in particular, we shall explore a few important coding schemes that have been devised to represent decimal numbers. In general, the set of bit string used to represent a set of objects is called a *code* and each such bit string in the set is called a *code word*. Several special-purpose binary codes have already been developed over the years for performing specific functions. They each have particular advantages and characteristics. The assignment of code words to objects can be given either by an algebraic expression or by an assignment table. One of the goals in coding is to standardize a set of universal codes that can be used by all.

To code 2^n distinct quantities in binary, a minimum of n bits are required. Since there is no constraint on the maximum number of bits and if efficient codes are not a requirement, there can be numerous choices for binary coding. For example, the decimal numbers can be coded with ten bits, and each decimal number could be assigned a bit combination of a single 1 and nine 0s. In this particular coding scheme, the digit 6 can be coded, for example, as 0000001000 and, similarly, the digit 2 is coded as 0010000000. Consider a different coding scheme for the decimal number 10. We could write its equivalent in binary as 1010, or the individual decimal characters could be coded separately as 0001 and 0000. Herein, each decimal digit is represented by its 4-bit binary equivalent. Although this means of expressing the number in binary form requires multiple strings of four bits, this method is useful because it provides ease of conversion. This scheme could be used to express more complex numbers. For example, 657 can be readily expressed as 0110 0101 0111 instead of by its binary equivalent 1010010001_2. This most frequently used decimal code is called the *binary coded decimal* (BCD) code. Recall that finding the binary equivalent of 657_{10} will require a nontrivial effort.

BCD is also known as the 8-4-2-1 code since 8, 4, 2, and 1 are the weights of the corresponding bits as in binary. It is what we refer to as a weighted code. In general, the number represented by a weighted code word is found by adding the weights corresponding to each 1 in the code word. The BCD weight, in particular, makes use of the binary representations of decimal digits 0_{10} through 9_{10}. The bit combinations corresponding to the binary numbers greater than 1001 are not used in BCD. To stress its distinctiveness from its binary equivalent, therefore, the BCD code of 657, for example, is thus expressed as 011001010111_{BCD}.

One of the questions that need to be answered is "How many different ways can we construct codes for the ten decimal numbers?" The number of different ways to assign ten out of sixteen 4-bit code words is given by (16!)/(16!6!). However, there can be up to 10! ways to assign each different choice to the 10 digits. Accordingly, there can be up to (16!10!)/(16!6!) or 29,059,430,400 possible 4-bit decimal codes. Table 1.9 shows some of the more common weighted codes that may be used to code each decimal digit to form a number. The list includes a few other codes besides BCD: 4221, 2421, $84\overline{2}\overline{1}$, and $631\overline{1}$. The latter four codes are also known as *self-complementing BCD* codes. What is special about these codes is that the 1's complement of the code word in any of these coding schemes yields the 9's complement of the number itself. We shall discuss their applications later in this section.

Depending on the applications, digital systems are often equipped with *encoders* and *decoders* for translating one code to another. It is important to note that a string of bits

TABLE 1.9

Examples of Binary Codes

Decimal	BCD	2421	4221	842$\overline{1}$	631$\overline{1}$
0	0000	0000	0000	0000	0011
1	0001	0001	0001	0111	0010
2	0010	0010	0010	0110	0101
3	0011	0011	0011	0101	0111
4	0100	0100	1000	0100	0110
5	0101	1011	0111	1011	1001
6	0110	1100	1100	1010	1000
7	0111	1101	1101	1001	1010
8	1000	1110	1110	1000	1101
9	1001	1111	1111	1111	1100

TABLE 1.10

Unweighted Codes

Decimal	XS3	Cyclic	Gray
0	0011	0000	0000
1	0100	0001	0001
2	0101	0011	0011
3	0110	0010	0010
4	0111	0110	0110
5	1000	0100	0111
6	1001	1100	0101
7	1010	1110	0100
8	1011	1010	1100
9	1100	1000	1101

may not always represent a decimal number; the string of digits may represent some other information as specified by the coding scheme. Some digital systems store information in BCD form and also perform BCD arithmetic operations. These systems, if they were to be forced to use binary adders, may contribute to the following three problems:

1. The use of binary adders may result in 4-bit groups that may not represent legitimate BCD codes.
2. The BCD sum even when it is formed of legitimate code words may not be same as the correct answer.
3. Finding the 9's complement of BCD values is not a simple task.

The aforementioned problems are typically overcome by using a code called *Excess-3, or XS3* for short. In this code, as listed in Table 1.10, a binary number three greater than the decimal digit represents each decimal digit. Conversely, an XS3 number can be converted to its BCD equivalent readily by adding 1101 to it and by discarding the carry, if any, from the sum. For example, the XS3 number 0111 (representing 4_{10}) is converted to its BCD equivalent by first performing the addition $0111 + 1101$, which gives 10100. By discarding the carry-out we obtain 0100 (the correct BCD code for 4_{10}). The XS3 code is not a weighted code but it has the advantage that when binary addition is performed on two decimal operands, their sum produces a carry bit from the most significant bit position under exactly similar circumstances as those in a decimal addition.

Digital Design: Basic Concepts and Principles

Consider adding 12973_{10} and 23834_{10} in both BCD and XS3 code. By representing these two 5-digit numbers first in BCD form, they are added as follows:

Carry-out	00110		0	0	1	0	0
Addend	12973	=	0001	0010	1001	0111	0011
Augend	23834	=	0010	0011	1000	0011	0100
Sum	36807	=	0011	0110	0001	1010	0111

We notice that in decimal addition, two carry-outs are generated respectively out of digit positions 1 and 2. However, when these same numbers were represented in BCD form, only one carry-out is generated (i.e., from position 2). Besides, BCD code has no code such as 1010. Now let us consider this same addition but in XS3 code.

Carry-out	00110		0	0	1	1	0
Addend	12973	=	0100	0101	1100	1010	0110
Augend	23834	=	0101	0110	1011	0110	0111
Sum	36807	=	1001	1100	1000	0000	1101

In XS3-based addition, we notice though that two carries were indeed generated from positions 1 and 2. Unfortunately, some of the resulting sum bit strings are also invalid here. In either case, therefore, the sums obtained using binary addition will have to be corrected to get error-free result.

An examination of the binary addition of BCD numbers will convince us that as long as the BCD sum is 9 (i.e., 1001) or less, the addition always yields the correct sum. If and when the sum exceeds decimal 9, the result has to be adjusted by adding decimal 6 (i.e., 0110) to it. Let us illustrate the concept by considering binary addition in the following three examples:

Decimal	BCD	Decimal	BCD	Decimal	BCD
6	0110	8	1000	9	1001
+2	0010	+5	0101	+8	1000
	1000		1101		10001

The leftmost example leads to the correct sum while the other two examples result in error. The example in the middle, in particular, also leads to an invalid BCD code. However, if we use binary arithmetic to add 0110 to the two wrong results, we get the correct sums since $1101 + 0110 = 0001\ 0011 \rightarrow 13_{10}$ and $10001 + 0110 = 0001\ 0111 \rightarrow 17_{10}$. This correction takes care of both wrong sums and invalid code.

The XS3 code, in particular, can be easily complemented to accommodate storage of negative numbers. It, therefore, lends itself to be used effectively in 2's complement subtraction. It so happens that the 9's complement of an XS3-coded decimal digit may be obtained readily by complementing each bit of the XS3 code. Consider the number 4_{10}, for example, that is represented as 0111 in XS3. The 9's complement of 4_{10} is 5_{10} that is represented as 1000 in XS3. We note that 1000 can be obtained from 0111 and vice versa simply by complementing bits. The list of self-complementing code includes the 4221, 2421, $842\bar{1}$, and $631\bar{1}$ codes that were listed in Table 1.9. These codes are all self-complementing BCD codes. Like the BCD code, these codes are all weighted.

Sometimes it may be necessary for a designer to avoid a number system where two or more bits change value between successive numbers. We shall consider some of those situations in Chapter 10. When such is the case, *Cyclic code* that uses four unweighted bits to represent decimal numbers may be used. They have this unique property that the successive code words differ only in one bit position. Table 1.10 lists two such unweighted codes: the typical cyclic code and the Gray code. The *Gray*, or *distance*, code is very similar to the cyclic code except that it also has a reflective property. By considering the first 2^n Gray codes where n is a positive integer, this reflective property of the code will become obvious. For example, the first 16 Gray codes are 0000, 0001, 0011, 0010, 0110, 0111, 0101, 0100, 1100, 1101, 111, 1110, 1010, 1011, 1001, and 1000. The MSD of the first eight code words and that of the second eight code words are different. All other bit values are "reflected" about the midpoint. For 4-bit Gray code, the mid-point is between 7 and 8. Consider for example, the Gray code for 6, which is 0101. This implies that the code for 9 should be the same except for the MSD. And, indeed, Gray code for 9 is 1101.

There are two convenient ways to formulate a Gray code with any arbitrary number of bits. The first scheme follows directly from its reflective property. It is given as follows:

1. A 1-bit Gray code has two code words, 0 and 1.
2. The first 2^n code words of a $n + 1$-bit Gray code equal the code words of an n-bit Gray code, written in the same order with a leading 0 appended.
3. The last 2^n code words of a $n + 1$-bit Gray code equal the code words of an n-bit Gray code, written in the reverse order with a leading 1 appended.

Thus, a 2-bit Gray code, for example, can be constructed easily from a 1-bit Gray code. Since the code words of 1-bit Gray code are 0 and 1, respectively, the least significant bits of the four code words of a 2-bit Gray code are 0, 1, 1, and 0. By appending a leading 0 to the least significant bits of the first two code words and by appending a leading 1 to the least significant bits of the last two code words, we obtain 00, 01, 11, and 10 for the 2-bit Gray code. Similarly, 3-bit Gray code is generated by appending 0 to the left of 00, 01, 11, and 10 and 1 to the left of 10, 11, 01, and 00. The resulting 3-bit Gray codes are, therefore, 000, 001, 011, 010, 110, 111, 101, and 100.

The second scheme to construct an n-bit Gray code follows directly from the corresponding n-bit binary code word. The Gray code can be formulated from the corresponding binary number using the following rule: Bit j of the Gray code word is 0 if bits j and $j + 1$ of the corresponding binary code word are the same; otherwise bit j is a 1. When $j + 1 = n$, bit $j + 1$ of the binary code word is considered to be 0. Accordingly, when determining a Gray code word corresponding to decimal 9 (i.e., 1001), for example, the four bit pairs $j + 1$ and j of the binary code word are 01, 10, 00, and 01. The bits of the corresponding Gray code word are therefore 1, 1, 0, and 1. The code word for 9 is thus 1101. Table 1.10 shows that this formulation works for all 4-bit Gray code words.

Digital computers process not only numeric information, but also alphabetic characters, punctuation marks, and other operations as well as control characters. To handle such alphanumeric items, binary codes may be used to represent them. The two most popular alphanumeric codes are American Standard Code for Information Interchange (ASCII) and Extended BCD Interchange Code (EBCDIC). They are used extensively to represent characters that come from the keyboard. It quite often happens in computer systems that the codes used for its inputs and the codes used for its outputs are different. In any case, however, a *table look-up* process is used to convert one form of number/character representation to another.

The alphanumeric items encountered in a digital computer include 26 uppercase letters, 26 lowercase letters, 10 decimal digits, 32 printable special characters, and several unprintable control data. The alphanumeric character set thus consists of less than 128 items. Since

TABLE 1.11

ASCII Code

	0000	0001	0010	0011	0100	0101	0110	0111
0000	NUL	DLE	SP	0	@	P		p
0001	SOH	DC1	!	1	A	Q	a	q
0010	STX	DC2	"	2	B	R	b	r
0011	ETX	DC3	#	3	C	S	c	s
0100	EOT	DC4	$	4	D	T	d	t
0101	ENQ	NAK	%	5	E	U	e	u
0110	ACK	SYN	&	6	F	V	f	v
0111	BEL	ETB	'	7	G	W	g	w
1000	BS	CAN	(	8	H	X	h	x
1001	HT	EM	)	9	I	Y	i	y
1010	LF	SUB	*	:	J	Z	j	z
1011	VT	ESC	+	;	K	[	k	{
1100	FF	FS	,	<	L	\	l	|
1101	CR	GS	−s	=	M	]	m	}
1110	SO	RS	.	>	N	^	n	~
1111	SI	US	/	?	O	_	o	DEL

NUL: Null; ETX: End of text; ACK: Acknowledge; HT: Horizontal tab; FF: Form feed; SI: Shift in; DC2: Device control 2; NAK: Negative ACK; CAN: Cancel; ESC: Escape; RS: Record separator; SP: Space; SOH: Start of heading; EOT: End of transmission; BEL: Bell; LF: Line feed; CR: Carriage return; DLE: Data link escape; DC3: Device control 3; SYN: Synchronize; EM: End of medium; FS: File separator; US: Unit separator; STX: Start of text; ENQ: Enquiry; BS: Backspace; VT: Vertical tab; SO: Shift out; DC1: Device control 1; DC4: Device control 4; ETB: End transmission block; SUB: Substitute; GS: Group separator; DEL: Delete.

$2^7 = 128$, a 7-bit ASCII code is used to represent all these characters as shown in Table 1.11. Note that the 34 control characters appear abbreviated in Table 1.11. Control information are of three types: format effecters, information separators, and communications controllers. The format effecters are used to control the printing layout while the information separators separate data into divisions such as paragraphs, and pages, and the communication controllers are used to guide transmission of texts between terminals. The 7-bits of ASCII are designated often by bit notations b_6 through b_0, respectively. However, since most digital computers are capable of manipulating 8-bit numbers, ASCII characters may as well be written as an 8-bit code, provided the most significant bit b_7 is set to a 0. The first column of Table 1.11 lists the least significant four bits while the first row lists the more significant four bits. Using Table 1.11, the text string "Hello!," for example, is represented by the string "0100 1000 0110 0101 0110 1100 0110 1100 0110 1111 0010 0001."

EBCDIC uses eight bits for each of the characters. It accounts for all symbols as ASCII but the bit assignments for each of the characters are different. EBCDIC bit assignments are made in such a way that both $b_7b_6b_5b_4$ and $b_3b_2b_1b_0$ range from 0000 through 1001 as in BCD.

1.8 Error-Detecting and Correcting Codes

As digital codes are manipulated within a system, there is always a possibility, albeit small, that a random-noise pulse will change a 0 to a 1 or a 1 to a 0. The chance of having more than one bit changed simultaneously is even smaller. It is possible, however, to code the data so that the occurrence of an error can be detected once the damaged data has been examined.

The simplest approach is to add an extra bit, called a *parity bit*, to each of the number codes. If the coded data including the parity bit has an even number of 1s, the code is said to have an *even parity*. On the other hand, if the coded word including the parity bit has an odd number of 1s, the code is said to have an *odd parity*. Following this approach, we can keep track of bits in a BCD code, for example. In a parity-justified code, the parity bit makes the number of 1s in the 5-bit number either odd or even, respectively giving rise to an odd parity and even parity BCD codes. For example, BCD value of 8 can be encoded as 10001 in even format and as 10000 in odd format by placing the proper parity bit on the right side of 1000. Similarly, for example, BCD value of 3 will become 00110 and 00111 respectively in even and odd format. Even though the two parity schemes are equivalent, odd parity is generally preferred since it ensures at least one 1 in the code word. If any one and only one bit of the code word gets changed, we would notice immediately that the code word has its parity changed. However, if two bits get changed simultaneously, the parity codes will fail to detect such errors. The parity bit for the BCD may be added arbitrarily, for example, on the extreme right.

The *biquinary* (also referred to as 5043210) code listed in Table 1.12 is a 2-of-7 code, meaning that only two bits are 1 in a 7-bit code word. This code also allows for error detection. If the code used is biquinary, a simple inspection of its bits will reveal the occurrence of an error. Biquinary is a special type of the m-out-of-n codes, where all valid code words have n bits, of which m bits are 1s and $n - m$ bits are 0s. Table 1.11 lists also the 2-out-of-5 code, which is unweighted. In 2-out-of-5 code, decimal 0 in particular is unweighted while all of the other decimal numbers are weighted.

Residue codes can be designed for an integer N in terms of an arbitrary quotient m for error detection as well. A residue r is the remainder obtained after a division operation such that $N = (I \times m) + r$ and $0 < r < m$. The quantity r is the residue, which is also referred to as "N modulo m." The residue, for example, of the decimal number 6578 with respect to $m = 7$ is 2 since $6578 = (939 \times 5) + 2$. In residue codes, the data bits define the number N and the residue appended to the data bits are used as check bits. The residue code of 6578 is obtained by concatenating 1100110110010 (binary equivalent of 6578) and 010 (binary equivalent of 2). The residue code of 6578 is thus 1100110110010010.

The error detection and correction capabilities of a code are measured often by its Hamming distance. *Hamming distance* refers to the number of bit changes necessary to convert one code to another. Accordingly, it is same as the number of bits in which two distinct code words differ. The minimum Hamming distance is just the minimum number of bits that must be changed to convert one code to another. In general, the minimum Hamming

TABLE 1.12

Error Detecting Codes

Decimal	Odd Parity	Even Parity	2-Out-of-5	Biquinary
0	00001	00000	01001	0100001
1	00010	00011	00011	0100010
2	00100	00101	00101	0100100
3	00111	00110	00110	0101000
4	01000	01001	01010	0110000
5	01011	01010	01100	1000001
6	01101	01100	10001	1000010
7	01110	01111	10010	1000100
8	10000	10001	10100	1001000
9	10011	10010	11000	1010000

distance d of a code is related to its error detection and error correction capabilities:

$$d = C + D + 1 \tag{1.18}$$

where C and D are respectively the number of to-be-corrected bit errors and to-be-detected bit errors. Typically, D is equal or greater than C since no errors can be corrected without being detected. To correct k errors the minimum Hamming distance must not be smaller than $2k + 1$. A total of k errors would produce a code word k distance away from the correct code word. To be able to correct this error, no other k errors should be able to produce this code word. The new code word, therefore, should be at a distance of at least $k + 1$ from any other code word. Accordingly, the minimum distance between two code words should be $2k + 1$. In codes with $d = 1$, such as in BCD, two valid code words may differ only in one bit. A single error in such a code word may thus produce only another valid code word. Consequently, codes with $d = 1$ is not suited for detecting single errors. On the other hand, if $d = 2$, a single-bit error in a code word will convert it to an invalid code word. These codes are thus able to detect single errors. However, if two bits were to change simultaneously, the errors may cause a code word to generate another valid code word. An example of such codes are parity-coded BCDs that can detect single errors but are unable to detect double errors. Each code in an m-out-of-n scheme is at least distance two away from the other code words. Consequently, these codes can be used to also detect single errors.

The main idea behind error-detecting codes is to add check (or parity) bits to the data bits so that if an error ever occurs in the data it can be detected. On the other hand, the error-correction codes can take us a step further. They can be used to not just detect error(s) but also identify their exact location(s). A minimum distance of four will provide for both single-error correction and double-error detection. A minimum distance of five would allow double-error correction. Next, we shall examine a code for which both error detection and error correction are straightforward.

In order to achieve single-bit error correction, we must append p parity bits, $x_p, \ldots, x_2, x_1$, to the m-bit data such that we can determine which one, if any, of the $(m + p)$ bits is in error. The number of parity bits required for unambiguous single-bit error correction is determined by solving the Hamming relationship:

$$2^p \geq m + p + 1. \tag{1.19}$$

The resulting error correcting code is referred to as the Hamming code. It uses multiple parity bits placed at specific locations of the coded word. The general rules for generating parity bits are as follows:

1. Parity bits are placed at those bit positions that correspond to the powers of 2, that is, at bit positions 1, 2, 4, 8, 16, and so on, of the coded word. The data bits are placed in order at other locations, that is, 3, 5, 6, 7, 9, 10, and so on.

2. Each parity bit individually takes care of only a few bits and not all of the coded word. To determine the bits of the coded word that are accounted for by a parity bit, every bit position is expressed in binary form. A parity bit would check only those bit positions, including itself, that have a 1 in the same location of their binary representation as in the binary representation of the parity bit.

Example 1.20

Determine the Hamming code of 101101 using an even parity scheme.

Solution

The number of data bits is 6 and, therefore, $p = 4$. The resulting Hamming code should thus form a 10-bit code word. The bit position values are expressed first in binary. The data bits (101101) are positioned respectively at bit locations 3, 5, 6, 7, 9, and 10.

The parity bits (even scheme in this case) are then determined by studying the binary representations of the bit positions and by checking off each of the parity bits against its corresponding data bits as follows: Parity 1, for example, checks bit positions 1, 3, 5, 7, and 9 since binary representation of all these bit locations have a 1 in the LSD of their binary representations. The data bits present at all four of the data locations (3, 5, 7, and 9) are 0, 1, 0, and 1, respectively. Since there are two 1s in the data bits, Parity 1 at location 0001 should be a 0 (indicated by a bold 0) to maintain an even parity.

Position	10	9	8	7	6	5	4	3	2	1
Binary	1010	1001	1000	0111	0110	0101	0100	0011	0010	0001
Data bits	1	0		1	1	0		1		
Parity 1		0		1		0		1		**0**
Parity 2	1			1	1			1	**0**	
Parity 3				1	1	0	**0**			
Parity 4	1	0	**1**							

Once we have determined all of the four parity bits, we can write out the 10-bit Hamming code as 1011100100.

Hamming code allows for both error detection and error correction. The *general detection algorithm* consists of the following steps:

Step 1. Starting with bit 1, check parity on each parity bit and the bits for which it provides parity.

Step 2. If the test indicates the preservation of assumed parity, a 0 is assigned to the test result. A failed test is assigned a 1.

Step 3. The binary number formed by the score of parity tests indicates the location of the bit in error. The last test score is interpreted as the MSD of this number and the first test score is assigned its LSD. Every other test score is assigned bit position in order between these two digits.

Example 1.21

Assume that the Hamming code obtained in Example 1.20 had its tenth bit position changed. By using the correction algorithm as above, show that the Hamming code is actually able to locate the error.

Solution

On the basis of the given assumption, the damaged data appears as 0011100100. We may now test the state of parity for Parity 1, Parity 2, Parity 3, and Parity 4. For each passed test (if the bits were to maintain even parity), we assign a 0 and for each failed test (when the bits do not preserve even parity), we assign a 1.

Position	10	9	8	7	6	5	4	3	2	1	Test
Binary	1010	1001	1000	0111	0110	0101	0100	0011	0010	0001	
Data bits	0	0		1	1	0		1			
Parity 1		0		1		0		1		0	Pass
Parity 2	0			1	1			1	0		Fail
Parity 3				1	1	0	0				Pass
Parity 4	0	0	1								Fail

The position of the damaged data bit is thus given by the number formed of the successive test results: Fail Pass Fail Pass → 1010, that is, position 10. So the scheme indeed identifies the bit in error.

1.9 CAD System

From the very early days of electronic computing, computers were used to analyze and design circuits. This is because circuit design often involves optimization of (or balancing with) many competing design requirements to be calculated with complex computation formulas. Computers typically resolve these difficult computations more efficiently (accurate and faster) than most experienced circuit designers. Computer software used to aid circuit design is commonly known as *computer-aided design* (CAD) system.

For example, in 1960 General Electric's Haning and Mayes reported that computers were used in interpretation of Boolean Equations, partition of circuits for plug-in cards, generating interconnects between circuit cards, generating circuit schematics, and so on, as well as design and building new computers. In 1961, Kaskey, Prywes and Lukoff of the UNIVAC Project in Pennsylvania reported the use of computers in all aspects of designing, building, and maintaining the famous UNIVAC computers.

The most famous and longest legacy CAD application for circuit design and analysis is perhaps the SPICE program, which grew out from several student projects led by Professor Donald O. Pederson in late 1950s at the UC-Berkeley campus. Over the years, the SPICE program has been enhanced in many ways and many companies now offer SPICE-like CAD products in the *electronic design automation* (EDA) market.

With *integrated circuit* (IC) designs surpassing 125-million transistors on a single silicon chip, CAD systems help engineers to manage the complex design-to-manufacturing process, and shorten the *time-to-market* (TTM) as well as postdesign and postmanufacturing processes. A typical IC design methodology flow is shown in Figure 1.4, wherein there are numerous CAD applications used at every design and engineering step.

In this book, we will introduce and use VHDL/Verilog® HDL with Xilinx WebPACK™ ISE Logic Design software for *field-programmable-gate-array* (FPGA) platforms to enhance students' learning experience. The software and a quick start handbook can be downloaded, free of charge, at www.xilinx.com.

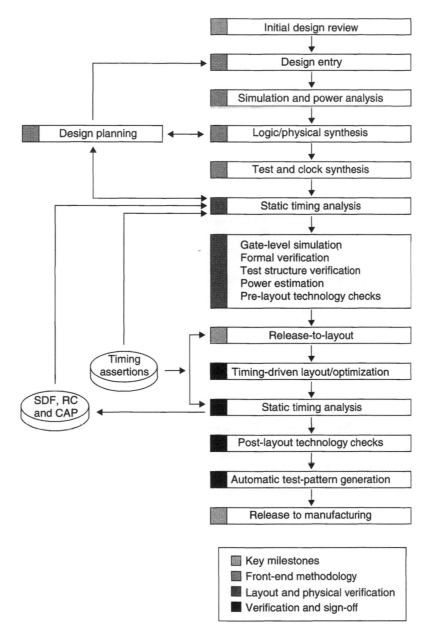

FIGURE 1.4
IBM Blue Logic ASICs design flow ©IBM (at http://www-306.ibm.com/chips/products/asics/methodology/ 5/26/2004).

1.10 Summary

In this chapter, the various number systems were explored. Schemes for transforming numbers from one system to another were discussed and also demonstrated. The techniques for representing negative numbers were also identified, and these representations were used in binary arithmetic. Two unconventional number systems (negabinary and signed-digit) were also explored even though they will not be used anywhere else in the book.

These schemes, however, are actively being considered for use in hybrid electro-optical computing systems. Finally, we introduced various binary codes and demonstrated how some of these could be used. In particular, we studied and discussed error-detection as well as error-correction codes.

Bibliography

Floyd, T., *Digital Fundamentals*. 8th edn. Englewood Cliffs, NJ. Prentice Hall, 2003.

Gerner, H.L., Theory of Computer Additions and Overflows, *IEEE Trans. Comp.*, **C-27**, 297, 1978.

Hamming, R.W., *Coding and Information Theory*. Englewood Cliffs, NJ. Prentice Hall, 1980.

Hwang, K., *Computer Arithmetic*, New York, NY. Wiley, 1979.

Johnson, E.J. and Karim, M.A., *Digital Design: A Pragmatic Approach*. Boston, MA. PWS-Kent Publishing, 1987.

Karim, M.A. and Awwal, A.A.S., *Optical Computing: An Introduction*. New York, NY. John Wiley & Sons, 1992.

Katz, R.H., *Contemporary Logic Design*. Boston, MA. Addison Wesley, 1993.

Mowle, F.J.A., *A Systematic Approach to Digital Logic Design*. Reading, MA. Addison-Wesley, 1976.

Nelson, V.P.P., Nagle, H.T., and Carroll, B.D., *Digital Logic Circuit Analysis and Design*. Englewood Cliffs, NJ. Prentice Hall, 1995.

Rajeraman, V. and Radhakrishnan, T., *Introduction to Digital Computer Design*. Englewood Cliffs, NJ. Prentice-Hall, 1983.

Smith, K.C., Multiple-Valued Logic: A Tutorial and Appreciation, *Computer*, **37** (4) 17–27, 1988.

Swartzlander, E.E., Digital Optical Arithmetic, *Applied Optics*, **25**, 3021–3032, 1986.

Wakerley, J., *Error Detecting Codes, Self-Checking Circuits and Applications*. New York, NY. North-Holland, 1978.

Waser, S. and Flynn, M.J., *Introduction to Arithmetic for Digital Systems Designers*. New York, NY. Holt, Rinehart & Winston, 1982.

Problems

1. Convert the following base-R number to their base-10 equivalents:

    (a) $(1101.01)_2$ (b) $(6432.00)_8$ (c) $(1010.10)_2$

    (d) $(7767.64)_8$ (e) $(2312.30)_4$ (f) $(1ABF.23)_{16}$

    (g) $(3123.12)_4$ (h) $(F011.20)_{16}$ (i) $(4567.03)_8$

    (j) $(1101.01)_2$ (k) $(11011.11)_2$ (l) $(777666)_8$

    (m) $(23123.12)_4$ (n) $(AACE42)_{16}$ (o) $(32112.31)_4$

    (p) $(FF2310)_{16}$ (q) $(43526.71)_8$ (r) $(A001.1C)_{16}$

2. Convert the following base-10 numbers to their equivalent numbers in (i) base-2, (ii) base-4, (iii) base-8, and (iv) base-16:

    (a) $(12546.23)_{10}$ (b) $(43002.01)_{10}$ (c) $(43562.675)_{10}$

    (d) $(11909.91)_{10}$ (e) $(432760.02)_{10}$ (f) $(4321.023)_{10}$

    (g) $(3240.64)_{10}$ (h) $(17110.75)_{10}$ (i) $(12325.53)_{10}$

    (j) $(532.42)_{10}$ (k) $(5945.35)_{10}$ (l) $(4575.95)_{10}$

3. Perform the following additions:

   (a) $(1011011)_2 + (0011011)_2$ (b) $(54672.453)_8 + (11222.435)_8$
   (c) $(A453.FF23)_{16} + (A43.22F)_{16}$ (d) $(1011.1101)_2 + (1000.0011)_2$

4. Realize the subtractions indicated below using (i) the radix complement and (ii) the diminished-radix complement method. Note that the numbers within the parentheses are all magnitudes only:

   (a) $(10101.101)_2 - (11110.011)_2$ (b) $(11010.001)_2 - (10101.11)_2$
   (c) $(45627.723)_8 - (2564.234)_8$ (d) $(5634.234)_8 - (45367.723)_8$
   (e) $(AAC2F.4FE)_{16} - (9B854.11F)_{16}$ (f) $(19854.11F)_{16} - (AA82F.34E)_{16}$

5. Realize the subtraction $423_{10} - 526_{10}$ using (a) negabinary, and (b) signed-binary number system. Compare these subtractions with that using 2's complement arithmetic.

6. Given that n-bit word X is the 2's complement of n-bit word Y, prove that Y is also the 2's complement of X.

7. Show that the numerical value A of an all-integer 1's complement number $(a_{n-1}a_{n-2}\ldots a_1a_0)$ is given by

$$A = -a_{n-1}(2^{n-1} - 1) + \sum_{j=0}^{n-2} a_j 2^j$$

8. Truncate the 6-bit fractional fixed-point numbers between $+1$ and -1 to form (a) 2-bit, (b) 3-bit, (c) 4-bit, and (d) 5-bit fractions and then plot the truncated values against the actual values of the fractions for all three forms of binary representations. Summarize your findings.

9. Determine the BCD code for the following decimal numbers:

   (a) 87645   (b) 342145.32   (c) 14523.25

10. Determine the following codes for the decimal numbers listed in Problem 9:

    (a) XS3   (b) Gray   (c) Cyclic
    (d) Biquinary   (e) 2-out-of-5   (f) 4221

11. Determine Hamming codes for the following data bits:

    (a) 11010111   (b) 10011110   (c) 11110011111
    (d) 100111101   (e) 11101010101   (f) 101010101010

    Assume an even parity scheme.

12. Determine if the following Hamming codes have damaged data in it. Provide the corrected data for (i) even parity scheme and (ii) odd parity scheme:

    (a) 1010111010011   (b) 011010101101   (c) 0110011101011
    (d) 01110101110111   (e) 0101111011100   (f) 10011001100110

13. Show how the BCD code words representing the decimal digits can be encoded as a Hamming code using an odd parity scheme.

14. Repeat Problem 13 but for the XS3 code words.

2

Boolean Algebra

2.1 Introduction

Design of digital circuits is typically approached from an understanding of mathematical formalism pertinent to binary systems. This particular formalism is commonly known as Boolean algebra. Claude Shannon proposed this particular algebra, by extending the works of algebra of logic that was initially devised by George Boole, for analyzing and designing discrete binary systems. *Boolean algebra* is a mathematical system that defines three primary logical operations: AND, OR, and NOT, on sets of binary logical variables. Boole had used his algebraic manipulations for describing the logical relationships of natural language. This reference to natural language is very relevant here since we are also interested in translating a word statement of the function of the desired digital system to a mathematical description. Boolean algebra serves as the basis for moving from a verbal description of the logical function to an unambiguous mathematical description. This unambiguous representation allows us to design logic circuits using a given library of logic components.

Boolean algebra is finite but richer in scope than the ordinary algebra and, accordingly, it leads to simpler optimization schemes. Complex logical functions can be simplified using Boolean algebraic rules. Correspondingly, the design process leads to logic circuits that are both simplified and inexpensive. The properties of Boolean algebra need to be understood first before we could learn to design digital systems efficiently. This chapter will acquaint you with Boolean algebra and provide necessary tools for handling complex logical functions.

2.2 Logic Operations

The logic functions introduced here are the allowed operations in Boolean algebra, which is explored later in Section 2.4. An understanding of these logic operations is vital since they are used in translating a word statement of a logical problem to a digital logic network. These logic functions operate on variables that are allowed to assume only two values, 1 or 0. To define a complex logic function of several variables, typically only three logic functions (i.e., AND, OR, and NOT) are needed. There are a few other logic functions that can be derived from AND, OR, and NOT that will also be considered in this section. The digital electronic circuits corresponding to the logic functions are alternatively known as logic gates or simply *gates*. Note that *gate* refers to an actual electronic device, and *function* or *operation* refers to a logic operator.

Typically, a logic function may be described by means of a table, known commonly as a *truth table*. Such a table usually lists the output values for all possible combinations of the input values. Accordingly, in the following subsections, all logic operations are defined using truth tables.

2.2.1   AND Function

The *AND function*, defined by $f(A, B) = A \cdot B \equiv AB$, is a logical function that produces a 1 if both A and B are 1, and is a 0 otherwise. The dot used for the AND operation is frequently omitted if the function is obvious. This definition of a 2-input AND logic may be extended to that for more than two input variables. The output is a 1 as long as all of the inputs are 1, and is a 0 otherwise. The AND logic operation often resembles the multiplication process; however, it must not be confused with the arithmetic multiplication operation.

The AND function is best illustrated by the analogy shown in Figure 2.1a where A and B are two switches connected in series, and the output is indicated by the photoemitter f. Herein, the open switch is analogous to logic 0, and the closed switch is analogous to logic 1. When all the switches are closed the photoemitter is turned on, indicating an output of 1. For all other switch combinations, the photoemitter remains turned off. The truth table and the logic symbol for the AND function are shown respectively in Figure 2.1b and c.

2.2.2   OR Function

The *OR function*, defined by $f(A, B) = A + B$, is a function that is a 1 when either A or B or both are 1. The + sign used here stands for logical OR operation and not addition. The OR definition may also be extended to include more than two input variables. The output is a 0 as long as all of the inputs are 0, and is a 1 otherwise.

The OR logic, like AND, can be illustrated by an analogy involving switches where the switches are connected in parallel. This analogous circuit with a photoemitter is shown in Figure 2.2a. The photoemitter gets turned on (i.e., logic 1) in this configuration as long as at least one of the switches is closed (i.e., logic 1), and it remains turned off when

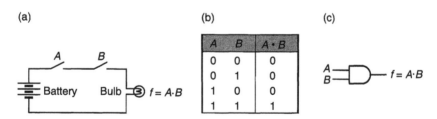

FIGURE 2.1
Two-input AND function (a) analogous circuit, (b) truth table, and (c) logic symbol.

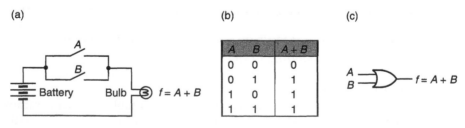

FIGURE 2.2
Two-input OR function (a) analogous circuit, (b) truth table, and (c) logic symbol.

(a)

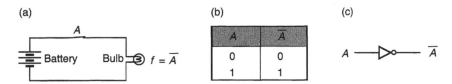

$f = \overline{A}$

(b)

A	$\overline{A}$
0	0
1	1

(c)

$A \longrightarrow \!\!\!\triangleright\!\circ \longrightarrow \overline{A}$

FIGURE 2.3
NOT function (a) analogous circuit, (b) truth table, and (c) logic symbol.

(a)

A	B	$\overline{AB}$
0	0	1
0	1	1
1	0	1
1	1	0

(b)

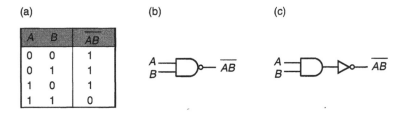

$\overline{AB}$

(c)

$\overline{AB}$

FIGURE 2.4
Two-input NAND function (a) truth table, (b) logic symbol, and (c) equivalent logic.

all the switches are open (i.e., logic 0). Figure 2.2b and c show the truth table and the logic symbol for the OR function.

2.2.3 NOT Function

The *NOT function*, defined by $f(A) = \overline{A}$ *or* A', is a function that is a 1 when A is a 0. The bar indicates the operation and is called interchangeably the *not*, *complement*, or *inversion* of A. Figure 2.3a shows the analogous circuit where normally the switch A remains closed. When it is not operated, the photoemitter remains turned on. When the switch is operated, the circuit is broken, causing the photoemitter to be turned off. The bubble on the logic symbol illustrated in Figure 2.3c typically indicates an inversion. The NOT function is the only meaningful logic function that has a single input. Figure 2.3b and c show the truth table and the logic symbol of the NOT function.

2.2.4 NAND Function

The *NAND function*, defined by $f(A, B) = \overline{AB}$, is a function that is a 0 when all of the inputs are 1 and it is a 1 for all other input combinations. NAND is thus identical to the AND function that has its output inverted. Accordingly, NAND logic is realized by also having an AND function followed by a NOT function. Figure 2.4 shows the truth table, logic symbol, and an equivalent circuit for a NAND function. It will be shown later that NAND of a given set of inputs is also equivalent to first realizing NOT of the inputs and then having these inverted inputs processed with an OR function.

2.2.5 NOR Function

The *NOR function*, defined by $f(A, B) = \overline{A + B}$, is a function that is a 1 when all of the inputs are 0 and a 0 for all other input combinations. NOR is thus identical to the OR function that has its output inverted. The NOR function is realizable by combining an OR gate and a NOT gate. Figure 2.5 shows the truth table, logic symbol, and equivalent logic for the two-input NOR functions. It will also be shown later that NOR of a given set of inputs is the same as

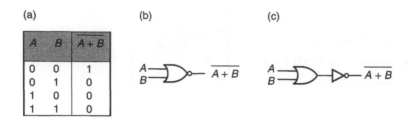

FIGURE 2.5
Two-input NOR function (a) truth table, (b) logic symbol, and (c) equivalent logic.

A	B	$\overline{A+B}$
0	0	1
0	1	0
1	0	0
1	1	0

(a)   (b)   (c)

A	B	$A \oplus B$	$A \otimes B$
0	0	0	1
0	1	1	0
1	0	1	0
1	1	0	1

FIGURE 2.6
Two-input XOR and XNOR functions (a) truth table and (b) logic symbols.

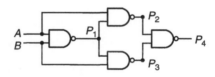

FIGURE 2.7
Circuit consisting of only NAND logic gates.

first realizing NOT of the inputs and then having these inverted inputs processed with an AND function.

2.2.6 Exclusive-OR and Exclusive-NOR Functions

The *exclusive-OR function* (XOR for short), defined by $f(A, B) = A \oplus B$, is a function that has an output of 1 if an odd number of its inputs are 1. The *exclusive-NOR* (XNOR for short) is just the opposite of the exclusive-OR and is denoted by $f(A, B) = A \otimes B$. Figure 2.6 shows the truth table and the logic symbols for both of these logic functions.

All these latter four functions (i.e., NAND, NOR, XOR, and XNOR) can be realized by suitably combining the other three logic functions. In fact, either only NAND or only NOR functions can be successively combined in certain sequence to generate both XOR and XNOR functions. The XOR and XNOR functions occur so frequently in digital design that devices have been designed to perform these functions as well as the others just discussed.

Consider the logic circuit of Figure 2.7. Since the logic elements used in this circuit are all NAND gates, we shall use NAND logic concept to first investigate the nature of the logic output at point P_1, and then at points P_2 and P_3, and finally at point P_4. The output at point P_1 (which is equivalent to NAND of A and B) is determined as follows.

P_1 and A appear as inputs to the second NAND logic and, similarly, P_1 and B appear as inputs to the third NAND logic. P_4 can be obtained then by performing NAND of P_2 and P_3. Consequently, we can now determine P_2 and P_3 and, finally, P_4 as follows.

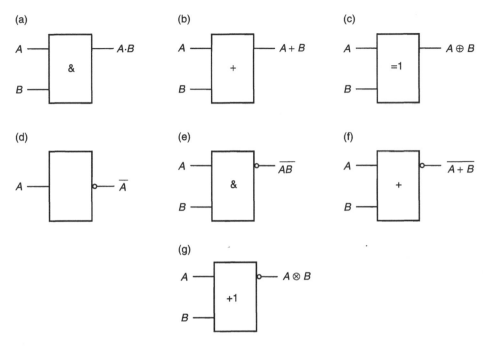

FIGURE 2.8
IEEE Symbols (a) AND, (b) OR, (c) XOR, (d) NOT, (e) NAND, (f) NOR, and (g) XNOR functions.

By comparing P_4 values with XOR output column of Figure 2.6a, we find that the two columns are identical for all combinations of the inputs. Consequently, we may safely conclude that the logic circuit of Figure 2.7 functions exactly as an XOR function. This equivalency illustration thus suggests that for most digital problems, there is more than one acceptable solution. Typically, the designers seek to compare the various alternatives to determine an optimum solution. In later sections, we shall investigate techniques for determining such optimal designs.

Figures 2.1 through 2.6 show distinctly different logic symbols corresponding to AND, OR, NOT, NAND, NOR, XOR, and XNOR logic elements respectively. However, besides these visibly distinct logic symbols, logic elements are also represented by rectangule shaped symbols. One such convention has resulted in what we refer to now as the Institute of Electrical and Electronics Engineers (IEEE) logic symbols. These symbols have an advantage that they are not only uniform but can also be more easily drawn on a video monitor. As shown in Figure 2.8, each of the logic elements is distinguished by what appears within the rectangles. With symbols &, >1 (or +), and =1 (or ⊕) appearing within a rectangle, they refer respectively to AND, OR, and XOR logic. Inputs to the function are customarily fed from the left of the rectangle and the outputs typically emerge from the right. Both inputs and outputs can also appear inverted. When a bubble is placed next to the input or output, it refers to an inversion. Thus by using a bubble either on the input or output, we are able to also denote NOT, NOR, NAND, and XNOR functions.

2.3   Logic Functions from Truth Table

Design problems are typically introduced in the form of a verbal description. These descriptions are often vague and misleading in many respects. As a first step, therefore, the design engineer translates the verbal description into a truth table form. The design problem is completely specified by a truth table and, therefore, it is a tool that allows the designer to examine the combination of all possible inputs and the corresponding outputs. In logic design, construction of the truth table is often overlooked, but it should be the first step in organizing a problem. We will introduce the concept of designing with truth tables.

It will be proper to introduce at this time also the concepts of literal, minterm, and maxterm for describing different combinations of inputs. Both a variable x and its complement $\bar{x}$ in an expression are known as literals. Although both x and $\bar{x}$ pertain to the same variable they are considered to be different literals. A *minterm* is that particular combination of input literals, which, when fed to an AND gate, yields a 1. For a function of three inputs A, B, and C, for example, if $A = 0$, $B = 1$, and $C = 1$, then the corresponding minterm is $\bar{A}BC$ since the AND of $\bar{A}$, B, and C would yield a 1. A *maxterm*, on the other hand, is that combination of the input literals, which, when fed to an OR gate would yield a 0. The maxterm corresponding to $A = 0$, $B = 1$, and $C = 0$ is, therefore, $A + \bar{B} + C$. No minterm or maxterm can include both the literals of any input. The minterm $\bar{A}BC$ is the complement of the maxterm $A + \bar{B} + \bar{C}$. This equivalency can be verified easily once the next section has been covered.

Consider designing a special-purpose democracy circuit that accepts three inputs: A, B, and C. The circuit outputs a 1 when a majority of the inputs are 1. The action of this or any circuit can be described completely by considering all possible input combinations and their expected outputs. Accordingly, a truth table may be constructed to describe the desired output function. In general, to accommodate N input variables the truth table will have a total of 2^N rows each for one particular combination of the inputs. Table 2.1 completely specifies this democracy function. For illustration, it lists both minterms and maxterms.

Both minterms and maxterms are sometimes referred to by the decimal equivalents of the input variables. Minterm 5, for example, corresponds to $A\bar{B}C$ while maxterm 3 corresponds to $A + \bar{B} + \bar{C}$. Considering $f(A, B, C)$ as the output of this democracy function, we can write a logic equation by noting which particular input combinations leads to an output of 1 and which yields an output of 0. The algebraic expression describing the resulting Boolean function can be formed by using an OR operator to combine only those minterms for which the function is equal to 1. This form of function is called the *sum-of-product* (SOP) form. For the function under consideration, it is given by

$$f(A, B, C) = \bar{A}BC + A\bar{B}C + AB\bar{C} + ABC \qquad (2.1)$$

TABLE 2.1

Truth Table for a Democracy Function

Decimal Equivalents	A	B	C	Output	Minterms	Maxterms
0	0	0	0	0	$\bar{A}\bar{B}\bar{C}$	$A + B + C$
1	0	0	1	0	$\bar{A}\bar{B}C$	$A + B + \bar{C}$
2	0	1	0	0	$\bar{A}B\bar{C}$	$A + \bar{B} + C$
3	0	1	1	1	$\bar{A}BC$	$A + \bar{B} + \bar{C}$
4	1	0	0	0	$A\bar{B}\bar{C}$	$\bar{A} + B + C$
5	1	0	1	1	$A\bar{B}C$	$\bar{A} + B + \bar{C}$
6	1	1	0	1	$AB\bar{C}$	$\bar{A} + \bar{B} + C$
7	1	1	1	1	ABC	$\bar{A} + \bar{B} + \bar{C}$

Equation 2.1, the SOP function, takes on the value 1 for four variable combinations: 011, 101, 110, and 111. For every other input combinations, the four AND terms included in Equation 2.1 would be a 0 and, consequently, $f(A, B, C)$ becomes a 0. Alternatively, the SOP function given in Equation 2.1 can be expressed as follows:

$$f(A, B, C) = \Sigma m(3, 5, 6, 7) \tag{2.2}$$

where 3, 5, 6, and 7 are the decimal values of the minterms that cause the function to be a 1. The symbol Σ refers to sum (i.e., OR operation) and m refers to the minterms.

An alternative form by which this same function f can be expressed is referred to as the *product-of-sum* (POS) form. It is obtained by using an AND operator to combine all those maxterms for which the function is equal to 0. The POS form of the democracy function is given by,

$$f(A, B, C) = (A + B + C)(A + B + \bar{C})(A + \bar{B} + C)(\bar{A} + B + C) \tag{2.3}$$

This POS form can be also expressed in a concise form:

$$f(A, B, C) = \Pi M(0, 1, 2, 4) \tag{2.4}$$

where 0, 1, 2, and 4 are the decimal equivalent values of those maxterms that yield a 0. The symbol Π refers to product (i.e., AND operation) and M refers to the maxterms. The reader should be aware that the terms "sum of products" and "product of sums" are both misnomers. Instead, they should have been "OR of ANDs" and "AND of ORs" respectively. As with many other things in life, these more appropriate names did not become much popular.

The circuits that implement the democracy function follow directly from Equations 2.1 and 2.3 and are obtained as shown in Figure 2.9. In the logic circuit of Figure 2.9a, for example, we first employ three NOT gates since we need to generate three different complemented literals ($\bar{A}$, $\bar{B}$, and $\bar{C}$). The appropriate literal combinations are then fed to the AND gates. Since there are four minterms and three variables in the expression, the circuit includes four 3-input AND gates. The four AND outputs are finally fed to an OR gate. This last OR gate must be capable of taking in four inputs. By similar reasoning of how Equation 2.3 was derived, we can obtain the logic circuit of Figure 2.9b. It also requires three NOT gates for obtaining the complemented literals. But in the next level, four 3-input OR gates are used to generate four OR terms, each of three literals. And, finally, the OR outputs are fed to a 4-input AND gate to generate the logic function.

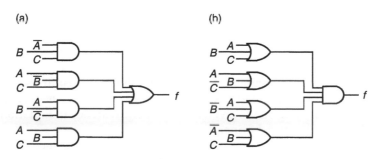

FIGURE 2.9
Logic circuits for democracy function (a) SOP form and (b) POS form.

(a)   (b)

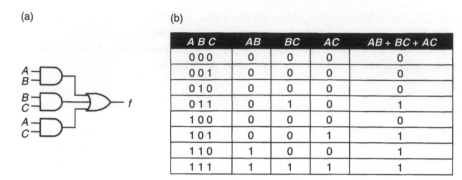

A B C	AB	BC	AC	AB + BC + AC
0 0 0	0	0	0	0
0 0 1	0	0	0	0
0 1 0	0	0	0	0
0 1 1	0	1	0	1
1 0 0	0	0	0	0
1 0 1	0	0	1	1
1 1 0	1	0	0	1
1 1 1	1	1	1	1

FIGURE 2.10
Simplified democracy function (a) logic diagram and (b) truth table.

The circuits just designed are more complex than necessary. The two circuits are about equivalent as far as their complexities are concerned. This is not surprising, since the truth table pertaining to this democracy function had equal numbers of 0s and 1s in its output column. When the number of 1s is less than or equal to the number of 0s for the output, typically the SOP form leads to a simpler circuit; otherwise the POS form is used. To make the simplest possible logic circuit, accordingly, a designer should look at the truth table to determine whether or not SOP form is to be preferred. Boolean algebra, yet to be introduced, is used to simplify both SOP and POS forms of any function. The SOP form of the function given in Equation 2.1, for example, can be reduced to

$$f(A, B, C) = AB + BC + AC \qquad (2.5)$$

Its circuit implementation is shown in Figure 2.10a. Verify for yourself that the simplified logic function will also yield a 1 for the same combination of input values. The truth table shown in Figure 2.10b clearly shows that the logic circuit of Figure 2.10a functions exactly as those shown in Figure 2.9. A designer will undoubtedly prefer the logic circuit of Figure 2.10a since it not only involves less number of logic gates but each of the OR and AND gates also has fewer gate inputs. Fewer gates and fewer gate inputs readily translate to less expensive circuits. Moreover, the input signals now have fewer logic gates to propagate through. This obviously implies significant speed enhancement since speed is a function often of the maximum number of logic gates through which the signals propagate. In summary, the logic circuit of Figure 2.10a is not only less expensive but also faster and simpler. In this book, we shall be learning techniques that will help simplify our design solutions. The next section on Boolean algebra is the starting point.

A Boolean function is said to be in its *canonical* form when it is expressed in terms of either a sum of minterms or a product of maxterms. The canonical forms are unique for each function. From the definitions of the two canonical forms, we can derive a general principle for converting from one canonical form to another. To convert from one canonical form of an n-input Boolean function (having m decimal equivalent entries) to its other canonical form, one needs to interchange Σ and Π signs and list all of the $2^n - m$ decimal equivalent numbers that were excluded from the given form. The equivalency of Equations 2.2 and 2.4 reflects the validity of this general principle.

The order of the variables in the canonical notation is important since it determines the order of the input bits, which, in turn determines the values of the minterms and maxterms.

Consider, for example, the SOP functions $f_1(A,B,C)$ and $f_2(B,A,C)$ both of which are formed of identical minterms 1, 2, 5, and 7.

$$f_1(A, B, C) = \Sigma m(1,2,5,7)$$

$$= \bar{A}\bar{B}C + \bar{A}B\bar{C} + A\bar{B}C + ABC \qquad (2.6)$$

$$f_2(B, A, C) = \Sigma m(1,2,5,7)$$

$$= \bar{B}\bar{A}C + \bar{B}A\bar{C} + B\bar{A}C + BAC$$

$$= \bar{A}\bar{B}C + A\bar{B}\bar{C} + \bar{A}BC + ABC \qquad (2.7)$$

Two of the AND terms are common between Equations 2.6 and 2.7; but the other two AND terms are uncommon. Clearly, therefore, f_1 and f_2 are not identical even though the minterm lists appeared to be the same. Defining the function $g(A, B, C)$ which is same as f_2, we find

$$g(A, B, C) = f_2(B, A, C)$$

$$= \bar{A}\bar{B}C + A\bar{B}\bar{C} + \bar{A}BC + ABC$$

$$= \bar{A}\bar{B}C + \bar{A}BC + A\bar{B}\bar{C} + ABC$$

$$= \Sigma m(1,3,4,7) \qquad (2.8)$$

Although equivalent, Equations 2.7 and 2.8 show the difference in their minterm lists. This difference is caused by the difference in the ordering of input variables.

Given a canonical form of an n-input Boolean function f, if we only list those decimal equivalent numbers that were excluded from the given form, we generate the complement function $\bar{f}$. But if in addition we also change the sign symbol (either Σ to Π or Π to Σ), we generate the function f itself but in its other canonical form. In other words, for example, while $f_2(A, B, C) = \Sigma m(1,2,5,7) = \Pi M(0,3,4,6)$, $\bar{f}_2 = \Sigma m(0,3,4,6) = \Pi M(1,2,5,7)$.

For a logic circuit that has n different inputs, the corresponding truth table may consist of up to 2^n different minterms. Furthermore, the outputs corresponding to an n-input logic function can be organized in 2^{2^n} different ways. Consequently, a 4-input logic function, for example, may result in 65,536 different output functions. Table 2.2 lists all the functions but of two variables A and B.

Of the 16 functions listed in Table 2.2, only 12 are unique. Two of the functions (F_0 and F_{15}) yield 0 and 1 respectively while two other functions (F_3 and F_5) are transfer functions. Accordingly, the implementation of these four functions requires no circuitry. Two of the twelve functions (F_{10} and F_{12}) are complement functions. In particular, there are four binary operations (two inhibition and two implication operations) that can be realized using other logic gates but are never used in computer design just because they are very difficult to implement as individual components. Neither of these is commutative or associative, and both are, therefore, not desirable in logic design. The eight functions that are implemented using single logic function include two complement operations, F_1, F_6, F_7, F_8, F_9, and F_{14}. The latter six logic functions correspond to AND, XOR, OR, NOR, XNOR, and NAND functions of A and B, respectively.

TABLE 2.2

Boolean Functions of Two Variables

Function	AB = 00	AB = 01	AB = 10	AB = 11
$F_0 = 0$	0	0	0	0
$F_1 = AB$	0	0	0	1
$F_2 = A\bar{B}$	0	0	1	0
$F_3 = A$	0	0	1	1
$F_4 = \bar{A}B$	0	1	0	0
$F_5 = B$	0	1	0	1
$F_6 = A\bar{B} + \bar{A}B$	0	1	1	0
$F_7 = A + B$	0	1	1	1
$F_8 = \overline{A + B}$	1	0	0	0
$F_9 = AB + \bar{A}\bar{B}$	1	0	0	1
$F_{10} = \bar{B}$	1	0	1	0
$F_{11} = A + \bar{B}$	1	0	1	1
$F_{12} = \bar{A}$	1	1	0 .	0
$F_{13} = \bar{A} + B$	1	1	0	1
$F_{14} = \overline{AB}$	1	1	1	0
$F_{15} = 1$	1	1	1	1

2.4 Boolean Algebra

As shown in the last section, a canonical form of a logic function can usually be reduced to a simpler form that leads to a less expensive, less complicated and faster logic circuit. Such function complexity reduction is accomplished often by means of Boolean algebra, although it was not initially developed specifically for the purpose of function simplification. For a set of logic elements, Boolean algebra can be defined in terms of a set of operators and a number of axioms or postulates that are taken as true without the need for any proof. A binary operator defined on the set of elements is a rule that assigns to each pair of elements in the set another unique element that is also included in the set.

To appreciate the usefulness of Boolean algebra, we need to briefly consider the principles of ordinary algebra first. Some of the more common postulates used in defining formalisms for ordinary algebra are as follows:

Closure: A set is closed with respect to an operator ∇ if and only if for every A and B in set S, $A \nabla B$ is also a member of set S.

Associativity: An operator ∇ defined on a set S is said to be associative if and only if for all A, B, and C in set S, $(A \nabla B) \nabla C = A \nabla (B \nabla C)$.

Commutativity: An operator ∇ defined on a set S is said to be commutative if and only if for all A and B in set S, $A \nabla B = B \nabla A$.

Distributivity: If ∇ and $\square$ are two operators on a set S, the operator ∇ is said to be distributive over $\square$ if for all A, B, and C in set S, $A \nabla (B \square C) = (A \square B) \nabla (A \square C)$.

Identity: With an operator ∇ defined on set S, the set S is said to have an identity element if and only if for an element A in set S, $A \nabla B = B \nabla A = B$ for every B in set S.

Inverse: With an operator ∇ and the identity element I defined on set S, the set S is said to have an inverse element if and only if for every A in set S, there exists an element B in set S such that $A \nabla B = I$.

Just as the case with ordinary algebra, the formulation of Boolean algebra is based on a set of postulates, known commonly as *Huntington's postulates*. Boolean algebraic structure is defined on a set of elements $S = \{0, 1\}$ with two binary operators, $+$ and $\cdot$, and satisfies the following postulates:

Postulate 1. The set S is closed with respect to the operators: $+$ (i.e., OR) and $\cdot$ (i.e., AND). Closure is self-evident in tables of Figures 2.1b and 2.2b. Both the AND and OR output are elements of set S.

Postulate 2. The set S has identity elements (a) 1 with respect to $+$ (i.e., OR) and (b) 0 with respect to $\cdot$ (i.e., AND). Using the tables of Figures 2.1b and 2.2b, we note that

(a) $A + 0 = A$ and $A \cdot 1 = A$

(b) $A \cdot 0 = 0$ and $A + 1 = 1$

which shows that both 0 and 1 are identity elements.

Postulate 3. The set S is commutative with respect to the operators: $+$ (i.e., OR) and $\cdot$ (i.e., AND). It follows directly from the symmetry of the tables in Figures 2.1b and 2.2b since $A + B = B + A$ and $A \cdot B = B \cdot A$.

Postulate 4. The operator $+$ (i.e., OR) is distributive over $\cdot$ (i.e., AND) and similarly $\cdot$ (i.e., AND) is distributive over $+$ (i.e., OR). This can be demonstrated by verifying both sides of the logical equation $A \cdot (B + C) = (A \cdot B) + (A \cdot C)$ for all possible cases of the variables.

$A\,B\,C$	$B + C$	$A \cdot (B + C)$	$A \cdot B$	$A \cdot C$	$A \cdot B + A \cdot C$
0 0 0	0	0	0	0	0
0 0 1	1	0	0	0	0
0 1 0	1	0	0	0	0
0 1 1	1	0	0	0	0
1 0 0	0	0	0	0	0
1 0 1	1	1	0	1	1
1 1 0	1	1	1	0	1
1 1 1	1	1	1	1	1

Similarly, we can also show by tabular method that $A + B \cdot C = (A + B) \cdot (A + C)$. Note that this is not the case in ordinary algebra.

$A\,B\,C$	$B \cdot C$	$A + B \cdot C$	$f_1 = A + B$	$f_2 = A + C$	$f_1 \cdot f_2$
0 0 0	0	0	0	0	0
0 0 1	0	0	0	1	0
0 1 0	0	0	1	0	0
0 1 1	1	1	1	1	1
1 0 0	0	1	1	1	1
1 0 1	0	1	1	1	1
1 1 0	0	1	1	1	1
1 1 1	1	1	1	1	1

Postulate 5. For every A in set S, there exists a complement element $\bar{A}$ in set S such that $A + \bar{A} = 1$ and $A \cdot \bar{A} = 0$.

There are several differences between ordinary algebra and Boolean algebra. In ordinary algebra, for example, + (i.e., OR) is not distributive over · (i.e., AND). Ordinary algebra, on the other hand, applies to the set of infinite real numbers while Boolean algebra applies to a set of finite elements. In Boolean algebra there is no room for subtraction or division operations since there are no inverses with respect to + or ·. On the plus side, however, complements are available in Boolean algebra but not in ordinary algebra. Although there are major differences between the two, Boolean algebra and ordinary algebra are alike in many respects. Note that Huntington's list of postulates does not include associativity since it can be derived from other postulates. Many in the field often include associativity among the postulates but that is unnecessary.

As is true for other axiomatic systems, each Huntington's postulate has a corresponding dual. Briefly explained, the *principle of duality* states that if and when a given logic expression is valid, the dual of the same logic expression is also valid. The Huntington's dual or mirror image is obtained by replacing each 0 with a 1, each 1 with an 0, each AND with an OR, and each OR with an AND. For example, the dual expression of the democracy function given by

$$\bar{A}BC + A\bar{B}C + AB\bar{C} + ABC = AB + BC + CA$$

is

$$(\bar{A} + B + C) \cdot (A + \bar{B} + C) \cdot (A + B + \bar{C}) \cdot (A + B + C) = (A + B) \cdot (B + C) \cdot (A + C)$$

This principle of duality along with the idea of taking a complement will be used later in the development of DeMorgan's theorems.

Boolean algebra is established on a set of two binary operators, AND and OR. However, since one of the postulates (i.e., Postulate 5) establishes the complement operator, Boolean algebra is frequently defined in terms of two or more elements subject to an equivalence relation "=" and three binary operators OR, AND, and NOT. Boolean theorems may now be derived using the Huntington's postulates. These theorems can be proven easily using the postulates stated above. While some of these proofs are left as an end-of-the-chapter problem (see Problem 1), a few of them are worked out so that you may get used to the mechanics of Boolean reductions. The theorems are listed as follows:

THEOREM 1
The Law of Idempotency.

For all A in set S,

(a) $A + A = A$ and
(b) $A \cdot A = A$

The part (a) follows directly from Postulate 2(b) since

$$A + A = A \cdot (1 + 1) = A \cdot 1 = A$$

Alternatively, we can verify this theorem by proceeding as follows:

$$A + A = (A + A) \cdot 1 \qquad \text{using Postulate 2(a)}$$
$$= (A + A) \cdot (A + \bar{A}) \qquad \text{using Postulate 5}$$
$$= A \cdot A + A \cdot \bar{A} \qquad \text{using Postulate 4}$$
$$= A + 0 \qquad \text{using Postulate 5}$$
$$= A \qquad \text{using Postulate 2(a)}$$

Part (b), on the other hand, can be justified as follows:

$$A \cdot A = A \cdot A + 0 \qquad \text{using Postulate 2(a)}$$
$$= A \cdot A + A \cdot \bar{A} \qquad \text{using Postulate 5}$$
$$= A \cdot (A + \bar{A}) \qquad \text{using Postulate 4}$$
$$= A \cdot 1 \qquad \text{using Postulate 5}$$
$$= A \qquad \text{using Postulate 2(a)}$$

THEOREM 2
The Law of Absorption.

For all A and B in set S,

(a) $A + (A \cdot B) = A$ and
(b) $A \cdot (A + B) = A$

Part (a) of this theorem can be proven as follows:

$$A + AB = A \cdot 1 + AB \qquad \text{using Postulate 2(a)}$$
$$= A \cdot (1 + B) \qquad \text{using Postulate 4}$$
$$= A \cdot 1 \qquad \text{using Postulate 2(b)}$$
$$= A \qquad \text{using Postulate 2(a)}$$

Part (b) of the theorem likewise can be justified as follows:

$$A \cdot (A + B) = A \cdot A + A \cdot B \qquad \text{using Postulate 4}$$
$$= A + A \cdot B \qquad \text{using Theorem 1(b)}$$
$$= A \qquad \text{using Theorem 2(a)}$$

THEOREM 3
The Law of Identity.

For all A and B in set S, if (a) $A + B = B$ and (b) $A \cdot B = B$ then $A = B$.

By substituting condition (b) into the left-hand side of condition (a), we get

$$A + A \cdot B = B$$

But according to Theorem 2(a)

$$A + A \cdot B = A$$

Therefore, $A = B$.

THEOREM 4
The Law of Complement.

For all A in set S, $\bar{A}$ is unique.
Assume that there are two distinct elements $\bar{a}_1$ and $\bar{a}_2$ that satisfy Postulate 5, that is,

$$A + \bar{a}_1 = 1, \quad A + \bar{a}_2 = 1, \quad A \cdot \bar{a}_1 = 0, \quad A \cdot \bar{a}_2 = 0$$

Then,

$$
\begin{aligned}
\bar{a}_2 &= 1 \cdot \bar{a}_2 && \text{using Postulate 2(a)} \\
&= (A + \bar{a}_1) \cdot \bar{a}_2 && \text{using Postulate 5} \\
&= A \cdot \bar{a}_2 + \bar{a}_1 \cdot \bar{a}_2 && \text{using Postulate 4} \\
&= 0 + \bar{a}_1 \cdot \bar{a}_2 && \text{using Postulate 5} \\
&= A \cdot \bar{a}_1 + \bar{a}_1 \cdot \bar{a}_2 && \text{using Postulate 5} \\
&= (A + \bar{a}_2) \cdot \bar{a}_1 && \text{using Postulate 4} \\
&= 1 \cdot \bar{a}_1 && \text{using Postulate 5} \\
&= \bar{a}_1 && \text{using Postulate 2(a)}
\end{aligned}
$$

Therefore, A has a unique complement in set S.

THEOREM 5
The Law of Involution.

For all A in set S, $(\bar{A})' = A$.
Since,

$$A + \bar{A} = 1 \quad \text{and} \quad A \cdot \bar{A} = 0 \qquad \text{(by Postulate 5)}$$

Then the complement of $\bar{A}$ is A. Therefore, $(\bar{A})' = A$.

THEOREM 6
The Law of Association.

For every A, B, and C in set S,

(a) $A + (B + C) = (A + B) + C$ and
(b) $A \cdot (B \cdot C) = (A \cdot B) \cdot C$

Let

$$D = \{A + (B + C)\} \cdot \{(A + B) + C\}$$

$$= A\{(A + B) + C\} + (B + C)\{(A + B) + C\} \qquad \text{using Postulate 4}$$

$$= A(A + B) + AC + \{B[(A + B) + C] + C[(A + B) + C]\} \qquad \text{using Postulate 4}$$

$$= A + AC + \{B(A + B) + BC + C[(A + B) + C]\} \qquad \text{using Theorem 2}$$

$$= A + \{B + BC + C(A + B) + CC\} \qquad \text{using Theorem 2}$$

$$= A + \{B + C(A + B) + C\} \qquad \text{using Theorem 2}$$

$$= A + (B + C) \qquad \text{using Theorem 2}$$

But

$$D = \{A + (B + C)\}(A + B) + \{A + (B + C)\}C \qquad \text{using Postulate 4}$$

$$= \{[A + (B + C)]A + [A + (B + C)]B\} + AC + (B + C)C \qquad \text{using Postulate 4}$$

$$= \{[A + (B + C)]A + [A + (B + C)]B\} + AC + C \qquad \text{using Theorem 2}$$

$$= \{[A + (B + C)]A + (AB + B)\} + C \qquad \text{using Theorem 2}$$

$$= \{AA + (B + C)A + B\} + C \qquad \text{using Theorem 2}$$

$$= \{A + (B + C)A + B\} + C \qquad \text{using Theorem 2}$$

$$= (A + B) + C \qquad \text{using Theorem 2}$$

Therefore, $A + (B + C) = (A + B) + C$

THEOREM 7
Law of Elimination.

For all A and B in set S,

(a) $A + \bar{A}B = A + B$
(b) $A \cdot (\bar{A} + B) = A \cdot B$

Part (a) of this theorem can be proved as follows:

$$A + \bar{A}B = (A + \bar{A})(A + B) \qquad \text{using Postulate 4}$$

$$= 1 \cdot (A + B) \qquad \text{using Postulate 5}$$

$$= A + B \qquad \text{using Postulate 2(a)}$$

This theorem is very much like that of absorption in that it can be employed to eliminate extra literal from a Boolean function.

THEOREM 8
DeMorgan's Law.

For all A and B in set S,

(a) $\overline{A + B} = \bar{A} \cdot \bar{B}$
(b) $\overline{A \cdot B} = \bar{A} + \bar{B}$

This theorem implies that a function may be complemented by changing each OR to an AND, each AND to an OR, and by complementing each of the variables. Let us prove part (a) only, first, for a two variable function.

Let $x = A + B$, then $\bar{x} = \overline{A + B}$. If $x \cdot y = 0$ and $x + y = 1$, then $y = \bar{x}$ per Theorem 4. Thus to prove part (a) of DeMorgan's theorem, we need to set $y = \bar{A} \cdot \bar{B}$ and evaluate $x \cdot y$ and $x + y$.

$$x \cdot y = (A + B) \cdot (\bar{A} \cdot \bar{B})$$

$$= A\bar{A}\bar{B} + B\bar{A}\bar{B} \qquad \text{using Postulate 4}$$

$$= 0 + 0 \qquad \text{using Postulate 5}$$

$$= 0 \qquad \text{using Postulate 2(a)}$$

$$x + y = (A + B) + \bar{A} \cdot \bar{B}$$

$$= A + B + \bar{A}\bar{B}$$

$$= A + (B + \bar{A}\bar{B}) \qquad \text{using Theorem 6}$$

$$= A + (B + \bar{A}) \qquad \text{using Theorem 7}$$

$$= B + (A + \bar{A}) \qquad \text{using Theorem 6}$$

$$= B + 1 \qquad \text{using Postulate 2(b)}$$

$$= 1$$

Therefore, $\overline{A + B} = \bar{A} \cdot \bar{B}$.
This theorem may be generalized for all $A, B, \dots,$ and Z in set S as follows:

(a) $\overline{A + B + \cdots + Z} = \bar{A} \cdot \bar{B} \cdots \cdot \bar{Z}$
(b) $\overline{A \cdot B \cdots \cdot Z} = \bar{A} + \bar{B} + \cdots + \bar{Z}$

The consequence of these two theorems can be summarized as follows: The complement of any logic function is obtained by replacing each variable with its complement, each AND with an OR, and each OR with an AND, each 0 with an 1 and each 1 with a 0.

THEOREM 9
The Law of Consensus.

For all A, B, and C in set S,

(a) $AB + \bar{A}C + BC = AB + \bar{A}C$
(b) $(A + B)(\bar{A} + C)(B + C) = (A + B)(\bar{A} + C)$

Part (a) of this theorem can be proved as follows:

$$
\begin{aligned}
AB + \bar{A}C + BC &= AB + \bar{A}C + 1 \cdot BC &&\text{using Postulate 2(a)} \\
&= AB + \bar{A}C + (A + \bar{A}) \cdot BC &&\text{using Postulate 5} \\
&= AB + AC + AB\bar{C} + \bar{A}B\bar{C} &&\text{using Postulate 4} \\
&= AB(1 + C) + \bar{A}C(1 + B) &&\text{using Postulate 4} \\
&= AB + \bar{A}C &&\text{using Postulate 2(b)}
\end{aligned}
$$

This theorem is used in both reduction and expansion of Boolean expressions. The key to using this theorem is to locate a literal in a term, its complement in another term, and associated literal or literal combination in both of these terms and only then the included term (the consensus term) that is composed of the associated literals can be eliminated.

THEOREM 10
The Law of Interchange.

For all A, B, and C in set S,

(a) $AB + \bar{A}C = (A + C) \cdot (\bar{A} + B)$
(b) $(A + B) \cdot (\bar{A} + C) = AC + \bar{A}B$

Part (a) of the theorem can be justified as follows:

$$
\begin{aligned}
AB + \bar{A}C &= (AB + \bar{A}) \cdot (AB + C) &&\text{using Postulate 4} \\
&= (A + \bar{A}) \cdot (B + \bar{A}) \cdot (A + C) \cdot (B + C) &&\text{using Postulate 4} \\
&= 1 \cdot (\bar{A} + B) \cdot (A + C) \cdot (B + C) &&\text{using Postulate 5} \\
&= (\bar{A} + B) \cdot (A + C) \cdot (B + C) &&\text{using Postulate 2(a)} \\
&= (A + C) \cdot (\bar{A} + B) &&\text{using Theorem 9}
\end{aligned}
$$

THEOREM 11
The Generalized Functional Laws.

The AND/OR operation of a variable A and a multivariable composite function that is also a function of A is equivalent to AND/OR operation of A with the composite function wherein A is replaced by 0:

(a) $A + f(A, B, \ldots, Z) = A + f(0, B, \ldots, Z)$
(b) $A \cdot f(A, B, \ldots, Z) = A \cdot f(0, B, \ldots, Z)$

The basis of this theorem is Theorem 1 and Postulate 2(a). Since $A = A + A = A + 0$, the variable A within the function in part (a) may be replaced by 0. For all $A, B, \ldots,$ and Z

in set S,

(a) $f(A, B, \ldots, Z) = A \cdot f(1, B, \ldots, Z) + \bar{A} \cdot f(0, B, \ldots, Z)$

(b) $f(A, B, \ldots, Z) = [A + f(0, B, \ldots, Z)] \cdot [\bar{A} + f(1, B, \ldots, Z)]$

The latter two versions of the theorem can be proved by making use of the other two versions, and Postulates 2(a) and 5(a).

The generalized functional laws become useful in designing a particular digital circuit called *multiplexer* or *data selector*. This circuit in turn can be used for implementing almost any Boolean function. These are amazingly powerful since they allow writing a Boolean function so that a selected variable and its complement appear only once. More will be said about this in Chapter 4. The following examples illustrate the use of the various Boolean theorems we have already introduced in this section.

Example 2.1

Obtain the overflow logic function in terms of (a) only sign bits and (b) only carries (in and out of sign bit).

Solution

(a) If the sign bit of the sum is different from that of both addend and augend, an overflow has occurred. Consider two n-bit numbers A and B when added together yields the sum S. Let A_{n-1}, B_{n-1}, and S_{n-1} represent the sign bits of A, B, and S, respectively. The truth table for the overflow logic function f is shown in Figure 2.11. Whenever the sign bit of both A and B are similar but different than that of S, the overflow has occurred.

Since there are only two 1s in the output column, it is wise to obtain an SOP expression for f.

Therefore,

$$f = \bar{A}_{n-1}\bar{B}_{n-1}S_{n-1} + A_{n-1}B_{n-1}\bar{S}_{n-1}$$

(b) The second method involves an examination of both carry-in and carry-out of the sign bit. Presence of different values for these two carries is indicative of an overflow. Given that we denote carry-in as C_{in} and carry-out as C_{out}, the truth table for f is obtained as shown in Figure 2.12.

Accordingly, we can use the SOP form to obtain

$$f = \bar{C}_{in}C_{out} + C_{in}\bar{C}_{out} = C_{in} \oplus C_{out}$$

Alternatively, we can make use of the POS form as,

$$f = (C_{in} + C_{out})(\bar{C}_{in} + \bar{C}_{out})$$
$$= C_{in}\bar{C}_{in} + C_{in}\bar{C}_{out} + C_{out}\bar{C}_{in} + C_{out}\bar{C}_{out}$$
$$= C_{in} \oplus C_{out}$$

A_{n-1}	B_{n-1}	S_{n-1}	f
0	0	0	0
0	0	1	1
0	1	0	0
0	1	1	0
1	0	0	0
1	0	1	0
1	1	0	1
1	1	1	0

FIGURE 2.11
Overflow logic function.

C_{in}	C_{out}	F
0	0	0
0	1	1
1	0	1
1	1	0

FIGURE 2.12
Truth table for Example 2.1.

$A\,B\,C$	$A+B$	$\bar{A}+C$	$B+C$	$(A+B)(\bar{A}+C)(B+C)$
0 0 0	0	1	0	0
0 0 1	0	1	1	0
0 1 0	1	1	1	1
0 1 1	1	1	1	1
1 0 0	1	1	0	0
1 0 1	1	1	1	1
1 1 0	1	0	1	0
1 1 1	1	1	1	1

FIGURE 2.13
Truth table for left side of Theorem 9(b) for Example 2.2.

Example 2.2
Verify Theorem 9(b) using truth tables.

Solution
Theorem 9(b) states that

$$(A + B)(\bar{A} + C)(B + C) = (A + B)(\bar{A} + C)$$

The truth table for generating the left-hand side of the above Boolean equation is obtained as shown in Figure 2.13. For each combination of the inputs, we first obtain three OR outputs and then we AND them together in the last column.

Similarly the truth table for generating the right-hand side of the above Boolean equation is obtained next as shown in Figure 2.14. Here only two ORs are derived first, which are then "AND"ed as listed in the last column.

Since entries of the two columns $(A+B)(\bar{A}+C)(B+C)$ in Figure 2.13 and $(A+B)(\bar{A}+C)$ in Figure 2.14 are alike for all possible minterms, the theorem is valid.

ABC	$A+B$	$\bar{A}+C$	$(A+B)(\bar{A}+C)$
0 0 0	0	1	0
0 0 1	0	1	0
0 1 0	1	1	1
0 1 1	1	1	1
1 0 0	1	0	0
1 0 1	1	1	1
1 1 0	1	0	0
1 1 1	1	1	1

FIGURE 2.14
Truth table for right side of Theorem 9(b) for Example 2.2.

Example 2.3

A student uses DeMorgan's theorem to complement $\bar{x} + yz$ as $x\bar{y} + \bar{z}$. Verify if this conclusion is valid.

Solution

Using truth table, as shown in Figure 2.15, we may compare both the function and its complement.

On examination, we find that for the input conditions $xyz = 000$ and 010, $\overline{(\bar{x} + yz)}$ is not equal to $x\bar{y} + \bar{z}$. Therefore, the DeMorgan's theorem has not been used properly by the student. A closer examination of the function reveals that

$$\bar{f} = (\bar{x} + yz)' = x(y' + z') \neq x\bar{y} + \bar{z}$$

Example 2.4

Use Boolean algebra to show that the logic circuits shown in Figures 2.9a and 2.10a are equivalent.

Solution

The logical function corresponding to the circuit of Figure 2.9a is given by

$$f(A, B, C) = \bar{A}BC + A\bar{B}C + AB\bar{C} + ABC$$

$$= \bar{A}BC + ABC + A\bar{B}C + ABC + AB\bar{C} + ABC \quad \text{using Theorem 1}$$

$$= (\bar{A} + A)BC + AC(\bar{B} + B) + AB(\bar{C} + C) \quad \text{using Theorem 6}$$

$$= 1 \cdot BC + AC \cdot 1 + AB \cdot 1 \quad \text{using Postulate 5}$$

$$= BC + AC + AB \quad \text{using Postulate 2(a)}$$

This expression for $f(A, B, C)$ is the same as Equation 2.5 that was used for realizing the circuit of Figure 2.10a. Hence, the two logic circuits are equivalent.

x y z	$\bar{x}$	yz	$\bar{x}+yz$	$\bar{y}$	$x\bar{y}$	$\bar{z}$	$x\bar{y}+\bar{z}$	$(\bar{x}+yz)'$
0 0 0	1	0	1	1	0	1	1	0
0 0 1	1	0	1	1	0	0	0	0
0 1 0	1	0	1	0	0	1	1	0
0 1 1	1	1	1	0	0	0	0	0
1 0 0	0	0	0	1	1	1	1	1
1 0 1	0	0	0	1	1	0	1	1
1 1 0	0	0	0	0	0	1	1	1
1 1 1	0	1	1	0	0	0	0	0

FIGURE 2.15
Truth table for Example 2.3.

Example 2.5
Simplify

$$f = (AB + \bar{A}C)(B\bar{A} + A\bar{C} + B\bar{C}) + (\bar{A}B + A\bar{C})(A\bar{C} + B\bar{C})$$

Solution

$$f = (AB + \bar{A}C)(B\bar{A} + A\bar{C} + AB\bar{C} + \bar{A}B\bar{C}) + (\bar{A}B + A\bar{C})(A\bar{C} + B\bar{C})$$
$$= (AB + \bar{A}C)[\bar{A}B(1 + \bar{C}) + A\bar{C}(1 + B)] + (\bar{A}B + A\bar{C})(A\bar{C} + B\bar{C})$$
$$= (AB + \bar{A}C)(\bar{A}B + A\bar{C}) + (\bar{A}B + A\bar{C})(A\bar{C} + B\bar{C})$$
$$= (\bar{A}B + A\bar{C})(A\bar{C} + B\bar{C} + AB + \bar{A}C)$$
$$= \bar{A}B(A\bar{C} + B\bar{C} + AB + \bar{A}C) + A\bar{C}(A\bar{C} + B\bar{C} + AB + \bar{A}C)$$
$$= (0 + \bar{A}B\bar{C} + 0 + \bar{A}BC) + (A\bar{C} + AB\bar{C} + AB\bar{C} + 0)$$
$$= \bar{A}B\bar{C} + \bar{A}BC + A\bar{C} + AB\bar{C}$$
$$= \bar{A}B(\bar{C} + C) + A\bar{C}(1 + B)$$
$$= \bar{A}B + A\bar{C}$$

Example 2.6
Use Theorem 11 to transform the function $f(A,B,C) = AB + A\bar{C} + \bar{A}C$ to its canonical
SOP form.

Solution
The theorem is employed three times to extract all three variables, A, B, and C
respectively.

$$f(A,B,C) = AB + A\bar{C} + \bar{A}C$$
$$= A \cdot f(1,B,C) + \bar{A} \cdot f(0,B,C)$$
$$= A(B + \bar{C}) + \bar{A}C$$
$$= B \cdot [A(1 + \bar{C}) + \bar{A}C] + \bar{B} \cdot [A(0 + \bar{C}) + \bar{A}C]$$
$$= B[A + \bar{A}C] + \bar{B}[A\bar{C} + \bar{A}C]$$
$$= AB + \bar{A}BC + A\bar{B}\bar{C} + \bar{A}\bar{B}C$$
$$= C \cdot (AB + \bar{A}B + \bar{A}\bar{B}) + \bar{C} \cdot (AB + A\bar{B})$$
$$= ABC + \bar{A}BC + \bar{A}\bar{B}C + AB\bar{C} + A\bar{B}\bar{C}$$
$$= \Sigma m(1,3,4,6,7)$$

Example 2.7

Show for a 3-input logic circuit that $\Sigma m(0,3,4,6) = \Pi M(1,2,5,7)$.

Solution

$$\Pi M(1,2,5,7) = (A+B+\bar{C})(A+\bar{B}+C)(\bar{A}+B+\bar{C})(\bar{A}+\bar{B}+\bar{C})$$

$$= (AA+A\bar{B}+AC+BA+B\bar{B}+BC+\bar{C}A+\bar{C}B+\bar{C}C)$$
$$\times (\bar{A}\bar{A}+\bar{A}\bar{B}+\bar{A}\bar{C}+B\bar{A}+B\bar{B}+BC+\bar{C}\bar{A}+\bar{C}B+\bar{C}\bar{C})$$

$$= (A+A\bar{B}+AC+BA+0+BC+\bar{C}A+\bar{C}B+0)$$
$$\times (0+\bar{A}\bar{B}+\bar{A}\bar{C}+B\bar{A}+0+BC+\bar{C}\bar{A}+\bar{C}B+\bar{C})$$

$$= (A+A\bar{B}+AC+BA+BC+\bar{C}A+\bar{C}B)$$
$$\times (\bar{A}\bar{B}+\bar{A}\bar{C}+B\bar{A}+BC+\bar{C}\bar{A}+\bar{C}B+\bar{C})$$

$$= \{A(1+\bar{B}+C+B+\bar{C})+BC+\bar{C}B\}$$
$$\times \{\bar{A}(\bar{B}+\bar{C}+B)+(B+\bar{A}+\bar{B}+1)\bar{C}\}$$

$$= \{A(1)+BC+\bar{C}B\}\{\bar{A}(1+\bar{C})+(1)\bar{C}\}$$

$$= \{A+BC+\bar{C}B\}\{\bar{A}+\bar{C}\}$$

$$= A\bar{A}+\bar{A}BC+\bar{A}\bar{C}B+A\bar{C}+BC\bar{C}+\bar{C}B\bar{C}$$

$$= 0+\bar{A}BC+\bar{A}\bar{C}B+A\bar{C}+0+B\bar{C}$$

$$= \bar{A}BC+\bar{A}\bar{C}B+A(1)\bar{C}+(1)B\bar{C}$$

$$= \bar{A}BC+\bar{A}B\bar{C}+A(B+\bar{B})\bar{C}+(A+\bar{A})B\bar{C}$$

$$= \bar{A}BC+\bar{A}B\bar{C}+AB\bar{C}+A\bar{B}\bar{C}+AB\bar{C}+\bar{A}B\bar{C}$$

$$= \bar{A}BC+\bar{A}B\bar{C}+AB\bar{C}+A\bar{B}\bar{C}$$

$$= \Sigma m(0,3,4,6)$$

Each time we connect an input to a circuit, it involves a wire connected to a pin of an integrated circuit. Every wire and every pin we use increases the overall cost of the circuit. Thus the primary objective of reducing or simplifying a function by the use of Boolean algebra is to achieve reduction in the number of literals in the given function. The number of literals in a function of n variables can reach a maximum of $2n$. Fortunately, the number of literals can be reduced by other usable techniques as well. These techniques are the subject of the next chapter.

2.5 Summary

In this chapter, the logic operations and logic gates were introduced and methods of generating the logic functions from their truth tables were presented. Finally, the algebraic techniques of manipulating and reducing logical expressions were developed and demonstrated. These Boolean algebra basics will be used in subsequent chapters to develop schemes for designing digital systems and understanding their applications.

Bibliography

Floyd, T., *Digital Fundamentals*. 8th edn. Englewood Cliffs, NJ. Prentice Hall, 2003.

Hwang, K., *Computer Arithmetic*, New York, NY. Wiley, 1979.

Johnson, E.J. and Karim, M.A., *Digital Design: A Pragmatic Approach*. Boston, MA. PWS-Kent Publishing, 1987.

Karim, M.A. and Awwal, A.A.S., *Optical Computing: An Introduction*. New York, NY. John Wiley & Sons, 1992.

Katz, R.H., *Contemporary Logic Design*. Boston, MA. Addison Wesley, 1993.

Mowle, F.J.A., *A Systematic Approach to Digital Logic Design*. Reading, MA. Addison-Wesley, 1976.

Waser, S. and Flynn, M.J., *Introduction to Arithmetic for Digital Systems Designers*. New York, NY. Holt, Rinehart & Winston, 1982.

Problems

1. Verify the following Boolean theorems: (a) Theorem 6(b), (b) Theorem 7(b), (c) Theorem 8(b), and (d) Theorem 10(b).

2. Prove the following canonical equations for a 4-variable function $f(A, B, C, D)$:

   (a) $\Sigma m(2, 5, 8, 10, 12–14) = \Pi M(0, 1, 3, 4, 6, 7, 9, 11, 15)$
   (b) $\Sigma m(2–5, 9–12, 15) = \Pi M(0, 1, 6–8, 13, 14)$
   (c) $\Pi M(2–5, 9–12, 15) = \Sigma m(0, 1, 6–8, 13, 14)$
   (d) $\Pi M(2, 5, 8, 10, 12–14) = \Sigma m(0, 1, 3, 4, 6, 7, 9, 11, 15)$

3. Prove the following canonical equations for a 3-variable function $f(A, B, C)$:

   (a) $\Sigma m(2, 5, 7) = \overline{\Sigma m(0, 1, 3, 4, 6)}$
   (b) $\Pi M(2, 5, 7) = \overline{\Pi M(0, 1, 3, 4, 6)}$
   (c) $\Sigma m(0, 4, 67) = \overline{\Sigma m(1 - 3, 5)}$
   (d) $\Pi M(0, 4, 6, 7) = \overline{\Pi M(1 - 3, 5)}$

4. Convert the following function $f(A, B, C)$ to (i) canonical SOP form and (ii) canonical POS form:

   (a) $AB + A\bar{C} + \bar{A}C$
   (b) $A(A + \bar{C})$
   (c) $B\bar{C} + \bar{B}C$
   (d) $AB + \bar{A}\bar{B}$

5. Take complement of the following Boolean functions:

   (a) $AB + A\bar{C} + \bar{A}C$
   (b) $A(A + \bar{C})$
   (c) $B\bar{C} + \bar{B}C$
   (d) $AB + \bar{A}\bar{B}$

6. Reduce the following logical functions to a minimum number of literals:

   (a) $ABC + \bar{A}D + \bar{B}D + CD$ (b) $ABC + BCD + A\bar{B}\bar{D} + \bar{A}BC\bar{D} + \bar{A}\bar{B}\bar{D}$

   (c) $(A + BC)(\bar{D} + BC)(\bar{A} + \bar{D})$ (d) $\bar{A} + \overline{AB}(A + \bar{C} + \bar{A}C)$

   (e) $\bar{A}(\bar{B} + \bar{C})A + B + +\bar{C})$ (f) $(\bar{A} + B + \bar{B}\bar{C})(A + \bar{B}C) + \bar{D}$

   (g) $\overline{\overline{CD} + A} + ACD + AB$ (h) $AB + \bar{A}B + AB\bar{C}D$

   (i) $(A \oplus B) + BC(A + B)$ (j) $(A \oplus B) + BC(A + B)$

7. Verify the following logic equations:

   (a) $A \oplus B = \bar{A} \oplus \bar{B}$ (b) $A \oplus B \oplus AB = A + B$

   (c) $\overline{A \oplus B} = A \oplus B \oplus 1$ (d) $BC + \bar{A}C = \bar{A}BC$

8. Construct a truth table for converting a 4-variable binary number into the corresponding 4-variable Gray code. Use Boolean algebra to obtain a logic circuit for binary-to-Gray conversion.

9. Obtain the logic circuit for converting 4-bit Gray code into their corresponding 4-bit binary numbers.

10. Obtain a logic circuit for converting four-bit BCD numbers to their XS3 equivalents.

11. Design a 3-input logic circuit that produces a high output whenever the majority of its inputs are high.

12. Design a 3-bit comparator circuit that compares $A_2A_1A_0$ with $B_2B_1B_0$ and gives a high output whenever $A_2A_1A_0 > B_2B_1B_0$.

13. Design a 2-level AND–OR logic circuit that converts a 4-bit sign-and-magnitude number to its equivalent 2's complement format.

14. Show that $B = C$ when $AB = AC$ and $A + B = A + C$.

15. Show $(\overline{AB \oplus AC})(A + B) \oplus (A + C)) = \overline{B \oplus C}$.

16. Find the complements of the following logical functions:

    (a) $AB\bar{C} + A\bar{B}C + \bar{A}C$

    (b) $AB \oplus C \oplus \bar{B}\bar{C}$

    (c) $(B\bar{C} + \bar{A}D)(\bar{B}\bar{D} + AC)(AB + CD)$

    (d) $\bar{A}B\bar{C} + A\bar{B}C$

17. The complementary metal-oxide semiconductor (CMOS) technology allows for the availability of a transmission gate TG, as shown in Figure 2.P1, that functions as follows: When $C = 1$, there is a closed path between X and Y and when $C = 0$, there is an open path between X and Y. Design a 2-input XOR logic device using such TGs.

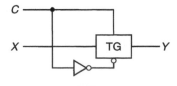

FIGURE 2.P1

18. Complement the following logical expressions and simplify the result so that the only complemented terms are individual variables.

(a) $A[B + D(C + \bar{A})]$
(b) $A(B + C) + \bar{A}B$

19. Show that XOR is associative:

$$A \oplus (B \oplus C) = (A \oplus B) \oplus C$$

20. Prove that AND is distributive over XOR:

$$A \cdot (B \oplus C) = (A \cdot B) \oplus (A \cdot C)$$

21. Prove that the complement of XOR function is equal to its dual.

22. A Boolean function F is referred to as self-dual if the dual of the function is equal to the function itself. Show that there are $2^{2^{n-1}}$ self-dual functions of n variables.

3

Minimization of Logic Functions

3.1   Introduction

Typically, we must have a circuit format in mind before a criterion for the simplest logic circuit can be defined. When implementing a logic function, it is often desirable that we keep the number of logic gates as well as the number of literals to a minimum. Boolean expressions with fewer logic operators imply fewer gates and fewer gate inputs per gate, thus enabling us to design inexpensive logic circuits. At the same time, we need to be cognizant of the fact that minimizing the number of operators may help us to minimize the cost, but that may cause larger propagation delays. But to minimize the overall propagation delay, one needs to reduce the number of operators on each critical path from input to output. Boolean functions can be simplified by algebraic means, but the process is often tedious and the designer cannot be certain that the manipulations have produced the minimum logical function. Depending on the sequence in which the Boolean theorems are used, we may end up with different reduced forms. A much easier and faster method of minimization is considered in this chapter. This method involves plotting function minterms on a two-dimensional map. This mapping scheme identifies all of the cases for a given set of input variables of the form $A + \bar{A} = 1$. When such minterm groups have been identified, the redundant variables can be eliminated, thus yielding a simplified function for the output. Following the rules that are yet to be identified in this chapter, algebraic manipulations will be traded for graphical methods. The graphical technique, which was proposed by Veich and later modified by Karnaugh, allows minimized 2-level functions to be obtained with very little effort. This particular map is referred to as *Karnaugh map*, or K-map, for short.

It must be noted though that designers may often use devices other than gates for realizing complex logic functions. For example, read-only-memory as well as programmable logic devices can be used to generate multiple functions of multiple-input variables without regard to much minimization. Consequently, the traditional demand for absolute minimization has been diminished somewhat in recent years by the introduction of such devices. However, there are many occasions when it is absolutely necessary to reduce complex logic functions. The choice between a simpler logic circuit and a faster logic circuit is generally a matter of judgement. Higher speed often means higher cost. The cost, on the other hand, is a composite function of the number of logic gates, the number of logic layers, and the number of inputs per gate in each of these layers. The exact ratio between the cost of a logic gate and the cost of logic gate input depends on the type of logic gate being used. In most cases, however, the cost of an additional logic gate will be several times that of an additional gate input on an already existing logic gate. When speed, for example, is the central issue, either a sum-of-product (SOP) or a product-of-sum (POS) expression is desirable for the most simplified logic function. For certain circuit technology, one may not assign any cost to inverting the variables. In such cases, both complemented and uncomplemented literals are assumed to be available when needed throughout the logic network.

Unfortunately, K-map-based scheme is neither suited for solving problems involving more than six input variables nor easily programmable. In this chapter, therefore, we also present a minimization algorithm, known as Quine-McCluskey's (Q-M) tabular reduction technique, that is reasonably straightforward. It is possible to come up with more than one valid minimization of a function if K-maps are used because of the different ways the maps may be interpreted. The Q-M technique, in comparison, uses a systematic method to conduct an exhaustive search to determine all possible combinations of logically adjacent minterms. Typically, the technique converges to the same optimum solution for the logic function. However, the technique is often time consuming, especially for functions having multiple-input variables. But since it follows a systematic and algorithmic approach, it is possible to write software programs based on the Q-M method to allow computer-aided design of logic circuits. The Q-M technique can also be used to optimize multiple-output logic functions for circuits that have one or more common input variables.

3.2   Karnaugh Map

A K-map may be viewed as a pictorial two-dimensional form of a truth table and hence there exists a one-to-one mapping between the two. The truth table has one row for a single minterm while the K-map has one cell per minterm. Likewise, there is also a one-to-one correspondence between truth table rows and K-map cells if and when maxterms are utilized. To understand its relevance consider the Boolean cubes shown in Figure 3.1 for the number of inputs $n = 1, 2, 3$, and 4. Each vertex (also referred to as a 0-subcube) in an n-cube represents a minterm of an n-variable Boolean function. Each n-input Boolean function is represented visually in an n-cube by marking up those vertices corresponding to which the function is 1. The unmarked vertices correspond to the function values of 0. In a 3-cube, for example, the vertex 101 will correspond to minterm 5. Figure 3.2 shows the cube representation corresponding to the 3-variable democracy function considered earlier in Section 2.3. The vertices corresponding to the minterms are indicated with large black dots.

Two 0-subcubes of a logical function may be combined to form a 1-subcube if they differ in the value of only one input variable. In the function of Figure 3.2, for example, we have three 1-subcubes, each consisting of a pair of 0-subcubes. These 1-subcubes are $\{011, 111\}$, $\{110, 111\}$, and $\{101, 111\}$. In general, m-subcube of an n-cube (where $m < n$) is referred to as a set of 2^m vertices (i.e., 0-subcubes) in which $n - m$ of the variable values are same at every vertex, while the remaining m variable values will take on 2^m possible combinations of 1s and 0s. Consequently, if a Boolean function has a value of 1 at each one of the vertices of an m-subcube, the 2^m minterms included in that subcube are represented by a single product term formed of $n - m$ common literals. In case of the 3-variable democracy function shown in Figure 3.2, therefore, the three 1-subcubes may be represented respectively by single product terms of two literals, BC, AB, and AC. Not surprisingly, we had already obtained the reduced democracy function to be $AB + AC + BC$ using Boolean algebra in Equation 2.5.

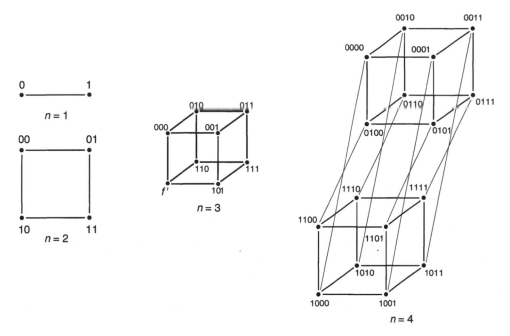

FIGURE 3.1
Boolean cubes for $n = 1, 2, 3,$ and 4.

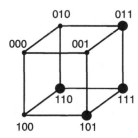

FIGURE 3.2
Boolean cube representing a 3-variable democracy function.

Consider that $A_1, A_2, \ldots,$ and A_{n-m} are the input variables whose values are the same for each vertex of an m-subcube, provided that $X_1, X_2, \ldots,$ and X_{n-m} are the literals such that $X_i = A_i$ whenever $X_i = 1$ and $X_i = \overline{A_i}$ whenever $X = 0$. Then the single product term that represents this m-subcube is given by

$$X_1 X_2 \ldots X_{n-m}(\overline{A_{n-m+1}}\, \overline{A_{n-m+2}} \ldots \overline{A_n} + \overline{A_{n-m+1}}\, \overline{A_{n-m+2}} \ldots A_n$$

$$+ \cdots + A_{n-m+1} A_{n-m+2} \ldots A_n) = X_1 X_2 \ldots X_{n-m}$$

The logical expression within the parentheses is equal to unity since it represents the SOP equivalent of all possible minterms of m variables. Very much like what has already been found to be true in the case of the democracy function, logical functions can be expressed in standard SOP form. Each of the product terms in it corresponds to a subcube and each minterm is included in at least one subcube. In order to minimize a logical function, the designer seeks to reduce the number of OR functions as well as the number of inputs to the AND functions. By choosing as few subcubes as possible we minimize the number of OR operations and by choosing the largest possible subcubes, we minimize the number of inputs that need to be ANDed together.

In general, a subcube is referred to as a *prime implicant* if it is not included within another subcube. An *essential prime implicant*, on the other hand, is that subcube which includes a minterm that is not included in any other subcube. During minimization, the prime implicants can be identified and selected visually from the cube representing the logic function. But with n exceeding 4, we find that this visualization process may not be that simple. Karnaugh devised a graphical scheme for representing minterms in a single plane. The K-map is a set of 2^n squares arranged in an ordered two-dimensional array. Each of the squares on the map corresponds to only one of the 2^n possible minterms. Each square thus corresponds to one vertex of an n-cube. If the minterm contributes to a function value of 1, a 1 is entered inside the square to otherwise a 0 is entered in it. The possible K-maps for functions of up to four variables are shown in Figure 3.3. Each of the squares on the K-map, called a *cell*, is also identified by the decimal value of the corresponding minterm.

In the K-maps of Figure 3.3, we identify each of the cells with its respective decimal designation for reference. During function minimization process, however, we need to be careful so that we do not confuse these decimal designations with the cell entries. Each of the map columns is assigned variable values in such a way that as one scans the values from left to right or from right to left, from cell to cell, only one variable changes. Each cell is thus adjacent to the next. In addition, the cells on the right end are only one unit distance from (adjacent to) the cells on the left end. Likewise, rows are also designated using variable values so that as one scans the variable values from top to bottom or from

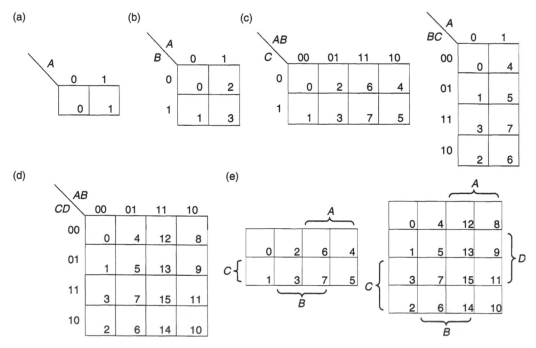

FIGURE 3.3
Karnaugh maps for (a) $f(A)$, (b) $f(A, B)$, (c) $f(A, B, C)$, (d) $f(A, B, C, D)$, and (e) $f(A, B, C)$ and $f(A, B, C, D)$. ·

bottom to top from cell to cell, only one variable changes. Cells at the top of the map are adjacent to cells along the bottom of the map. In summary, K-maps are more useful than truth tables since they can also show adjacency and can therefore visually aid in identifying possible subcubes.

The rows and columns of 4-variable K-maps, for example, are labeled by 2-bit Gray code numbers such as 00, 01, 11, and 10. Alternatively, however, K-map columns and rows can also be labeled by input variables. While each input variable is common among half of the cells, its complement is common among the remaining half of the cells as shown in Figure 3.3e. The two K-maps of Figure 3.3e are equivalent to those shown in Figure 3.3c and d respectively. A 4-variable K-map consists of sixteen cells, each of which corresponds to a different minterm of a 4-variable function. Each cell of a K-map is adjacent to four other cells that differ in the value of one variable. Since the K-maps are all labeled using Gray codes, the cells on the top (left) boundary are considered adjacent to the cells on the bottom (right) boundary and vice versa. The key feature of a K-map is that each cell on the map is logically adjacent to the cell that is physically adjacent to it. On the other hand, two cells are considered physically adjacent when minterms corresponding to the cells differ by a single variable.

Figure 3.4a shows a truth table for 2-input Boolean functions AND, OR, XOR, and XNOR. Their corresponding K-maps are shown then in Figure 3.4b through e. The AND function of A and B is trivial in the sense that there is only one minterm that contributes to an output of 1. The OR function of A and B, on the other hand, involves the 0-subcubes {01}, {10}, and {11} as shown in the K-map of Figure 3.4c. The 0-subcubes {01} and {11} are adjacent to each other and, therefore, can be combined to form a 1-subcube. This particular 1-subcube corresponds to the reduced product term B since B is common in these two 0-subcubes.

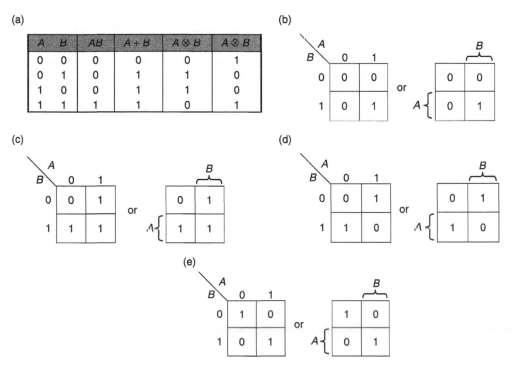

FIGURE 3.4
(a) AND, OR, XOR, and XNOR truth table and (b)–(e) the corresponding K-maps.

Likewise, the 0-subcubes {10} and {11} can form a 1-subcube that contributes to the reduced product term A. Neither of these 1-subcubes is included within the other and, therefore, both these are prime implicants. Since the 0-subcube {01} is not included in the second 1-subcube and since 0-subcube {10} is not included in the first 1-subcube, both of these 1-subcubes are also essential prime implicants. The K-map of OR function of A and B thus leads to the reduced form $A + B$. This result is not surprising since we can also derive this reduced form using Boolean algebra as follows:

$$A\bar{B} + \bar{A}B + AB = A\bar{B} + (\bar{A} + A)B$$
$$= A\bar{B} + B$$
$$= A + B$$

The other two K-maps shown respectively in Figure 3.4d and e provide a different situation. The 0-subcubes {01} and {10} of the XOR K-map and the 0-subcubes {00} and {11} of the XNOR K-map cannot be combined to form a 1-subcube since they are nonadjacent. In general, two 0-subcubes are considered to be adjacent, or neighbors, when they can be labeled with two successive Gray codes. In the case of XOR, we see that two changes are necessary to convert {01} to {10} and vice versa. Accordingly, the 0-subcubes of XOR function are nonadjacent and, therefore, will not lead to any function simplification. This is also true for XNOR function since two changes will be necessary to convert {00} to {11} and vice versa. It may be safe now to generalize that for any arbitrary logical function, two 0-subcubes that are adjacent are grouped to form a 1-subcube and expressed as the product of the literal(s) common in both of the 0-subcubes.

Minimization of a logical function using K-maps consists of four basic steps: (a) map generation, (b) prime implicant generation, (c) selection of essential implicants, and (d) identification of minimal cover. Generating a K-map directly from a truth table is rather simple since for each minterm a 1 is entered in the corresponding cell if that minterm contributes to a 1. A 0 is entered in each of the remaining cells. If instead the function is given in standard or reduced form, it is expanded into its canonical form first. Alternatively, a function in standard form can be introduced in the K-map by entering $n - m$ 1's. A 1 is entered, if not already entered, in each one of the $n - m$ subcubes. See Example 3.3 for an application of this principle. Each one of these $n - m$ subcubes corresponds to an m-literal product term included in the standard form. The process for identification of m-subcube (or groups) requires that we utilize the following rules:

Rule 1. If all entries in an n-variable K-map are 1s, the group size is 2^n and the corresponding reduced function is a 1.

Rule 2. For n variables, the largest nontrivial group size is 2^{n-1}. The other allowable group sizes can be of size $2^{n-2}, 2^{n-3}, \ldots,$ and 2^0.

Rule 3. In order for a group to be valid, it must be possible to start at a cell and travel from cell to adjacent cell through each cell in the group without passing through any cell twice and finally return to the starting cell.

Using the rules for forming valid groups, the prime implicants are generated next as follows:

1. For an n-variable K-map, determine the group, if there is any, which has 2^{n-1} 0-subcubes. In case there is such a group, it is identified by circling all of those 0-subcubes.

2. Next the group, if there is any, of size 2^{n-2} 0-subcubes is determined. Groups that are completely contained within an already-determined larger group are just ignored from further consideration. Only those groups that include additional yet-to-be-included 0-subcube(s) will be considered.

3. This process is continued until the time we may have 0-subcubes that may not be grouped with any other 0-subcubes to form a 1-subcube. In the process of grouping, each 1 on the map must be included in at least one group.

4. The isolated 0-subcubes appear as an n literal product term in the reduced function.

After we have generated the prime implicants, the designer attempts to identify next those prime implicants that can be characterized as essential. The essential prime implicants are those prime implicants that have at least one 0-subcube that is not included in any other prime implicant. As will be demonstrated later, a *cover* describing a logical function is obtained by performing an OR of all its prime implicants. These prime implicants together must include each one of the minterms describing the function. The essential implicants are invariably included in the list of the covering terms for generating the minimized logical function. In its final step, the *minimal cover* is identified. It is a combination of the least possible number of prime implicants selected in such a way that each 0-subcube is included in at least one prime implicant. One of the approaches by which the minimal cover is identified involves identifying the prime implicant that includes the maximum number of 0-subcubes that have not been included in any other prime implicant. If it turns out that more than one such prime implicant include the same number of 0-subcubes that have not been included in any other prime implicant, then only one of these prime implicants is randomly chosen. This process is repeated until all of the 0-subcubes have been covered.

For each of the groups to be included within the minimal cover, a reduced product term (i.e., AND) is generated from the literals that are constant in each one of the 0-subcubes of the group. For x such groups, there will be x product terms. For an n-variable function, a group consisting of 2^m 0-subcubes contributes to a product term of $n - m$ literals. It must be understood that it is possible to obtain more than one reduced logical functions.

The minimization rules are best understood by considering the K-maps of Figure 3.5. There are only two 0-subcubes, 1 and 5, for example, on the third 3-variable map of Figure 3.5a. These 0-subcubes may be combined to form a single product term using Boolean algebra as follows:

$$\bar{A}\bar{B}C + A\bar{B}C = \bar{B}C(\bar{A} + A) = \bar{B}C$$

The two 0-subcubes are adjacent, and the variables $\bar{B}$ and C are constant within this group-of-two. Similarly, in the fourth K-map of Figure 3.5a, the 0-subcubes are adjacent, and thus the function reduces to AB. The K-map groups can also be formed of the 0-subcubes that are positioned at the edges. Examples of few such K-maps are also shown in Figure 3.5. The first K-map of Figure 3.5b reduces to $\bar{B}$; the bottom left K-map of Figure 3.5c reduces to $\bar{B}C$, and the first K-map of Figure 3.5d reduces to B.

The bottom right K-map of Figure 3.5c consists of 0-subcubes 0, 2, 8, and 10 that are located at the corners. Since the top row is adjacent to the bottom row and since the left-most column is adjacent to the right-most column, these four 0-subcubes are considered neighbors. A logical function that includes these 0-subcubes can be reduced using Boolean

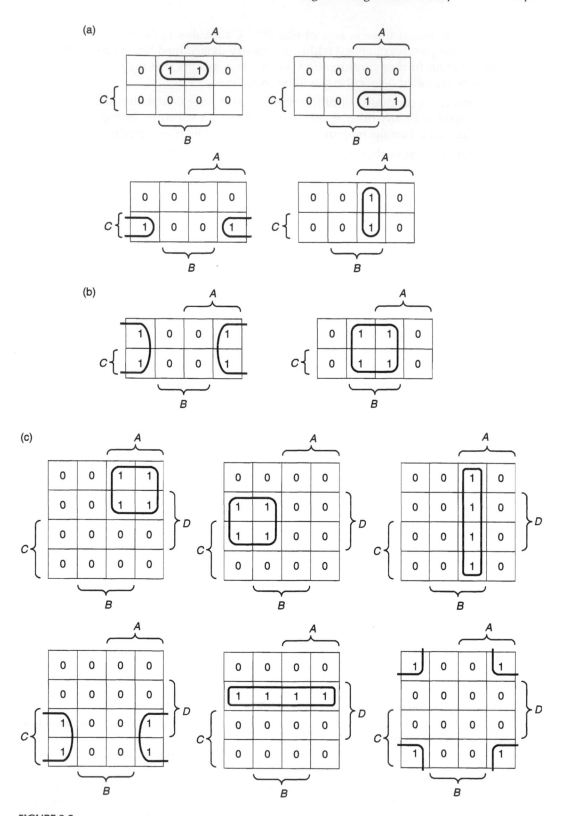

FIGURE 3.5
Examples of K-maps (a) a 1-subcube in a 3-variable K-map, (b) a 2-subcube in a 3-variable K-map, (c) a 2-subcube in a 4-variable K-map, and (d) a 3-subcube in a 4-variable K-map.

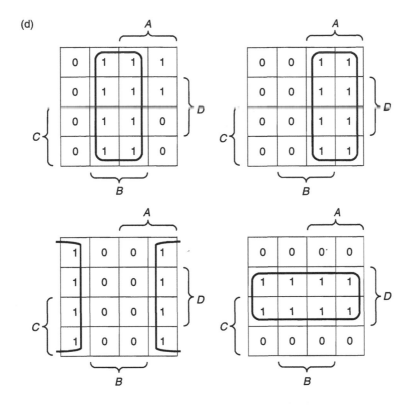

FIGURE 3.5
Continued.

algebra as follows:

$$\bar{A}\bar{B}\bar{C}\bar{D} + \bar{A}\bar{B}C\bar{D} + A\bar{B}\bar{C}\bar{D} + A\bar{B}C\bar{D} = \bar{A}\bar{B}\bar{D}(\bar{C} + C) + A\bar{B}\bar{D}(\bar{C} + C)$$

$$= \bar{A}\bar{B}\bar{D} + A\bar{B}\bar{D}$$

$$= \bar{B}\bar{D}(\bar{A} + A) = \bar{B}\bar{D}$$

These four corner 0-subcubes, thus, form a valid 2-subcube. The maps are considered continuous in both directions so that both top and bottom edges and left and right edges are considered to be touching, forming a donut-shaped contour. The remaining K-maps of Figure 3.5 show other more obvious groupings. Example 3.1 demonstrates the application of mapping scheme when there is more than one adjacent group.

Example 3.1
Obtain the logical equation for a digital circuit that outputs a 1 each time the decimal equivalent of four variables, *A*, *B*, *C*, and *D*, is other than a prime number.

Solution
This 4-variable problem can be described using a truth table of 2^4 rows. The resulting truth table is obtained as shown in Figure 3.6. The function can also be expressed in its canonical form as

$$f(A, B, C, D) = \Sigma m(0, 4, 6, 8, 9, 10, 12, 14, 15)$$

A	B	C	D	f
0	0	0	0	1
0	0	0	1	0
0	0	1	0	0
0	0	1	1	0
0	1	0	0	1
0	1	0	1	0
0	1	1	0	1
0	1	1	1	0
1	0	0	0	1
1	0	0	1	1
1	0	1	0	1
1	0	1	1	0
1	1	0	0	1
1	1	0	1	0
1	1	1	0	1
1	1	1	1	1

FIGURE 3.6
Truth table for Example 3.1.

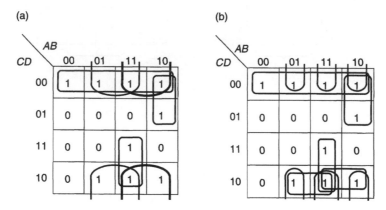

FIGURE 3.7
K-Map for $f(A, B, C, D) = \Sigma m(0, 4, 6, 8, 9, 10, 12, 14, 15)$ (a) an optimum grouping and (b) an example of nonoptimal grouping.

The nine 0-subcubes are entered as nine 1s in a 4-variable K-map as shown in Figure 3.7a. From visual inspection, we see that the largest possible group could be of size 4. Here, the minterms 0, 4, 8, and 12 (corresponding to the 0-subcubes {0000}, {0100}, {1000}, and {1100}) can be combined to form a 2-subcube. Likewise, minterms 4, 6, 12, and 14 (corresponding to the 0-subcubes {0100}, {0110}, {1100}, and {1110}) and minterms 8, 10, 12, and 14 (corresponding to the 0-subcubes {1000}, {1010}, {1100}, and {1110}) can be combined to yield two 2-subcubes. Next, we look for possible groups of two (i.e., 1-subcubes) that are not completely included within any one of these 2-subcubes. We see that the minterms 14 and 15 (corresponding to 0-subcubes {1110} and {1111}) as well as the minterms 8 and 9 (corresponding to 0-subcubes {1000} and {1001}) can be combined to form 1-subcubes. In this particular example, we see that there is no 0-subcube that could not be included in at least one 1-subcube.

The K-map groups as shown in Figure 3.7a are as follows:

1. Minterm set {0, 4, 8, 12}, that is, 0-subcube set {0000, 0100, 1000, 1100} results in $\bar{C}\bar{D}$

2. Minterm set $\{4, 6, 12, 14\}$, that is, 0-subcube set $\{0100, 0110, 1100, 1110\}$ results in $B\bar{D}$

3. Minterm set $\{8, 10, 12, 14\}$, that is, 0-subcube set $\{1000, 1010, 1100, 1110\}$ results in $A\bar{D}$

4. Minterm set $\{14, 15\}$, that is, 0-subcube set $\{1110, 1111\}$ results in ABC

5. Minterm set $\{8, 9\}$, that is, 0-subcube set $\{1000, 1001\}$ results in $A\bar{B}\bar{C}$

Each of these prime implicants is considered essential. Each of these prime implicants alone provides cover respectively for the minterms 0, 6, 10, 15, and 9. Therefore, the minimized SOP function for the digital system is given by

$$f = \bar{C}\bar{D} + B\bar{D} + A\bar{D} + ABC + A\bar{B}\bar{C}$$

However, if the designer fails to see, for example, the last two 2-subcubes, one may instead conclude that there exists seven possible choices for minterm pairs: 8 and 9, 4 and 6, 6 and 14, 10 and 14, 12 and 14, and 8 and 10. Such nonoptimal groupings for this problem are shown in the K-map of Figure 3.7b. Typically, the designer would prefer to have as few groups as possible and yet be able to include each 0-subcube in at least one group. The important point is that all 0-subcubes will have to be covered. These nonoptimal K-map groups are

(i) Minterm set $\{0, 4, 8, 12\}$, that is, 0-subcube set $\{0000, 0100, 1000, 1100\}$ results in $\bar{C}\bar{D}$

(ii) Minterm set $\{8, 9\}$, that is, 0-subcube set $\{1000, 1001\}$ results in $A\bar{B}\bar{C}$

(iii) Minterm set $\{4, 6\}$, that is, 0-subcube set $\{0100, 0110\}$ results in $\bar{A}B\bar{D}$

(iv) Minterm set $\{6, 14\}$, that is, 0-subcube set $\{0110, 1110\}$ results in $BC\bar{D}$

(v) Minterm set $\{10, 14\}$, that is, 0-subcube set $\{1010, 1110\}$ results in $AC\bar{D}$

(vi) Minterm set $\{12, 14\}$, that is, 0-subcube set $\{1100, 1110\}$ results in $AB\bar{D}$

(vii) Minterm set $\{8, 10\}$, that is, 0-subcube set $\{1000, 1010\}$ results in $A\bar{B}\bar{D}$

(viii) Minterm set $\{14, 15\}$, that is, 0-subcube set $\{1110, 1111\}$ results in ABC

Studying the various K-map groups, one can conclude that the groups (i), (ii), and (viii) are essential prime implicants. This is because only group (i) covers minterm 0, only group (ii) covers minterm 9, and only group (viii) covers minterm 15. These three essential prime implicants together cover the minterms 0, 4, 8, 9, 12, 14, and 15. The essential prime implicants are always included within the minimal cover. To cover the remaining two minterms (i.e., 6 and 10), however, a certain combination of the other five pairs (but not necessarily all of these five pairs) will have to be included in the reduced function. There are two possibilities that will involve combining only two prime implicants: (a) groups (iii) and (vii) and (b) groups (iv) and (v).

Therefore, the minimized SOP function for the digital system is either

$$f = \bar{C}\bar{D} + A\bar{B}\bar{C} + ABC + \bar{A}B\bar{D} + A\bar{B}\bar{D}$$

or

$$f = \bar{C}\bar{D} + A\bar{B}\bar{C} + ABC + BC\bar{D} + AC\bar{D}$$

Each of these two latter functions requires five AND and one 5-input OR operations. The optimized grouping shown in Figure 3.7a also requires five AND and one 5-input OR operations. The critical difference then is in the number of inputs feeding the AND gates.

In the case of Figure 3.7a, three of the five AND gates require 2 inputs each while in the case of Figure 3.7b only one AND gate can be of 2-input type. Such difference may appear trivial for many of the applications, but it may make a substantial difference in other situations. However, in comparison, the unminimized function would have required nine 4-input AND and one 9-input OR operations. Our simplification process thus results in marked improvement both in terms of the number of AND operations and in the number of inputs to OR operation.

We have so far considered the minimization cases involving four or fewer variables. In the event of a 4- or 6-variable logic function, we shall need more than one 4-variable map. In the 4-variable map, each cell is adjacent to four other cells: top, bottom, left, and right. To show similar adjacencies in a 5-variable K-map, however, one needs two 4-variable K-maps that are positioned next to each other. In this arrangement, the second 4-variable K-map is overlaid on top of the first 4-variable K-map so as to create a 3-dimensional visualization. Only then each of the thirty-two cells is seen to have four adjacent cells. This consideration becomes obvious by noting the most significant digit of the 5-variable logic function. The first sixteen of the thirty-two 5-variable 0-subcubes is always a 0, and that of the remaining sixteen 0-subcubes is a 1. Thus in one of the two 4-variable K-maps the most significant variable is a 0 in all of its 0-subcubes and that in the other 4-variable K-map is a 1 in all of its 0-subcubes. Similarly, for a 6-variable problem, we could split the sixty-four 0-subcubes into four groups of sixteen. One group of sixteen is distinguished from the other by its more significant two bits. Consequently, just as the 5-variable K-map consisted of two 4-variable K-maps, a 6-variable K-map may consist of four 4-variable K-maps. For a 6-variable K-map, the right half and the left half as well as the top half and the bottom half are considered adjacent. These mapping schemes are shown in Figure 3.8.

In a 5-variable map each 0-subcube can have up to five possible neighbors. For example, in Figure 3.8a the 0-subcubes {00011}, {01001}, {01010}, {01111}, and {11011} (i.e., minterms 3, 9, 10, 15, and 27) are all neighbors of the 0-subcube {01011} (i.e., minterm 11). The 0-subcubes that are 1 at identical position on the two 4-variable K-maps are also considered adjacent. Some of the valid groupings in a 6-variable map are, for example, (0, 16, 32, 48), (5, 7, 13, 15, 21, 23, 29, 31, 37, 39, 45, 47, 53, 55, 61, 63), (0, 1, 2, 3, 16, 17, 18, 19), and (0, 2, 8, 10, 16, 18, 24, 26, 32, 34, 40, 42, 48, 50, 56, 58).

Example 3.2
Simplify $f(A, B, C, D, E) = \Sigma m$ (0, 1, 4, 5, 6, 11, 12, 14, 16, 20, 22, 28, 30, 31).

Solution
The minterms are plotted and grouped in a 5-variable K-map as shown in Figure 3.9. These minterms may be grouped as follows:

     Minterm set {4, 6, 12, 14, 20, 22, 28, 30} yielding $C\bar{E}$

     Minterm set {0, 1, 4, 5} yielding $\bar{A}\bar{B}\bar{D}$

     Minterm set {0, 4, 16, 20} yielding $\bar{B}\bar{D}\bar{E}$

     Minterm set {30, 31} yielding $ABCD$

     Minterm 11 yielding $\bar{A}B\bar{C}DE$

Therefore, the minimized SOP function is given by

$$f(A, B, C, D, E) = C\bar{E} + \bar{A}\bar{B}\bar{D} + \bar{B}\bar{D}\bar{E} + ABCD + \bar{A}B\bar{C}DE$$

(a)

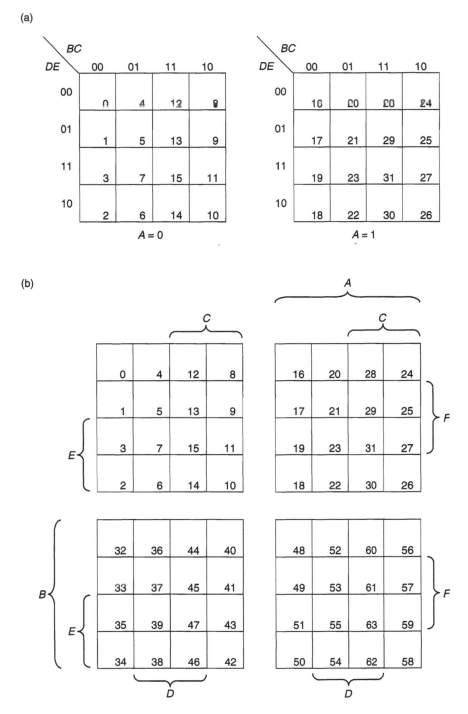

FIGURE 3.8
(a) 5-variable K-map and (b) 6-variable K-map.

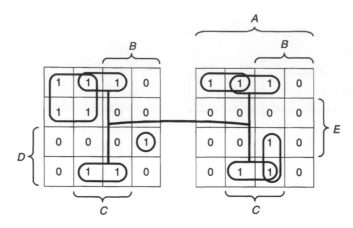

FIGURE 3.9
K-map for Example 3.2.

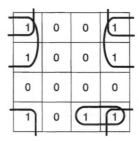

FIGURE 3.10
K-map for Example 3.3.

Example 3.3
Simplify the Boolean function $f = ABC\bar{D} + \bar{A}\bar{B}C + A\bar{B}\bar{C} + \bar{B}C\bar{D}$ using a K-map.

Solution
The decimal equivalent of the 0-subcubes for this 4-variable function are obtained as follows:

$$ABC\bar{D} \Rightarrow 14$$

$$\bar{A}\bar{B}C \Rightarrow 0,1 \text{ since } \bar{A}\bar{B}C = \bar{A}\bar{B}C\bar{D} + \bar{A}\bar{B}CD$$

$$A\bar{B}\bar{C} \Rightarrow 8,9 \text{ since } A\bar{B}\bar{C} = A\bar{B}\bar{C}\bar{D} + A\bar{B}\bar{C}D$$

$$\bar{B}C\bar{D} \Rightarrow 2,10 \text{ since } \bar{B}C\bar{D} = \bar{A}\bar{B}C\bar{D} + A\bar{B}C\bar{D}$$

These seven minterms (0, 1, 2, 8, 9, 10, 14) may now be plotted and grouped in a 4-variable K-map as shown in Figure 3.10.
   Therefore,

$$f = \bar{B}\bar{D} + \bar{B}\bar{C} + AC\bar{D}$$

Note that in this particular problem, no two minterms were covered by two or more product terms. In practice, there may be situations when product terms can result from common minterms. In either case, K-map simplification process, if carried out properly, may lead to a simpler logic function implementation.

3.3 Incompletely Specified Functions in K-Maps

Most logic circuit designs begin with a verbal statement of the problem. The next step in the design process is constructing a truth table for the to-be-determined function(s). Up to this point, we have considered truth tables with output values for all possible input conditions specified. Oftentimes, however, there are practical reasons for which a function may not be completely specified. Sometimes the circuit we are designing is a part of a larger circuit in which certain inputs occur only under circumstances such that the output of the smaller circuit may not influence the overall circuit. Accordingly, in this section we shall consider using K-map for implementing logical functions that are incompletely specified. In such cases, practical constraints may exclude a particular group of minterms from contributing to the output function, which makes their occurrence in the function strictly optional. Note that this does not imply that the circuit would not develop some output if the forbidden input occurred. Assuming that such occurrences are rare, oftentimes designers thus do not care if these minterms are included in the function since they would not occur. These optional minterms are called the *don't-care* minterms. For example, consider the BCD inputs that were discussed in Section 1.7. The 4-bit BCD code with weights 8-4-2-1 uses only 10 of 16 possible combinations. Six of the combinations {1010}, {1011}, {1100}, {1101}, {1110}, and {1111} are not just used in BCD. Constraints of this type often become very helpful since the don't-care minterms may be included in the K-map as either a 0 or a 1, depending on which leads to the most minimum logic function. Don't-care conditions also arise where all input combinations do occur for a given digital circuit, but the output is required to be 1 or 0 only for certain combinations. On the other hand, if we express a logical function in terms of its maxterms, don't-care minterms are usually expressed in the corresponding maxterm form, in which case they are called don't-care maxterms.

On K-maps, the don't-care minterms are entered with a "-" symbol. During the grouping procedure, those don't-cares leading to the formation of larger prime implicants when they are grouped with other minterms are considered as producing an output of 1. In other words, only those don't cares that aid in simplification are included. The other don't cares are treated as though they are 0s in the K-map. We implicitly assign a value of 1 to those don't cares that are included in at least one prime implicant. A minterm list for a function with don't cares has the form $f = \sum m(\ldots) + d(\ldots)$ where don't-care minterms are listed in $d(\ldots)$. The maxterm list for this same function will have the form $\Pi M(\ldots) \bullet D(\ldots)$ where the don't-care maxterms are listed in $D(\ldots)$. A note of caution: groups of only don't cares are never used. Example 3.4 illustrates the benefit of having don't cares in a logic function.

Example 3.4
Design a logic circuit that outputs the 9's complement of a BCD digit.

Solution
The truth table for the 9's complement logic circuit is obtained as shown in Figure 3.11. The BCD digits are introduced through four input lines A, B, C, and D.

The four output functions corresponding to the BCD inputs are expressed as follows:

$$F_3(A,B,C,D) = \Sigma m(0,1) + d(10\text{--}15)$$

$$F_2(A,B,C,D) = \Sigma m(2,3,4,5) + d(10\text{--}15)$$

$$F_1(A,B,C,D) = \Sigma m(2,3,6,7) + d(10\text{--}15)$$

$$F_0(A,B,C,D) = \Sigma m(0,2,4,6,8) + d(10\text{--}15)$$

To each of the above expressions, we have included all six don't cares. On the basis of whether their inclusions will lead to the formation of larger groups, we may treat some of these don't cares as minterm-like. The four K-maps for these output functions are obtained as shown in Figure 3.12.

We see that with inclusion of a selected set of don't cares, the minterms lead to the following reduced expressions for the output functions:

$$F_3(A,B,C,D) = \Sigma m(0,1) = \bar{A}\bar{B}\bar{C}$$

$$F_2(A,B,C,D) = \Sigma m(2,3,4,5) + d(10,11,12,13) = B \oplus C$$

$$F_1(A,B,C,D) = \Sigma m(2,3,6,7) + d(10,11,14,15) = C$$

$$F_0(A,B,C,D) = \Sigma m(0,2,4,6,8) + d(10,12,14) = \bar{D}$$

In this K-map processing only those don't cares are included that lead to larger grouping. In the K-map corresponding to F_3, the don't cares were of no help. But for the other three functions don't cares contribute to significant simplification. If we had not taken advantage of these don't cares, these three functions would have instead reduced to

$$F_2(A,B,C,D) = \Sigma m(2,3,4,5) = \bar{A}\bar{B}C + \bar{A}B\bar{C}$$

$$F_1(A,B,C,D) = \Sigma m(2,3,6,7) = \bar{A}C$$

$$F_0(A,B,C,D) = \Sigma m(0,2,4,6,8) = \bar{A}\bar{D} + \bar{B}\bar{C}\bar{D}$$

We see that by making use of don't cares, we are able to eliminate literal $\bar{A}$ from the Boolean expressions F_2, F_1, and F_0. Furthermore, from F_0 the larger AND term got eliminated. It must be understood though that if the logic circuit were to be fed now with unallowable BCD digits, the logic circuit may yield undesirable outputs.

Example 3.5

Obtain a 6-input digital circuit that gives an output, y_0, given by (a) x_1 if $R = 1, S = 1$; (b) x_{-1} if $R = 0, S = 1$; and (c) x_0 if $R = \text{-}, S = 0$ where $R, S, x_0, x_1,$ and x_{-1} are all circuit inputs.

Solution

Typically, a 5-variable truth table needs to include a total of 32 combinations of the inputs. However, one may not have to construct such a truth table since the output does not necessarily depend on all of the input variables simultaneously. The reduced truth table can be obtained then as shown in Figure 3.13 where some of the inputs are treated as don't cares.

In a case such as this, there may not be any need to plot them all in a K-map. The equation for y_0 can be obtained readily from the truth table.

$$y_0 = RSx_1 + \bar{R}Sx_{-1} + \bar{S}x_0$$

The resulting logic circuit is obtained as shown in Figure 3.14. This circuit may be referred to as a shifter since it can be used for shift-left (when $S = 1$ and $R = 0$), shift-right (when $S = 1$ and $R = 1$), or no-shift (when $S = 0$) operations.

ABCD	$F_3F_2F_1F_0$
0 0 0 0	1 0 0 1
0 0 0 1	1 0 0 0
0 0 1 0	0 1 1 1
0 0 1 1	0 1 1 0
0 1 0 0	0 1 0 1
0 1 0 1	0 1 0 0
0 1 1 0	0 0 1 1
0 1 1 1	0 0 1 0
1 0 0 0	0 0 0 1
1 0 0 1	0 0 0 0

FIGURE 3.11
Truth table for Example 3.4.

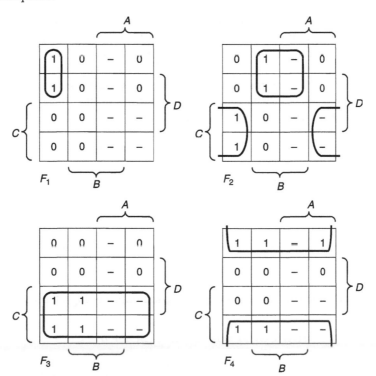

FIGURE 3.12
K-maps for Example 3.4.

R	S	x_1	x_0	x_{-1}	y_0
1	1	0	-	-	0
1	1	1	-	-	1
0	1	-	-	0	0
0	1	-	-	1	1
-	0	-	0	-	0
-	0	-	1	-	1

FIGURE 3.13
Reduced truth table for Example 3.5.

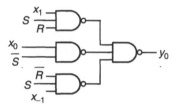

FIGURE 3.14
Logic circuit for Example 3.5.

3.4   K-Maps for POS Form of Functions

At times, it becomes convenient to implement the POS form of a function rather than its traditional SOP form. Such an instance may occur, for example, when it is more economical to implement the complement of a function than the function itself. When such is the case, however, it is not necessary to redraw the K-maps for finding the complement of a function. The typical K-map may be used but only the 0s may be grouped to obtain the complement function. The rules for grouping the 0s are the same as those used earlier in the grouping of 1s. When the resulting reduced SOP expression for the complement function is complemented, an expression logically equivalent to the POS form of the function will be obtained. This technique is known as the complementary approach.

Consider the SOP and POS forms of the function $f(A, B, C, D) = \sum m(0, 2, 4, 8, 10, 14)$. The function is plotted on two different K-maps, as shown in Figure 3.15. Map(a) is used for determining the SOP form and map(b) is used for determining the POS form. By grouping the 1s in map(a), one may obtain the SOP form of the function as

$$f(A, B, C, D) = \bar{B}\bar{D} + AC\bar{D} + \bar{A}\bar{C}\bar{D}$$

and by grouping the 0s in map(b), one may obtain $\bar{f}$, instead of f, as

$$\bar{f}(A, B, C, D) = D + \bar{A}BC + AB\bar{C}$$

The complement function $\bar{f}$ now may be complemented using DeMorgan's theorem to yield the function f.

$$f(A, B, C, D) = \bar{D}(A + \bar{B} + \bar{C})(\bar{A} + \bar{B} + C)$$

The logic circuits corresponding to the SOP and POS forms are shown in Figure 3.16.

The technique just demonstrated for determining the POS form of Boolean function can be summarized as follows. We treat 0s in the K-map as though they were 1s. These 0s are then grouped just as we had grouped 1s in case of the SOP-based K-map scheme. The

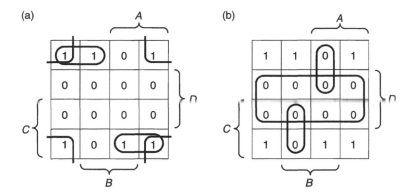

FIGURE 3.15
K-Maps for $f(A,B,C,D) = \sum m(0,2,4,8,10,14)$: (a) SOP form and (b) POS form.

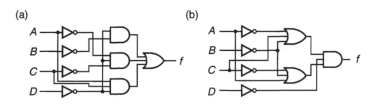

FIGURE 3.16
Logic Circuit for $f(A,B,C,D) = \sum m(0,2,4,8,10,14)$: (a) SOP form and (b) POS form.

reduced expression is finally complemented using DeMorgan's theorem to yield the desired function but in POS form. In minimization of logic circuits, it is best to explore both SOP and POS forms before a decision is made as to which form will be better to implement.

An equivalent way to view simplification of maxterms is to translate the groups of 0s directly into an OR form rather than into an AND form. In such a case, these ORed variables are then ANDed together to produce a reduced POS function. The procedure for reading maxterms is as follows:

1. Variables that are included within a prime implicant are complemented. If input x is a 1 for all maxterms within that prime implicant, then it would be read as $\bar{x}$. On the other hand, if it were a 0 for all maxterms within that prime implicant, then it would be read as x.

2. Each such literal contribution from within the prime implicant is then ORed.

3. These reduced OR terms are then ANDed to yield the reduced function in the POS form.

We see that the SOP form of logic circuit as shown in Figure 3.16a requires a total of 8 gates and 15 inputs and the POS form of logic circuit as shown in Figure 3.16b requires a total of 7 gates and 13 gate inputs. Usually the cost of a circuit is directly proportional to the weighted sum of the total number of gate inputs and number of discrete logic gates. If the relative weights were to be considered the same, at this point the POS form appears to be more economical than the SOP form even though in the given function there are fewer 1s than 0s. When we consider practical problems, such as the availability of certain gate types, you will see that the decision to choose between a logic circuit and its alternative may not be so clear-cut.

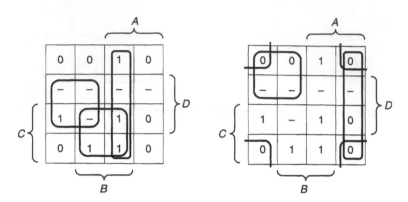

FIGURE 3.17
K-maps for Example 3.6.

Example 3.6

Compare the minimized SOP forms of the function $f(A, B, C, D) = \Sigma m(3, 6, 12, 14, 15) + d(1, 5, 7, 9, 13)$ by grouping (a) 1s and (b) 0s.

Solution

The function can be expressed in terms of maxterms as well since,

$$f(A, B, C, D) = \Sigma m(3, 6, 12, 14, 15) + d(1, 5, 7, 9, 13)$$
$$= \Pi M(0, 2, 4, 8, 10, 11) \bullet D(1, 5, 7, 9, 13)$$

The grouping of 1s and 0s can be carried out as shown in the K-maps of Figure 3.17. We make efficient use of don't cares for optimum solution. From the groupings of 1s, one obtains

$$f = AB + BC + \bar{A}D$$

But then from the grouping of 0s, we obtain

$$f = (\bar{A} + B)(A + C)(B + D)$$
$$= (\bar{A} + B)(AB + AD + BC + CD)$$
$$= \bar{A}BC + \bar{A}CD + AB + ABD + BC + BCD$$
$$= \bar{A}BC + \bar{A}CD + BC(1 + D) + AB(1 + D)$$
$$= \bar{A}BC + \bar{A}CD + BC + AB$$
$$= (\bar{A}C + A)B + BC + \bar{A}CD$$
$$= AB + BC + \bar{A}CD$$

We see that the two forms of f are not necessarily identical. This occurs since we have used the don't cares differently in minimizing the functions. For example, the don't cares at cells 1 and 5 are interpreted as 1s when deriving the SOP expression but they are used as 0s when deriving the POS expression.

3.5 Map-Entered Variables

Karnaugh map scheme is usable in minimizing Boolean functions of up to six input variables. Often to increase K-map's capacity to handle a larger number of input variables, one may make use of a scheme called *map-entered variables* (MEVs). The scheme increases the effective size of K-map. In the traditional K-map scheme, each of its cells represents a minterm, a maxterm, or a don't-care term. In a MEV K-map, a cell instead of housing 1 or 0, may house also an input variable or a Boolean function of one or more variables. In case at least one K-map cell houses an input variable or a Boolean function of one or more variables, the columns and rows of the K-map are all labeled by all but these MEVs. A MEV K-map having 2^n cells thus is able to handle up to $n + m$ input variables where m is the number of map-entered variables.

We shall first explore ways to construct a MEV K-map for a given function. Then we shall be introducing rules for determining the corresponding prime implicants. Consider the 4-variable logic function $f(A, B, C, D) = \Sigma m(0, 1, 2, 3, 12, 13, 15) + d(4, 5)$. We can readily determine its minimized form using a 4×4 K-map. However, we shall try here to construct a 3-variable MEV K-Map for dealing with this same function. This implies that one of the four input variables will need to be treated as an MEV. Typically, the MEVs are chosen from the lesser significant variables or the variables that appear the least in the unminimized logical expression. For simplicity, accordingly, we shall consider input D as a MEV. Figure 3.18a shows the truth table corresponding to this 4-variable logic function. Two minterm columns are included in the table: M for an MEV K-map and m for the regular K-map. Since we intend to use only one variable as an MEV, eight cells of the MEV K-map should be able to cover all sixteen cells of the regular K-map. Each of the MEV K-map cells is thus equivalent to two corresponding cells in the regular K-map.

The rules for transferring truth table entries to an equivalent MEV K-map are as follows:

1. If for all of the standard minterms covered by an MEV cell, the output is a 1 (or a 0, or a don't care), then a 1 (or 0, or don't care) is entered in that MEV cell.

(a)

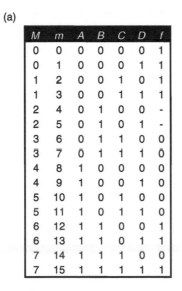

M	m	A	B	C	D	f
0	0	0	0	0	0	1
0	1	0	0	0	1	1
1	2	0	0	1	0	1
1	3	0	0	1	1	1
2	4	0	1	0	0	-
2	5	0	1	0	1	-
3	6	0	1	1	0	0
3	7	0	1	1	1	0
4	8	1	0	0	0	0
4	9	1	0	0	1	0
5	10	1	0	1	0	0
5	11	1	0	1	1	0
6	12	1	1	0	0	1
6	13	1	1	0	1	1
7	14	1	1	1	0	0
7	15	1	1	1	1	1

(b)

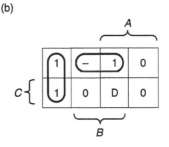

FIGURE 3.18
Example of an MEV K-map scheme (a) truth table with D as a MEV and (b) the corresponding MEV K-map.

2. If for all of the standard minterms covered by an MEV cell, the output and the MEV are identical (or complements), then the MEV (or MEV complement) is entered in that MEV cell.

3. If for standard minterms covered by an MEV cell, the output variable is a don't care in one case and 1 (or 0) in the other, then a 1 (or 0) is entered in that MEV cell.

Accordingly, the MEV K-map corresponding to the truth table of Figure 3.18a is obtained as shown in Figure 3.18b. For each value of M, we compare the values in both columns D and f. Aforementioned rules are then employed to determine the MEV K-map entries. For example, since for $M = 7$ both values of D and f are identical, we enter D in cell 7 of the MEV K-map.

Once the MEV K-map has been constructed for a particular function, we are then ready to determine the resulting prime implicants. The grouping and the corresponding product terms are then determined as follows:

Step 1. Form as large a group as possible of only the 1s, if necessary, by also including the don't cares. The corresponding product term is determined by taking AND of all variables that are common to the group.

Step 2. By treating all the 1s that have been already included in a group as don't cares, form as large a group as possible of only the common MEVs, if necessary, by also including don't cares. The corresponding product term is determined by taking an AND of all variables that are common to the group as well as the MEV.

In accordance with Step 1, therefore, the K-map of Figure 3.18b leads to the following valid groups:

$$\text{Cell set } \{0,1\} \rightarrow \text{(formed of two 1s)} \rightarrow \bar{A}\bar{B}$$

$$\text{Cell set } \{2,6\} \rightarrow \text{(formed of a don't care and a 1)} \rightarrow B\bar{C}$$

Then in the next step, we can treat all these 1s as simply don't cares. The grouping of the MEVs leads to the following additional group:

$$\text{Cell set } \{6,7\} \rightarrow \text{(formed of a don't care and } D) \rightarrow AB(D)$$

Therefore, the minimized function is $f(A, B, C, D) = \Sigma m(0, 1, 2, 3, 12, 13, 15) + d(4, 5) = \bar{A}\bar{B} + B\bar{C} + ABD$. We could have reached the same minimized function using a standard 4-variable K-map as well.

The use of MEV permits the K-map dimensions to be compressed. Thus, it becomes possible to reduce a Boolean function that may have more than six input variables. The MEV K-map is not necessarily restricted to only a single MEV. When we limit MEV to only a single variable, one can construct the MEV K-map readily from the truth table. In particular, if we were to use the least significant variable as an MEV, the construction of MEV K-map becomes relatively easy since it follows directly from the comparison of the last two columns of the truth table. However, an MEV can also be a logical function of multiple-input variables. Unfortunately, for such MEV K-map, a truth table may not easily lend itself to its construction.

3.6   Hazards

We have assumed so far that the optimum approach to designing a logic circuit is to generate a minimum SOP or POS expression and then implement it using logic gates. In this process, we have always assumed that variable values are constant during evaluation. However, in practice, variables are free to change. Such changes often increase the potential for incorrect circuit output if corrective measures have not been taken. Moreover, Boolean algebra does not account for propagation delays along the various signal paths of an actual logic circuit. Practical logic gates have inherent propagation delays, that is, their outputs change only a finite time after the inputs have changed. Accordingly, there are many circumstances when a minimal gate realization may not always be a desirable condition. When the inputs undergo transitions in a combinatorial logic circuit, some or all the outputs may change in response. Consequently, in minimized logic circuits, however, an output may change more than once for a single change in an input variable. These logic circuits are commonly referred to as having circuit *hazard*. The resulting momentary output changes are often referred to as *glitches*. Glitches are caused by the transient behavior of signal paths that may have different propagation delays. The propagation delays of most logic gates are very short depending on the technology and configuration. However, the propagation delay may not be exactly the same for even two similar logic devices.

Typically, hazard-related output malfunctions eventually cease to exist in a purely combinatorial circuit and may not pose a serious problem to the operation of the circuit. However, hazard begins to contribute to serious problems when logic circuits are operated at the limit of the propagation delay paths. Consider for example, the worst case propagation path for a given circuit amounts to a delay of Δt. That implies that data inputs may not be changed in that circuit at a rate faster than once every Δt duration. The circuit that is behaving predictably at some particular data rate may easily get into a hazardous range because of, for example, change in temperature. In later chapters it will be shown that the combinatorial output is sometimes used to drive a class of circuits, referred to as sequential circuits, highly sensitive to $1 \rightarrow 0$ and $0 \rightarrow 1$ transitions. In case of the sequential circuits, therefore, the presence of combinatorial hazard is very critical.

Timing diagram is an important tool that can be used to understand hazards. Often, many of the circuit inputs change with considerable frequency, and it may not be too easy to determine the resulting output that should occur. *Timing diagrams*, in general, are used to describe the time behavior of electronic systems. Conventionally, a timing diagram consists of the waveforms pertaining to the inputs and the output(s) as a function of time. Typically, time is assumed to increase along the horizontal axis from left to right and the vertical axis corresponds to the values of the signals involved. In the case of digital logic circuits, only logical values are used along the vertical axis since our primary interest is in the functional behavior of the circuit.

There are two types of hazards: static and dynamic. A *static hazard* is an error condition for which a single variable change causes a momentary output change when no output change should occur. Typically, the circuit hazard is referred to as static when the initial and final outputs are the same. A static hazard is known as static-1 hazard for which the output is a 1 but temporarily becomes a 0 producing a transient pulse. The other type when the transient causes a momentary 0 is known as static-0 hazard. Static-1 hazard occurs more commonly in AND–OR (i.e., SOP) circuits while static-0 hazard occurs in OR–AND (i.e., POS) circuits. On the other hand, a *dynamic hazard* is an error condition that occurs when, in response to an input change, the output changes several times before settling into its new steady-state value. In the case of dynamic hazard, the final and initial output values differ. Typically, a network that is free of static hazard is also free of dynamic hazards. The natures of the outputs for these hazard conditions are shown in Figure 3.19.

In designing logic circuits, we seek to have circuits that operate properly under all possible conditions. There are two sources that contribute to these hazards. A *function-induced hazard* (or, simply, function hazard) is a hazard that is caused by the function itself. A *logic-induced hazard* (logic hazard) is caused by the particular logic circuit implementing the

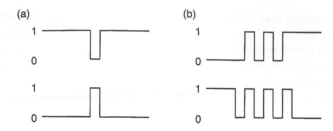

FIGURE 3.19
Hazards (a) static type and (b) dynamic type.

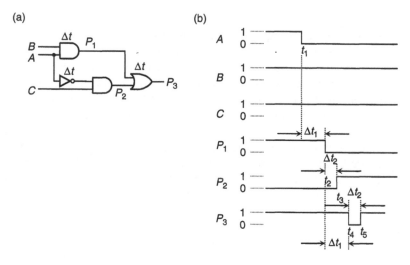

FIGURE 3.20
Example of a static hazard (a) logic circuit and (b) timing diagram.

logic function. We can eliminate a function hazard by not allowing multiple inputs to change at the same time. Logic hazards, however, can exist even when there is a single input.

To illustrate a static hazard, for example, consider a logic circuit shown in Figure 3.20a, where it is logic-induced. As is customary, the circuit behavior can be visualized through its associated timing diagram shown in Figure 3.20b. This particular hazard is a result of nonideal operation of physical devices. Owing to the presence of capacitance (among other things) in the electronics used to implement the logic functions, it can be shown that a small delay of a few nanoseconds exists between input changes and corresponding output response. This time required for the output of a gate to change in response to a change in the input is referred to as *propagation delay*. However, we first consider the case where each of the logic gates has the same propagation delay Δt.

Figure 3.20a shows a 3-input logic circuit having four logic gates. We are assuming, for example, that the propagation delays of both AND and OR gates are identical (say, Δt_1). We assume further that Δt_1 is greater than the propagation delay Δt_2 of the NOT gate. With input A changing from 1 to 0 at t_1 while B and C are both constant and 1, P_1 changes from 1 to 0 at t_2, a Δt_1 time later as shown in Figure 3.20b. But because of this same change in A, P_2 changes from 0 to 1 at $t_3 = t_2 + \Delta t_2 = t_1 + \Delta t_1 + \Delta t_2$. The change in P_1 in turn causes a change in f (from 1 to 0) at $t_4 = t_2 + \Delta t_1$ and similarly the change in P_2 causes a follow-on change in f (from 0 to 1) at $t_5 = t_4 + \Delta t_2$. In an ideal circuit where the logic gates have no propagation delay, f would have been 1 all the time. However, in reality, the change in P_2 follows behind P_1 by a time that is equivalent to the propagation delay of the NOT gate. Accordingly, $f = P_1 + P_2$ becomes 0 momentarily since for that finite duration both P_1 and P_2 are 0. Basically, the glitch occurred because there were two separate paths from input A to the output P_2, each having different number of delay elements. The path through P_1 involves two logic gates and that through P_2 involves three logic gates. This difference in gate delay causes the output to see A go low before $\bar{A}$ goes high.

Using a K-map that corresponds to the logic circuit shown in Figure 3.20a, one will see that the canonical form of this logic function is given by $f(A,B,C) = \Sigma m(1,3,6,7)$. Accordingly, the output should be a 1 as long as the minterms corresponding to the current input conditions remain in the cells 1, 3, 6, and 7. The logic circuit of Figure 3.20a corresponds to the grouping of 1s of such a K-map. The two pairs {1,3} and {6,7} when expressed in SOP form yield the equivalent logic $f(A,B,C) = AB + \bar{A}C$. The timing diagram of Figure 3.20b just discussed represents the input conditions that change from $ABC = 111$ to $ABC = 011$, that is, when A changes from 1 to 0. During this change the logic circuit transitions from the group represented by AB to the group represented by $\bar{A}C$, which is from cell 7 to cell 3. Owing to the inherent characteristic of the NOT gate, both P_1 and P_2 become momentarily 0, which forces the output to be a 0. This momentary drop in the output is the direct outcome of what we refer to as a static hazard. More precisely, this static-1 hazard has been caused since there are two 0-subcubes that differ only in one variable but are not covered by a common product term in an SOP implementation. Likewise, a static-0 hazard is caused when there are two or more maxterm groups that differ only in one variable but are not covered by a common sum term in a POS implementation.

The definition of static hazard suggests ways to construct hazard-free logic circuits. Such hazard exists because two different product terms are used to cover two adjacent minterms. Thus to eliminate such hazard condition, the designer must include an additional product term (corresponding to the consensus term) covering at least those two minterms. This additional prime implicant is redundant but it is the price one has to pay to overcome circuit hazard. In the case of the logic circuit that we just analyzed, this prime implicant formed of the group {3,7} leads to BC term and it remains invariant to the changes in A. Consider now the modified logic circuit shown in Figure 3.21a that has an additional AND gate. The inclusion of this additional 1-subcube (by means of an added AND gate) removes the hazard condition that was discussed earlier. In the corrected logic, the product term BC remains 1 all throughout and thus the resulting output is 1 when the circuit moves from cell 7 to cell 3. This is illustrated in the timing diagram of Figure 3.21b. Similarly, to eliminate a static-0 hazard from an OR–AND logic circuit, the designer will have to include OR gates so as to accommodate bridging adjacent prime implicants.

Additional examples of logic groupings that result in circuit hazards are shown in Figure 3.22. They can be recognized as contributing to static hazards by noting the map positions where one group is exited and another is entered with no other group covering the cells being traversed. As shown in the example just considered, such hazards are eliminated by adding to the reduced SOP expression an additional product term that covers the transition in question. A POS-based circuit, on the other hand, if improperly designed, may contribute to a glitch of the type $0 = >1 = >0$. The latter static-0 hazard is thus eliminated by including additional OR terms to cover hazardous boundaries between the adjacent 0s of the corresponding K-map. In reality, the SOP and the corresponding POS expressions are dual. The hazards in the circuit corresponding to Figure 3.22a is eliminated by replacing the pair {8,9} with the group {0,1,8,9} and including the group {1,5,9,13}. Similarly, the static hazards in Figure 3.22b, for example, can be removed by including two additional pairs, {8,12} and {4,6}.

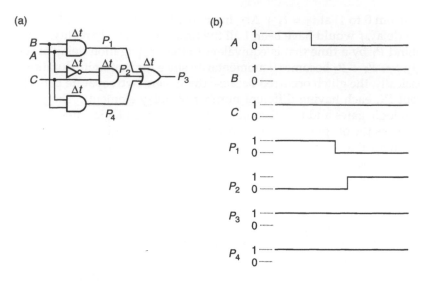

FIGURE 3.21
(a) A hazard-free logic circuit and (b) its timing diagram.

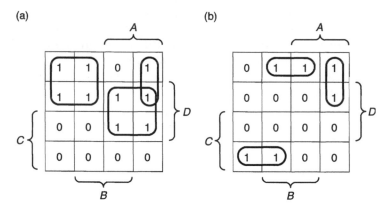

FIGURE 3.22
Examples of K-map groupings that contribute to circuit hazards.

In general, static hazards are caused by two complementary signals that become equal for a finite time because of different delays on different circuit paths. If, in addition, two signals that always have the same value become different for a finite time, the corresponding circuit then has a dynamic hazard associated with it. Same variable value propagating through different circuit paths with different propagation delays may contribute to a dynamic hazard. Consider the circuit of Figure 3.23a, which is nothing but that of Figure 3.20a with only an added AND and an OR gate. On the basis of what we have seen in the timing diagram of Figure 3.20b, P_3 contributes to a static-1 hazard at t_4, $2\Delta t_1$ time after A has changed from 1 to 0. P_4, on the other hand, is generated at about the same time as P_2. The final output which is an AND of P_3 and P_4 thus goes through an oscillatory motion once before taking on its value of 1. There are different ways to remove this dynamic hazard. One way is to eliminate the static hazard by adding a redundant primary implicant, as was done in the solution posed in Figure 3.21. Another valid way of handling dynamic hazard would involve introducing a delay element on P_4 between the OR gate and the last AND gate

(a)

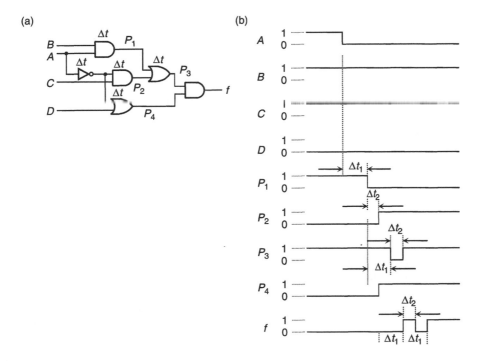

(b)

FIGURE 3.23
A case of dynamic hazard (a) logic circuit and (b) its timing diagram.

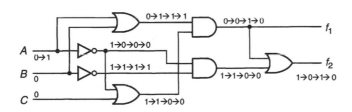

FIGURE 3.24
An ill-formed 3-input 2-output logic circuit.

such that this delay element can postpone the change P_4 from 0 to 1 until the glitch has disappeared.

Consider the ill-formed circuit of Figure 3.24 that has three inputs A, B, and C and two outputs, f_1 and f_2. Note that there are multiple paths in this circuit with different propagation delays between the inputs and the outputs. For simplicity of discussion, we may consider that all of these gates have identical propagation delays and that initially ABC is set to 000. Now suppose A is switched from 0 to a 1. Since the longest circuit path goes through four logic gates, it may be appropriate to study four successive signal transitions. The successive transitions are mentioned next to the outputs of each of the logic gates. It can be seen that f_1 exhibits a static-0 type of hazard since it goes through $0 \Rightarrow 0 \Rightarrow 1 \Rightarrow 0$ transitions. On the other hand, f_2 exhibits a dynamic hazard since it involves $1 \Rightarrow 0 \Rightarrow 1 \Rightarrow 0$ type of transitions.

Static-0 type hazard is generally associated with a POS form of function. An investigation will reveal that the 4-variable f_1 has been derived in two steps. First, $\bar{f}_1$ was obtained (as in the case of Figure 3.15b) from the K-map by grouping two sets of 0s (located at cells 0 through 3 and those at cells 8, 9, 12, and 13). $\bar{f}_1$ was next complemented to give f_1. The groups of 0s, however, had a hazard condition similar to those that existed between the groups of 1s as in Figure 3.22. This condition is thus overcome by grouping the 0s located at cells 0, 1, 8, and 9. This contributes to an OR of $\bar{B}\bar{C}$ and $\bar{f}$. Accordingly, the static-0 hazard associated with $f_1 = (A + B)(\bar{A} + C)$ can be eliminated by including an additional OR gate such that the modified logic function is now $(A + B)(\bar{A} + C)(B + C)$. Surprisingly, such a modification for f_1 can also eliminate the dynamic hazard associated with f_2. Typically, static hazards needed only two different delay paths. Dynamic hazards, on the other hand, must have three or more propagation delay data paths in order for them to occur.

In this section, we have worked with circuit hazards that cause a logic circuit to malfunction during single variable changes. If only a single input is allowed to change at a given time, and the effect of this change is allowed to propagate through the logic circuit before any other additional inputs are changed, then we can indeed implement a Boolean function that is free of static and dynamic hazards. The simple but expensive solution, however, is to realize the function using all of its prime implicants. The resulting logic circuit will be free of hazards irrespective of variations in propagation delays among the various gates implementing the circuit.

The hazard cases caused by the multivariable changes are simply beyond the scope of this book. There are occasions when a circuit must have all hazard conditions removed and other occasions when the hazards will not adversely affect the circuit operation. They may not be so critical in combinatorial circuits since the outputs reach the proper steady state in a short while. This scenario is acceptable when circuit speed is not an issue. In any event, if the variable change causing the hazard does not ever occur, or if it is determined that the spurious outputs do not affect the operations, typically no corrective circuitry is necessary. However, when the output of a combinatorial circuit is being used as an input to a sequential circuit (as it will be shown in Chapter 6) that responds to input changes, then the hazard conditions may be extremely critical. In those cases, glitches can be mistaken for valid signals that may cause the sequential circuit to undergo changes in its states. Accordingly, in sequential circuits, one ends up using a synchronizing clock to control variable changes.

Interestingly, there are logic circuits that take advantage of the existence of hazards. Figure 3.25 shows one such circuit that is commonly used to detect the leading and trailing edge of a randomly varying input signal. The duration of the output pulse generated by such circuits is a function of the propagation delay of the inverters. In many cases, where the propagation delay is not enough to make the output pulse transparent, even numbers

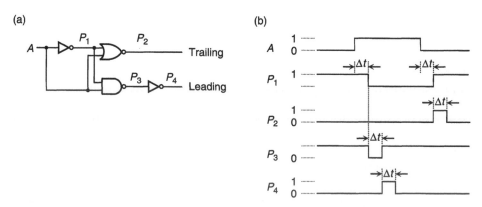

FIGURE 3.25
Leading and trailing edge detectors (a) circuits and (b) timing diagram.

of cascaded NOT gates (in head-to-tail configuration) may be placed next to each of these inverters.

3.7 Single-Output Q-M Tabular Reduction

We have already seen that K-map-based minimization scheme is somewhat ad hoc as it requires the designer to visually recognize groups formed of 0-subcubes. As soon as the number of input variables exceeds six, the K-map scheme becomes unwieldy since it becomes increasingly difficult to recognize valid groups with so many 4-variable K-maps involved in describing the logic function. The tabular reduction technique that will be introduced in this section is a viable scheme for handling a larger number of input variables as opposed to the K-map. It is not dependent on visual pattern recognition. Moreover, since the tabular reduction technique searches exhaustively for the prime implicants, it is suitable for adaptation to a computer-based minimization scheme. Such an alternative is much desirable since otherwise the tabulation scheme is quite tedious as well as very monotonous. This tabular design method was first proposed by W. V. Quine and later improved by E. J. McCluskey, and therefore is frequently referred to as the *Quine-McCluskey's (Q-M) tabular reduction technique*. The tabular method of function simplification involves four steps: (a) setting up the problem in its SOP canonical form, (b) determining the prime implicants, (c) identifying the essential prime implicants, and (d) determining the minimal cover. But to clarify some of the pertinent steps of the Q-M tabular reduction method, we explore next the concepts that govern formation of valid groups.

A reduced SOP function can always be translated back to obtain the original function identifying each of the component minterms or maxterms. This expanded form of a function is typically referred to as the *expanded SOP or POS* or canonical form. Each of these canonical terms corresponds to a different entry in the K-map of the function.

Example 3.7
Obtain the canonical form of the function $f(A, B, C, D) = A\bar{B}\bar{C} + BD + \bar{A}\bar{B}C$.

Solution
$$f(A, B, C, D) = A\bar{B}\bar{C}(D + \bar{D}) + (A + \bar{A})BD + \bar{A}\bar{B}C(D + \bar{D})$$

$$= A\bar{B}\bar{C}D + A\bar{B}\bar{C}\bar{D} + ABD + \bar{A}BD + \bar{A}\bar{B}CD + \bar{A}\bar{B}C\bar{D}$$

$$= A\bar{B}\bar{C}D + A\bar{B}\bar{C}\bar{D} + AB(C + \bar{C})D + \bar{A}B(C + \bar{C})D + \bar{A}\bar{B}CD + \bar{A}\bar{B}C\bar{D}$$

$$= A\bar{B}\bar{C}D + A\bar{B}\bar{C}\bar{D} + ABCD + AB\bar{C}D + \bar{A}BCD + \bar{A}B\bar{C}D + \bar{A}\bar{B}CD + \bar{A}\bar{B}C\bar{D}$$

$$= \Sigma m(2, 3, 5, 7, 8, 9, 13, 15)$$

As shown in Example 3.7, we can express a logical function in several ways. For every reduced form, we can determine a unique canonical form:

$$f(A, B, C, D) = A\bar{B}\bar{C} + BD + \bar{A}\bar{B}C \text{ reduced SOP form}$$

$$= \Sigma m(2, 3, 5, 7, 8, 9, 13, 15) \text{ canonical minterm form}$$

If we were to plot the minterms in a K-map, we shall see that these eight minterms can be grouped as follows:

$$\{8,9\} \qquad\qquad \ggg A\bar{B}\bar{C}$$

$$\{5,7,13,15\} \qquad\qquad \ggg BD$$

$$\{2,3\} \qquad\qquad \ggg \bar{A}\bar{B}C$$

Consider the pair of minterms 8 and 9 that also reduces to $A\bar{B}\bar{C}$ using Boolean algebra as follows:

$$A\bar{B}\bar{C}\bar{D} + A\bar{B}\bar{C}D = A\bar{B}\bar{C}(\bar{D}+D) = A\bar{B}\bar{C}$$

If we are to consider now their binary representations, we shall see that the algebraic minimization is similar to the cellular manipulation as well:

$$A\bar{B}\bar{C}\bar{D} + A\bar{B}\bar{C}D = A\bar{B}\bar{C}$$

$$\{1000\} + \{1001\} => \{100\text{-}\}$$

where uncomplemented literal and complemented literal are represented respectively by 1, and 0. The symbol "-" is not an indication of the value of a variable but is rather a placeholder for a variable that has been eliminated from the expression. The binary representations of the minterms 8 and 9 differ in only one bit position, which, in this case, has a positional weight of 1. Accordingly, the pair $\{8,9\}$ formed of the two minterms, 8 and 9, yields the product term $A\bar{B}\bar{C}$ (obtained by taking AND of the literals common to all of the component minterms), which is also a derivative of its cellular term $\{100\text{-}\}$. Likewise, the pair formed by combining the minterms 2 and 3 gives $\bar{A}\bar{B}C$ in product form or $\{001\text{-}\}$ in its cellular form. The minterms 5, 7, 13, and 15 can be grouped as $\{5,7,13,15\}$, which yields BD in AND form or $\{\text{-}1\text{-}1\}$ in cellular form.

We now have a form to express a reduced function in terms of the minterms or maxterms that are included in each of the implicant groups. The function of Example 3.6, for example, may also be expressed as

$$f(A,B,C,D) = 8,9\{100\text{-}\} + 5,7,13,15\{\text{-}1\text{-}1\} + 2,3\{001\text{-}\}$$

Each of the minterm groups can be traced backward to generate the AND term from which they have been derived. This AND term is derived by expressing the minterms in binary form and then taking an AND of the appropriate set of literal(s) that are common to all of these binary representations. For example, the minterms included in 8,9$\{100\text{-}\}$ can be expressed respectively as $A\bar{B}\bar{C}\bar{D}$ and $A\bar{B}\bar{C}D$. When the literal corresponding to positional weight of 1 is eliminated from either of these two product terms, we obtain $A\bar{B}\bar{C}$ as the reduced product term equivalent to 8,9(100-). Similarly, the product term BD is obtained by eliminating the literals corresponding to the positional weights of 2 and 8.

In order to determine if a cellular group can be formed of two 0-subcubes, both the following conditions must be met:

Condition 1. The decimal equivalent values of the component 0-subcubes must differ by a power of 2, that is, 1, 2, 4, 8, and so on.

Condition 2. Two 0-subcubes can form a cellular group only if the number of 1s in their binary representations differ by one. The decimal value of one component 0-subcube with k ones in its binary representation is subtracted from that with $k+1$ ones. The positive difference of these decimal values, if it is a power of 2, indicates the positional weight of the particular variable that can be eliminated to obtain the resulting AND term. These AND terms are used to denote the cellular group. If the resultant difference is negative, then no grouping can be made even if the difference is a power of 2.

On the basis of an understanding of the above conditions, the Q-M technique for minimizing a single logical function may be elaborated by the following steps. Example 3.7 that follows will provide one with necessary explanations for each of the steps. It is suggested that these steps be read and then each step re-read as the student goes through the steps of Example 3.7.

Step 1. Convert each of the 0-subcubes (i.e., minterms) to its binary form and then partition the 0-subcubes into different sets in accordance with the total count of 1s present in their binary representations.

Step 2. Construct all possible 1-subcubes by grouping two of the appropriate 0-subcubes. The resulting 1-subcubes will have one don't care in their cellular representation.

Step 3. Construct all possible 2-subcubes by grouping two of the appropriate 1-subcubes. This process is continued until no further higher-order subcubes may be produced. An n-subcube will have n don't cares in its cellular representation.

Step 4. Identify those subcubes that could not be used in the formation of higher-order subcubes. Such subcubes are the prime implicants. Each of the prime implicants correspondingly represents all possible subcubes with no smaller-order subcubes being part of a larger-order subcubes.

Step 5. Identify those prime implicants that are necessary to provide a minimum cover of all the 0-subcubes. In other words, each 0-subcube of the original canonical function must be present in at least one subcube.

In Step 1, we sort the 0-subcubes in ascending order by their count of 1s. They are partitioned into sets of 0-subcubes each having a different count of 1s. We next identify 0-subcubes that differ only in one literal. This is accomplished by comparing the 0-subcubes in one partitioned set with 0-subcubes of only the next partitioned set. In order for them to be logically adjacent, two 0-subcubes must differ in exactly one literal. Therefore the binary representation of one of these 0-subcubes must have either one more or one fewer 1 bit than the other 0-subcube. The partitioning thus allows the designer to easily identify 1-subcubes in Step 2. The binary representation of the resulting 1-subcube contains a don't-care symbol "-" in the position of the eliminated literal. Next, in Step 3, we compare the 0-subcubes present in the just-formed 1-subcubes with those present in the next set of 1-subcubes to see if one can form a 2-subcube. This process is continued until all the 0-subcubes have been grouped into one of the n-subcubes. Any subcube not forming a higher-order subcube is a prime implicant by definition.

When the number of inputs increases significantly, it becomes very difficult for one to manually use Q-M scheme for identifying the various higher-order subcubes and, finally, the prime implicants. To produce a higher-order subcube, numerous lower-order subcubes may have to be identified first. There must be at least n pairs of neighboring $(n-1)$-subcubes to yield a single n-subcube. Thus for a multivariable Boolean function that generates an n-subcube where n is very large, the manual search process is not trivial. In practice, designers use software versions of the Q-M tabular reduction algorithm for handling such Boolean functions.

Typically, the minimum cover for a particular SOP canonical function is determined by constructing an implication table. The rules for constructing the implication table are

0-subcubes	ABCD	Count of 1s
2	0010	
4	0100	1
8	1000	
6	0110	
9	1001	
10	1010	2
12	1100	
13	1101	3
15	1111	4

FIGURE 3.26
Partitioned 0-subcubes for Example 3.8.

summarized as follows:

1. Use 0-subcubes as its column headings.
2. Use the prime implicants as its row labels. It is recommended that we separate the prime implicants by their order. If there are two prime implicants that cover one or more of the same 0-subcubes, the higher-order prime implicant is preferred.
3. A check mark, "√" is entered in the table to indicate that the 0-subcube in question is covered by one or more of the corresponding prime implicants.
4. Essential implicants are identified next. A prime implicant is referred to as essential if it is the only cover for one or more of the 0-subcubes. All 0-subcubes that are covered by an essential implicant are then removed from further consideration.
5. The secondary essential implicants are determined next from the remaining prime implicants by making judicious choices. A prime implicant is preferred over another if (a) it covers more of the remaining 0-subcubes, or (b) it is higher in order than the other. This process is continued until all the 0-subcubes have been considered.

One way to identify the essential implicant(s) is to construct an implication table in which each row represents one prime implicant and each column represents a 0-subcube. A x is placed at the intersection of a row and column to indicate that the 0-subcube represented by that column is included in the prime implicant listed as the corresponding row label. After the table has been completed, if we can locate a single x in any particular column, we may conclude that the prime implicant corresponding to that single x (represented now with x circled, i.e., $\otimes$) is indeed an essential implicant. Next example illustrates the Q-M tabular reduction technique in more detail.

Example 3.8
Use Q-M technique to reduce $f(A, B, C, D) = \Sigma m(2, 4, 6, 8, 9, 10, 12, 13, 15)$.

Solution
Step 1. The 0-subcubes are partitioned as shown in Figure 3.26.

Step 2. We try to identify valid 1-subcubes by studying these partitioned 0-subcubes. Two 0-subcubes may form a pair only if the numbers of 1s present in their binary representations differ by one. But in order for the pair to be valid, the decimal equivalents of the candidate 0-subcubes must differ by a power of 2. Starting with the top section, we can match 2 and 6 because they differ by $6 - 2 = 2^2$ to form a 1-subcube. These two

0-subcubes differ in only a single literal. Binary representations of these two 0-subcubes differ from one another only at the location corresponding to positional weight 2^2. The reduced product term is obtained by interpreting the resulting cellular value of {0-10} as $\bar{A}C\bar{D}$. Here "-" corresponds to the positional weight 2^2. This cellular value of 0-10 is listed in the top section of the 1-subcube column shown in Figure 3.27. For bookkeeping purposes $\sqrt{}$ mark is placed next to both of the 0-subcubes, indicating that both of the 0-subcubes have been included in at least one higher-order subcube. Continuing in this manner, 2 and 10 form another valid 1-subcube {2,10} equivalent to the cell {-010}, and so on.

Although an n-subcube may be included in more than one $n+1$-subcube, the n-subcube in question is checked off only once. For example, the 0-subcube 2, which has been used twice, is marked off only once. Between the three 0-subcubes in the top section and the four 0-subcubes in the next section, there are seven 1-subcubes: -

$$
\begin{array}{lll}
2,6 & \ggg & \{2,6\} \gg \{0\text{-}10\} \\
2,10 & \ggg & \{2,10\} \gg \{\text{-}010\} \\
4,6 & \ggg & \{4,6\} \gg \{01\text{-}0\} \\
4,12 & \ggg & \{4,12\} \gg \{\text{-}100\} \\
8,9 & \ggg & \{8,9\} \gg \{100\text{-}\} \\
8,10 & \ggg & \{8,10\} \gg \{10\text{-}0\} \\
8,12 & \ggg & \{8,12\} \gg \{1\text{-}00\}
\end{array}
$$

This accounts for all possibilities between the top two sections of the 0-subcube column. The seven 1-subcubes are listed in the next column in its topmost section. This group of 1-subcubes formed between the top two sections of the 0-subcube column is different from those 1-subcubes listed in the section below it.

Next, we attempt to form 1-subcubes, if possible, by combining 0-subcubes listed in the second section with 0-subcubes listed in the third section. There are two possible 1-subcubes between these two sets of 0-subcubes:

$$
\begin{array}{lll}
9,13 & \ggg & \{9,13\} \gg \{1\text{-}01\} \\
2,13 & \ggg & \{12,13\} \gg \{110\text{-}\}
\end{array}
$$

Likewise, between the third section 0-subcubes and the fourth section 0-subcubes, there is only one 1-subcube possible, {13,15}. In all these cases the difference between the 0-subcubes is always a positive power of 2. In addition, the 0-subcube from the n-th section must always be smaller than the one from the $(n + 1)$-th section. Consequently, the 0-subcubes 8 and 6, for example, could not be combined to form a 1-subcube even though the difference between 8 and 6 is 2. As a verification of this conclusion, we could see a K-map where we shall find that the cells corresponding to 8 and 6 are nonadjacent. The corresponding binary representations 1000 and 0110 imply that they are different in more than one bit. In the table of Figure 3.27, we see that all of the 0-subcubes have been used in at least one 1-subcube. Accordingly, none of these 0-subcubes is a candidate for becoming a prime implicant.

Next we proceed to identify 2-subcubes from the list of already identified 1-subcubes, as shown in Figure 3.28. A valid 2-subcube may be formed by combining 1-subcubes from two neighboring sections if and only if (a) don't cares in both 1-subcubes are located at identical bit position and (b) each of the 0-subcubes included in the 1-subcube differs from those of the other 1-subcube by a power of 2.

By comparing the 1-subcube column entries, we find that {8,9} and {12,13} (respectively represented by {100-} and {110-} from the first and second sections) can be combined to form a valid 2-subcube. Formation of this 2-subcube meets all of the following constraints:

(a) The don't care for the two 1-subcubes occupies the same bit position (the least significant bit in this case)

(b) The 0-subcube included in the first 1-subcube is smaller than the corresponding 0-subcube included in the second 1-subcube since $8 < 12$ and $9 < 13$

(c) The difference in the decimal value of the corresponding 0-subcubes is a power of 2 since $12 - 8 = 13 - 9 = 4 = 2^2$.

The resulting 2-subcube is {8,9,12,13}. Its cellular form is {1-0-} since it is formed of {100-} and {110-}. Upon examination, we shall also find that {8,12} and {9,13} can be combined to also produce the same 2-subcube. In this latter case, the 2-subcube is formed since (a) the don't care for both is also at the same bit position (next to the most significant bit in this case), (b) $8 < 9$ and $12 < 13$; and (c) $9 - 8 = 13 - 12 = 1 = 2^0$. For every 2-subcube identified, there will always be two pairs of 1-subcubes that will produce this 2-subcube. Consequently, both these 1-subcubes are checked off. By comparing the second and the third sections under the 1-subcube column, we conclude that there is no possibility of forming another 2-subcube. This is more true since in order to form a 2-subcube, there must be at least two 1-subcubes present in each of the two sections. In general, to form an n-subcube there must be at least $n - 1$ $(n - 1)$-subcubes present in each of the two neighboring sections. In this particular example, if there was this possibility of forming a 2-subcube by combining 1-subcubes from second and third sections, such a 2-subcube would have been entered in the next section.

Step 3. The 2-subcubes of the neighboring sections are compared next for possible 3-subcubes. In this example, however, no such comparison can be made since there is only one entry in this column. Besides, even if there were entries distributed in several of its sections, there must be at least three 2-subcubes in each of at least two neighboring sections for them to contribute to a valid 3-subcube.

Step 4. All subcubes without $\sqrt{}$ mark are the prime implicants. The prime implicant list for this function is obtained as follows:

$$
\begin{array}{ll}
PI_1 & \{8,9,12,13\} \\
PI_2 & \{2,6\} \\
PI_3 & \{2,10\} \\
PI_4 & \{4,6\} \\
PI_5 & \{4,12\} \\
PI_6 & \{8,10\} \\
PI_7 & \{13,15\}
\end{array}
$$

Step 5. Once the prime implicants have been determined, an implication is constructed next as shown in Figure 3.29a. It is used to determine which and how many of the prime implicants are necessary to assure that all of the minterms are covered. The presence of a ⊗ mark in one or more of its columns indicates that the corresponding prime implicant is essential. From Figure 3.29a, we find that PI₁ and PI₇ are essential since each alone covers the 0-subcubes 9 and 15 respectively. These two essential prime implicants will inevitably be included in the final logical expression. These two essential prime implicants together also cover the 0-subcubes 8, 12, and 13. For further consideration, we can construct a reduced implication table as shown in Figure 3.29b. This table has only 2, 4, 6, and 10 (the rest of the 0-subcubes) as its columns.

The next step is to select from among the remaining prime implicants, one or more of them together to cover the remaining 0-subcubes.

We notice that the five remaining prime implicants all have the same order since they all have equal number of don't cares. Accordingly, if we had to choose any one of these prime implicants over any other prime implicant there should be no problem as long as it covers more of the remaining 0-subcubes. However, we would like to also include as few prime implicants as possible for the logical function since that will guarantee a simpler logic circuit. By observation, we may conclude that if we were to include in addition PI₃ and PI₄ all of the 0-subcubes will be covered. All other combinations will require that we include at least three of these five prime implicants. The reduced function is thus obtained by including the two essential implicants, and PI₃ and PI₄:

$$f(A, B, C, D) = PI_1 + PI_7 + PI_3 + PI_4$$

$$- 1\text{-}0\text{-} + 11\text{-}1 \mid 010 + 01\text{-}0$$

$$= A\bar{C} + ABD + \bar{B}C\bar{D} + \bar{A}B\bar{D}$$

The steps that we undertook to simplify implication table (Figure 3.29a and b) can now be generalized to include the following rules:

(a) If a 0-subcube is covered by only one prime implicant, then that column and all other columns covered by that prime implicant may be removed from further consideration. The corresponding prime implicant is an essential implicant and it must appear in the reduced expression for the logical function.

(b) A column that *covers* another column may be eliminated. If two or more columns have identical entries, then all but any one of these columns may be removed from consideration.

(c) A row that is *covered* by another row may be removed from further consideration. When identical rows are present, all but one of the rows may be eliminated.

Even after sufficient simplifications, it is sometimes difficult to decide on the optimum cover. Oftentimes this step involving the selection of the minimum cover is heuristic and therefore not guaranteed to yield an optimum solution. A technique, referred to as *Petrick's Algorithm*, uses an algebraic process to identify all possible covers of a function. This scheme involves setting a logic function that describes the possible choices. Once the essential implicants have already been identified and removed from the implication table, a POS logic expression is posed as follows:

1. For each remaining 0-subcube, construct an OR expression of all the prime implicants that cover this particular 0-subcube.
2. Construct an AND of all the ORs from step 1.

The resulting POS expression is next converted to its equivalent SOP form using distributive law. Each product term in the resulting reduced expression represents one possible cover. The cover that is associated with the least cost is selected as the optimum solution. Typically, the cost is a function of the number of prime implicants and the number of literals present in each prime implicant.

For the implication table of Figure 3.29b, for example, the minimal cover can be obtained from the equation:

$$
\begin{aligned}
\text{Cover} &= (PI_2 + PI_3) \bullet (PI_4 + PI_5) \bullet (PI_2 + PI_4) \bullet (PI_3 + PI_6) \\
&= (PI_2PI_4 + PI_3PI_4 + PI_2PI_5 + PI_3PI_5) \bullet (PI_2PI_3 + PI_4PI_3 + PI_2PI_6 + PI_4PI_6) \\
&= PI_2PI_4PI_2PI_3 + PI_3PI_4PI_2PI_3 + PI_2PI_5PI_2PI_3 + PI_3PI_5PI_2PI_3 + PI_2PI_4PI_4PI_3 \\
&\quad + PI_3PI_4PI_4PI_3 + PI_2PI_5PI_4PI_3 + PI_3PI_5PI_4PI_3 + PI_2PI_4PI_2PI_6 + PI_3PI_4PI_2PI_6 \\
&\quad + PI_2PI_5PI_2PI_6 + PI_3PI_5PI_2PI_6 + PI_2PI_4PI_4PI_6 + PI_3PI_4PI_4PI_6 \\
&\quad + PI_2PI_5PI_4PI_6 + PI_3PI_5PI_4PI_6 \\
&= PI_2PI_4PI_3 + PI_3PI_4PI_2 + PI_2PI_5PI_3 + PI_3PI_5PI_2 + PI_2PI_4PI_3 + PI_3PI_4 \\
&\quad + PI_2PI_5PI_4PI_3 + PI_3PI_5PI_4 + PI_2PI_4PI_6 + PI_3PI_4PI_2PI_6 + PI_2PI_5PI_6 \\
&\quad + PI_3PI_5PI_2PI_6 + PI_2PI_4PI_6 + PI_3PI_4PI_6 + PI_2PI_5PI_4PI_6 + PI_3PI_5PI_4PI_6 \\
&= PI_2PI_4PI_3 + PI_2PI_5PI_3 + PI_3PI_4 + (PI_2 + 1)PI_3PI_5PI_4 + PI_2PI_4PI_6(1 + PI_3) \\
&\quad + PI_2PI_5PI_6(1 + PI_3) + PI_2PI_4PI_6(1 + PI_5) + (1 + PI_5)PI_3PI_4PI_6 \\
&= PI_2PI_4PI_3 + PI_2PI_5PI_3 + PI_3PI_4 + PI_3PI_5PI_4 + PI_2PI_4PI_6 + PI_2PI_5PI_6 \\
&\quad + PI_2PI_4PI_6 + PI_3PI_4PI_6 \\
&= (PI_2 + 1 + PI_5 + PI_6)PI_3PI_4 + PI_2PI_4PI_6 + PI_2PI_5PI_6 + PI_2PI_5PI_3 \\
&= PI_3PI_4 + PI_2PI_4PI_6 + PI_2PI_5PI_6 + PI_2PI_5PI_3
\end{aligned}
$$

This just completed algebraic manipulations, therefore, indicate that there are four possible choices of prime implicants that will cover the minterms. Only one cover involves two prime implicants while the other three covers include a combination of three prime implicants. Each of these possible covers should be examined individually to determine which choice would require the minimum number of gates and gate inputs. The Petrick's algorithm may be used without even simplifying the implication table but then algebraic manipulation may be somewhat lengthy and tedious. It is best to make all obvious prime implicant choices before employing the Petrick's algorithm.

In Section 3.3 it was shown that inclusion of don't cares provide an additional opportunity for function minimization. The same opportunity exists when the Q-M technique is

0-subcubes			1-subcubes	
2	0010	√	{2,6}	0-10
4	0100	√	{2,10}	-010
8	1000	√	{4,6}	01-0
			{4,12}	-100
			{8,9}	100-
			{8,10}	10-0
			{8,12}	1-00
6	0110	√	{9,13}	1-01
9	1001	√	{12,13}	110-
10	1010	√		
12	1100	√		
13	1101	√	{13,15}	11-1
15	1111	√		

FIGURE 3.27
0-Subcubes for Example 3.8.

0-subcubes			1-subcubes			2-subcubes		
2	0010	√	{2,6}	0-10	PI$_2$	8,9,12,13	1-0-	PI$_1$
4	0100	√	{2,10}	-010	PI$_3$			
8	1000	√	{4,6}	01-0	PI$_4$			
			{4,12}	-100	PI$_5$			
			{8,9}	100-	√			
			{8,10}	10-0	PI$_6$			
			{8,12 }	1-00	√			
6	0110	√	{9,13 }	1-01	√			
9	1001	√	{12,13}	110-	√			
10	1010	√						
12	1100	√						
13	1101	√	{13,15}	11-1	PI$_7$			
15	1111	√						

FIGURE 3.28
1-Subcubes for Example 3.8.

(a)

PI	2	4	6	8	9	10	12	13	15
PI$_1$				x	⊗		x	x	
PI$_2$	x		x						
PI$_3$	x				x				
PI$_4$		x	x						
PI$_5$	x					x			
PI$_6$				x	x				
PI$_7$							x	⊗	

(b)

PI	2	4	6	10
PI$_2$	x		x	
PI$_3$	x			x
PI$_4$		x	x	
PI$_5$		x		
PI$_6$				x

FIGURE 3.29
Implication table for Example 3.8.

employed for minimizing logical functions. If the to-be-reduced logical function includes don't cares, the tabular reduction is carried out exactly as before with two important exceptions. The don't cares are included in all of the steps except that (a) the higher-order subcube formed only of don't cares is not treated as a prime implicant and (b) the don't cares are not used as column headings of the implication table. Example 3.9 illustrates the procedure.

Example 3.9

Simplify the logical function $f(A, B, C, D) = \Sigma m(1, 3, 13, 15) + d(8, 9, 10, 11)$ using Q-M technique.

Solution

The don't cares 8, 9, 10, and 11 are included also as 0-subcubes as shown in Figure 3.30.

We see that even though we made 1-subcubes of only don't cares, {8,9}, {8,10}, {9,11}, and {10,11}, we carried them through to 2-subcube column. We tried to see if one or more of these 1-subcubes could be used to form a 2-subcube that may cover a non-don't-care 0-subcube. In this case, we do see, however, that {9,11} has actually contributed to the formation of 2-subcube {9,11,13,15}. This 2-subcube in addition to covering don't cares also covers minterms 13 and 15.

Since {8,9,10,11} is a prime implicant but formed exclusively of the don't cares, it will not be included in the covering function and, therefore, it will not be used in the implication table. Also, the implication table must not include any of the don't cares as its columns. Accordingly, for this problem, the implication table is obtained as shown in Figure 3.31.

It is obvious that both of the PIs are essential implicants. The reduced function is thus obtained as

$$f(A, B, C, D) = \text{PI}_1 + \text{PI}_3$$

$$= \text{-0-1} + \text{1--1}$$

$$= \bar{B}D + AD$$

0-subcubes			1-subcubes			2-subcubes		
1	0001	√	{1,3}	00-1	√	{1,3,9,11}	-0-1	PI$_1$
8	1000	√	{1,9}	-001	√	{8,9,10,11}	10--	PI$_2$
			{8,9}	100-	√			
			{8,10}	10-0	√			
3	0011	√	{3,11}	-011	√	{9,11,13,15}	1--1	PI$_3$
9	1001	√	{9,11}	10-1	√			
10	1010	√	{9,13}	1-01	√			
			{10,11}	101-	√			
11	1011	√	{11,15}	1-11	√			
13	1101	√	{13,15}	11-1	√			
15	1111	√						

FIGURE 3.30
Minimization for Example 3.9.

PI	1	3	13	15
PI$_1$	x	x		
PI$_3$			x	x

FIGURE 3.31
Implication table for Example 3.9.

3.8 Multiple-Output Q-M Tabular Reduction

In the design of more complex systems, for example, very large-scale integrated (VLSI) circuits, it becomes necessary to generate multiple-output functions that may have all or some of the input variables common between some of them. The real estate size occupied by such circuits has a significant impact on the VLSI design. It is usual to expect having the maximum amount of logic in the minimum possible space. Although one can reduce each of the output functions independently by either K-map or Q-M scheme, considerable savings can be realized if we could share hardware between these output functions. Typically, therefore, we seek to optimize these multiple functions together so that they may share a significant part of a logic circuit. Here, the emphasis is on deriving product terms that can be shared among the various outputs. We have so far described a logical function to be cost-effective on the basis of the following premises. First, a higher-order subcube will always cost less to realize than a lower-order subcube. Second, a cover of fewer prime implicants leads to a less expensive circuit than that of more prime implicants. However, as we begin to deal with multiple-output functions, we may have to be a bit more precise in defining cost on the basis of the need of the technology involved.

The determination of shared product terms among different Boolean functions is a rather complicated task. In any event, if the optimization process using either K-maps or Q-M tabular reduction scheme were to lead to identifying a common product term(s) between these multiple-output functions, we may design a multiple-output logic circuit with common ports. In either case, the first step is finding the prime implicants. The second step involves selecting a subset of these prime implicants to provide optimal cover. We shall first explore a K-map-based effort at reducing multiple-output logic functions.

Consider the output functions:

$$f_1(A, B, C) = \Sigma m(2, 5, 6, 7)$$

$$f_2(A, B, C) = \Sigma m(2, 4, 5, 6)$$

The K-maps corresponding to these two output functions are obtained as shown in Figure 3.32a. Correspondingly, the simplified SOP expressions are

$$f_1(A, B, C) = AC + B\bar{C}$$

$$f_2(A, B, C) = A\bar{B} + B\bar{C}$$

Without any special effort, we see that the two K-maps lend themselves to a situation whereby the reduced functions are able to have a common product term, $B\bar{C}$.

Let us next consider the following two functions:

$$f_3(A, B, C) = \Sigma m(3, 5, 7)$$

$$f_4(A, B, C) = \Sigma m(4, 5, 6)$$

The corresponding K-maps plotted in Figure 3.32b can be used to obtain the following reduced functions:

$$f_3(A, B, C) = \{3, 7\} + \{5, 7\} = \{-11\} + \{1-1\} = BC + AC$$

$$f_4(A, B, C) = \{4, 6\} + \{4, 5\} = \{1-0\} + \{10-\} = A\bar{C} + A\bar{B}$$

In this example, however, no common product term is obvious. Comparing the canonical forms, we see though that only the minterm 5 is common between the two functions.

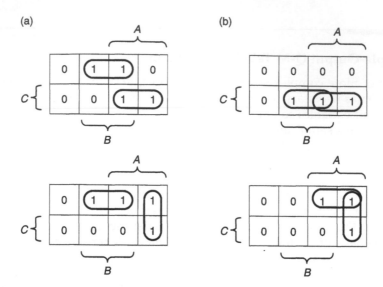

FIGURE 3.32
K-maps for multiple-output logic circuits.

The 1-subcubes {5, 7} and {4, 5} have respectively contributed to the product terms AC and $A\bar{B}$. If the intention is to share logic circuit between these two functions, we can attempt to first generate the 0-subcube 5 and then OR it with the pertinent 1-subcube to obtain the outputs:

$$f_3(A, B, C) = \{3, 7\} + \{5\} = \{\text{-}11\} + \{101\} = BC + A\bar{B}C$$

$$f_4(A, B, C) = \{4, 6\} + \{5\} = \{1\text{-}0\} + \{101\} = A\bar{C} + A\bar{B}C$$

It is clear though that consideration of multiple outputs together does not always guarantee a minimum number of logic elements. In this example just considered, however, the first option involving f_3 and f_4 would have required the use of six logic gates: four 2-input AND gates and two 2-input OR gates. The second option, on the other hand, requires the use of five logic gates: two 2-input AND gates, one 3-input AND gate, and two 2-input OR gates.

The rules for carrying out Q-M-based multiple-output logic minimization problem are similar to those of the single-output minimization with the following exceptions:

1. Each 0-subcube is labeled with distinctive symbols to indicate its association with one or more output functions.
2. Lower-order subcubes may be combined to form higher-order subcubes only if they belong to the same one or more outputs. The resulting higher-order subcube carries with it the symbol or symbols common to the lower-order subcubes.
3. A lower-order subcube is checked off if and only if *all* symbols associated with it are also associated with the other lower-order subcube that combines with it to form a higher-order subcube.

A multiple-output minimization problem that illustrates these additions to the single-output Q-M algorithm is given in Example 3.10.

0-subcubes				1-subcubes				2-subcubes			
0	0000	$\alpha\beta$	√	{0,2}	00-0	$\alpha\beta$	PI_2				
				{0,8}	-000	α	PI_3				
1	0010	$\alpha\beta\gamma$	PI_9	{2,6}	0-10	γ	PI_4				
4	0100	γ	√	{2,10}	-010	$\beta\gamma$	PI_5	{4,5,6,7}	01--	γ	PI_1
8	1000	$\alpha\gamma$	PI_{10}	{4,5}	010-	γ	√				
				{4,6}	01-0	γ	√				
				{8,10}	10-0	γ	PI_6				
5	0101	$\alpha\gamma$	√	{5,7}	01-1	$\alpha\gamma$	PI_7				
6	0110	γ	√	{6,7}	011-	γ	√				
10	1010	$\beta\gamma$	√								
12	1100	β	PI_{11}								
7	0111	$\alpha\beta\gamma$	PI_{12}	{7,15}	-111	β	PI_8				
15	1111	β	√								

FIGURE 3.33
Determining prime implicants for Example 3.10.

Example 3.10
Use Q-M technique to show the design of an optimum digital circuit that implements the following three logical functions:

$$f_\alpha(A, B, C, D) = \Sigma m(2, 7, 8) + d(0, 5)$$
$$f_\beta(A, B, C, D) = \Sigma m(0, 2, 7, 10) + d(12, 15)$$
$$f_\gamma(A, B, C, D) = \Sigma m(2, 4, 5, 6) + d(7, 8, 10)$$

Solution
The prime implicants are determined as shown in the table of Figure 3.33. Consider the 0-subcubes, 0 and 2, that are combined to form the 1-subcube {0,2}. The resulting 1-subcube is denoted with $\alpha\beta$ since those are the two symbols common between the two 0-subcubes. However, we check off only 0 but not 2 since $\alpha\beta$ is included in $\alpha\beta\gamma$ and $\alpha\beta\gamma$ is not included in $\alpha\beta$. Also to be noticed is that the 1-subcube {0,8} carries with it the symbol α since only α is common between the 0-subcubes. However, neither $\alpha\beta$ is completely included within $\alpha\gamma$ nor $\alpha\gamma$ is completely included within $\alpha\beta$. Accordingly, we check neither 0 nor 8 even though they have combined to form a 1-subcube. The only 2-subcube that can be identified is formed of two 1-subcubes from the second and third sections. The 1-subcubes {4,5} and {5,7} as well as {4,6} and {5,7} lead to this same 2-subcube {4,5,6,7}. It is important to note that even though {4,6}, for example, got checked off, the partner 1-subcube {5,7} did not get checked off. This is because the first has associated with it the symbol γ but the latter has associated with it the symbol $\alpha\gamma$.

We construct next an implication table (as shown in Figure 3.34) that will include all 0-subcubes other than the don't cares as columns. Also, we do not include those prime implicants (PI_6, PI_{11}) that are formed of only don't cares.

We see that PI_1, PI_2, and PI_5 are essential prime implicants. These essential prime implicants do cover several 0-subcubes that can now be removed from further consideration. With the removal of these 0-subcubes, we find that PI_5 and PI_9 are not required. Accordingly, we obtain a reduced implication table as shown in Figure 3.35.

In this reduced implication table there are three 1-subcubes and two 0-subcubes. The best cover can be obtained by including PI_{12} and PI_3 where PI_{12} covers two output functions. Although there are two prime implicants covering the 0-subcube 8, PI_3 is preferred over PI_{10} since PI_3 is a 1-subcube while PI_{10} is only a 0-subcube. Therefore, the resulting reduced Boolean equations are

$$
\begin{aligned}
f_\alpha(A,B,C,D) &= \Sigma m(2,7,8) + d(0,5) \\
&= PI_2 + PI_{12} + PI_3 \\
&= \{00\text{-}0\} + \{0111\} + \{\text{-}000\} \\
&= \bar{A}\bar{B}\bar{D} + \bar{A}BCD + \bar{B}\bar{C}\bar{D} \\
f_\beta(A,B,C,D) &= \Sigma m(0,2,7,10) + d(12,15) \\
&= PI_2 + PI_5 + PI_{12} \\
&= \{00\text{-}0\} + \{\text{-}010\} + \{0111\} \\
&= \bar{A}\bar{B}\bar{D} + \bar{B}C\bar{D} + \bar{A}BCD \\
f_\gamma(A,B,C,D) &= \Sigma m(2,4,5,6) + d(7,8,10) \\
&= PI_1 + PI_5 \\
&= \{01\text{-}\text{-}\} + \{\text{-}010\} \\
&= \bar{A}B + \bar{B}C\bar{D}
\end{aligned}
$$

PI	2 α	7 α	8 α	0 β	2 β	7 β	10 β	2 γ	4 γ	5 γ	6 γ
PI_1									⊗	x	x
PI_2	x			⊗	x						
PI_3			x								
PI_4								x			x
PI_5					x		⊗	x			
PI_7		x								x	
PI_8					x						
PI_9	x				x			x			
PI_{10}			x								
PI_{12}		x			x						

FIGURE 3.34
Implication table for Example 3.10.

PI	7 α	8 α	7 β
PI_3		x	
PI_7	x		
PI_8			x
PI_{10}		x	
PI_{12}	x		x

FIGURE 3.35
Reduced implication table for Example 3.10.

As has been shown earlier in this section certain simplifications may often be obvious to the designer without having to employ the Q-M technique. However, with more variables and more outputs this would not be the case. The Q-M multiple-output minimization algorithm just illustrated increases the possibility of having a shared logic circuit and is therefore, more desirable. It must be stressed though that in determining the minimal cover, the designer must have a clear view of the cost function. The cost often is a function of the available inventory, or the current state of the technology, and so on. What may be considered optimal under one scenario may not remain so optimal under a different cost assumption.

3.9   Summary

This chapter summarizes the principles of logic simplification and shows how such simplification translates to a cost-efficient design of combinatorial logic circuits. K-map scheme has been demonstrated to be suitable for minimizing both SOP and POS functions of up to six variables. We have also demonstrated a map-entered scheme to increase the capacity of K-map to handle functions of more than six input variables. The concepts of combinatorial hazards were also introduced and techniques were shown for eliminating such circuit hazards. Finally, the Q-M tabular reduction method was used not only to handle more than six input variables, but also to design multiple-output logic circuits more effectively. This latter algorithmic method and its variations can be utilized to develop computer-aided circuit minimization scheme.

Bibliography

Floyd, T., *Digital Fundamentals*. 8th edn. Englewood Cliffs, NJ. Prentice Hall, 2003.

Johnson, E.J. and Karim, M.A., *Digital Design: A Pragmatic Approach*. Boston, MA. PWS-Kent Publishing, 1987.

Karim, M.A. and Awwal, A.A.S., *Optical Computing: An Introduction*. New York, NY. John Wiley & Sons, 1992.

Karnaugh, M., The Map Method for Synthesis of Combinational Logic Circuits, *Trans. AIEE*. **72**, 593, 1953.

Katz, R.H., *Contemporary Logic Design*. Boston, MA. Addison Wesley, 1993.

McCluskey, E.J., Minimization of Boolean Functions, *Bell Syst. Tech. J.*, **35**, 1417, 1956.

Mowle, F.J.A., *A Systematic Approach to Digital Logic Design*. Reading, MA. Addison-Wesley, 1976.

Nelson, V.P.P., Nagle, H.T., and Carroll, B.D., *Digital Logic Circuit Analysis and Design*. Englewood Cliffs, NJ. Prentice Hall, 1995.

Problems

1. Use K-map scheme to obtain the minimized SOP expression for the following functions:

   (a) $f(A, B, C) = \Sigma m(0, 2, 3, 5, 7)$

   (b) $f(A, B, C) = \Sigma m(0, 2, 4) + d(5, 7)$

   (c) $f(A, B, C, D) = \Sigma m(0, 3, 6, 7, 8, 9, 13, 15)$

   (d) $f(A, B, C, D) = \Sigma m(1, 5, 8, 12) + d(3, 7, 10, 11, 14, 15)$

   (e) $f(A, B, C, D) = \Sigma m(0, 1, 6\text{--}9, 12\text{--}15)$

   (f) $f(A, B, C) = \Sigma m(0, 2, 3, 4, 5, 7)$

   (g) $f(A, B, C, D) = \Sigma m(0\text{--}2, 4\text{--}6, 8, 9, 12)$

   (h) $f(A, B, C, D) = \Sigma m(1, 3\text{--}5, 7\text{--}9, 11, 15)$

   (i) $f(A, B, C, D, E) = \Sigma m(0, 2, 8, 10, 16, 18, 24, 26)$

   (j) $f(A, B, C, D, E) = \Sigma m(4\text{--}7, 9, 11, 13, 15, 25, 27, 29, 31)$

   (k) $f(A, B, C, D, E, F) = \Sigma m(0, 2, 4, 6, 8, 10, 12, 14, 16, 18, 24, 32, 40) + d(20\text{--}23, 48, 56)$

For each of the above minimizations, identify the hazard locations in the map and include appropriate product terms to eliminate the circuit hazard.

2. Obtain the most minimum hazard-free POS expression for the following functions:

   (a) $f(A, B, C, D) = \Sigma m(1, 5, 8, 12) + d(3, 7, 10, 11, 14, 15)$

   (b) $f(A, B, C) = \Sigma m(0, 2, 3, 4, 5, 7)$

   (c) $f(A, B, C, D) = \Sigma m(5, 6, 7, 8, 9) + d(10\text{--}15)$

   (d) $f(A, B, C, D) = \Sigma m(1\text{--}4, 9) + d(10\text{--}15)$

   (e) $f(A, B, C, D, E) = \Sigma m(0, 4\text{--}7, 9, 14\text{--}17, 23) + d(12, 29\text{--}31)$

   (f) $f(A, B, C, D, E) = \Sigma m(0, 4, 5, 7\text{--}9, 11\text{--}13, 15)$

   (g) $f(A, B, C, D) = \Sigma m(2, 3, 8, 9, 10, 11, 12, 13, 14, 15)$

3. Use K-map scheme to obtain the minimized POS expression for the following functions:

   (a) $f(A, B, C) = \Pi M(0, 2, 3, 5, 7)$

   (b) $f(A, B, C) = \Pi M(0, 2, 4) + d(5, 7)$

   (c) $f(A, B, C, D) = \Pi M(0, 3, 6, 7, 8, 9, 13, 15)$

   (d) $f(A, B, C, D) = \Pi M(1, 3, 8, 10, 12, 13, 14, 15)$

   (e) $f(A, B, C, D) = \Pi M(0, 1, 6\text{--}9, 12\text{--}15)$

   (f) $f(A, B, C, D, E) = \Pi M(5, 7, 8, 21, 23, 26, 30) \bullet D(10, 14, 24, 28)$

   (g) $f(A, B, C) = \Pi M(0, 2, 3, 4, 5, 7)$

   (h) $f(A, B, C, D) = \Pi M(0\text{--}2, 4\text{--}6, 8, 9, 12)$

   (i) $f(A, B, C, D) = \Pi M(1, 3\text{--}5, 7\text{--}9, 11, 15)$

   (j) $f(A, B, C, D, E) = \Pi M(0, 2, 8, 10, 16, 18, 24, 26)$

   (k) $f(A, B, C, D, E) = \Pi M(4\text{--}7, 9, 11, 13, 15, 25, 27, 29, 31)$

   (l) $f(A, B, C, D, E, F) = \Pi M(0, 2, 4, 6, 8, 10, 12, 14, 16, 18, 24, 32, 40) \bullet D(20\text{-}23, 48, 56)$

For each of the above minimizations, identify the hazard locations in the map and include sum terms to eliminate the circuit hazard.

4. Obtain the most minimum hazard-free SOP expression for each of the canonical expressions given in Problem 3.

5. For the following logical functions, obtain the most reduced logic functions. Use a MEV K-map by treating the least significant variable as a MEV.

   (a) $f(A, B, C, D, E) = \Sigma m(0, 2, 8, 10, 16, 18, 24, 26)$

   (b) $f(A, B, C, D, E) = \Sigma m(4\text{--}7, 9, 11, 13, 15, 25, 27, 29, 31)$

(a)

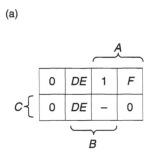

(b)

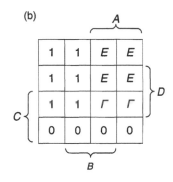

(c)

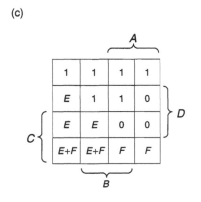

(d)

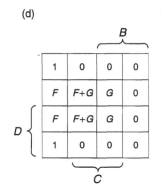

(e)

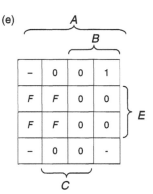

FIGURE 3.P1

(c) $f(A,B,C,D,E,F) = \Sigma m(0,2,4,6,8,10,12,14,16,18,24,32,40) + d(20\text{--}23,48,56)$

(d) $f(A,B,C,D,E) = \Sigma m(0,4\text{--}7,9,14\text{--}17,23) + d(12,29\text{--}31)$

6. For each of the Boolean functions listed in Problem 5, obtain the most reduced logic function using MEV K-maps where the two least significant variables are used as MEV. (Hint: For each MEV minterm, the standard minterms and the corresponding function output can be compared to determine the corresponding MEV.)

7. Find the reduced logic function for the MEV K-maps shown in Figure 3.P1.

8. Use Q-M tabular reduction technique to obtain the minimized SOP form for the following logical functions:

   (a) $f(A,B,C,D) = \Sigma m(0,1,2,3,6,7,8,9,14,15)$

   (b) $f(A,B,C,D) = \Sigma m(1,3,13,15) + d(8,9,10,11)$

   (c) $f(A,B,C,D,E) = \Sigma m(1,3\text{--}7,10\text{--}15,18\text{--}23,26,27)$

   (d) $f(A,B,C,D,E,F) = \Sigma m(7,12,22,23,28,34,37,38,40,42,44,46,56,58,60,62)$

   (e) $f(A,B,C,D) = \Sigma m(1,5,7,9,13,15) + d(8,10,11,14)$

   (f) $f(A,B,C,D) = \Sigma m(0,2,4,8,10,14) + d(5,6,7,12)$

   (g) $f(A,B,C,D,E) = \Sigma m(5,7,9,12\text{--}15,20\text{--}23,25,29,31)$

   (h) $f(A,B,C,D,E) = \Sigma m(0,4,8\text{--}15,23,27) + d(20\text{--}22,31)$

   (i) $f(A,B,C,D,E,F) = \Sigma m(0,2,4,5,7,8,10\text{--}12,24,32,36,40,48,56) + d(14\text{--}17,46,61\text{--}63)$

9. Determine the canonical form (in terms of minterms) for the following functions:

   (a) $f(A, B, C, D) = (\bar{A} + \bar{D})(C + D)(\bar{B} + \bar{D})$
   (b) $f(A, B, C, D) = (A + B + \bar{D})(\bar{A} + D)(\bar{A} + \bar{B})$
   (c) $f(A, B, C, D, E) = AB + \bar{C}D + DE$
   (d) $f(A, B, C, D) = BD + A\bar{B}C\bar{D}$
   (e) $f(A, B, C, D) = (\bar{A} + \bar{C})(A + \bar{B})(B + \bar{D})$
   (f) $f(A, B, C, D, E) = A\bar{B}C\bar{D} + \bar{C}D + \bar{A}\bar{B}DE + A\bar{B}DE + B\bar{C}E$

10. Determine the canonical form (in terms of maxterms) for the functions listed in Problem 8.

11. Determine the minimal SOP form for the following multiple-output logic systems:

    (a)

    $$f_x(A, B, C, D) = \Sigma m(2, 4, 10, 11, 12, 13)$$
    $$f_y(A, B, C, D) = \Sigma m(4, 5, 10, 11, 13)$$
    $$f_z(A, B, C, D) = \Sigma m(1, 2, 3, 10, 11, 12)$$

    (b)

    $$f_x(A, B, C, D) = \Sigma m(0, 2, 9, 10) + d(1, 8, 13)$$
    $$f_y(A, B, C, D) = \Sigma m(2, 8, 10, 11, 13) + d(3, 9, 15)$$
    $$f_z(A, B, C, D) = \Sigma m(1, 3, 5, 13) + d(0, 7, 9)$$

    (c)

    $$f_x(A, B, C, D, E) = \Sigma m(0, 1, 2, 8, 9, 10, 13, 16\text{–}19, 24, 25)$$
    $$f_y(A, B, C, D, E) = \Sigma m(2, 3, 8\text{–}11, 13, 15\text{–}19, 22, 23)$$
    $$f_z(A, B, C, D, E) = \Sigma m(0, 1, 3, 5, 7, 9, 13, 16, 17, 22, 23, 30, 31)$$

12. Find a cost-effective implementation of $f = (A + B)(A + B + C)(A + C)$ using K-map.

13. A digital circuit is to have a single output and four inputs: $A, B, C,$ and D. The output is a 1 whenever the decimal equivalent of $(ABCD)_2$ is divisible by either 3 or 5. Design the logic circuit.

14. Design an SOP logic circuit that converts XS3 input code to its equivalent BCD code.

15. Design a POS logic circuit that converts XS3 input code to its equivalent BCD code.

16. Compare the SOP and POS logic circuits for realizing the function $f(A, B, C, D) = \Sigma m(12, 13, 15) + d(2, 4, 7, 9, 11)$.

17. Study the POS logic circuit corresponding to the function $f(A, B, C, D) = (\bar{A} + B + C)(D + \bar{C})(\bar{B} + \bar{D})$. Determine all of the static-0 hazards associated with this circuit. Using timing diagrams, show the effect of these glitches. What modification will be necessary to obtain a glitch-free logic circuit?

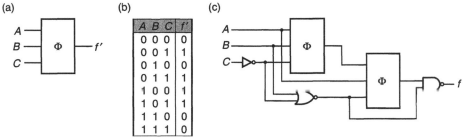

FIGURE 3.P2

18. The truth table of a 3-input digital device, shown in Figure 3.P2a, is given in the table of Figure 3.P2b. Obtain the truth table for the circuit shown in Figure 3.P2c and then obtain its equivalent minimal logic circuit.

19. Design a multiple-output logic circuit that will multiply two 2-bit binary numbers.

4

Logic Function Implementations

4.1 Introduction

The logic circuit design is usually a multistep process. The realization of logical function from a word statement of the logical problem, and the subsequent minimization of the logical function is not the end of the design. In Chapters 2 and 3, we have come to understand how various visual as well as algorithmic schemes may be used for deducing reduced logic function either in the sum-of-product (SOP) or in the product-of-sum (POS) format. These function formats can be translated easily into either a familiar AND–OR or OR–AND pattern of logic circuits. Unfortunately, it may not be possible to have an arbitrary n-input logic gate, for example, especially for large n. Digital integrated circuits (ICs) have inherent practical limitations that may directly affect the implementation of logic circuits. Some of the issues pertaining to ICs include the fan-in and fan-out limitations and the fact that ICs are more frequently available in a particular form than, for example, in the AND and OR form.

The logic gates that are easier to realize with electronic components or the ones that are available in one's inventory eventually determine as to how a particular logic circuit may be implemented. Consequently, it is important for the designer to be familiar with the techniques for translating a reduced SOP or POS function to an acceptable form so that other types of more frequently available logic gates may be used to implement the logic circuit. Another design consideration is speed of the logic circuit. Calculating the total propagation delay requires determining the longest delay path and then accounting for the worst-case propagation delay. This worst-case propagation delay determines the speed with which new data can be introduced to the circuit inputs.

There are three basic approaches to implementing a digital logic circuit: with standard SSI, MSI, and LSI components, with full-custom very large-scale integrated (VLSI) devices, or with the semicustom devices. Small-scale integrated (SSI) circuits contain up to an equivalent of ten 2-input logic gates. Medium-scale integrated (MSI) circuits contain up to an equivalent of 100 logic gates and large-scale ICs contain over 1000 logic gates. VLSI circuits may contain an equivalent of thousands of logic gates in it. Standard SSI and MSI functions are simple in that a circuit can be assembled quickly with readily available off-the-shelf parts. It is quite possible though that the total parts count and, thus, cost per logic gate can be too high as well as the speed may be too low for the resulting design. The alternative may be to consolidate all of the logic subfunctions into one or more custom or semicustom devices. The full-custom design, on the other hand, involves the physical electrical component layouts and carefully controlled design of the necessary interconnections. It can lend itself to highly optimized logic circuit but the involved design process is both very expensive and time-consuming.

In the beginning of this chapter, we limit ourselves to SSI circuit components, but in the later sections we consider logic circuits that may be realized using a standard combinatorial MSI circuit module called a data selector or multiplexer (MUX) and decoder. Typically, when using a MUX to implement a logic function, some of the input variables are used as its input selectors while the remaining variables are fed to this same module as its data inputs. The MUX unit can be organized to serve as a logic circuit that produces correct output for each combination of the inputs. Accordingly, the designer needs to figure out how the input variables (other than those used as selector inputs) or their combinations are to be fed to the module as its data inputs. The key to this design process is that the MUX itself cannot be redesigned, but the inputs can be preprocessed, if needed, using other logic functions before they are fed into the MUX. Likewise, a decoder could not be changed either but its outputs can be suitably postprocessed to generate the desired circuit outputs. Once that mapping of the inputs has been identified, we can use the same MUX or decoder to implement a whole host of different Boolean functions. Use of semicustom devices, on the other hand, reduces design time by utilizing predesigned gate arrays and programmable logic devices (PLDs). In that case, a designer needs to specify how to interconnect the gates on the array.

This chapter dwells on designing logic circuits using gate arrays and PLDs. The read-only memories (ROMs) that are actually nonprogrammable are discussed as a way of introducing the PLDs. These VLSI circuits are frequently used to implement complex logic functions in a cost-effective manner. Typically, the PLDs are prefabricated ICs that include flexible interconnection layers. A PLD circuit design process includes formulating logic functions, translating them to an acceptable PLD format, and then installing them into the PLD using a programmer. In general, PLDs are of three types: programmable read-only memory (PROM), programmable logic array (PLA), and programmable array logic (PAL—a registered trademark of Monolthic Memories, Inc.).

This chapter will explore the possibilities of using only one type of logic gate, gate arrays, or one of the predesigned circuit modules—MUX, decoder, ROM, PROM, PAL, or PLA—for implementing logic functions. Once we have mastered these ideas, we should be able to implement logic functions using either any one type of basic logic gates or one of the predesigned circuit modules unless there is a particular reason for doing otherwise.

Besides the bilateral electronic devices, there exists a different type of switching device, called the threshold element. Threshold elements may not be easily fabricated but in spite of that there is a renewed interest in this device because of its potential application in some special-purpose logic circuits. A section of this chapter is thus devoted to the synthesis of logic circuits using threshold elements.

4.2   Functionally Complete Operation Sets

A functionally complete operation set refers to a set of logic functions that can be used as building blocks for realizing any arbitrary combinatorial logic expressions. There are six such functionally complete operation sets: {AND, OR, NOT}, {AND, NOT}, {OR, NOT}, {NAND}, {NOR}, and {XOR, AND}. In the earlier chapters, we have already encountered logic circuits that use only the first functionally complete operation set, that is, AND, OR, and NOT. Two of the sets are particularly significant since they include only a single logic function, that is, either NAND or NOR. Accordingly, it is possible to design any arbitrary logic circuit using either only NAND or only NOR logic gates. Not surprisingly, many logic circuits are built using either only NAND or only NOR gates. These two logic operations

TABLE 4.1

Typical Gate Configurations and Characteristics

Logic Gate	Number of Inputs	Number of Transistors	Delay (in ns)
NOT	1	2	1
AND	2	6	2.4
	3	8	2.8
	4	10	3.2
OR	2	6	2.4
	3	8	2.8
	4	10	3.2
NAND	2	4	1.4
	3	8	1.8
	4	10	2.2
NOR	2	4	1.4
	3	8	1.8
	4	10	2.2
XOR	2	14	4.2
XNOR	2	12	3.2
AOI	4 (2 ANDs)	8	2
	6 (3 ANDs)	12	2.4
	6 (2 ANDs)	12	2.2
OAI	2 (2 ORs)	8	2.0
	6 (3 ORs)	12	2.2
	6 (2 ORs)	12	2.4

are also known as universal logic operations since one may not need any other logic gates for implementing a function.

Some IC technologies implement logic functions more cheaply than others. Transistor–transistor logic (TTL) realizes logic functions using NANDs while complementary metal-oxide semiconductor (CMOS) logic and emitter-coupled logic (ECL) do so but by using NORs. Each of these logic families is made of numerous subfamilies. Some consume less power and therefore are used in power-sensitive circuit designs; others switch fast and are therefore used in high-speed circuits. Table 4.1 lists some of the more typical logic gate configurations and their pertinent characteristics. The number of transistors needed to generate these logic gates are also listed as this factor determines the gate cost. This table also includes a few multiple-operator gates: AND–OR–INVERT (AOI) and OR–AND–INVERT (OAI). They are often used to respectively implement SOP and POS forms of functions with far less cost (since they require comparatively fewer transistors) and smaller propagation delays. An m-input AOI gate with r ANDs, for example, functions by feeding m/r number of inputs into each one of the r ANDs and then by taking an OR of the r AND outputs. An m-input OAI gate with r ORs, for example, functions by feeding m/r number of inputs into each one of the r ORs and then by taking an AND of the r OR outputs.

It is highly recommended that one uses either only NAND gates or only NOR gates to design logic circuits since then the designer has to maintain only a smaller inventory of discrete logic gates. Besides, NAND and NOR gates both have a smaller propagation delay when compared to other logic gates. NAND and NOR operations are dual of each other. This simplification follows directly from Boolean theorems. Typically, the dual of a Boolean expression is obtained by replacing each of the ANDs with ORs, ORs with ANDs, 0s with 1s, and 1s with 0s. Six of these NAND- and NOR-related dual properties are as follows:

NAND	NOR
$\overline{a \cdot 0} = 1$	$\overline{a + 1} = 0$
$\overline{a \cdot 1} = \bar{a}$	$\overline{a + 0} = \bar{a}$
$\overline{a \cdot a} = \bar{a}$	$\overline{a + a} = \bar{a}$
$\overline{a \cdot b} = \bar{a} + \bar{b}$	$\overline{a + b} = \bar{a} \cdot \bar{b}$
$\overline{\bar{a} \cdot \bar{b}} = a + b$	$\overline{\bar{a} + \bar{b}} = a \cdot b$
$\overline{\overline{a \cdot b}} = a \cdot b$	$\overline{\overline{a + b}} = a + b$

Accordingly, AND, OR, NOT, and NAND (or NOR) logic are implementable using only NOR (or NAND) logic gates. But since XOR and XNOR logic functions are also expressible in either a SOP or a POS form, all of the logic functions may be generated using either only NAND or only NOR logic. The NAND- and NOR-based logic functions are all shown in Figure 4.1.

One of the methods to transform an arbitrary logic function either to its NAND-only form or its NOR-only form is the *brute force* scheme. According to this, each of the basic logic operations that are used to construct the more complex logic function is replaced by its corresponding NAND or NOR equivalent circuit shown in Figure 4.1.

Restriction on the maximum number of inputs that may be fed to the logic gates may not cause any major problem as long as the designer judiciously uses the law of involution and DeMorgan's theorems during such transformation. While using brute force technique, or sometimes only DeMorgan's theorem, the designer may end up having two or more NOT logic next to each other in a circuit path. This allows for an opportunity whereby each pair of adjacent NOT gates between any two neighboring nodes can be eliminated since $\bar{\bar{f}} = f$. However, if there exists a circuit node between the NOT gates, their simultaneous elimination may cause changes in other signals along circuit paths connected to the node in question. In manipulating logic circuits, however, it may be necessary to move NOT gates around sometimes across a node. But in order to get the logic circuit to behave in accordance with the purpose of its original design, therefore, we may have to make further changes in the logic circuit. Figure 4.2 illustrates the method of moving NOT gates across a node without altering the overall functional characteristics of the circuit.

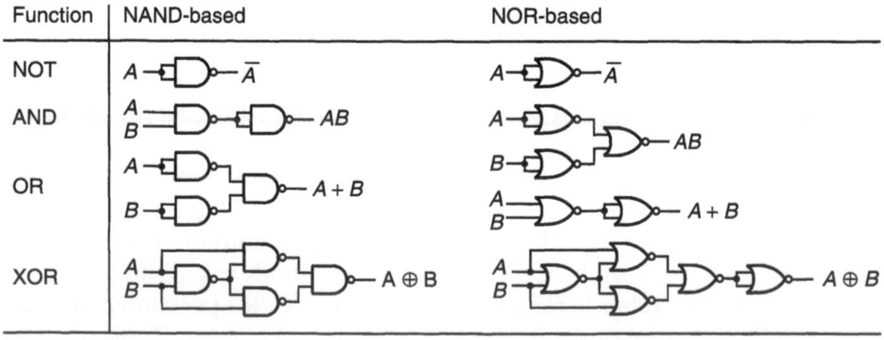

FIGURE 4.1
Basic logic functions using NANDs and NORs.

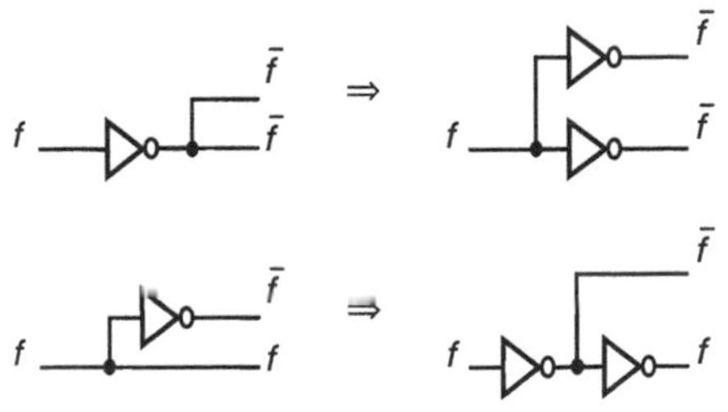

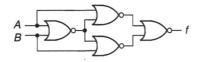

FIGURE 4.2
Rules for moving NOT gates across a junction.

FIGURE 4.3
Logic circuit for Example 4.1.

Example 4.1
Design a NOR-based logic circuit for realizing $\overline{A \oplus B}$.

Solution
There are two different approaches for realizing an XNOR logic: (i) by feeding the output of a NOR-based XOR logic to the input of a NOR-based NOT logic gate, and (ii) by implementing either the SOP or POS form of XNOR function using NOR-based AND, OR, and NOT logic gates. The resulting logic circuit corresponding to the first approach is obtained as shown in Figure 4.3.

The second approach may lead to several different logic circuits.

1. $\overline{A \oplus B}$ can be expressed in the SOP form as $AB + \bar{A}\bar{B}$. Each of these product terms can be realized using NOR-based logic functions as shown in Figure 4.4.

2. Realizing that $\bar{A}\bar{B} = \overline{A + B}$, one may use a NOR gate to generate $\bar{A}\bar{B}$. The resulting logic circuit may be obtained as shown in Figure 4.5.

3. The function may be also expressed in the POS form as $\overline{A \oplus B} = (A + \bar{B})(\bar{A} + B)$. Consequently, a NOR-based logic circuit can be organized as shown in Figure 4.6 for realizing the function. However, the circuit of Figure 4.6 can be reduced further since $X = \bar{\bar{X}}$. The resulting reduced logic circuit is shown in Figure 4.7.

The logic circuits shown in Figures 4.3 through 4.7 require respectively four, eight, six, nine, and five NOR gates. Typically, the maximum propagation delay for a circuit corresponds to the number of maximum gate levels. For the logic circuits just discussed, the number of gate levels are respectively 3, 5, 4, 5, and 3. Note that we can also obtain the circuit of Figure 4.5 directly from that of Figure 4.4 by eliminating two successively placed NOT gates from two of the circuit paths.

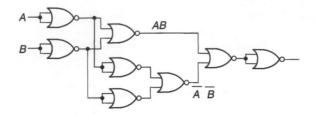

FIGURE 4.4
A NOR-based logic circuit for Example 4.1.

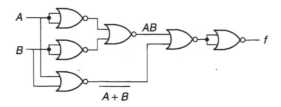

FIGURE 4.5
Logic circuit for Example 4.1.

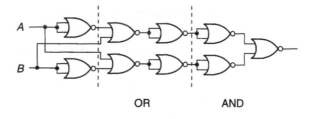

FIGURE 4.6
A NOR-based logic circuit for Example 4.1.

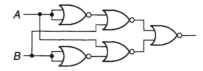

FIGURE 4.7
A reduced logic circuit for Example 4.1.

Example 4.2

By introducing pairs of NOT gates and using DeMorgan's theorem, convert the logic circuit of Figure 4.8 to an equivalent one that uses only NOR gates.

Solution

For simplicity, we may use bubbles to represent NOT gates. Accordingly, f can be realized using the circuit of Figure 4.9.

Finally, noting the DeMorgan's relationships $\bar{x}\bar{y}\bar{z} = \overline{(x+y+z)}$ and $\bar{x}\bar{y} = \overline{(x+y)}$, the logic circuit of Figure 4.9 can be modified to that of Figure 4.10.

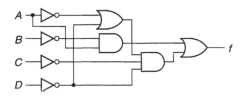

FIGURE 4.8
Given logic circuit for Example 4.2.

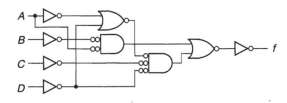

FIGURE 4.9
Logic circuit for f.

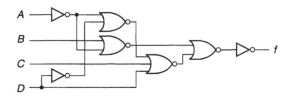

FIGURE 4.10
An alternative logic circuit for f.

Examples 4.1 and 4.2 illustrate that the brute force technique does not necessarily guarantee that the resulting NAND-only logic circuit is going to be any faster than the original logic circuit.

4.3   NAND- and NOR-Only Implementations

In Chapter 3, we have already shown that SOP and POS forms of minimized logic expressions lead to circuits that may be implemented respectively by AND–OR and OR–AND circuits. In this section, we show that such functions can be realized using either only NANDs or only NORs without increasing the number of gate levels. In fact, any Boolean function in SOP form can be implemented using two-level NAND gates while that in POS form can be implemented using two levels of NOR gates. It must be noted that such an implementation is based on the assumption that double-rail inputs are available. *Double-rail* or *dual-rail* input to a logic circuit indicates that both the input variable and its complement are available. If that is not the case, that is, when the logic circuit inputs are of *single-rail* type, one instead needs an additional gate level (of NOT gates) to generate the complements.

NAND-only circuit implementation makes use of the fact that complementing a function twice successively returns the function back to its original form. This implementation process is realized in two steps:

1. The function is complemented first by complementing each of the AND terms and replacing each of the OR operations with an AND operation.

2. The complement function is then complemented once to recover the original function.

Realizing NAND output of two inputs, for example, is equivalent to first complementing the two inputs and then obtaining OR of the two complemented inputs. This statement follows directly from DeMorgan's theorem.

Conversely, given a NAND-only logic circuit, its equivalent logical expression is derived by repeatedly applying DeMorgan's theorem. Typically, the gate level containing the logic gate that produces the function output is referred to as the first level of a logic circuit. The preceding gate level is referred to as the second level, and the level preceding the second level is referred to as the third level, and so on. Circuits with more than just two levels are often needed where there are gate fan-in limits. Given a multilevel NAND circuit, its equivalent Boolean expression is determined in accordance with the following rules:

Rule 1. NAND gates function as OR gates in odd-numbered levels and as AND gates in even-numbered levels.

Rule 2. Input to NAND gate in odd-numbered levels, if it has not gone through a preceding level of NAND gate, appears complemented in the final logical expression. Inputs to NAND gate at even-numbered levels, on the other hand, appear as simply uncomplemented in the final logical expression.

Example 4.3

Using only NAND logic gates, implement the function given by

$$f(A, B, C) = \Sigma m(1, 2, 3, 5, 6)$$

Solution

$$
\begin{aligned}
f(A, B, C) &= \bar{A}\bar{B}C + \bar{A}B\bar{C} + \bar{A}BC + A\bar{B}C + AB\bar{C} \\
&= (\bar{A} + A)\bar{B}C + \bar{A}B(\bar{C} + C) + AB\bar{C} \\
&= \bar{B}C + \bar{A}B + AB\bar{C} \\
&= \bar{B}C + (\bar{A} + A\bar{C})B \\
&= \bar{B}C + (\bar{A} + \bar{C})B \\
&= \bar{B}C + \bar{A}B + B\bar{C}
\end{aligned}
$$

Since,

$$\bar{f} = (B'C)'(A'B)'(BC')'$$

Therefore,

$$f = [(B'C)'(A'B)'(BC')']'$$

The final circuit, therefore, is obtained as shown in Figure 4.11. The logic circuit requires three 2-input NAND gates and one 3-input NAND gate provided that all inputs are also available in complemented form. In case the inputs are single-rail type, NAND gates with all inputs tied together may be used in level 3 for generating the necessary complements.

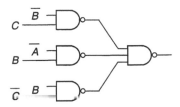

FIGURE 4.11
Logic circuit for Example 4.3.

A function expressed in SOP form may be implemented also using only NOR gates. However, in order to accomplish this feat, the function in question is first converted to its equivalent POS form. The NOR-only implementation of SOP expressions thus involves the following design steps:

1. Plot the given SOP function on a K-map.
2. Obtain the complemented function by grouping only the 0s.
3. Use DeMorgan's theorem to expand each of the resulting AND terms.
4. Complement the complemented function.

In case the inputs are dual-rail type, NOR-only realizations of SOP functions result in 2-level logic circuits.

A NOR output of two or more inputs is equivalent to first taking the complement of the inputs and then taking an AND of the complemented inputs. Therefore, DeMorgan's theorem may be used to interpret also the NOR-only logic circuits. Given a multilevel NOR logic circuit, its equivalent logical expression is determined in accordance with the following rules:

Rule 1. NOR gates function as AND gates in odd-numbered levels and as OR gates in even-numbered levels.

Rule 2. Input to NOR gate in odd-numbered levels, if it has not gone through a preceding level of NOR gate, appears complemented in the final Boolean expression. Input to NOR gate in even-numbered levels appear simply as uncomplemented in the final logical expression.

Example 4.4
Implement the following logical function using only NOR gates.

$$f(A, B, C, D) = \Sigma m(1, 5\text{--}9, 12\text{--}14)$$

Solution
The 4-variable K-map of Figure 4.12 shows the plotted minterms. By grouping all the 0s, we obtain

$$\bar{f} = \bar{B}C + \bar{A}\bar{C}\bar{D} + ACD$$
$$= (B + \bar{C})' + (A + C + D)' + (\bar{A} + \bar{C} + \bar{D})'$$

Therefore,

$$f = [(B + \bar{C})' + (A + C + D)' + (\bar{A} + \bar{C} + \bar{D})']'$$

The NOR-only circuit is, therefore, obtained as shown in Figure 4.13. It requires one 2-input NOR gate and three 3-input NOR gates, provided the inputs are also available in their complemented form.

Table 4.2 summarizes the schemes for converting SOP and POS forms of logic circuits to their equivalent NAND-only and NOR-only logic circuits. The SOP form can be converted to a NAND-only circuit simply by replacing each gate with a NAND gate. On the other hand, it can be converted to a NOR-only circuit by replacing each gate with a NOR gate and inserting a NOR-based inverter on both input and output ports. Converting a POS form to its equivalent NOR- and NAND-only circuit follows similar transformation rules. Since it

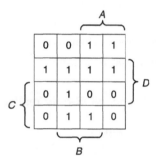

FIGURE 4.12
K-map for Example 4.4.

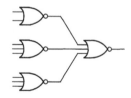

FIGURE 4.13
Logic circuit for Example 4.4.

TABLE 4.2

Schemes for Converting SOP and POS Forms of Logic Circuits

Form	Standard Implementation	NAND-Based Implementation	NOR-Based Implementation
SOP			
POS			

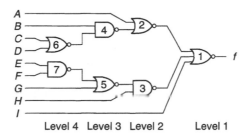

FIGURE 4.14
Given logic circuit for Example 4.5.

is desirable not to have extra gate levels, however, one would rarely convert a SOP form to a NOR-only logic circuit or a POS form to a NAND-only logic circuit.

Example 4.5

Determine the Boolean expression for the output of the multilevel logic circuit shown in Figure 4.14.

Solution

This is a 4-level circuit. Next, we identify each of the logic gates with its corresponding gate level. The Boolean operations associated with these seven logic gates are as follows:

(a) Gate 1 (in level 1) performs an AND operation where input I is considered complemented.

(b) Gates 2 and 3 (in level 2) respectively perform OR and AND operations.

(c) Gates 4 and 5 (in level 3) respectively perform OR and AND operations, and the inputs B and G are considered complemented.

(d) Gates 6 and 7 (in level 4) perform OR and AND operations respectively.

Therefore,

$$f(A, B, C, D, E, F, G, H, I) = (A + (\bar{B} + (C + D))) \cdot (H \cdot (\bar{G} \cdot (E \cdot F))) \cdot \bar{I}$$

$$= (A + \bar{B} + C + D)EF\bar{G}H\bar{I}$$

This particular form is conceptually acceptable since its implementation requires us to have one 4-input OR gate and one 6-input AND gate. If we were to expand it instead to its SOP form, for example, its implementation will require us to have four 6-input AND gates and one 4-input OR gate. The expression for output of this logic circuit may be determined also by starting at level 4 and then by repeatedly applying DeMorgan's theorem, when needed, to determine respective outputs.

4.4   Function Implementation Using XORs and ANDs

In the last section, we have explored the opportunities provided by two functionally complete operation sets {NAND} and {NOR}. Consequently, we were able to implement arbitrary logic functions using either only NAND gates or only NOR gates. In this section, however, we shall be discussing the functionally complete operation set {XOR, AND}. As with other functionally complete sets, by definition, any arbitrary logical function should be realized also using only XOR and AND logic gates.

From one perspective, XOR gate is logically more powerful than, for example, NAND gate. For an *n*-input NAND gate, only one of 2^n-input combinations can cause a change in the output. This limited logic discrimination capability is what forces a NAND-only logic circuit to require a larger number of such gates. XOR gate, on the other hand, is comparatively less interconnection sensitive. Accordingly, only a few of them may be enough to implement a complex logic circuit. Figure 4.1, for example, shows that a 2-input XOR logic is equivalent to having four specially connected 2-input NAND gates. For illustration, the K-maps of several XOR functions are shown in Figure 4.15. These K-maps incidentally have a special characteristic in that they all have equal numbers of 1s and 0s on each half.

Implementation of an arbitrary Boolean function using only XOR and AND logic gates requires that we understand the various XOR properties. A number of these pertinent

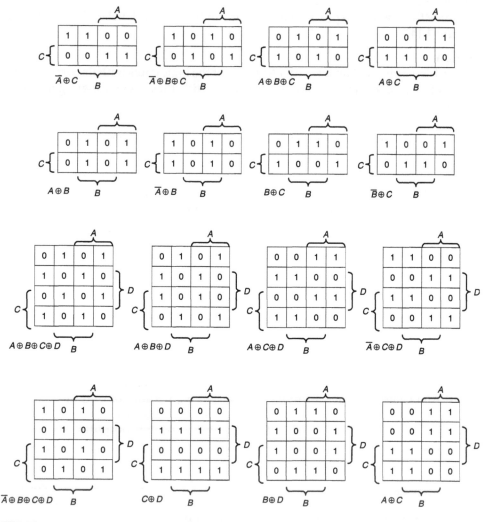

FIGURE 4.15
Examples of XOR functions.

XOR-related theorems are as follows:

1. $A \oplus A = 0$
2. $A \oplus \bar{A} = 1$
3. $A \oplus 0 = A$
4. $1 \oplus \bar{A} = A$ and $1 \oplus A = \bar{A}$
5. $A \oplus B = B \oplus A$ (commutative property)
6. $A + B = A \oplus B \oplus AB = A \oplus \bar{A}B$
7. $A(B \oplus C) = AB \oplus AC$
8. $(A \oplus B) \oplus C = A \oplus (B \oplus C)$ (associative property)

Theorem 4 above, in particular, suggests that an XOR gate can be used as a programmable inverter. If one of the inputs of a 2-input XOR gate is a 1, the corresponding output is equivalent to complement of the other input. If this control input is instead a 0, the other XOR input is simply transferred to the XOR output. In general, the complement of a purely XOR function of multiple inputs is obtained either by complementing the function or by complementing an odd number of input variables. This is possible since the combination of XOR and AND logic gates can be used to implement AND, OR, and NOT operations. Since A OR B is also equivalent to $A \oplus B \oplus AB$ (according to Theorem 6 above), a 2-input OR logic can be formulated using a 2-input AND logic gate and a 3-input XOR logic gate. The complement of a variable, on the other hand, is obtained by feeding the given variable and a 1 to a 2-input XOR logic gate (according to Theorem 4).

A multi-input XOR function can be changed to another equivalent multi-input XOR function by either having an even number of input variables complemented, or by first complementing an odd number of input variables and then complementing the entire function. For example,

$$A \oplus B \oplus C \oplus D = \bar{A} \oplus \bar{B} \oplus \bar{C} \oplus \bar{D} = A \oplus B \oplus \bar{C} \oplus D$$
$$= A \oplus \bar{B} \oplus C \oplus \bar{D} = \overline{A \oplus \bar{B} \oplus C \oplus D}$$

In order to realize an arbitrary Boolean function using XOR and AND logic gates, we first convert the given function to its canonical SOP form. For example, a 2-input Boolean function can be expressed in terms of its canonical terms as follows:

$$f(A, B) = c_0 \bar{A}\bar{B} + c_1 \bar{A}B + c_2 A\bar{B} + c_3 AB$$

where c_i is either 0 or 1. This canonical form will remain unchanged if we replace each of its OR gates with an XOR gate since at any one time only one minterm in an SOP expression can take the value of 1. Using complement property of XOR logic, where needed, we can then rewrite the Boolean equation as

$$\begin{aligned}
f(A, B) &= c_0 \bar{A}\bar{B} + c_1 \bar{A}B + c_2 A\bar{B} + c_3 AB \\
&= c_0 \bar{A}\bar{B} \oplus c_1 \bar{A}B \oplus c_2 A\bar{B} \oplus c_3 AB \\
&= c_0 (1 \oplus A)(1 \oplus B) \oplus c_1 (1 \oplus A)B \oplus c_2 A(1 \oplus B) \oplus c_3 AB \\
&= c_0 (1 \oplus A \oplus B \oplus AB) \oplus c_1 (B \oplus AB) \oplus c_2 (A \oplus AB) \oplus c_3 AB \\
&= c_0 \oplus (c_0 \oplus c_2)A \oplus (c_0 \oplus c_1)B \oplus (c_0 \oplus c_1 \oplus c_2 \oplus c_3)AB \\
&= d_0 \oplus d_1 A \oplus d_2 B \oplus d_3 AB
\end{aligned}$$

where $d_0 = c_0$, $d_1 = c_0 \oplus c_2$, $d_2 = c_0 \oplus c_1$, $d_3 = c_0 \oplus c_1 \oplus c_2 \oplus c_3$. Thus, any 2-input Boolean function can be implemented using only XOR and AND gates. An XOR-related function such as the one just derived that includes only uncomplemented literals is said to be in *Reed–Muller canonical form*. In general, any arbitrary n-input function can be expressed in Reed–Muller canonical form:

$$f(A_1, A_2, \ldots, A_n) = c_0 \oplus c_1 A_1 \oplus c_2 A_2 \oplus \cdots \oplus c_n A_n \oplus c_{n+1} A_1 A_2 \oplus c_{n+2} A_1 A_3$$
$$\oplus \cdots \oplus c_x A_1 A_2 \ldots A_n$$

where $x = 2^n - 1$.

Example 4.6

Convert $f(A, B, C) = A\bar{B} + \bar{A}B + AB\bar{C}$ to its Reed–Muller canonical form so that it can be realized using only XOR and AND logic gates.

Solution

$$f(A, B, C) = A(\bar{B} + \bar{C}) + \bar{A}B$$
$$= A(\bar{B} + \bar{C}) \oplus \bar{A}B \oplus A(\bar{B} + \bar{C}) \cdot \bar{A}B$$
$$= A(\bar{B} + \bar{C}) \oplus \bar{A}B \oplus (A\bar{B} + A\bar{C}) \cdot \bar{A}B$$
$$= A(\bar{B} + \bar{C}) \oplus \bar{A}B \oplus 0$$
$$= A(\bar{B} + \bar{C}) \oplus \bar{A}B$$
$$= A(\bar{B} \oplus \bar{C} \oplus \bar{B}\bar{C}) \oplus \bar{A}B$$
$$= A((1 \oplus B) \oplus (1 \oplus C) \oplus (1 \oplus B)(1 \oplus C)) \oplus (1 \oplus A)B$$
$$= A(1 \oplus B \oplus 1 \oplus C \oplus (1 \oplus B \oplus C \oplus BC)) \oplus (B \oplus AB)$$
$$= A(1 \oplus B \oplus C \oplus B \oplus C \oplus BC) \oplus (B \oplus AB)$$
$$= A(1 \oplus 0 \oplus 0 \oplus BC) \oplus (B \oplus AB)$$
$$= A \oplus ABC \oplus B \oplus AB$$

Reed–Muller canonical form guarantees that the resultant expression is realizable using only XOR and AND gates. However, as we saw in Example 4.6, the solution may require us to use multiple levels of XOR logic gates. If and when too many levels of XOR gates are needed, the designer may then like to settle for a compromise. Consider the XOR K-maps of Figure 4.15. Since the 1s in them are typically equally distributed, they are often a little harder to group. Thus, a designer may prefer to use XOR gates for a function only when its K-map is either that of an XOR K-map or when it is nearly similar to one of the XOR K-maps. A technique known as bridging may be employed to implement such a function.

The *bridging technique* is a way to bend the characteristics of a given Boolean function to match that of a known function. In our case, for example, this known function could be that of an XOR gate. Typically, when the function is relatively unreduced due to having logically separated minterms, a given function may be bridged using known XOR functions that exhibit similar patterns of logically separated minterms. As long as the to-be-implemented function bears the distribution characteristic of an XOR function, it can be realized using

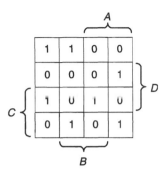

FIGURE 4.16
K-map for Example 4.7.

TABLE 4.3

$f = gX + Y$

f	g	X	Y
0	0	—	0
0	1	0	0
1	0	—	1
1	1	d	d

one or more XOR gates. The bridging technique can be used to express the function in terms of a closely resembling XOR function. The technique consists of the following steps:

1. Compare K-map of the to-be-implemented function f with that of a known XOR K-map g.
2. Construct K-maps of X and Y so as to bridge f and g in accordance with Boolean expression: $f = gX + Y$.
3. Determine the reduced form of X and Y from their K-maps and then realize the logic circuit.

The next example illustrates the bridging technique. The technique is very general and it does not have to bridge only XOR functions. It can be used for generating a higher-level Boolean function when a like function of the same variables already exists. See Problem 30 at the end of the chapter.

Example 4.7
Use bridging to implement $f(A, B, C, D) = \Sigma m(0, 3, 4, 6, 9, 10, 15)$ with an XOR gate.

Solution
The K-map for the given function is obtained as shown in Figure 4.16. By comparing the K-map of Figure 4.16 with those of Figure 4.15, it is obvious that its closest match occurs with $f = A \oplus B \oplus C \oplus D$. The two K-maps are different, however, at three cell locations, 4, 5, and 12. The two K-maps are different, however, at three cell locations, 4, 5, and 12.

To meet the constraint $f = gX + Y$, the functions f and g can be bridged in accordance with the entries of Table 4.3. The ds in Table 4.3 indicate that either X or Y or both must equal 1. In other words, d value for both X and Y may not simultaneously be a 0 when $f = g = 1$. This use of d, however, allows for many choices for the selection of X and Y. Following the dictates of Table 4.3, the two K-maps for f and g can be bridged as shown in Figure 4.17.

It can be seen that if all ds in the K-map for X are set equal to 0, then $f = Y$. This is contrary to what we expect to find in bridging. When $f = Y$, no bridging is needed. Since we do not want to generate f from an identical function, this particular choice for d is disregarded. Alternatively, if all ds in X are set equal to 1, then we can minimize the K-maps for X and Y to obtain $X = \bar{B} + C$ and $Y = \bar{A}\bar{C}\bar{D}$. Therefore,

$$f(A, B, C, D) = (\bar{A} \oplus B \oplus C \oplus D)(\bar{B} + C) + \bar{A}\bar{C}\bar{D}$$

The resultant logic circuit is obtained as shown in Figure 4.18. This bridged circuit is certainly better than the logic circuit that would have been obtained by minimizing the given logic function. If a designer had to use only K-map, for example, the given function would have reduced to

$$f(A, B, C, D) = \bar{A}\bar{C}\bar{D} + \bar{A}B\bar{D} + \bar{A}\bar{B}CD + A\bar{B}\bar{C}D + A\bar{B}C\bar{D} + ABCD$$

As shown in Figure 4.1, it takes a number of NAND or NOR gates to realize the equivalent of an XOR logic. Accordingly, the drawback of using XOR-based logic circuits is that their propagation delay is more than that based on other logic gates. Another difficulty is that typically while a nontrivial number of inputs can be fed simultaneously to AND, OR, NAND, and NOR gates, this may not be that easy with an XOR logic gate. Consider the NAND-based logic circuit of Figure 4.19b. This circuit consisting of seven NAND gates

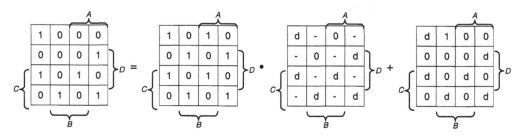

FIGURE 4.17
K-map bridging for Example 4.7.

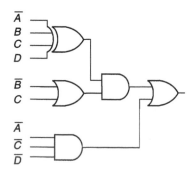

FIGURE 4.18
Logic circuit for Example 4.7.

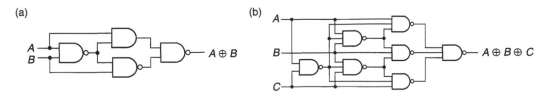

FIGURE 4.19
NAND-based XOR logic having (a) 2 inputs and (b) 3 inputs.

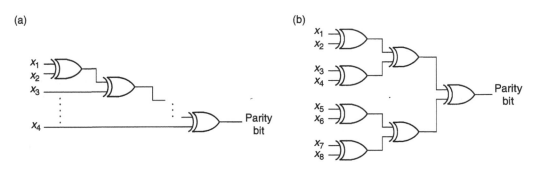

FIGURE 4.20
An 8-input XOR logic in (a) cascade form and (b) tree format.

(six 3-input NAND gates and one 2-input NAND gate) can realize a 3-input XOR logic. For this reason, it is preferable often to use only 2-input XOR logic gates. For a multi-input XOR logic, one uses 2-input XOR logic gates either in cascade as shown in Figure 4.20a or in the tree configuration format as shown in Figure 4.20b. In both cases, the circuit delay increases with the increase in the number of XOR inputs. The case of the cascade form is not desirable if we were to be interested in speed. Even the tree format is not purely parallel. Thus if an entirely parallel operation is necessary, XOR gates with more than two inputs each may not be used.

4.5 Circuit Implementation Using Gate Arrays

One of the hardest decisions a designer has to make is to choose between a simpler circuit and a faster circuit. It is often a matter of relative judgment. Each gate level adds to the delay in the production of signal at the circuit's output. Reduction of the number of gate levels unfortunately does not always allow the circuit to be any simpler. Thus, faster circuits invariably are more expensive as they tend to increase the size of chip real estate. These are typically two-level SOP or POS type, provided dual-rail inputs are available to the circuit. In the context of VLSI chip design, cost for dual-rail routing often exceeds that of the inverters, in which case one may prefer to use single-rail inputs. The exact ratio between the cost of a logic gate and that of gate input depends on the type of logic gate and technology. In most cases, however, the cost of an additional gate is several times that of an additional input on an already existing logic gate.

When gate cost and gate-input cost drive decision-making, as in SSI, MSI, and LSI level designs, the primary motivation behind simplification is elimination of logic gates. An SOP form of logical function is regarded as optimized when there exists (a) no other equivalent Boolean expression involving fewer AND gates, and (b) no other equivalent

Boolean expression involving the same number of AND gates but a smaller number of literals. With the availability of VLSI technology, however, the definition of cost function has changed somewhat. The cost of a VLSI chip is measured by its silicon area that is proportional to the number of transistors or gates. Two particular approaches using VLSI chips are discussed in this chapter. In this section, in particular, we address the requirements of semicustom VLSI circuits and a later section is devoted to programmable VLSI designs.

In semicustom approach, VLSI designs are primarily pursued using gate array technologies. Typically, the gate arrays consist of a large number of identical logic gates that are prefabricated. The logic gates present in a gate array have a fixed number of inputs. The number of inputs in a logic gate is referred to as its fan-in. This always is a limiting factor in circuit design. An n-input logic function can always be realized using either n-input NAND-only or n-input NOR-only gates. Accordingly, in order to use a gate array of m-input gates for implementing an n-input logic function where $m < n$, the designer will need to first transform the given n-variable logic function. This transformation process, referred to as technology mapping, is essential so that the given n-variable logic function is implementable using only m-input gates.

Technology mapping of a minimized Boolean function (in SOP or POS forms) typically consists of the following phases: decomposition, conversion, and retiming. Decomposition is the process of representing the minimized function as a collection of several subfunctions. It typically increases the number of gate levels but it decreases the fan-in of the logic gates. The conversion, on the other hand, is the phase when the decomposed functions are transformed, for example, to either NAND- or NOR-only equivalent logic circuits. During retiming, the designer identifies the circuit paths that contribute to the longest propagation delay. Often times, the designer may need to proceed through one more round of conversion in an attempt to decrease propagation delay of the critical circuit paths identified during retiming.

Consider the case of Example 4.5 where the resulting SOP form requires that we use 6-input AND gates. If we were constrained to use, for example, only 2-input AND gates, it will be necessary to decompose this 6-input AND function. For example, Figure 4.21a shows a 5-level solution while Figure 4.21b shows a 3-level solution. Either of these tree arrangements requires five 2-input AND gates but the latter is a lot faster than the former. Decomposition process thus leads us to a solution such as that shown in Figure 4.21b that may next need to be taken through the conversion phase.

The decomposition phase of technology mapping amounts to decomposing each of the n-input AND (OR) gates into a tree of m-input AND (OR) gates. The resulting gate tree has $p \equiv \text{Int}(\log_m n + 1)$ levels and $\text{Int}[(n - 1)/(m - 1) + 1]$ m-input gates where $\text{Int}(x)$ refers to the integer value of x. In level p ($\neq 1$) of this gate tree, there are $\text{Int}(n/m)$ gates. The outputs of these logic gates together with the remaining $[n - n(\text{Int}(n/m))]$ inputs from level p serve

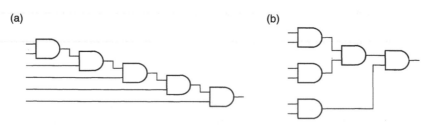

(a)   (b)

FIGURE 4.21
6-Input AND logic using (a) 5-level format; and (b) 3-level format.

as the inputs to gates in level $p - 1$. This process can be repeated for each of the p levels to identify the gate tree.

Example 4.8

Decompose a 12-input OR gate so that it can be implemented using a gate tree consisting of 3-input OR gates.

Solution

For the given problem,

$$\text{Number of OR gate levels} = \text{Int}(\log_3 12 + 1) = 3 \quad \text{and}$$
$$\text{Number of OR gates} = \text{Int}[(12 - 1)/(3 - 1) + 1] = 6$$

The number of gates and gate inputs for these three levels are obtained next as follows:

Level	Number of Inputs	Number of Gates
3	12	$\text{Int}(12/3) = 4$
2	$4 + (12 - 3\,\text{Int}(12/3)) = 4$	$\text{Int}(4/3) = 1$
1	$1 + (4 - 3\,\text{Int}(4/3)) = 2$	$\text{Int}(2/3 + 1) = 1$

In level 1, in particular, there should be two inputs. If we had used the standard formula for determining the number of logic gates in level 1, we would have got 0 as the answer. Assuming there is no other gate levels beyond this level, we need one last OR gate to process these 2 inputs.

The designer will still need to determine the combination of gate inputs at all gate levels. Each of these gate organizations is associated with a different propagation delay. Figure 4.22 shows one possible logic tree of five 3-input OR gates and one 2-input OR gate for implementing this 12-input OR operation.

At times, the designer may use a factoring operation in manipulating the SOP representation of the functions. *Factoring* is the transformation of a function in the SOP form to an equivalent form with parentheses and having a minimum number of literals. Typically, it uses the distributive law of Boolean algebra. Factoring is used often to reduce the number of literals in large AND terms. This is accomplished by selecting a literal that is common to the maximum possible number of AND terms. Accordingly,

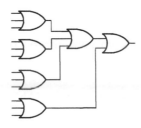

FIGURE 4.22
Logic circuit for Example 4.8.

products terms can be partitioned into two sets: one set consisting of those AND terms that have the common literal; and the other of rest of the AND terms. If the common literal is factored out from the first set, a new SOP expression is obtained. We may look also for a common literal among the AND terms of the second set. This factoring process is continued until they cannot be factored any more. Consider, for example, the logical function $f(A, B, C, D) = A\bar{B} + A\bar{C} + A\bar{D}$. Assuming that inverted inputs are available, this logical function can be realized using a 2-level AND-OR circuit formed of three 2-input AND gates and one 3-input OR gate. However, if we use factoring operation, this same function can be expressed as $f(A, B, C, D) = A(\bar{B} + \bar{C} + \bar{D}) = A\overline{BCD}$, which in turn can be realized using a 2-level circuit formed of one 3-input NAND gate and one 2-input AND gate. Besides, we could accommodate only single-rail inputs.

A factored form of a representation of a logical function is not necessarily unique. Thus, the designers may have to explore all possible choices of common literal before settling on a particular circuit configuration. At times, factoring also requires the introduction of redundancy at intermediate levels.

The decomposition of AND and OR gates is followed next by the conversion step. The gates are then replaced with either only NAND gate or only NOR gate using DeMorgan's law and involution theorem. While converting to NAND-only form, we simply replace each AND with a NAND gate followed by an inverter and each OR with a NAND gate with inverters at its inputs. Likewise, while converting to NOR-only form, we replace each OR with a NOR gate that is followed by an inverter and each AND with a NOR gate with inverters at its inputs. During this process, we eliminate any double inverters that were introduced during conversion.

Often the designers are also faced with the circumstance that they should use custom gate libraries to implement logic functions. The conversion, therefore, becomes a rather complex process especially since it may also involve AOI or OAI gates introduced in Table 4.1. Each logic gate present in a to-be-converted circuit schematic can be replaced by one of these multiple-operator logic gates. On the other hand, either an AOI or an OAI can also replace a number of gates present in a to-be-converted circuit schematic. Technology mapping for custom gate libraries when it involves, for example, multiple-operator gates requires the designer to group the gates present in the to-be-converted circuit schematic. By examining various such grouping patterns, the designer tries to have one or more of these groups replaced by one of the logic gates from the library, and at the same time have a converted logic circuit that has associated with it the minimum cost or delay or both.

It is obvious that speed consideration typically forces one to use standard SOP or POS form. On the other hand, cost consideration forces us to use maximally factored logic expressions requiring multiple gate levels. Accomplishing cost reduction and speed improvement simultaneously is thus a complex problem. A preferred design approach thus has the following goal. The delay minimization process is driven so as to reduce the delay on the critical circuit path. Cost minimization is then geared toward improving cost of all noncritical circuit paths as long as the resulting delay of the noncircuit paths do not exceed that of the critical path. Accordingly, in technology mapping efforts involving any arbitrary custom gate library, the following design steps may be followed:

Step 1. Map the given circuit schematic into a NAND- or NOR-only circuit schematic to determine the critical path.

Step 2. Minimize the propagation delay associated with the critical path.

Step 3. Minimize the cost of noncritical paths.

Example 4.9

Use Table 4.1 as your custom gate library to implement the Boolean function $f(A, B, C, D) = (\bar{A} + \bar{B})(AC + D)$ so that the resulting logic circuit is the fastest and also cost effective.

Solution

This design can be approached from two different angles: (a) its minimal SOP or POS expression and (b) the given form. We shall explore both these approaches.

The given logical expression can be rewritten as

$$f(A, B, C, D) = (\bar{A} + \bar{B})(AC + D)$$
$$= \bar{A}D + A\bar{B}C + \bar{B}D$$

The resulting AND–OR form of the logic circuit may be obtained as shown in Figure 4.23a. For ease of comparison, the number of transistors needed to construct logic gates have been listed within each of the corresponding gate symbols. Each of their respective propagation delays is proportional to the number listed just above their corresponding gate symbols. The NAND-only version, for example, of the circuit of Figure 4.23a can be readily obtained as shown in the circuit of Figure 4.23b. The worst-case propagation delays for these two logical circuits are respectively 6.6 and 4.6 ns. The gate cost, on the other hand, is proportional to 32 and 28 respectively. Therefore, we improve by nearly 33% in speed but do not gain as much in cost.

If the designer had instead begun from the given logical expression, the equivalent OR–AND logic circuit is obtained as shown in the circuit of Figure 4.23c for which the delay (7.2 ns) gets worse and the cost factor does not improve much. Clearly, this solution is not desirable. However, when we convert the logic circuit of Figure 4.23c to its equivalent NOR-only logic circuit cost goes down drastically from 30 to 22 and the speed factor is improved as it changes from 7.2 to 5.2 ns. The logic circuit of Figure 4.23d lends itself to further conversions on critical paths. One choice that may involve multiple-operator gates leads to the use of a 4-input AOI (with 2 ANDs) as shown in Figure 4.23e. The cost factor for the resulting logic circuit gets better but the speed gets a little worse. An alternative choice leads to the introduction of a 4-input OAI (with 2 ORs) as shown in the circuit of Figure 4.23f. Compared to the previous circuit, this latter logic circuit has a dramatically improved speed (3.8 ns instead of 5.4 ns) while the cost factor is the same. Next, we make one final attempt at seeing if we can improve the cost on no-critical circuit paths. Clearly, NOR and the two preceding NOT gates can be replaced by a NAND gate followed by a NOT gate as shown in Figure 4.23g. This conversion leads to significant improvement on cost without further changes in the delay. This final logic circuit thus is the most optimized since the propagation delay now is only 3.8 ns and the cost factor is only 18.

4.6 Logic Function Implementation Using Multiplexers

A rather well-known MSI is a *multiplexer* (MUX for short). It is also known as a *data selector*, a digital equivalent of an analog selector switch. It is a specially connected logic circuit of order n that has up to 2^n data inputs, n selector inputs (S), and an output. Commercial multiplexer units are typically limited to values of n from 1 through 4. An additional input

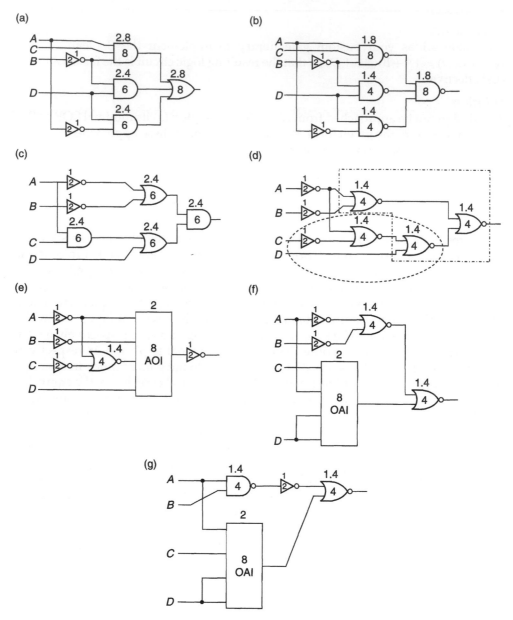

FIGURE 4.23
Logic circuit options for Example 4.9.

(referred to as an enable input) is also available that allows one to cascade multiple low-order multiplexers to obtain higher-order multiplexers. The multiplexer is also referred to as a 1-of-2^n selector since only one of 2^n data inputs is determined by n selector inputs to appear at its output.

A block diagram of a multiplexer with eight data inputs $D_0, D_1, D_2, D_3, D_4, D_5, D_6$, and D_7, three selector inputs, S_0, S_1, and S_2, and one output f is shown in Figure 4.24a. The multiplexer is shown to have two additional lines: $\bar{f}$ for the complemented output and E for enabling the multiplexer. For a set of 2^n data inputs, the multiplexer has exactly n control lines. By applying appropriate signals to the selectors, therefore, any one of the inputs may be selected. For example, when $S_2 S_1 S_0 = 011$, the D_3 input is selected to appear at the

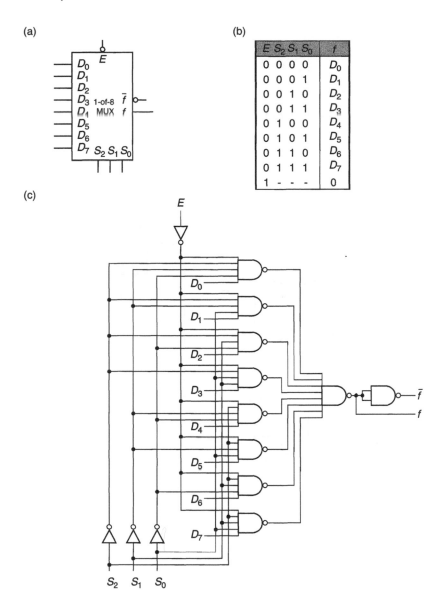

FIGURE 4.24
An 8-input multiplexer (a) block diagram, (b) function truth table, and (c) logic circuit.

output. When E is a 0, the multiplexer functions in accordance with its design objectives. Otherwise the multiplexer is disabled such that the output is a 0 regardless of the values of the selectors. This enable input allows several of these multiplexer devices to be cascaded together in a tree format to produce higher order MUX units. The operation of a multiplexer is best described by its truth table such as that listed in Figure 4.24b. Figure 4.24c shows the equivalent NAND-only logic circuit of the corresponding multiplexer. This follows directly from the truth table of Figure 4.24b.

Multiplexers are very useful devices in any application in which data must be directed from multiple originating points to a single destination point. This implies, however, that we may not be able to use it to simultaneously direct signals from two or more originating points. Signal from only a single signal-originating source will be able to make it to the

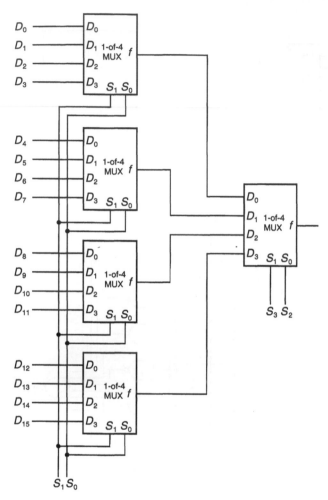

FIGURE 4.25
A 1-of-2^4 multiplexer using five 1-of-2^2 multiplexers.

destination point at one time. The signals from rest of the originating points will have to wait for their own turn. The order in which the signals appear at the output is decided by the values that are introduced at the selector inputs. If there are more signal originating sources than the number of data inputs in a multiplexer, then one will have to cascade more than one multiplexer. Figure 4.25 shows the typical scheme with which, for example, five 1-of-2^2 multiplexers are cascaded to obtain the equivalent of one 1-of-2^4 multiplexer. In this configuration, the two selector inputs belonging to the first level 1-of-2^2 multiplexer select one of the second level 1-of-2^2 multiplexer outputs. The selectors of the second level 1-of-2^2 multiplexers, on the other hand, selects any one of their respective 2^2 inputs.

The multiplexer circuit may be expressed formally by noting that any logic function $f(A_1, A_2, \ldots A_n)$ of n-inputs can be expanded with respect to any of its $n - 1$ variables as

$$f(A_1, A_2, \ldots A_n) = \bar{A}_1 \bar{A}_2 \ldots \bar{A}_{n-1} f(0, 0, \ldots, 0, A_n) + \bar{A}_1 \bar{A}_2 \ldots \bar{A}_{n-2} A_{n-1} f(0, 0, \ldots, 0, 1, A_n)$$

$$+ \bar{A}_1 \bar{A}_2 \ldots A_{n-2} \bar{A}_{n-1} f(0, 0, \ldots, 1, 0, A_n) + \cdots$$

$$+ A_1 A_2 \ldots A_{n-2} A_{n-1} f(1, 1, \ldots, 1, 1, A_n)$$

where for example, $f(x_1, x_2, x_3, \ldots, x_{n-1}, A_n)$ is a residue function of A_n with a value from the set $\{0, 1, A_n, \bar{A}_n\}$ such that x_i is either 0 or 1. The above equation describes a 1-of-2^{n-1} multiplexer. A_i is connected to the select line S_i and $f(x_1, x_2, x_3, \ldots, x_{n-1}, A_n)$ is fed to the data input line D_j where j is the decimal equivalent of $x_1 x_2 x_3 \ldots x_{n-1}$. Accordingly, multiplexers can be used to implement any arbitrary logical function. First, we show in a brute force manner how a multiplexer can be used to realize a logical function. This technique requires that a 1-of-2^n multiplexer is needed for realizing arbitrary functions of n-variables. The function values from a function's truth table are transformed directly to the inputs of the multiplexer. The n-input variables are treated as selector inputs of the multiplexer. Figure 4.26 shows the truth table for a 3-variable logical function and the corresponding 1-of-8 multiplexer implementation. For realizing two functions of the same variables, two multiplexers may be used as shown in the circuit of Figure 4.27.

Next, we shall show a more efficient utilization of multiplexer. In general, we shall find that a 1-of-2^n multiplexer can actually realize a logical function of $n + 1$ variables. This fact will be illustrated by implementing a 3-variable function using a 1-of-4 MUX and a 4-variable function using a 1-of-8 multiplexer respectively in Examples 4.10 and 4.11. A 1-of-8 multiplexer, for example, has three selector lines. While realizing a 4-variable function with this multiplexer, all but any one of the input variables are introduced as its selector inputs. For example, we may consider $A, B,$ and C inputs of a function $f(A, B, C, D)$

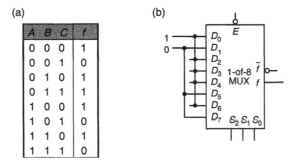

FIGURE 4.26
A 1-of-8 multiplexer for realizing $f(A, B, C) = \Sigma m(0, 2\text{--}4, 6)$ (a) truth table and (b) logic circuit.

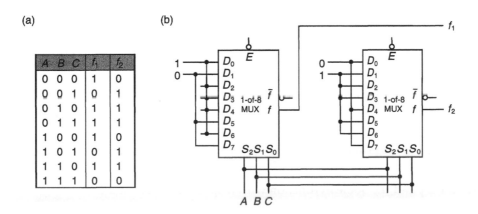

FIGURE 4.27
Two 1-of-8 multiplexers implementing $f_1(A, B, C) = \Sigma m(0, 2\text{--}4, 6)$ and $f_2(A, B, C) = \Sigma m(1\text{--}3, 5, 6)$ (a) truth table and (b) logic circuit.

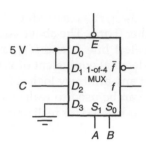

FIGURE 4.28
Logic circuit for Example 4.10.

as the three selector inputs, S_2, S_1, and S_0, respectively. In that case, the technique will need to determine the output function in terms of the fourth input, D, for all combinations of the selector inputs. The values so obtained are then introduced at the respective data inputs of the multiplexer.

Example 4.10

Using a 1-of-4 multiplexer, realize the logical function $f(A, B, C) = \bar{A} + \bar{B}C$.

Solution

We may arbitrarily choose A and B to be fed to the multiplexer as its selector inputs. To determine the appropriate data inputs, therefore, the function may be rewritten as,

$$f(A, B, C) = f(0, 0, C)\bar{A}\bar{B} + f(0, 1, C)\bar{A}B + f(1, 0, C)A\bar{B} + f(1, 1, C)AB$$

But $f(A, B, C) = \bar{A} + \bar{B}C$. Therefore, the four data inputs are obtained as follows:

$$f(0, 0, C) = 1$$
$$f(0, 1, C) = 1$$
$$f(1, 0, C) = C$$
$$f(1, 1, C) = 0$$

since $\bar{A}\bar{B} + \bar{A}B + A\bar{B}C = \bar{A}(\bar{B} + B) + A\bar{B}C = \bar{A} + A\bar{B}C = \bar{A} + \bar{B}C$. The resultant logic circuit is obtained as shown in Figure 4.28.

Example 4.11

Implement the logical function $f(A, B, C, D) = A\bar{C}\bar{D} + \bar{B}D$ using a 1-of-8 multiplexer.

Solution

We can expand the Boolean function with respect to $A, B, C,$ and D.

$$f(A, B, C, D) = A\bar{C}\bar{D} + \bar{B}D$$
$$= \bar{A}\bar{B}\bar{C}f_0 + \bar{A}\bar{B}Cf_1 + \bar{A}B\bar{C}f_2 + \bar{A}BCf_3 + A\bar{B}\bar{C}f_4 + A\bar{B}Cf_5 + AB\bar{C}f_6 + ABCf_7$$

where $f_0 \equiv f(0,0,0,D) = D, f_1 \equiv f(0,0,1,D) = D, f_2 \equiv f(0,1,0,D) = 0, f_3 = 0, f_4 = 1,$ $f_5 = D, f_6 = \bar{D},$ and $f_7 = 0.$ Figure 4.29a shows the implementation of the corresponding logical function. The information obtained by expanding the Boolean function can be derived also by representing the said function in a tabular form as shown in Figure 4.29b. The entries under the column $f(D)$ are determined by comparing the entries of column D with those of the output column f. For any combination of $A, B,$ and C, the following are true:

1. If $f = 1$ irrespective of D, then $f(D) = 1$
2. If $f = 0$ irrespective of D, then $f(D) = 0$
3. If $f = x$ when $D = x$, then $f(D) = D$
4. If $f = \bar{x}$ when $D = x$, then $f(D) = \bar{D}$

The MEV K-map for this problem is obtained as shown in Figure 4.29c where D is used as a MEV. The eight values of $f(D)$ are fed respectively into data inputs D_0 through D_7. Consequently, for any combination of the three selectors, $A, B,$ and C (respectively fed to $S_2, S_1,$ and S_0), the proper value of f would actually appear at the multiplexer output. However, if the input variables were to be fed to the selector inputs in any other order, the proper output will not be generated by the multiplexer. In that case, the list of minterms will have to be rederived first with the decimal equivalents of the product term formed of literals arranged in that same order. A table such as that in Figure 4.29b will have to be reconstructed to determine the new values of $f(D)$. Then and only then proper data input for the multiplexer can be determined.

Functions of n-variables can be realized using a multiplexer of fewer than $n-1$ selectors. When this is possible, larger functions can be realized with smaller multiplexers and logic gates external to it at times contributing to savings in cost and even increased speed. An n-variable Boolean function may be implemented with a multiplexer having m selectors if the number of minterms corresponding to the function is a multiple of 2^{n-m-1}. For example, the function $f(A,B,C,D) = A\bar{C}\bar{D} + \bar{B}D$ (used in Example 4.10) can be realized using a 1-of-8

(a)

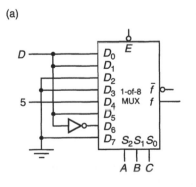

(b)

A B C D	f	f(D)
0 0 0 0	0	D
0 0 0 1	1	
0 0 1 0	0	D
0 0 1 1	1	
0 1 0 0	0	0
0 1 0 1	0	
0 1 1 0	0	0
0 1 1 1	0	
1 0 0 0	1	1
1 0 0 1	1	
1 0 1 0	0	D
1 0 1 1	1	
1 1 0 0	1	$\bar{D}$
1 1 0 1	0	
1 1 1 0	0	0
1 1 1 1	0	

(c)

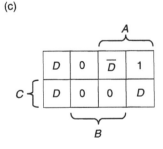

FIGURE 4.29
Logic realization using 1-of-8 multiplexer.

multiplexer since the number of minterms ($=6$) in the function (when $n = 4$) is a multiple of $2^{n-m-1} = 2^{4-3-1} = 1$. This same function is also implementable using a 1-of-4 multiplexer since 6 also is a multiple of $2^{n-m-1} = 2^{4-2-1} = 2$. The function obtained using a 1-of-8 multiplexer can be realized using 1-of-4 multiplexer where the selectors can be assigned any one of six input combinations: $\{A, B\}$, $\{A, C\}$, $\{A, D\}$, $\{B, C\}$, $\{B, D\}$, $\{C, D\}$. These six choices lead to their respective reduced functions as follows:

ABCD	f	ACBD	f	ADBC	f	BCDA	f	BDAC	f	CDAB	f
0000	0	0000	0	0000	0	0000	0	0000	0	0000	0
0001	1	0001	1	0001	0	0001	1	0001	0	0001	0
0010	0	0010	0	0010	0	0010	1	0010	1	0010	1
0011	1	0011	0	0011	0	0011	1	0011	0	0011	1
0100	0	0100	0	0100	1	0100	0	0100	1	0100	1
0101	0	0101	1	0101	1	0101	0	0101	1	0101	0
0110	0	0110	0	0110	0	0110	1	0110	1	0110	1
0111	0	0111	0	0111	0	0111	1	0111	1	0111	0
1000	1	1000	1	1000	1	1000	0	1000	0	1000	0
1001	1	1001	1	1001	0	1001	1	1001	0	1001	0
1010	0	1010	1	1010	1	1010	0	1010	1	1010	0
1011	1	1011	0	1011	0	1011	0	1011	0	1011	0
1100	1	1100	0	1100	1	1100	0	1100	0	1100	1
1101	0	1101	1	1101	1	1101	0	1101	0	1101	0
1110	0	1110	0	1110	0	1110	0	1110	0	1110	1
1111	0	1111	0	1111	0	1111	0	1111	0	1111	0

The corresponding logical function $f(A, B, C, D) = A\bar{C}\bar{D} + \bar{B}D$, therefore, can be expressed in six different ways:

1. $f(A, B, C, D) = \bar{A}\bar{B}(D) + \bar{A}B(0) + A\bar{B}(\bar{C} + D) + AB(\bar{C}\bar{D})$
2. $f(A, C, B, D) = \bar{A}\bar{C}(\bar{B}D) + \bar{A}C(\bar{B}D) + A\bar{C}(\bar{B} + \bar{D}) + AC(\bar{B}D)$
3. $f(A, D, B, C) = \bar{A}\bar{D}(0) + \bar{A}D(\bar{B}) + A\bar{D}(\bar{C}) + AD(\bar{B})$
4. $f(B, C, D, A) = \bar{B}\bar{C}(D + A) + \bar{B}C(D) + B\bar{C}(\bar{D}A) + BC(0)$
5. $f(B, D, A, C) = \bar{B}\bar{D}(A\bar{C}) + \bar{B}D(1) + B\bar{D}(A\bar{C}) + BD(0)$
6. $f(C, D, A, B) = \bar{C}\bar{D}(A) + \bar{C}D(B) + C\bar{D}(0) + CD(\bar{B})$

These options are all examples of function decomposition. The corresponding multiplexer solutions are obtained as shown in Figure 4.30. Of the choices we have, option 6 offers the best solution for single-rail inputs since it requires only one NOT gate. Option 3, in comparison, requires two NOT logic gates. Provided that dual-rail inputs are available, both options 3 and 6 are the most desirable solutions since we may not need anything but a single 1-of-4 multiplexer. Every one of the other solutions requires additional logic gates to preprocess the inputs. Although it is inconsequential here, the NOT gate that provides $\bar{B}$ must have a fan-out of at least 2. In case of many problems, such fan-out related issue could become critical enough in that we may have to use additional logic gates to compensate for fan-out limitations.

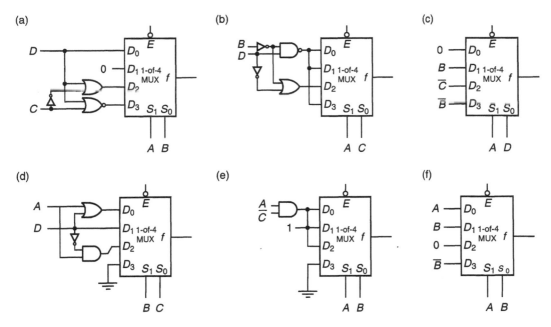

FIGURE 4.30
Various multiplexer options for $f(A,B,C,D) = A\bar{C}\bar{D} + \bar{B}D$.

It is very obvious that the availability of multiplexers provides a designer with numerous choices. The only way to find out which variables should be fed to the selector inputs to achieve the simplest logic circuit is to try out all possibilities. In addition, one or more of the selectors can be assigned complement of the function variables.

Example 4.12
Realize the logical function $f(A,B,C,D) = \sum m(3,4,8\text{--}10,13\text{--}15)$ using (a) a 1-of-4 multiplexer and assorted logic gates, if needed; and (b) 1-of-4 multiplexer in the first level and 1-of-2 multiplexer in level 2.

Solution

(a) The function may be expressed in its SOP form and then regrouped as follows:

$$f(A,B,C,D) = \bar{A}\bar{B}CD + \bar{A}B\bar{C}\bar{D} + A\bar{B}\bar{C}\bar{D} + A\bar{B}\bar{C}D$$

$$+ A\bar{B}C\bar{D} + AB\bar{C}D + ABC\bar{D} + ABCD$$

$$= \bar{A}\bar{B}(CD) + \bar{A}B(\bar{C}\bar{D}) + A\bar{B}(\bar{C}\bar{D} + \bar{C}D + C\bar{D})$$

$$+ AB(\bar{C}D + C\bar{D} + CD)$$

$$= \bar{A}\bar{B}(CD) + \bar{A}B(\bar{C}\bar{D}) + A\bar{B}(\bar{C} + \bar{D}) + AB(C + D)$$

The resulting multiplexer-based logic circuit is obtained as shown in Figure 4.31. As expected, function inputs are fed first to appropriate logic gates and their outputs in turn are fed to the multiplexer inputs.

(b) In order to use 1-of-2 multiplexer units in level 2 for preprocessing the inputs, each of the four functions $f(0,0,C,D)$, $f(0,1,C,D)$, $f(1,0,C,D)$ and $f(1,1,C,D)$ will also have to be decomposed, for example, in terms of C. We find that

$$f(A,B,C,D) = \bar{A}\bar{B}[\bar{C}(0) + C(D)] + \bar{A}B[\bar{C}(\bar{D}) + C(0)]$$
$$+ A\bar{B}[\bar{C}(1) + C(\bar{D})] + AB[\bar{C}(D) + C(1)]$$

Now a two-level multiplexer-based logic circuit can be used to generate this logical function. The 1-of-4 multiplexer used in level 1 is identical to that used in Figure 4.31. The second level 1-of-2 multiplxers essentially eliminates the need for discrete logic gates. Each of these multiplexers generates the same output as that of the corresponding logic gates of Figure 4.31. The resultant two-level multiplexer-based logic circuit is obtained as shown in Figure 4.32. However, upon further exploration we notice that

$$\overline{\bar{C}(0) + C(D)} = \bar{C}(1) + C(\bar{D})$$

and

$$\overline{\bar{C}(\bar{D}) + C(0)} = \bar{C}(D) + C(1)$$

This implies that two of the 1-of-4 multiplexer inputs are complements of the other two inputs. Fortunately, each of the multiplexers is equipped with an additional output for providing the complemented result. Consequently, two of the second level 1-of-2 multiplexers that were used in Figure 4.32 may not be needed. The resulting reduced multilevel multiplexer-based logic circuit is obtained as shown in Figure 4.33.

Binary arithmetic operations often require that the corresponding digital system is able to shift a string of binary digits a few bit either to the left or to the right. For example, if m bits of an n-bit number (where $m \leq n$) were to be shifted to the left, then m new bits (typically m 0s) will have to be shifted in from the right. A total of m shifts to the left, for example, is equivalent to having multiplied the preshifted number by 2^m. Similarly, m shifts to the right is equivalent to a division by 2^m. If the right shifted number is in 2's complement number form, however, m copies of the sign bit will have to be moved in from the left. During left-shift operation, on the other hand, the most significant bit must continue to store the

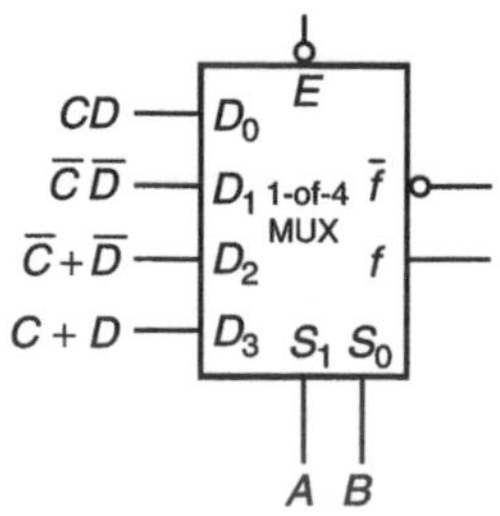

FIGURE 4.31
Logic circuit for Example 4.12.

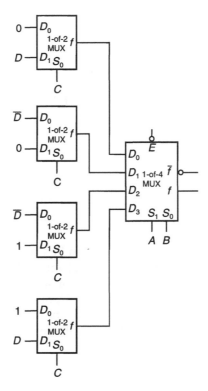

FIGURE 4.32
Logic circuit using 1-of-2 multiplexer for Example 4.12.

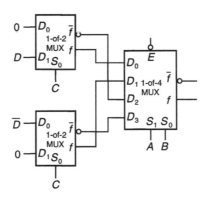

FIGURE 4.33
A reduced logic circuit for Example 4.12.

same sign bit value. The rotation operation is different in that it does not involve any newer bit. The bits that are shifted out from one side are simply shifted in from the other side.

Example 4.13

Using only 1-of-4 multiplexers, obtain a multibit shifter that takes in an $(n + 2)$-bit input, x, generates an n-bit output, y. The shifter output is determined by R and S as explained already in Example 3.5.

Solution

The output of the multibit shifter is determined as follows:

$$\text{If } R = 1 \quad \text{and} \quad S = 1, \quad y_i = x_{i-1} \text{ (equivalent to a left-shift)}$$

$$\text{If } R = 0 \quad \text{and} \quad S = 1, \quad y_i = x_{i+1} \text{ (equivalent to a right-shift)}$$

$$\text{If } S = 0 \quad \text{and} \quad R = -, \quad y_i = x_i \text{ (equivalent to no-shift)}$$

A 1-of-4 multiplexer can be employed to process each bit of x. The variables R and S can be fed as multiplexer selectors and the inputs x_i, x_{i-1}, and x_{i+1} as its data inputs. The resulting multibit shifter is obtained as shown in Figure 4.34. With E set to 0, the selector inputs can be manipulated to realize the required bit shift operations.

As shown in this last example, multiplexers can be used to realize shift operations. An universal n-bit shifter/rotator using 1-of-4 multiplxers is shown in Figure 4.35. The three selector inputs are used for determining the type of shift operations. When S_2 is a 0, the input appears at the output without any change. If, on the other hand, S_2 is a 1 either a shift or rotate operation is to be performed. The direction of the shift, in particular, depends on the value of S_1. When it is a 0, the bit string shifts to the left one bit and when it is a 1,

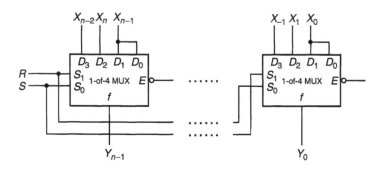

FIGURE 4.34
Multibit shifter for Example 4.13.

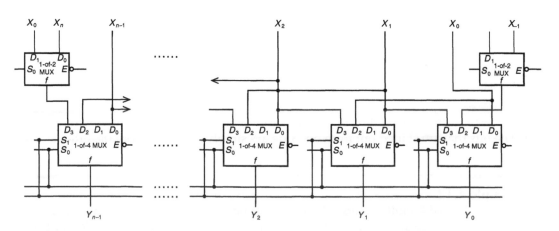

FIGURE 4.35
An universal shifter/rotator using 1-of-4 multiplexers (a) truth table and (b) logic circuit.

the bit string shifts to the right one bit. The third selector S_0 discriminates between shift and rotation operations. The shifter shown in Figure 4.35 has two 1-of-2 multiplexers at its two extreme bit positions. On the basis of the value of S_0, these two multiplexers select the appropriate bits during the shift and rotate operations. With $S_2 S_1 = 10$, the 1-of-2 multiplexer positioned on the extreme right shifts in a bit from the right (as in a left-shift operation) when $S_0 = 0$ but rotates in the most significant bit when $S_0 = 1$. The 1-of-2 multiplexer positioned on the extreme left, on the other hand, becomes effective when $S_2 S_1 = 11$. With $S_0 = 0$, it feeds in a bit from the left (as in a shift-right operation) but with $S_0 = 1$, it rotates in the least significant bit.

When we need to shift or rotate a given number more than 1 bit, the shifted number can be obtained using the circuit of Figure 4.35b but by passing the data bits multiple times. Each pass through the system amounts to a 1-bit shift. However, such repeatedly performed data transformation process is considered rather slow. A *barrel shifter* can be used to overcome this timing problem. An n-bit barrel shifter typically consists of $\log_2 n$ levels of 1-of-2 multiplexers. The j-th level of the multiplexers ($0 \leq j \leq \log_2 n - 1$) shifts 2^j positions. Figure 4.36 shows the logic circuit, for example, of an 8-bit barrel right rotator. In general, if we are to shift/rotate a string of numbers p bits (expressed as $p_{n-1} \ldots p_0$ in binary), each of the multiplexer selectors S_j is connected directly to p_j for all j in the range $0 \leq j \leq (\log_2 n - 1)$. This logic circuit consists of three levels of 1-of-2 multiplexers. For an n-bit barrel rotator, the circuit delay is approximately $\log_2 n$ times the delay of a single 1-of-2 multiplexer. On the other hand, the circuit cost is roughly $n \log_2 n$ times the cost of one 1-of-2 multiplexer.

As we have seen already, the multiplexers are very effective in implementing logical functions. The previous example also shows its capability to shift data. Figure 4.37 takes this idea further to show how strings of data can be routed to a particular destination. In Figure 4.37a, two stacks (4-bit each) of 1-of-2 multiplexers are used to route 8-bits of data from either X or Y to a destination by means of a single selector input. With $S = 0$, the least significant four bits of Y pass through the multiplexer on the right and the most significant four bits pass through the multiplexer on the left. With $S = 1$, however, eight bits of X appear at the output. Then in Figure 4.37b, by making use of the enable input, the same

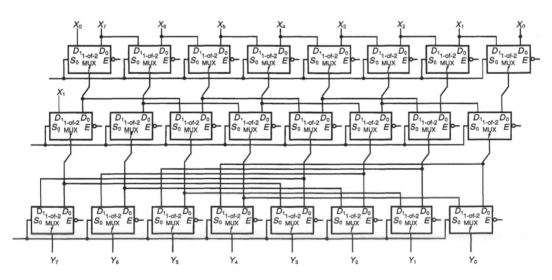

FIGURE 4.36
A 8-bit barrel right rotator.

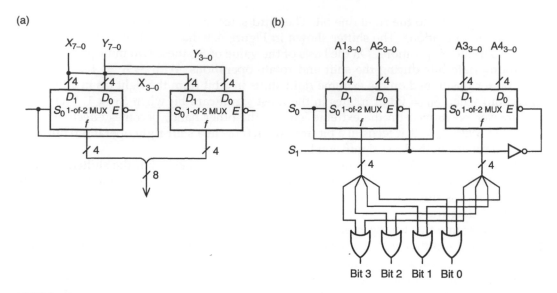

FIGURE 4.37
Data routing from (a) two 8-bit sources and (b) four 4-bit sources.

(a)   (b)

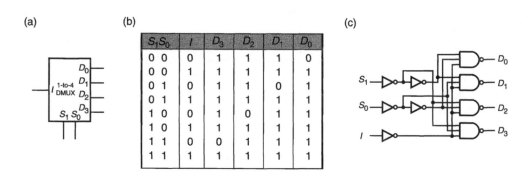

(c)

$S_1 S_0$	I	D_3	D_2	D_1	D_0
0 0	0	1	1	1	0
0 0	1	1	1	1	1
0 1	0	1	1	0	1
0 1	1	1	1	1	1
1 0	0	1	0	1	1
1 0	1	1	1	1	1
1 1	0	0	1	1	1
1 1	1	1	1	1	1

FIGURE 4.38
1-to-4 line demultiplexer (a) block diagram, (b) truth table, and (c) logic circuit.

two stacks of 1-of-2 multiplexers are used to route 4-bits of data from one of four sources. With $S_1 S_0 = 00, 01, 10$, and 11, the 4-bits of $A1$, $A2$, $A3$, and $A4$ appear respectively at the output.

4.7   Logic Function Implementation Using Demultiplexers and Decoders

A demultiplexer (DMUX for short) performs a function inverse of a multiplexer. It routes data on a single input to one of several outputs as determined by its selectors. A demultiplexer directs the signal to one of the 2^n outputs by means of n selector lines. The block diagram, truth table, and logic circuit of a 1-to-4 line demultiplexer circuit are shown in Figure 4.38. In this logic circuit, S_1 and S_0 serve as the demultiplexer selectors, I is the data input, and D_0 through D_3 are the outputs. Depending on the value of S_1 and S_0, the value of I may show up at any one of the outputs. For example, if $S_1 S_0 = 01$, D_1 takes the value of I, and so on.

(a) 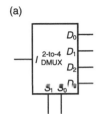 (b)

S_1 S_0	E	D_3	D_2	D_1	D_0
0 0	0	1	1	1	0
0 1	0	1	1	0	1
1 0	0	1	0	1	1
1 1	0	0	1	1	1
- -	1	1	1	1	1

(c)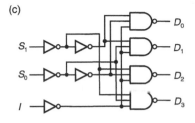

FIGURE 4.39
A 2-to-4 line decoder (a) block diagram, (b) truth table, and (c) logic circuit.

A decoder, on the other hand, produces a unique output corresponding to each input pattern. One can use a decoder to activate or enable any one of its n outputs. In fact, the demultiplexer just introduced discussed can be reorganized to obtain a decoder. For example, the 1-to-4 line demultiplexer of Figure 4.38 when rearranged can be made to function as a 2-to-4 line decoder as shown in Figure 4.39. S_1 and S_0 are redefined to serve as its data inputs and I is now redefined to be its enable input E. When $S_1 S_0 = 10$, for example, only the output line D_2 will be low, provided that $E = 0$. Thus, when the input E is 0, the decoder is enabled, and when E is 1 the decoder is disabled. When E is 0, all outputs are 1 except for the line corresponding to the decimal value of the input bits. Since each combination of inputs enables a unique output, a decoder can be made to function also as a minterm generator. Consequently, a decoder is one of the simplest and most useful multiple-output logic structures that may allow us to activate any one of many different modules/circuits, for example, by means of its different outputs. In that case, the operations of these peripheral circuits could be controlled by means of the decoder inputs. Note that in a decoder circuit, E may be altogether absent. With E present, however, decoder and demultiplexer circuits are nearly identical.

In general, an m-to-n line decoder has $m = \log_2 n$-input lines. There are times when several decoder/demultiplexer circuits may be cascaded together to form a larger decoder/demultiplexer. For example, m levels of 1-to-2 line decoders are needed to construct an m-to-n line decoder, each of which will decode only one address bit. In other words, the most significant address bit is decoded by one decoder, the next most significant bit by two decoders, and so on, until the least significant bit is decoded by $n/2$ decoders. The output of decoders at any one level enables the decoders at the next level. A 2-to-4 line decoder, on the other hand, can decode two address bits at a time. Thus, if we were to construct an m-to-n line decoder using 2-to-4 line decoders we shall need only half as many decoders as that constructed using 1-to-2 line decoders. Figure 4.40a shows how two 2-to-4 line decoders may be combined by means of their enable inputs. When $A_2 = 1$, only the top 2-to-4 line decoder is enabled and the lower one is disabled. Then when $A_2 = 0$, the top decoder is disabled and the bottom decoder is enabled. This decoder combination is equivalent to a 3-to-8 line decoder since any one of the eight outputs can be turned low by feeding appropriate values to the three address lines S_2, S_1 and S_0. Figure 4.40b shows an alternate logic circuit that uses a 1-to-2 line decoder in lieu of the NOT gate used in Figure 4.40a. Figure 4.40c shows a yet different variation but that which requires three levels of 1-to-2 line decoders. Another example of cascaded decoders is shown in Figure 4.41 where a 6-to-64 line decoder is designed using four 4-to-16 line decoders and a 2-to-4 line decoder. The 2-to-4 line decoder is used to enable only one of the four 4-to-16 line decoders.

The decoders that have been discussed thus far are often constructed using transistors in AND formation. The number of such transistors used in each of the gates is approximately equal to the number of gates and gate inputs. Typically, the number of gates present is of the

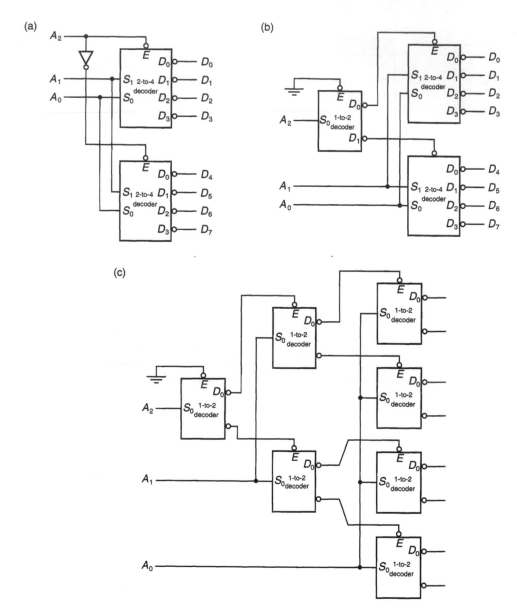

FIGURE 4.40
A 3-to-8 line decoder designed using (a) two 2-to-4 line decoders, (b) a 1-to-2 line decoder and two 2-to-4 line decoders, and (c) three levels of 1-to-2 line decoders.

order of the number of decoder outputs. Accordingly, the number of transistors required in a decoder circuit increase as 2^n with increasing inputs. Thus, one often prefers a decoder scheme where the sum of gates and gate inputs is as small as possible. In theory, decoder design employed in the circuit of Figure 4.39 can be extended to design any arbitrary n-to-2^n line decoder. However, in practice, for all possible technology there is a fan-in limitation that will be eventually reached. Consider that for the technology in question, there can be at most m inputs for a decoder. If we were to now realize equivalent of a decoder circuit that has more than n (i.e., greater than m) inputs, it can be designed by combining available decoders and follow-on logic gates. First, the inequality $p \leq (n/q) \leq m$ is solved for q (where q is a power of 2) for the largest possible value of p. We may then

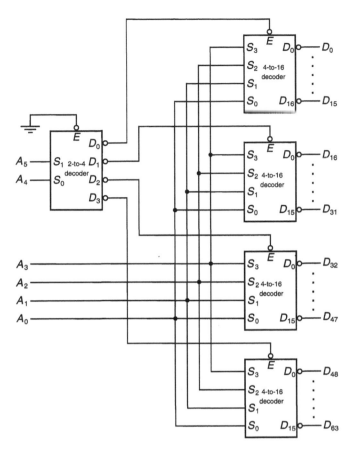

FIGURE 4.41
A 6-to-64 line decoder designed using four 4-to-16 line decoders and a 2-to-4 line decoder.

use q number of p-to-2^p line decoders to first form complement of minterms of disjoint subsets of p input variables. These minterm complements may then be combined using as many levels of OR gates as required to generate complement of minterms of all n input variables.

Example 4.14
Design a 12-to-2^{12} line decoder using a technology where the largest possible decoder that is available is a 5-to-2^5 line decoder.

Solution
For $m = 5$ and $n = 12$, there are two possible choices: (a) $p = 3$, and $q = 4$; and (b) $p = 1$ and $q = 8$. While the first choice would require us to use four 3-to-2^3 line decoders, the second choice would lead us to use eight 1-to-2 line decoders. Considering that it may not be productive to use 1-to-2 line decoders, we employ 3-to-2^3 line decoders as shown in Figure 4.42.

The 3-to-2^3 line decoders will generate complement of minterms formed of up to three variables. The next level of 128 ($= 2^3 \times 2^3 + 2^3 \times 2^3$) 2-input OR gates generate complement

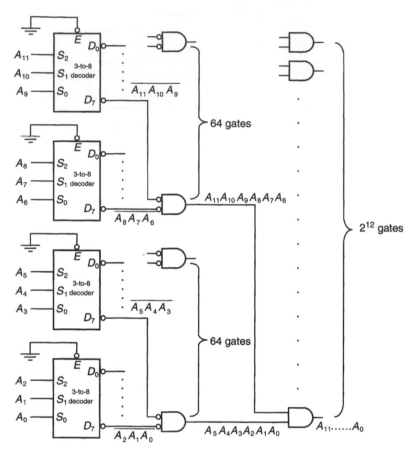

FIGURE 4.42
Logic circuit for Example 4.14.

of minterms of disjoint subsets of six input variables. Then the next level of 2^{12} OR gates generates complements of minterms of all 12 input variables. In this scheme, clearly the overall cost is determined more by the cost of the 2^{12} OR gates at the output. Provided the fan-in level would permit, one could have used instead 4-input OR gates and avoid having to use the intermediate-level OR gates.

The fan-in of logic gates often is a constraint when designing the class of decoders such as that shown in Figure 4.39. But as shown in Example 4.14, we see that fan-out limits can also pose serious problem in decoder having a large number of inputs. In Example 4.14, the intermediate-level OR gates will need to supply 64 of the output-level OR gates. Such fan-out problems, however, are solvable by considering a tree-type decoder configuration. Figure 4.43 shows a tree-type 4-to-16 line decoder. This particular decoder requires only 64 transistors for its fabrication while a regular 4-to-16 line decoder would have required 72 transistors. The tree-type decoder still has fan-out problem in that the variables introduced at a later level will still have to drive many logic gates. Besides, these decoders are relatively slow in speed.

A decoder network may use an alternative scheme called the balanced decoding scheme, as shown in Figure 4.40a. It requires only 56 transistors for its fabrication. The significance of such improvement becomes more obvious when we begin to consider larger decoders.

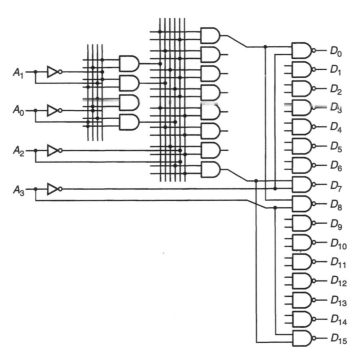

FIGURE 4.43
A 4-to-16 line tree-type decoder.

Switching speed of a balanced decoder is much better than that of a tree-type decoder since it involves fewer levels of logic gates. In tree-type 4-to-16 line decoders, $S_3 S_2 S_1 S_0$ are decoded in steps. First, S_1 and S_0 are decoded to identify four distinct outputs. Then with S_2 included, it is further decoded into eight distinct outputs. Finally, they are decoded into sixteen unique outputs by having included S_3. In the balanced scheme, in comparison, S_3 and S_2 are decoded into 4 distinct outputs while S_1 and S_0 are decoded into a different set of four distinct outputs. Finally, these two sets of four distinct outputs are decoded to yield a total of sixteen distinct outputs. The alternate logic circuit shown in Figure 4.44b requires still fewer gates but requires two 2-to-4 line decoders.

Very much like a multiplexer, a decoder can be used also to implement logical functions. Each of the decoder outputs corresponds to the complement of a particular product term formed of its input variables. Thus to realize a Boolean function, the decoder outputs corresponding to the set of minterms/maxterms may be accomplished in one of following ways. In the case of a decoder with active high inputs, either the outputs corresponding to the minterms in the given function are ORed or those corresponding to the maxterms are NORed. However, in the case of a decoder with active low outputs, either the outputs corresponding to the minterms forming the given function are NANDed or those corresponding to the maxterms are ANDed. The fact that decoders can implement logical equations in ways similar to a multiplexer is not unusual since we can easily construct a multiplexer using a decoder and a few assorted logic gates. To realize the equivalent of a multiplexer, the decoder outputs are first ANDed with the respective data input bits and the resulting AND outputs are all ORed next. In this scheme, the decoder inputs together serve equivalently as the selector inputs of the resulting multiplexer.

(a)

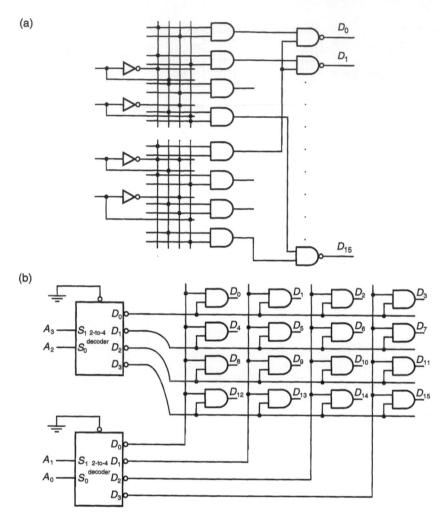

(b)

FIGURE 4.44
A 4-to-16 line decoder (a) balanced and (b) using 2-to-4 decoders.

Example 4.15
Implement the following set of logical equations using decoders:

$$F_1(A, B, C, D) = \Sigma m(2, 4, 10, 11, 12, 13)$$

$$F_2(A, B, C, D) = \Sigma m(4, 5, 10, 11, 13)$$

$$F_3(A, B, C, D) = \Sigma m(1, 2, 3, 10, 11, 12)$$

Solution
Since the three functions are all functions of the same four input variables, only one 4-to-16 line decoder is needed. The resulting logic circuit that will require in addition three NAND gates is obtained as shown in Figure 4.45. The input variables $A, B, C,$ and D are respectively connected to the inputs $S_3, S_2, S_1,$ and S_0.

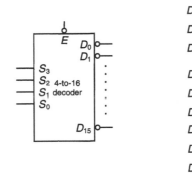

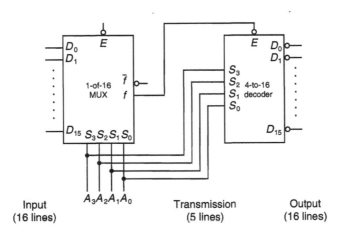

FIGURE 4.45
Logic circuit for Example 4.15.

FIGURE 4.46
Data routing using a multiplexer/demultiplexer combination.

Often times, a combination of multiplexers and demultiplexers is used for data routing. Figure 4.46 shows such a circuit formed of a 1-of-16 multiplexer and a 1-to-16 line demultiplexer that could, for example, be used to replace a cable of 16 lines. Instead of running a cable of sixteen lines, one could route only five lines (one data line and four selectors). While on the transmitting-end, the selectors are used to select one of sixteen signals and at the receiving-end, the selectors are used to demultiplex the signal back to sixteen parallel lines for further processing. It must be understood though that this system saving has come at an expense whereby at a time one and only one of the 16 signal channels may be in use.

A *bus* is used often to implement data routing. It typically uses decoders and tristate drivers such as that shown in Figure 4.47a. These drivers can have three output values—binary 0 and 1 and a high-impedance state. The high-impedance state can be interpreted as an open circuit. When the enable input E is a 1, the tristate output takes on the same value as its input. But when $E = 0$, the output becomes disconnected from the bus. Typically, a n-to-2^n line decoder and 2^n tristate devices are needed to implement a 2^n-input bus. Each driver is used for a source that is connected to the bus. Since only one source can be allowed to drive the bus at a given time, the enable lines of different drivers can be encoded into a source address via the use of a decoder. Figure 4.47b shows, for example, a 4-input bus.

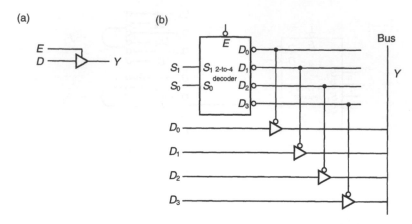

FIGURE 4.47
Bus implementation (a) a tristate device and (b) a 4-input bus.

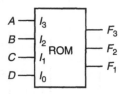

FIGURE 4.48
A black-box representation of the circuit shown in Figure 4.45.

In general, it is easy to design and modify a bus to either add new sources or remove old sources.

4.8 Logic Function Implementation Using ROMs

The decoder circuit of Figure 4.45 can be perceived in a different way to introduce a device that can be used also to implement logical functions. By putting the circuit of Figure 4.45 inside a box, we can visualize the decoder circuit as that shown in Figure 4.48. Such a representation is often called a *black box* representation where a user may not be required to know about the details of its logic components. In such a scenario, we might describe the black box of Figure 4.48, for example, by designating the inputs A, B, C, and D as its *address*, and F_1, F_2, and F_3 as what are stored in the black box. This is exactly similar to the operation of a ROM with which we can implement concurrently multiple logic functions each of which can be a function of the same set of input variables. A ROM can be considered as an array of storage cells with every cell having a value available to each of the multiple functions. Typically, a ROM is programmed by the manufacturer or by the designer if he or she has an available blank ROM chip and an appropriate programmer. The information *stored* in the ROM resides there permanently.

A ROM, as its name implies, holds fixed information that can only be read, not altered. The typical use of the ROM is to store binary information that may not be altered thereafter without undergoing a significantly involved process. A ROM circuit, thus, is identical to a logic network that has been designed, fabricated, tested, and encapsulated with only the inputs and the outputs available. ROMs have become an important part of many digital systems because of the ease with which complex functions can be readily implemented. The chip count of a logic circuit can be drastically reduced when one uses ROMs instead of assorted logic gates to implement multioutput logical functions.

The size of a ROM is determined by the number of storage cells (corresponding to the number of decoder minterms) and the number of bits stored in each cell (corresponding to the number of functions that can be realized). In general, n-variables (each corresponding to one address line) require 2^n storage cells. Each storage cell would have one bit (either 1 or 0) for each function of up to n-variables. Commercially available ROMs come in various sizes. In general, a ROM may store 2^n words of m bits each and, therefore, is referred to as a $2^n \times m$ ROM. It is associated with n address lines that are used to address each of its 2^n words. A 2K $\times$ 1 ROM refers to having 2048 storage locations one bit wide while a 2K $\times$ 8 ROM implies that it accommodates 2048 storage locations eight bit wide. The first ROM can be used to implement a function of up to eleven variables since $2^{11} = 2048$ while the second ROM can implement up to eight such functions. Depending on application, the circuit designer selects a ROM of adequate size to implement the desired function or functions. For most function generation applications, a ROM can be visualized as a set of decoders—each for one logical function of the input variables (i.e., addresses).

A $2^n \times m$ ROM is an array of memory cells organized into 2^n words of m bits each, as shown by the block diagram of Figure 4.49. This ROM corresponds to a logic network with m outputs, where each of the outputs may be associated with up to 2^n different minterms. A ROM may be equipped also with one or more chip-select lines to permit one to cascade smaller ROMs to form a ROM with more words (allowing implementation of functions of more variables).

A ROM can be implemented using only diodes, bipolar transistors, or MOS transistors. Although a diode matrix ROM no longer represents the current ROM technology, it serves as a simple model to show the basic concept. Figure 4.50 shows a simple diode ROM, where the row and column lines are interconnected (indicated by the presence of "·" at the intersection) via diodes placed at its intersections. Depending on the input values, A_2, A_1, and A_0, one of the eight decoder outputs is selected. The output D_x is asserted (by becoming low) whenever $(A_2A_1A_0)_2$ takes the value (x_{10}). Each decoder output is called a word line because it selects one row or word of the truth table stored in the diode ROM. The absence or presence of a diode at an intersection indicates that the corresponding row and column intersection is programmed with a 1 or 0. The output for the i-th address depends on the OR diodes connected to that line. For example, if $A_2A_1A_0 = 000$, the output will be $O_8O_7 \ldots O_1 = 10001001$. With $A_2A_1A_0 = 000$, a low is produced on the decoder output D_0.

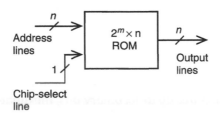

FIGURE 4.49
Block diagram of a ROM.

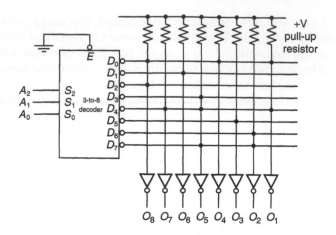

FIGURE 4.50
An 8 × 8 diode ROM.

This low and the pull-up resistors forward bias each of the diodes connected to this row and pull down the corresponding column outputs to a low. In the absence of a diode, the output is a 1. The ROM's size is described best in terms of the number of possible intersections in the matrix. The ROM of Figure 4.50, for example, is described as having $8 \times 8 = 64$ bits $= (1/16)$K bits.

Example 4.16
Define a ROM for realizing the square function $f(x) = x^2$ for $0 \le x \le 7$.

Solution
The square function truth table is obtained as follows:

x	$f(x)$ in Decimal	$f(x)$ in Binary
0	0	000000
1	1	000001
2	4	000100
3	9	001001
4	16	010000
5	25	011001
6	36	100100
7	49	110001

A 3-to-8 line decoder is sufficient to decode the numbers 0 through 7. Since the largest square number requires up to six binary bits, the diode matrix implementation of the corresponding x^2 ROM circuit will need to include up to six output lines as shown in Figure 4.51. The three select lines, A_0, A_1, and A_3, can be used to select any one of eight outputs.

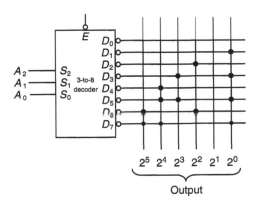

FIGURE 4.51
Logic circuit for Example 4.16.

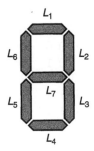

FIGURE 4.52
Seven-segment display device for Example 4.17.

Example 4.17

Define a ROM implementation to function as a seven-segment-to-BCD code converter. The seven-segment display device, shown in Figure 4.52, consists of seven specially arranged light-emitting diodes (LEDs) so that they may be used for displaying BCD codes.

Solution

The truth table for seven-segment-to-BCD code conversion is obtained as shown below:

$L_1 L_2 L_3 L_4 L_5 L_6 L_7$	$B_8 B_4 B_2 B_1$
1 1 1 1 1 1 0	0 0 0 0
0 1 1 0 0 0 0	0 0 0 1
1 1 0 1 1 0 1	0 0 1 0
1 1 1 1 0 0 1	0 0 1 1
0 1 1 0 0 1 1	0 1 0 0
1 0 1 1 0 1 1	0 1 0 1
0 0 1 1 1 1 1	0 1 1 0
1 1 1 0 0 0 0	0 1 1 1
1 1 1 1 1 1 1	1 0 0 0
1 1 1 0 0 1 1	1 0 0 1

Each one of these input combinations refers to how the seven LEDs are turned on so that they altogether represent a particular decimal digit. For example, when all LEDs but L_7 are turned on, the 7-segment display corresponds to a 0, the BCD code equivalent $(B_8B_4B_2B_1)$ of which is given by 0000. For displaying 1, however, there are two choices: L_2 and L_3 or L_5 and L_6. If one were to employ a ROM to program this truth table, it needs to have a size of $2^7 \times 4 = 512$ bits since there are seven inputs and four outputs in the truth table.

Alternatively, we may consider partitioning the above truth table by having, for example, only L_3 through L_7 as its inputs. If we disregard L_1 and L_2, however, we see that there are only seven unique combinations of L_3 through L_7. This suggests that if we were to construct a truth table with only L_3 through L_7 values as its inputs, the entries of this partitioned table (as shown below) may be decoded by only a 3-bit address (we use 001 through 111, for example) since $2^3 > 7$.

$L_3L_4L_5L_6L_7$	$X_2X_1X_0$
1 1 1 1 0	0 0 1
1 0 0 0 0	0 1 0
0 1 1 0 1	0 1 1
1 1 0 0 1	1 0 0
1 0 0 1 1	1 0 1
1 1 0 1 1	1 1 0
1 1 1 1 1	1 1 1

The advantage gained by having only three bits to uniquely represent the 5-bit inputs suggests that a reasonable improvement is possible if we cascade two separate ROMs. The first ROM will have five inputs $(L_3L_4L_5L_6L_7)$ but only three outputs $(X_2X_1X_0)$. The three outputs along with the two other inputs (L_1 and L_2) can then be fed to a second ROM to generate $(B_8B_4B_2B_1)$. The truth table for this second ROM will have to be carefully constructed so that its output becomes equivalent to that of the original truth table.

$L_1L_2X_2X_1X_0$	$B_8B_4B_2B_1$
1 1 0 0 1	0 0 0 0
0 1 0 1 0	0 0 0 1
1 1 0 1 1	0 0 1 0
1 1 1 0 0	0 0 1 1
0 1 1 0 1	0 1 0 0
1 0 1 1 0	0 1 0 1
0 0 1 1 1	0 1 1 0
1 1 0 1 0	0 1 1 1
1 1 1 1 1	1 0 0 0
1 1 1 0 1	1 0 0 1

The size of the first ROM is 96 bits $(= 2^5 \times 3)$. The second ROM size is 128 bits $(= 2^5 \times 4)$. Therefore, the total ROM size requirement is reduced from 512 bits to only 224 bits $(= 96 + 128)$. The resulting ROM implementation of the logic circuit is obtained as shown by the multilevel circuit of Figure 4.53.

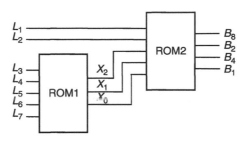

FIGURE 4.53
A ROM implementation for the logic circuit of Example 4.17.

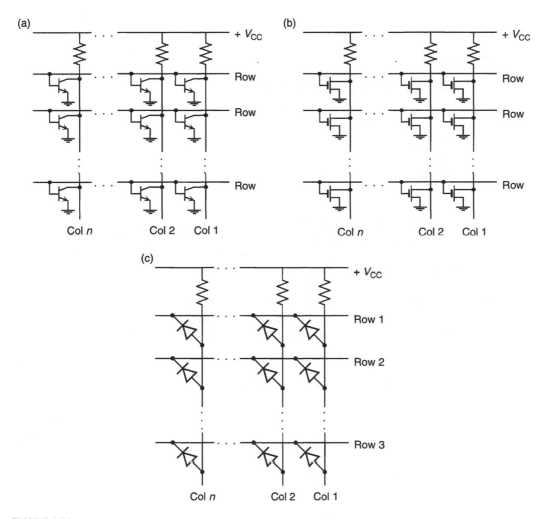

FIGURE 4.54
A $m \times n$ ROM using (a) bipolar devices, (b) MOS, and (c) diode.

Although the diode matrix concept serves to illustrate ROM principles, they are presently manufactured using bipolar and MOS transistors, as shown in Figure 4.54. The presence of a connection from a row line to either a transistor base or a MOSFET gate represents a logic 0, and the absence of such a connection represents a logic 1. Typically, MOS ROM is preferred over bipolar ROM since it makes more economic sense when there are a larger numbers of bits to be processed. However, access time pertaining to a bipolar ROM is much less than that for a MOS ROM.

There are two basic types of ROM: mask programmable and field programmable. For the mask programming type, the manufacturer considers customer-provided specifications and accordingly creates a mask used in the metallization phase of fabrication to connect transistors where required. Since it involves custom-made processings, mask programming is uneconomical if only a small number of the same ROM is to be built. In case of the field programmable ROMs, however, the customer programs the chip. The field programmable ROMs serve as the basis for a class of PLDs. We shall discuss them more in detail in the next section. The designer typically resorts to a ROM when the number of functions is large and the individual functions are complex.

4.9   Logic Function Implementation Using PLDs

Examples 4.16 and 4.17 illustrated how a ROM may be used to realize SOP logic expressions. In spite of their simplicity, however, a ROM is not preferable especially when we are not interested in large quantity of the same ROM. In such cases, the designers prefer an alternate module referred to as the PLDs. Like ROMs, PLDs are also made of an AND array followed by an OR array. The difference is that the user has an opportunity to program the PLDs in one's own laboratory to realize logic functions that might ordinarily take dozens of SSI circuit packages to implement.

PLDs are of three types: programmable read-only memory (PROM), programmable logic array (PLA), and programmable array logic (PAL—a registered trademark of Monolithic Memories, now part of Advanced Micro Devices, Inc.). In the PROM devices only the OR array is programmable, while in the PLA devices only the AND array is programmable and in PAL devices both of the arrays (AND and OR) are programmable. From a designer's perspective, a PROM is the least versatile of the PLDs. While PAL is only moderately versatile, the PLAs are the most versatile of the PLDs. As a general rule, PLAs are the most expensive, a little harder to program, and a little slower in speed. The PLA architecture allows a designer to implement complex logical functions. In fact, several commercially successful VLSI microprocessors use PLAs as their basic building blocks. In summary, the idea central to programming a PLD is that the user could program either the AND array, or the OR array, or both.

Consider the diode circuit shown in Figure 4.55a where the diode is connected to a power supply through a resistor. The voltage value at terminal A either forward biases or reverse biases the diode. If the diode is forward biased with a low voltage (i.e., binary 0) at A, the diode behaves as a closed switch, in which case B takes on the same voltage value as A (i.e., binary 0). If on the other hand, the diode is reverse biased by having high voltage

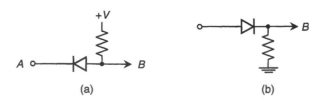

(a)   (b)

FIGURE 4.55
Diode connection with (a) power supply and (b) ground.

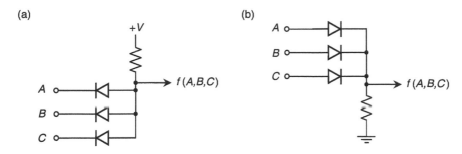

FIGURE 4.56
Diode configurations for (a) AND and (b) OR.

(i.e., binary 1) at A, the diode behaves as an open switch. Consequently, signal line B is pulled up by the resistor toward the power supply voltage V that forces B to take on the binary value of 1. Next consider the diode circuit shown in Figure 4.55b where the diode is grounded through a resistor. With A set to 1 (or 0), the diode is forward (or reverse) biased and therefore it can be treated as a closed (or open) switch and consequently B takes on the binary value of 1 (or 0). The diode circuit shown in Figure 4.55a can be used to construct AND gates and likewise that shown in Figure 4.55b can be used to construct OR gates. Figures 4.56a and b respectively show, for example, 3-input AND and OR gates. In case of the circuit shown in Figure 4.56a if any one of the inputs is a 0, the corresponding diode behaves as a closed switch thus forcing the output to become a 0. Whether or not the other inputs are 1 (with corresponding diodes being open) the output is invariant. On the other hand, if all inputs are 1, the diodes are all open and, therefore, the output becomes a 1. In the case of Figure 4.56b, if any one of the inputs is a 1, the corresponding diode is closed and consequently the output is a 1. The other diodes, if they were to remain open, do not affect the output. However, when all the inputs are 0, the diodes are all open and the resulting output is a 0. These concepts can be all summed up to construct AND and OR arrays. In most PLDs, the inputs are fed through buffers that generate both the true and complement of the input signals. Each gate in the AND array generates a minterm formed of the input variables. The device outputs are produced then by taking an OR of the appropriate minterms using an OR array. To make a device programmable, a metal fuse is placed in series with each diode between the diode and the output line. An intact fuse behaves as expected as a short circuit. A few of these examples are listed in Figure 4.57. Although we have illustrated these examples with diodes, PLDs are typically manufactured now using bipolar technology.

Why should one use the PLD when SSI and MSI devices are readily available? There are several reasons for this preference. First, use of PLDs reduces circuit real estate by reducing the total package count. Second, the reprogramming of PLDs allows for design changes to be made without redesigning the whole circuit. Finally, the use of PLDs increases the overall circuit reliability since it leads to fewer interconnections.

4.9.1 PROMs

A PROM is very much like a ROM consisting of a decoder section (corresponding to its AND array) followed by an OR array (made of diodes or transistors) that constitutes the encoder section. The decoder section comes as fixed but the OR array is readily programmable. The PROM units come with all the fuses intact, each contributing a 1 to the bits of the stored words. The user programs a PROM by blowing the fuses by having passed proper amount of current (via a PROM programmer) through the output terminals for each

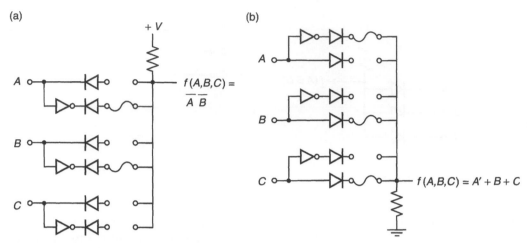

FIGURE 4.57
Examples of field programmable AND and OR arrays.

of the minterms. With its array of AND and OR gates, every combination of minterms of the input variables (addresses) can be formed. Traditionally, a PROM is used to store software in microprocessor-based systems in addition to implementing logic functions with it. The input variables are used as inputs to a PROM with the required output values being stored in the location corresponding to the input combination. The design of any particular circuit using a PROM device follows the same procedure as that used in the corresponding ROM device. Accordingly, concepts of circuit design using PROMs are not repeated in this section. The major advantage of using PROMs is that we do not need to conduct any logic minimization since this does not result in any savings. All possible AND terms (minterms) are already available (via the fixed AND array) by default. In fact the given expression is already reduced, and it will be necessary that we expand it back to its canonical form in order for us to be able to implement it using a PROM.

It should be noted that the hardware procedure for programming a PROM is irreversible and, once programmed, the truth table is permanent. They must be discarded and only new devices programmed to replace them. However, there is a particular type of PROM called erasable PROM (EPROM) which can be used to overcome this irreversibility problem. The OR array of an EPROM is programmed using a special programming voltage to trap electric charge in selected storage cell. When EPROMs are exposed to a special ultraviolet light through a quartz window for a specified time, the radiation discharges the internal gates that serve as fuses, restoring the OR array to its initial unprogrammed condition. These particular EPROMs are also referred to as ultraviolet erasable PROMs (UVEPROMs). There is another variety of EPROM, referred to as electrically erasable PROM (EEPROM), that uses electrical signal instead of ultraviolet light for erasure of stored data. When compared to PROMs, EPROM and EEPROM devices are more expensive per bit. In addition, most EPROM and EEPROMs are associated with longer propagation delays.

A nontrivial function may not include all possible minterms. For example, a PROM that processes twelve input variables requires a total of 4K byte (eight bits are called a *byte*) of memory. Such an arrangement was once needed for the Holerith code conversion circuit that has up to 12 input variables, but only 96 8-bit output combinations. This situation implied that 4000 of the 4096 bytes remain unused. In such situations implementing sparse functions (which have only a small number of 1s), PROMs do not provide very efficient solution. Such wastage of silicon area is usually minimized by the use of a PLA.

The complexity of a PROM device is determined by the number of equivalent diodes and fuses it contains. An n-input m-output PROM has $2n$ diodes (to generate both true and complement of the inputs) connecting the inputs to each AND term. Since there are 2^n possible AND functions of n-variables, the AND array is fabricated using $2n \times 2^n$ diodes. The OR array, on the other hand, includes $m \times 2^n$ diodes and fuses since each of the 2^n AND gates can be connected to each of the outputs. The total PROM cost is thus proportional to the cost of $(2n + m) \times 2^n$ diodes and $m \times 2^n$ fuses.

4.9.2 PLAs

A PLA consists of an array of AND-OR logic along with *inverters* that may be programmed to realize a desired output. In essence a PLA is highly flexible since it may be treated as having two separate PROMs: an AND PROM and an OR PROM, both programmable. Since the AND array (instead of a full decoder as in PROMs) is programmable, PLAs do not suffer from the limitation of PROMs that the AND array provides all possible product terms of input variables by default. Since both arrays are user programmable, OR gate can be made to access an arbitrary number of product terms. Like PROM, however, a PLA allows for product term sharing since it does not require any additional logic circuit. A typical $p \times q \times r$ PLA configuration is shown in Figure 4.58 where p, q, and r respectively represent the number of input variables, AND terms, and output functions. The p inputs are each internally buffered and also inverted to generate both true and complement values. The resulting $2p$ lines are available next as inputs to the AND gates. The AND terms are formed by appropriately taking an AND of one or more of p input variables. Each of the r OR gates can be programmed so as to provide sum of any or all of q AND terms. Thus, a PLA can implement r arbitrary logical functions of p variables each as long as the number of different AND terms in those r functions together does not exceed q.

When programmed, the programmed cell of a PLA matrix is indicated by inserting connections between the appropriate input lines and the AND lines and that between appropriate AND lines and OR lines. In addition, each of the PLA outputs is associated

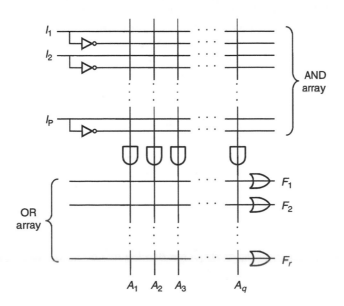

FIGURE 4.58
A $p \times q \times r$ PLA.

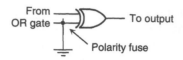

FIGURE 4.59
A programmable polarity fuse.

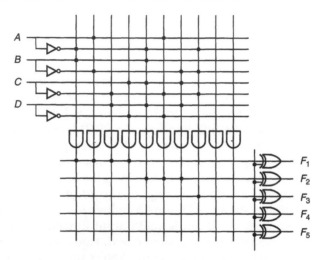

FIGURE 4.60
A PLA implementation.

with a programmable polarity fuse so that the PLA output can be individually programmed either true or complement. This is accomplished by feeding OR output to a 2-input XOR gate whose other input is connected to the ground through a fuse as shown in Figure 4.59. A blown polarity fuse corresponds to forcing the XOR input to become a 1 and consequently the XOR output takes on a value complement of the corresponding OR function. When the fuse remains intact, both the XOR output and the OR output become equivalent, in which case the XOR logic gate serves as a buffer. The programmable inversion of the outputs provides the added advantage. When the complement of a given function is simpler to realize than the function itself, then the complemented function is first programmed into the AND array and an inversion may be sought at the output. For example, if a given function had only a few 0s, the AND and OR arrays are programmed first to implement the complement of that function.

A PLA may be used also as a Boolean function generator in much the same way as a PROM. PLAs as well as PALs can be used to implement functions given in equation form as opposed to in tabular form. Figure 4.60 shows a PLA layout, for example, with five inputs, ten AND terms, and five outputs. The dots in the matrix of the top section can be thought of as AND function and those on the bottom section are interpreted as OR functions used to generate the outputs. The output functions shown in Figure 4.60 are as follows:

$$F_1 = \bar{A}B + A\bar{B} + \bar{C}D + C\bar{D}$$
$$F_2 = \bar{A}BCD + A\bar{C}\bar{D} + \bar{B}CD$$
$$F_3 = \bar{A}\bar{B}\bar{C}$$
$$F_4 = F_5 = 0$$

Note that the function F_3 is associated with an XOR gate for which the fuse is blown.

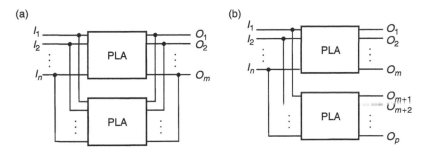

FIGURE 4.61
Scheme for PLA expansion of (a) product term and (b) output.

In using a PLA, the designer must ensure that the number of product terms in the given Boolean expression must not exceed the number of ANDs available in the PLA. Therefore, minimization of multiple-output Boolean functions (using either K-map scheme or Q-M tabular reduction technique) is very important in PLA-based logic circuits. The n-input m-output PLA that generates k AND terms consists of n buffer-inverter gates, k AND gates, m OR gates, and m XOR gates. The PLA in question will have $2n \times k$ fuses between the inputs and the AND array, $k \times m$ fuses between the AND and OR array, and m fuses associated with XOR gates. In practice, the designer should attempt to use a minimum number of AND terms for obtaining the functions. This is because a PLA has only a finite number of AND gates. The designer must always use care in choosing the minterms to be formed in the AND section of the PLA. In order to make efficient use of PLA, it is not necessary for each function to be minimized; the goal is to minimize the total number of minterms required to implement the set of all output functions.

It may be possible for one to incorporate a larger numbers of product terms and/or outputs by having more than one PLA cascaded. Some of these PLA expansion schemes are shown in Figure 4.61. By tying the outputs in parallel as in Figure 4.61a, the number of product terms can be increased. This configuration resembles the wiring of open-collector outputs. On the other hand, the word size can be increased by adopting a circuit configuration such as that shown in Figure 4.61b.

A PROM is only a special case of PLA. An n-input m-output PROM is equivalent to a PLA that has a fixed AND array (formed of 2^n n-input AND gates) that is capable of realizing 2^n possible input combinations of n-variables. The PLA is very much similar to a PROM except that the PLA does not generate all the minterms. In it, the decoder is replaced by an AND array that is programmed to generate only those minterms which are needed to generate the output in question. Typically, a PROM is preferable for those output functions which have a larger number of minterms. PLAs are used more often for the implementation of control logic, whereas ROMs and PROMs are used more frequently for constructing tables of coefficients, test vectors, startup programs, and other random data applications.

4.9.3 PALs

PALs serve as a lower-cost replacement for discrete logic devices, PROMs, and PLAs. It has a programmable AND array and a fixed OR array. In the fixed OR array, each output line is connected to a specific set of AND terms. It also allows the systems engineer to design his or her "own chip" using fusible links to configure AND gates to yield the desired logic functions. In comparison, both of the arrays of PLA are programmable, and PROM has a fixed AND matrix but a programmable OR array. Unlike PROM, in which all 2^n possible AND operations of n-variables are generated, a PAL generates only a limited number

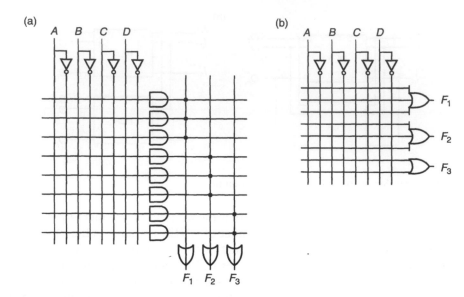

FIGURE 4.62
Representation type of a 4-input, 3-output PAL (a) nonstandard and (b) standard.

of AND terms. Therefore, the overall cost of PAL is considerably lower than that of a comparable PROM or PLA.

In PAL-based logic circuits, the AND array allows the designer to specify the product terms required and connect them to perform the required SOP logic functions. The PALs are available in a number of different part types each different from others in terms of how the OR gates have been organized. Specifying the OR gate connection, therefore, becomes a task of device selection rather than of programming. In general by using a PAL the user is able to improve the effectiveness of logic circuits by expediting and simplifying prototypes and board layouts. Figure 4.62 shows the PAL configurations of a 4-input, 3-output AND-OR circuit. Since in this case the OR array is fixed, the PAL representation shown in Figure 4.62b is preferred over that shown in Figure 4.62a. One would conclude that a PLA provides the most flexibility for implementing logic functions since the designer has complete control over all inputs and outputs. However, such flexibility makes a PLA rather expensive and somewhat formidable to comprehend. In comparison, a PAL combines much of the flexibility of a PLA with the low cost and easy programmability of the PROM.

When designing with PAL devices, the Boolean function will have to be modified to fit into the corresponding section of AND and OR arrays. Unlike in the PLAs, an AND term here cannot be shared by two or more OR gates. Accordingly, each output function will have to be individually minimized without investigating whether there are common AND terms between two or more output functions. If there are common AND terms between one or more functions, that AND term then will have to be generated separately for each of the output functions. The logic designer thus must ensure that the number of AND terms per output is enough for the worst-case number of AND functions in that application. However, if there is only a limited number of AND gates in a section, and if the number of terms in the function is too large, one may use one or more such sections to implement the function. In that case, it may be noted, the execution time increases linearly with the number of sections used.

To implement a set of logical functions using a PAL, their minimum SOP expressions are derived first. The primary PAL design objective is to minimize the number of AND terms in each of the SOP expressions, rather than the number of literals. Since each input variable and its complement are both available for the AND term, there is no cost advantage in reducing the number of literals from any of the AND terms. Again, since the AND terms cannot be shared (as in PROM and PLA), there is no need to employ any multiple-output minimization algorithm. Each of the SOP expressions should be minimized independently.

Consider, for example, a 6-input 4-output PAL device such that each of the outputs can be associated with only three AND gates. Consider further, the implementation of the following four Boolean functions, each of which is a function of the same four input variables A, B, C, and D.

$$f_1(A, B, C, D) = \sum m(3, 4, 5)$$

$$f_2(A, B, C, D) = \sum m(2, 12, 13)$$

$$f_3(A, B, C, D) = \sum m(3, 4, 5, 9, 14, 15)$$

$$f_4(A, B, C, D) = \sum m(7\text{--}15)$$

These four functions can be minimized using K-maps to yield

$$f_1(A, B, C, D) = \bar{A}B\bar{C} + \bar{A}BCD$$

$$f_2(A, B, C, D) = AB\bar{C} + \bar{A}BC\bar{D}$$

$$f_3(A, B, C, D) = \bar{A}B\bar{C} + ABC + \bar{A}BCD + A\bar{B}\bar{C}D$$

$$f_4(A, B, C, D) = A + BCD$$

If these four outputs were to be realized using the PAL just described, implementation of function f_3, in particular, is going to pose a serious problem. This function expression consists of four AND terms instead of three. In fact if we were to limit ourselves to using each of the four sections of the PAL only once, we just cannot generate f_3. However, by comparing f_3 and f_1, we find that

$$f_3(A, B, C, D) = f_1(A, B, C, D) + ABC + A\bar{B}\bar{C}D$$

Consequently, f_3 can be realized provided f_1 is treated as a single AND term. The resulting logic circuit, shown in Figure 4.63, uses f_1 as input for generating f_3. The function f_3 thus takes two times as much time as the other three outputs.

Example 4.18
Generate all of the functions of two variables (see Table 2.2 for reference) using (a) ROM, (b) PLA, and (c) PAL.

Solution
The 16 functions of two input variables can be realized using a $2^2 \times 16$ bit ROM as shown in Figure 4.64. Presence of each dot at the intersections of a 4×16 ROM matrix refers to a single minterm. The PLA and PAL implementation of these same sixteen functions is obtained as shown in Figures 4.65 and 4.66 respectively. It may be noted that presence of dots at all intersections of a row of the AND array corresponds to the absence of a minterm.

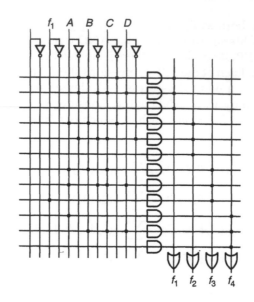

FIGURE 4.63
A PAL implementation of four functions.

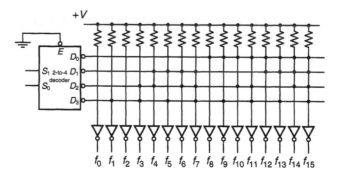

FIGURE 4.64
ROM-based implementation for Example 4.18.

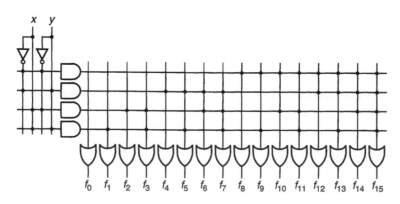

FIGURE 4.65
PLA-based implementation for Example 4.18.

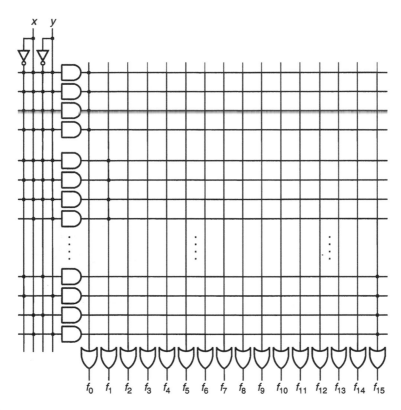

FIGURE 4.66
PAL-based implementation for Example 4.18.

TABLE 4.4

PLDs and Their Characteristics

PLD	Programmable	AND Sharing	Minimization Strategy
PROM	OR	Yes	Minimization does not affect the design
PLA	AND, OR	Yes	Minimization scheme must be executed to reduce the total number of minterms for the functions in question
PAL	AND	No	Each function should be independently minimized so that there are fewer AND terms rather than fewer number of literals

Table 4.4 highlights the differences between the three types of PLD devices. It classifies them in terms of which part of it (AND, or OR) is programmable, whether or not the AND terms (involved in the SOP form of the functions) can be shared between the functions, and the role of minimization schemes in implementing such functions.

4.9.4   PLD Highlights

Starting in late seventies, a new type of architecture on silicon started to gain commercial significance. Instead of placing standard logic devices on circuit boards, Ton Cline of Signetics (which was later purchased by Philips and then Xilinx) proposed a silicon

Inputs

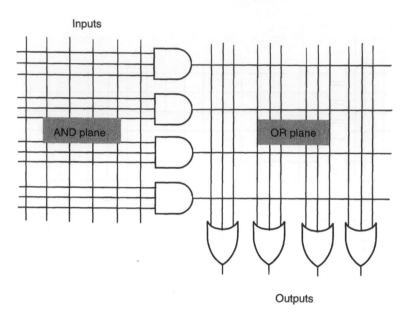

FIGURE 4.67
AND–OR planes architecture.

Inputs

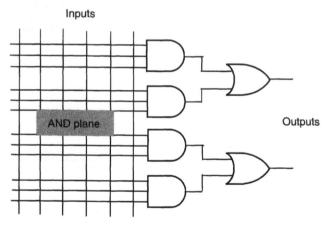

FIGURE 4.68
PAL topology with fixed OR plane.

architecture with two programmable planes. The two programmable planes provided any combination of "AND" and "OR" gates sharing of AND terms across multiple ORs, as illustrated in Figure 4.67. This architecture gave designers the ability to implement different interconnections in bigger devices—thus the birth of programmable logic.

Programmable Logic has involved in and branched out to many different silicon architectures, for example, early versions of PLA has both the AND and OR planes programmable, while PAL (Figure 4.68), has one of the two planes fixed to reduce the programming complexity and propagation delay but without the flexibility of the PLA. Other architectures followed, such as PLD, which is in the group often called simple PLD (SPLD).

Several programmable silicon technologies were commonly used in early programmable logic architectures. The oldest is referred to the use of *fuse* structures at programmable interconnections. Unwanted interconnects are disabled by applying a high-than-normal voltage across the fuse structures and breaking these undesired interconnections by blowing out the fuses. The *mask programming* technology controls the interconnections at the last step of the chip fabrication process by a semiconductor manufacturer. Desired interconnections are made at the metal layers on the basis of required functions. The process is costly because of the use of custom masking for a particular device. Therefore, mask programming is only economical for producing a large quantity of the same device. A third technology is referred to as *antifuse*. As the name suggests, it is the opposite of the fuse technology, with high-resistive structures initially placed to separate all interconnections. On the basis of required functions, desired connections are made at interconnects by applying a higher-than-normal power voltage across the high-resistive structures to become low resistive or conductive.

To provide more functionality for large digital systems requiring finite state machine (FSM) circuits, the architecture of a complex programmable logic device (CPLD) has three structure levels of interconnect, gates, and flip-flops, as illustrated in Figure 4.69. Desired inputs are propagated, by programming the necessary interconnections, to the next structure level to form AND terms. AND–OR formations are structured at the gates level, while flip-flops at the third structure level provide the synchronization control for FSM implementations.

To implement a logic function $F = a \cdot b \cdot c + \bar{b} \cdot \bar{c} \cdot d$, a PLD may be programmed as illustrated in Figure 4.70. Almost all PLD vendors provide users with device-programming software, which makes PLD implementations much faster and easier. With the software,

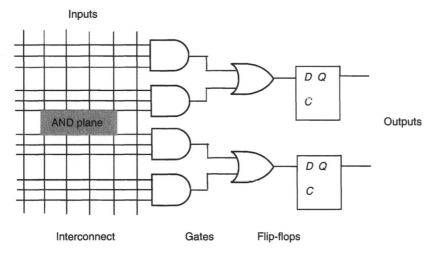

FIGURE 4.69
CPLD topology.

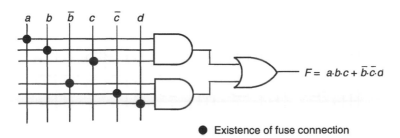

FIGURE 4.70
PAL configuration implementing $F = a \cdot b \cdot c + \bar{b} \cdot \bar{c} \cdot d$.

all that is required from users is to fill up a programming table that is used to configure the interconnects.

Another early programmable logic technology is often referred to as erasable-programmable read-only-memory (EPROM), with which EPROM devices can be reprogrammed for different function many times. This technology is very useful for initial product development and prototyping, where specification and implementation changes are made until designs are finalized. Table 4.5 summarizes the major characteristics of the early programmable logic technologies.

PLDs can also be used to implement ROM units. In this type of applications, PLD inputs are ROM read addresses and PLD outputs are ROM data outputs. Figure 4.71 illustrates the programmed PLD-configuration segment.

Modern PLDs have been dominated by a technology referred to as field programmable gate array (FPGA), with greater integration of individual programmable devices and interconnects on a single chip package. An early version of FPGA architecture is shown in Figure 4.71.

There are two basic types of FPGA: one-time programmable (OTP) and SRAM-based programmable. OTP uses the antifuse interconnect technology along with standard logic gates,

TABLE 4.5

Characteristics of Early Programmable Logic Technologies

Type	Speed	Programmability	Other Characteristics
PLA	Slow	Fully with both planes	High fuse count; sharing of AND terms across multiple ORs
PAL/SPLD	Fast	Only with one plane	Low fuses count; limited sharing of AND terms
CPLD	Fast	Only with the interconnects	Low fuses count; capable of handling wide and complex gating at high speed
EPROM	Slow		Erasable and programmable for many times

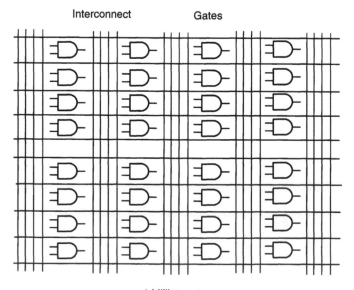

>1 Million gates

FIGURE 4.71
Architecture of an early version FPGA by Xilinx® (1985).

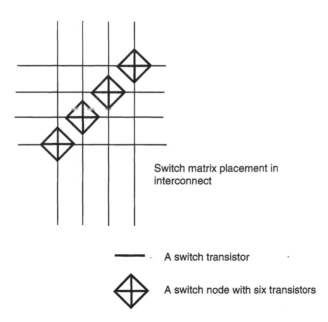

Switch matrix placement in
interconnect

———— - A switch transistor

A switch node with six transistors

FIGURE 4.72
SRAM-based switch matrix interconnect.

while SRAM-based uses SRAM memory to define interconnect as well as logic with look-up table (LUT). Obviously, the SRAM-based FPGA technology provides much flexibility to end users, especially for prototyping digital systems.

Key to the SRAM-based FPGA technology is the switch matrix transistors placed at inter-connect intersections, as illustrated in Figure 4.72. When a switch transistor is programmed to turn ON by the content in a memory bit in the SRAM, it would connect two wires either horizontally, or vertically, or a mix. As shown in Figure 4.72, six switch transistors are placed at each horizontal–vertical wire intersection.

There are four interconnect wires coming toward an intersection: coming from north, south, east, and west. The six switch transistors in each switch node connect northwest (as T1), southwest (as T2), northeast (as T3), southeast (as T4), east–west (as T5), and north–south (T6). To connect the wire from north with the wire from south, the memory bit in the SRAM used to program switch transistor T6 (for north-south) is set to a 1 (or ON). Multiple-switch transistors can be programmed to turn ON at each switch node for needed connections. For example, to connect wires from north, east, and south at a switch node, switch transistors T2 (for north–east) and T3 (for south–east) are turned ON by setting the memory bits in the SRAM to 1.

Similar methods are used to program logic gate functions in SRAM-based FPGA devices via LUTs. As with other programmable devices, vendors provide software to end users for programming the FPGA devices.

4.10 Logic Function Implementation Using Threshold Logic

This section deals with a special type of switching device called threshold elements. A single threshold element can realize a function that otherwise would have required dozens of conventional binary logic gates. Circuits formed of threshold elements generally require less

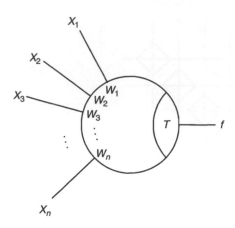

FIGURE 4.73
A n-input threshold element.

complex interconnections and fewer elements than the equivalent circuits involving conventional logic gates. However, neither the operation nor production of threshold elements is easy. In spite of early interest in such devices, they never became popular. But recent studies have shown that many of the threshold elements mimic the function of neurons. Human neurons are necessarily fast but they are capable of updating the outputs as newer experiences (information) are taken into consideration. Obviously, therefore, a surge of activities have now begun involving threshold elements with potential applications to neural networks, adaptive signal processing, adaptive control systems, learning automata, optical computing, and pattern recognition.

A threshold element may have n binary inputs, $X_1, X_2, \ldots, X_n$, and a single binary output f. The workings of a threshold element is, however, determined by a set of n weights so that weight W_j is associated with binary input X_j and a threshold value T. The weights as well as the threshold value may be real, finite, and either positive or negative. The binary output of the threshold element, shown in Figure 4.73, is given by

$$f = \begin{cases} 1 & \text{if } \sum W_i X_i \geq T \\ 0 & \text{if } \sum W_i X_i < T \end{cases} \tag{4.1}$$

where $W_i X_i$ is commonly referred to as the i-th inner product. To understand the behavior of these threshold elements, it will be better first to consider a couple of examples before resuming our discussion of threshold elements.

Example 4.19
Determine what Boolean function is being realized by the threshold element shown in Figure 4.74.

Solution
The truth table, as shown in Figure 4.75, is obtained by calculating the sum of inner products and then comparing the sum of inner products with the threshold value.

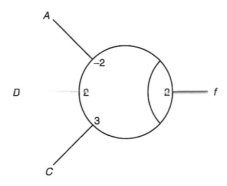

FIGURE 4.74
Threshold element for Example 4.19.

ABC	Sum of inner products	F
000	0	0
001	3	1
010	2	1
011	5	1
100	-2	0
101	1	0
110	0	0
111	1	0

FIGURE 4.75
Truthtable for Example 4.19.

Therefore,

$$f(A, B, C) = \sum m(1, 2, 3) = \bar{A}B + \bar{A}C.$$

The threshold element shown in Figure 4.74 is thus realizing a Boolean function that can also be implemented using a logic circuit formed of four conventional logic (one NOT, two 2-input ANDs, and one 2-input OR) gates.

Example 4.20
Obtain the threshold element that realizes the function $f(A, B, C) = \sum m(0, 2, 6)$.

Solution
The truth table corresponding to the given function is obtained as shown in Figure 4.76.
Therefore, the inequalities involving the weights and thresholds are determined as follows:

$$0 \geq T$$
$$W_3 < T$$
$$W_2 \geq T$$
$$W_2 + W_3 < T$$

A B C	f
0 0 0	1
0 0 1	0
0 1 0	1
0 1 1	0
1 0 0	0
1 0 1	0
1 1 0	1
1 1 1	0

FIGURE 4.76
Truth table for Example 4.20.

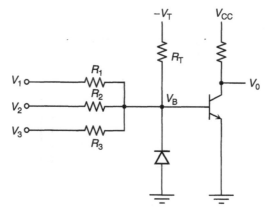

FIGURE 4.77
A 3-input threshold element.

$$W_1 < T$$
$$W_1 + W_3 < T$$
$$W_1 + W_2 \geq T$$
$$W_1 + W_2 + W_3 < T$$

From these inequalities, we conclude that T must be negative while it also has to be larger than both W_1 and W_3 and smaller than W_2. Again, since $W_1 + W_2 \geq T$, W_2 has to be positive. Accordingly, all of the aforementioned inequalities may be simultaneously satisfied, for example, by the following weight and threshold values:

$$W_1 = -2, \quad W_2 = 2, \quad W_3 = -4, \quad \text{and} \quad T = -1$$

A threshold element such as the one described in Figure 4.74 can be implemented using a resistor-transistor circuit such as that shown in Figure 4.77. When the three input resistances are nonidentical, the gate output V_0 is determined by the sum of inner products. For example, for the transistor to be turned on (by having output V_0 go low), V_B has to be 0 V (actually, 0.7 V in case the diode is made of silicon). Again, for the transistor to be turned off, V_B has to be negative. For V_B to be positive, therefore,

$$(V_1/R_1) + (V_2/R_2) + (V_3/R_3) \geq V_T/R_T \tag{4.2}$$

Otherwise, V_B is negative. Comparing Equations 4.1 and 4.2, we see that for this threshold element, $W_j = 1/R_j$ and the threshold value $T = V_T/R_T$.

One problem, however, becomes obvious that in mass-produced devices, the resistance values may end up differing significantly from their nominal values. Also, the voltage values may vary. Thus, the reliability of threshold elements is often questionable. However, this problem can be overcome by guaranteeing a large difference between the sum of the inner products for which f is a 1 and that for which f is a 0. But most importantly, threshold elements have not become sufficiently popular because of the lack of a simplified procedure with which one can conclude if a particular logic function is realizable using one or more threshold elements. The remaining part of this section will address some of these complex issues pertaining to threshold elements.

A threshold element satisfying the Boolean function $f(x_1, x_2, \ldots, x_n)$ may be specified by a weight–threshold vector given by

$$V = [W_1, W_2, \ldots, W_n : T] \tag{4.3}$$

and by its input variables $x_1, x_2, \ldots,$ and x_n. Accordingly, the threshold element of Example 4.20 is described by the weight–threshold vector $V = [-2, 2\text{--}4 : -1]$ and its inputs A, B, and C. The relationship, if there is any, between a Boolean function and its corresponding threshold element is described in terms of the characteristics of the function. A function $f(x_1, x_2, \ldots, x_n)$ is said to be positive in x_i if there exists an SOP or a POS expression for f in which x_i appears only in uncomplemented form. Likewise, f is said to be negative in x_i if there exists an SOP or a POS expression of f in which x_i appears only in its complemented form. Again if f is either positive or negative in x_i but not both, f is said to be "unate in x_i." And, finally, a function that is unate in each one of its input variables is said to be *unate*. A thorough mathematical investigation (not explored here, for simplicity) would confirm that the function corresponding to a valid weight–threshold vector is always unate. This does not, however, imply that all unate functions can be realized using threshold elements.

Consider, for example, the function $f(A, B, C) = \bar{A}B + \bar{A}\bar{B}C$. This function is unate since $\bar{A}B + \bar{A}\bar{B}C = \bar{A}(B + \bar{B}C) = \bar{A}(B + C) = \bar{A}B + \bar{A}C$ is unate in $\bar{A}, B,$ and C only. This particular logic function has already been used in Example 4.19 and was shown to be realizable using the weight–threshold vector $V = [-2, 2, 3 : 2]$.

The sets of input combinations contributing to an output of 1 are called "true vertices" while the remaining input combinations are referred to as "false vertices." For example, the true vertices corresponding to $f = \bar{A}B + \bar{A}C$ are $\{0,1,0\}$ and $\{0,1,1\}$ that are derived from $\bar{A}B$ and $\{0,0,1\}$ and $\{0,1,1\}$ that are obtained from $\bar{A}C$. Thus, there are only three true vertices that correspond to the logic function $f = \bar{A}B + \bar{A}C$, namely, $\{0,1,0\}$, $\{0,1,1\}$, and $\{0,0,1\}$. The remaining five vertices are false vertices. It is typical to order vertices, true or false, as follows. A vertex $\{a_1, a_2, \ldots, a_n\}$ is said to be higher in order than a second vertex $\{b_1, b_2, \ldots, b_n\}$ only if for all $j, a_j \geq b_j$. For example, the vertex $\{0,1,1\}$ has a higher rank when compared to the vertices $\{0,0,1\}$ or $\{0,1,0\}$. Interestingly, however, some vertices may not be comparable, for example, $\{0,0,1\}$ and $\{0,1,0\}$.

In the following, few of the most important theorems pertaining to the threshold elements are listed without any proof.

THEOREM 1
Provided there exists a weight–threshold vector $V_1 = [W_1, W_2, \ldots, W_n : T]$ that realizes the function $f(x_1, x_2, \ldots, x_i, \ldots, x_n)$, the logic function $f(x_1, x_2, \ldots, \bar{x}_i, \ldots, x_n)$ can be realized using the weight–threshold vector $V_2 = [W_1, W_2, \ldots, -W_i, \ldots, W_n : T - W_i]$.

THEOREM 2
Provided there exists a weight–threshold vector $V_1 = [W_1, W_2, \ldots, W_n : T]$ that realizes the function $f(x_1, x_2, \ldots, x_i, \ldots, x_n)$, the function $\bar{f}(x_1, x_2, \ldots, x_i, \ldots, x_n)$ can be realized using the weight–threshold vector $V_2 = [-W_1, -W_2, \ldots, -W_n : -T]$.

THEOREM 3
The weight–threshold vector, if it exists, corresponding to a Boolean function $f(x_1, x_2, \ldots, x_n)$ that is unate and also positive in each one of its variables, then can be determined by considering only minimal true vertices and maximal false vertices.

The step-by-step rules for identifying threshold element corresponding to a given logical function are listed below:

Step 1. Test the function for unateness. If the test fails, the function is not realizable using a threshold element.

Step 2. The unate function is made positive in each one of its variables by relabeling operations such as $\bar{X}_i$ as X_i.

Step 3. Determine all minimal true vertices (of the lowest rank) and maximal false vertices (of the highest two ranks).

Step 4. For each pair of minimal true vertex $\{a_1, a_2, \ldots, a_n\}$ and maximal false vertex $\{b_1, b_2, \ldots, b_n\}$, set up the inequality

$$a_1 W_1 + a_2 W_2 + \cdots + a_n W_n > b_1 W_1 + b_2 W_2 + \cdots + b_n W_n$$

Solve all these inequalities simultaneously for the weights.

Step 5. Determine the threshold value and, correspondingly, the weight–threshold vector.

Step 6. Convert the weight–threshold vector (using Theorem 1) to find the true weight–threshold vector by considering $Y_i \Rightarrow \bar{X}_i$ transformation.

On the other hand, if the function is not unate, one may split the function into two or more functions (using logical OR operation) such that each one of the decomposed functions is a unate. A separate threshold element can then be determined for each one of these decomposed functions.

Example 4.21
Obtain the threshold element, if there is any, for realizing $f(A, B, C, D) = AB + A\bar{B}C\bar{D}$.

Solution

$$AB + A\bar{B}C\bar{D} = A(B + \bar{B}C\bar{D}) = A(B + C\bar{D}) = AB + AC\bar{D}$$

This reduced form of the function confirms that the function is unate. The corresponding unate function that is completely positive is given by $f(A, B, C, D) = AB + ACD$. Its minimal true vertices are $\{1,1,0,0\}$ and $\{1,0,1,1\}$ while its maximal false vertices are $\{0,1,1,1\}$, $\{1,0,1,0\}$, and $\{1,0,0,1\}$. Therefore, the total number of to-be-solved inequalities is 6 ($= 2 \times 3$). They are obtained as follows:

$$\{1,1,0,0\} \quad \text{and} \quad \{0,1,1,1\} \Rightarrow W_1 + W_2 > W_2 + W_3 + W_4$$
$$\{1,1,0,0\} \quad \text{and} \quad \{1,0,1,0\} \Rightarrow W_1 + W_2 > W_1 + W_3$$
$$\{1,1,0,0\} \quad \text{and} \quad \{1,0,0,1\} \Rightarrow W_1 + W_2 > W_1 + W_4$$
$$\{1,0,1,1\} \quad \text{and} \quad \{0,1,1,1\} \Rightarrow W_1 + W_3 + W_4 > W_2 + W_3 + W_4$$
$$\{1,0,1,1\} \quad \text{and} \quad \{1,0,1,0\} \Rightarrow W_1 + W_3 + W_4 > W_1 + W_3$$
$$\{1,0,1,1\} \quad \text{and} \quad \{1,0,0,1\} \Rightarrow W_1 + W_3 + W_4 > W_1 + W_4$$

They can be reduced to yield the required constraints as given by

$$W_1 > W_3 + W_4$$
$$W_2 > W_3$$
$$W_2 > W_4$$
$$W_1 > W_2$$
$$W_4 > 0$$
$$W_3 > 0$$

These six inequalities can be summarized further to give

$$W_1 > W_3 + W_4 > 0$$
$$W_1 > W_2 > W_3 > 0$$
$$W_1 > W_2 > W_4 > 0$$

One of the solutions can be, for example, $W_1 = 4$, $W_2 = 3$, $W_3 = 2$, and $W_4 = 1$. Consider a pair of minimal true and maximal false vertices, $\{1,1,0,0\}$ and $\{0,1,1,1\}$, for example. The sum of inner products corresponding to $\{1,1,0,0\}$ is thus $4 + 3 = 7$ and that corresponding to $\{0,1,1,1\}$ is $3+2+1 = 6$. But since $7 \geq T \geq 6$, we may choose T to be 13/2, for example. The weight–threshold vector V_1 corresponding to $AB+ACD$ is thus $[4, 3, 2, 1: 13/2]$. Therefore, the weight–threshold vector V_2 corresponding to $AB + AC\bar{D}$ is given by

$$V_2 = [4, 3, 2, -1: 13/2 - 1]$$
$$= [4, 3, 2, -1: 11/2]$$

4.11 Logic Function Implementation Using Transmission Gates

Transmission gates are widely used in modern digital circuits because of the simplicity. A transmission gate (TG) consists of a pair of n-channel and p-channel transistors with their drains and sources tied together, while the gates are the control signal terminals. When the gate of the n-channel transistor is set to 1 and the gate of the p-channel transistor is set to 0, both transistors are turned ON and an input signal passes the transmission gate

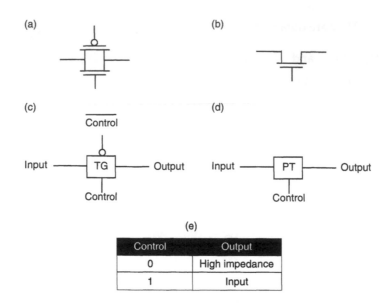

FIGURE 4.78
Transmission gate configurations. (a) CMOS configuration, (b) pass-transistor configuration, (c) symbol for TG, (d) symbol for PT, and (e) truth table for TG and PT.

through the drains to the sources. A simpler version of the transmission gate is called pass transistor (PT) (sometimes also called as pass gate or X-gate) which employs an n-channel (or p-channel) transistor. Figure 4.78 shows the configurations of transmission gate and pass gate with their symbolic representations.

One of the advantages with TG and PT implementation is simplicity. For example, an AND–OR implementation of an XOR gate would require two AND gates (typically using 12 CMOS transistors) and one OR gate (typically using six transistors), while a TG-based implementation would use only two gates with four transistors, as shown in Figure 4.79.

Another advantage of TG and PT implementation is that multiple logic outputs are allowed to be dotted on the same line such as a bus. Figure 4.80 shows a TG-based implementation of a 4-to-1 multiplexer. Outputs of basic logic gates such as AND, OR, NAND, and NOR are not allowed to be dotted together, since it would cause destructive bus-contention problems where one output is driving for a logic 1 while another is driving for a logic 0. Such conditions would cause bus burn out owing to draining of large currents.

One of the limiting factors with TGs and PTs implementations is the voltage drop when signals pass through these two types of components. Another is the higher internal capacitances in TG and PT configurations. Therefore, it is not recommended to cascade TGs and PTs in large numbers to minimize signal voltage drop. Table 4.6 summarizes the transmission characteristics.

One of the key steps in using TGs for logic implementation is the identification of pass variable(s) to replace the 1s and 0s in normal Karnaugh maps. Instead of grouping 1s, as one would with a normal Karnaugh map, any variable can be cast as a pass variable or control variable and grouped together. To illustrate the identification and use of pass variables and control variables, consider logic function $f(a,b,c) = a \cdot \bar{b} + b \cdot c$. Figure 4.81 shows the normal Karnaugh map and its modified version using pass variables, along with a TG-based implementation of f.

As shown in Figure 4.81a, the pass variable identified in the $abc = x0x$ group is variable a, with control variable $\bar{b}$ (since variable c has no effect for the group). Similarly, the pass variable identified in the $abc = x1x$ group is variable c, with control variable b.

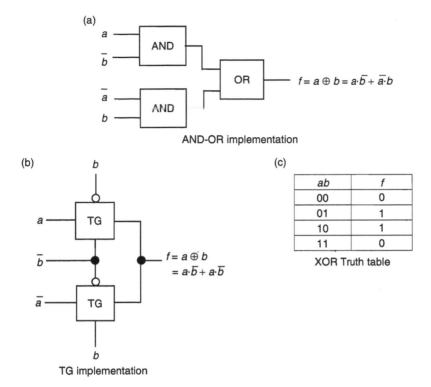

FIGURE 4.79
Implementations of XOR gate.

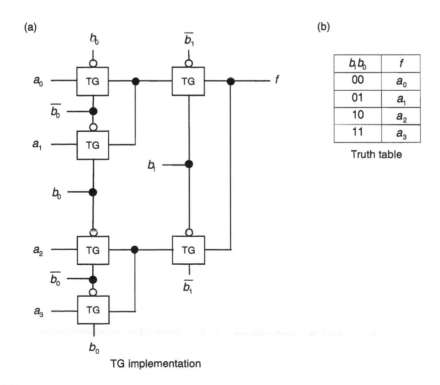

FIGURE 4.80
TG implementation of 4-to-1 multiplexer.

TABLE 4.6

Transmission Gate Characteristics

Device	Transmission of 1	Transmission of 0
CMOS transmission gate	Good	Good
N-channel pass transistor	Poor	Good
P-channel pass transistor	Good	Poor

(a) Karnaugh map for $f(a,b,c) = a \cdot \bar{b} + b \cdot c$

a	00	01	11	10
0	0	0	1	0
1	1	1	1	0

bc

(b) Modified Karnaugh map for $f(a,b,c) = a \cdot \bar{b} + b \cdot c$

a	00	01	11	10
0	a	a	c	c
1	a	a	c	c

bc

(c) TG-based schematic

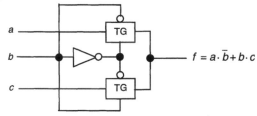

$f = a \cdot \bar{b} + b \cdot c$

FIGURE 4.81
Identification and grouping of pass and control variables for TG-based implementation.

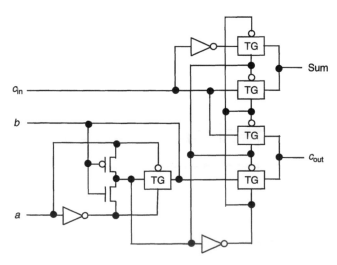

FIGURE 4.82
An optimized TG-based adder.

TABLE 4.7

Karnaugh Map for 4-to-1 Multiplexer

b_1	b_0	
	0	1
0	a_0	a_1
1	a_2	a_3

TABLE 4.8

Basic Functions Implemented with Multiplexer

Basic Function	a_0	a_1	a_2	a_3
AND (b_0, b_1)	0	0	0	1
OR (b_0, b_1)	0	1	1	1
NAND (b_0, b_1)	1	1	1	0
NOR (b_0, b_1)	1	0	0	0
XOR (b_0, b_1)	0	1	1	0

Used wisely, logic functions implemented with TGs and PTs can be very efficient. Figure 4.82 shows the logic schematic for an efficient TG-based adder. Furthermore, TG-based and PT-based implementations provide functional flexibility with very compact circuit sizes. Consider the example of TG-based 4-to-1 multiplexer as shown in Figure 4.80, with the Karnaugh map shown in Table 4.7. By assigning different combination of logic 1 and 0 to a_0, a_1, a_2, a_3, the same implementation can achieve different basic logic functions as illustrated in Table 4.8.

In many cases TGs can be replaced by PTs to further reduce the circuit size and complexity. However, such replacements should only be carried out after careful identification and examination of potential performance degradation in all possible input conditions. For example, in some circumstances CMOS TGs in a multiplexer configuration, such as the one shown in Figure 4.80, may be replaced with n-channel PTs, resulting in half of the original circuit size but with minimal performance impact.

4.12 Summary

In this chapter a number of practical implementation techniques were introduced for realizing arbitrary logical functions. In particular, circuits using SSI circuits (only NAND gates, only NOR gates, or only XOR gates and AND gates), MSI circuits (only multiplexers, and only decoders), ROMs, and PLDs were discussed. We have also introduced a bridging technique to handle XOR-like functions that are otherwise not reducible. This chapter also considers the topic of threshold logic, transmission gate, and the various issues related to their usage in realizing Boolean functions.

Bibliography

Comer, D.J., *Digital Logic and State Machine Design*. 3rd edn. New York, NY. Oxford University Press, 1994.

Floyd, T., *Digital Fundamentals*. 8th edn. Englewood Cliffs, NJ. Prentice-Hall, 2003.

Johnson, E.J. and Karim, M.A., *Digital Design: A Pragmatic Approach*. Boston, MA. PWS-Kent Publishing, 1987.

Karim, M.A. and Awwal, A.A.S., *Optical Computing: An Introduction*. New York, NY. John Wiley & Sons, 1992.

Katz, R.H., *Contemporary Logic Design*. Boston, MA. Addison Wesley, 1993.

Mowle, F.J.A., *A Systematic Approach to Digital Logic Design*. Reading, MA. Addison-Wesley, 1976.

Nagle, H.T., Jr., Carroll, B.D., and Irwin, J.D., *An Introduction to Computer Logic*, Englewood Cliffs, NJ. Prentice-Hall, 1975.

Nelson, V.P.P., Nagle, H.T., and Carroll, B.D., *Digital Logic Circuit Analysis and Design*. Englewood Cliffs, NJ. Prentice-Hall. 1995.

Peatman, J.B., *Digital Hardware Design*, New York, NY. McGraw-Hill, 1980.

Problems

1. Realize the following logical functions using only NAND gates:

    (a) $f(A, B, C, D) = \Sigma m(1, 3, 4, 6, 9, 11, 13, 15)$

    (b) $f(A, B, C, D) = \Sigma m(5, 6, 7, 10, 14, 15) + d(9, 11)$

    (b) $f(A, B, C, D, E) = \Sigma m(1, 4, 6, 8, 13, 14, 20, 21, 23)$

    (c) $f(A, B, C, D, E) = \Sigma m(0, 2, 3, 6, 9, 15, 16, 18, 20, 23, 26)$
    $$+ d(1, 4, 10, 17, 19, 25, 31)$$

    (d) $f(A, B, C, D, E, F) = \Sigma m(0, 2, 4, 7, 8, 16, 24, 32, 36, 40, 48)$
    $$+ d(5, 18, 22, 23, 54, 56)$$

    (e) $f(A, B, C, D, E) = \Sigma m(0, 1, 5, 14, 15, 21, 24, 26\text{--}28)$
    $$+ d(6, 11, 18, 19, 20, 29)$$

2. Implement each of the logical functions listed in Problem 1 using NOR-only logic circuits.

3. Implement the carry output function $C_4 = G_3 + P_3 G_2 + P_3 P_2 G_1 + P_3 P_2 P_1 G_0 + P_3 P_2 P_1 P_0 C_0$ using 3-input NAND gates and NOT gate if needed. Iterate the design to identify an equivalent logic circuit that will also be faster. Hint: Use Table 4.1.

4. Design a logic circuit that will take in 4-bit 2421 BCD code as its input and accordingly drive a 7-segment LED display (see Table 1.9 and Figure 4.52 for reference) using (a) only NAND gates, and (b) only NOR gates.

5. Prove the following identities:

    (a) $A + B = A \oplus B \oplus AB = A \oplus \bar{A}B$

    (b) $A(B \oplus C) = AB \oplus AC$

    (c) $A \oplus B \otimes C = \overline{A \oplus B \oplus C}$

    (d) $A \otimes B = A \oplus \bar{B} = \bar{A} \oplus B$

6. Use XOR and AND gates to implement the following logical functions:

    (a) $f(A, B, C, D) = \sum m(1, 2, 5, 7, 8, 10, 13, 14)$

    (b) $f(A, B, C, D) = \sum m(2, 3, 6, 7, 9, 11, 12, 13)$

    (c) $f(W, X, Y, Z) = \sum m(0, 2, 3, 6\text{--}8, 10, 13)$

    (d) $f(W, X, Y, Z) = \sum m(0, 6, 9, 10, 15)$

7. Bridge the functions listed in Problem 6 with the most resembling XOR functions (see Figure 4.15 for reference).

8. Implement the following logical functions using 3-input NAND gates:

   (a) $f(A,B,C,D) = \Sigma m(0,1,3,7,8,12) + d(5,10,13,14)$
   (b) $f(A,B,C,D) = \bar{A}B + \bar{B}CD + A\bar{B}\bar{D}$
   (c) $f(A,B,C,D) = AB(C+D) + CD(A+B)$
   (d) $f(A,B,C,D,E) = (\bar{A}B + \bar{A}C + \bar{A}DE)(\bar{B} + CD + \bar{A}E)$

9. Implement each of the functions listed in Problem 8 using 3-input NOR gates.

10. Generate NOT, AND, NAND, OR, and NOR logic using (a) only XOR gates; and (b) only XNOR gates.

11. Obtain a 2-input AND, NAND, OR, NOR, XOR, and XNOR logic functions using only logic units capable of generating (a) $\bar{A}B$, and (b) $A\bar{B}$.

12. Obtain $f(A,B,C,D) = A\bar{B}C + BD + A\bar{D}$ using an 1-of-4 MUX and assorted gates such that the selectors to the MUX are respectively (a) A and B, (b) B and D, and (c) C and A.

13. Obtain the $f(A,B,C,D) = A\bar{B}C + BD + A\bar{D}$ using an 1-of-8 MUX such that the selectors to the MUX are respectively (a) A, B, and C; and (b) B, C, and D; and (c) A, C, and D.

14. Using 1-of-2 MUX units construct a 1-of-8 MUX.

15. Using a 3-to-8 line decoder and assorted logic gates construct a 1-of-8 MUX. Compare this solution with that of Problem 14.

16. Using a single 1-of-4 MUX and as few 1-of-8 MUX as possible, design a 1-of-32 MUX.

17. Implement each of the Boolean functions listed in Problem 1 using 1-of-8 MUX and a few assorted logic gates, if necessary.

18. Implement each of the Boolean functions listed in Problem 1 using two levels of 1-of-4 MUXs.

19. Exhaust all possible input and selector combinations to determine the most optimum MUX (using 1-of-4 MUX) implementation for the following functions:

   (a) $f(A,B,C,D) = \Sigma m(1,3,4,6,9,11,13,15)$
   (b) $f(A,B,C,D) = \Sigma m(5,6,7,10,14,15) + d(9,11)$

20. Using 1-of-4 MUXs, obtain a two-level MUX circuit for each of the functions of Problem 1.

21. Implement $f(A,B,C,D) = \Sigma m(1,4,8,13)$ and $g(A,B,C,D) = \Sigma m(2,7,13,14)$ using 3-to-8 line decoders.

22. Design a 4-to-16 line decoder using only 2-to-4 line decoders.

23. Design a 5-to-32 line decoders using only 3-to-8 line decoders.

24. Design a 5-to-32 line decoder using a 2-to-4 line decoder and as few as possible of 3-to-8 line decoders.

25. Design a 6-to-64 line decoder using only 3-to-8 line decoders.

26. For each of the following circuits, determine the output in terms of its minterms (Figure 4.P1):

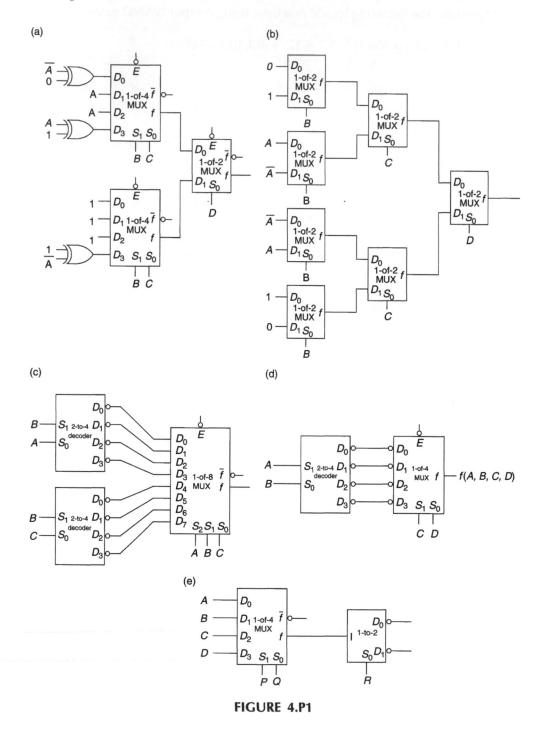

FIGURE 4.P1

27. Implement a multiplier of two 3-bit numbers $a_2a_1a_0$ and $b_2b_1b_0$ using a PROM.

28. Use a $10 \times 40 \times 5$ PLA to implement a logic circuit that receives 8-bit input and indicates whether or not the input contains all 0s, a single 1, or multiple 1s.

29. Implement the following multioutput functions using a cost-efficient PLA:

   (a) $F1(A, B, C, D, E) = \Sigma m(0\text{–}3, 6, 7, 20, 21, 26\text{–}28)$

   $F2(A, B, C, D, E) = \Sigma m(0, 1, 6, 7, 14\text{–}17, 19, 20, 24, 27)$

   $F3(A, B, C, D, E) = \Sigma m(0, 3, 8, 9, 16, 20, 26, 28, 30)$

   (b) $F1(A, B, C, D) = \Sigma m(0, 1, 2, 3, 6, 7)$

   $F2(A, B, C, D) = \Sigma m(0, 1, 2, 3, 8, 9)$

   $F3(A, B, C, D) = \Sigma m(0, 1, 6, 7, 14, 15)$

   (c) $F1(A, B, C, D) = \Sigma m(2, 10, 12, 14) + d(0, 7, 9)$

   $F2(A, B, C, D) = \Sigma m(2, 8, 10, 11, 13) + d(3, 9, 15)$

   $F3(A, B, C, D) = \Sigma m(2, 3, 6, 10) + d(3, 9, 15)$

30. Given the function $f(A, B, C, D) = \sum m(0, 4, 9, 10, 11, 12)$ and $f_1 = B \oplus D$, determine f_2 and f_3 such that $f = f_1 f_2 + \bar{f}_3$.

31. Use a ROM to design a 4-bit binary-to-Gray code converter.

32. Design a ROM circuit to drive a 7-segment LED display when driven with an XS3 code BCD input. Assume that the display requires an active low input to turn on the LED segments. Discuss the other implementation alternatives.

33. By implementing the following set of three Boolean equations in PLA and PAL format, discuss the savings and cost that one may have in each format by conducting simultaneous minimization of these functions.

   $F1(A, B, C, D) = \Sigma m(0, 2, 7, 10) + d(12, 15)$

   $F2(A, B, C, D) = \Sigma m(2, 7, 8) + d(0, 5, 13)$

   $F3(A, B, C, D) = \Sigma m(2, 4, 5) + d(6, 7, 8, 10)$

34. Design a 4-input logic network that squares each of the binary inputs using (a) a PROM, (b) a PLA, and (c) a PAL.

35. Design a 6-input logic network that generates the 2's complement of a positive number using (a) a PROM, (b) a PLA, and (c) a PAL.

36. Design a PAL circuit that compares two 4-bit unsigned binary numbers $A = a_3 a_2 a_1 a_0$ and $B = b_3 b_2 b_1 b_0$, and produces three outputs: X, if $A = B$; Y, if $A > B$; and Z, if $A < B$.

37. Bridge the carry-out function C_{i+1} as given in Table 1.5 with $A_i \oplus B_i$, where A_i and B_i respectively are the augend and addend.

38. Determine which of the following functions are realizable using a threshold element:

   (a) $A + \bar{B}\bar{C}$

   (b) $AB + CD$

   (c) $BC + CD + ABD$

   (d) $AB + BC$.

39. Determine the threshold element for realizing the following Boolean functions:

   (a) $\bar{A}B + \bar{A}D + B\bar{C} + \bar{A}\bar{C}$

   (b) $f(A, B, C, D) = \sum m(2, 3, 6, 7, 15)$

   (c) $f(A, B, C, D) = \sum m(10, 12, 14, 15)$

   (d) $f(A, B, C, D) = \sum m(3, 5, 7, 15)$.

40. Replace the circuit of Figure 4.P2 using a single threshold element.

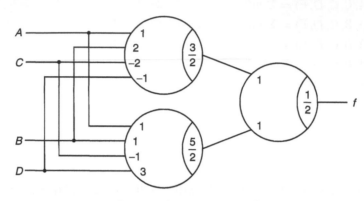

FIGURE 4.P2

41. Obtain the truth table for the following weight-threshold vectors corresponding to the inputs A_2, A_1, and A_0:

(a) [2, 2, 2: 5];

(b) [6, 6, 6, 6: 5] when its feeder inputs are respectively [−2, 3, 3: 5], [6, 2, −2: 5], [3, 2, 3: 5] and [3, 3, −2: 5];

(c) [6, 6, 6, 6: 5] when its feeder inputs are respectively [−2, −2, 6: 5], [−2, 6, −2: 5], [3, −2, 3: 5] and [3, 3, −2: 5].

Obtain a composite 3-input 3-output truth table where f_2, f_1, and f_0 are the outputs of the threshold circuits of parts (a), (b), and (c) respectively. Show that the composite truth table corresponds to the binary addition of $(A_2 A_1)_2$ and $(A_0)_2$.

5

Introduction to VHDL

5.1 Introduction

In the early 1980s, the U.S. Department of Defense's very high-speed integrated circuit (VHSIC) program supported the development of VHSIC hardware description language (VHDL). VHDL was then adopted as the IEEE Standard 1076-1987 and became the documentation, simulation, and verification medium for IC products, especially those with military applications. The wide use of VHDL (as well as its variations such as Verilog® HDL) throughout the industry has since made it one of the most important productivity-driving IC design tools available.

To meet all needs of the IC design industry and government agencies, VHDL has also evolved into a complex circuit design programming language since its introduction, with many books and manuals available in the market. In this chapter, we highlight many simple and basic features of VHDL for IC design. Readers are encouraged to download the free Xilinx WebPACK ISE Logic Design software and the Quick Start Handbook (at www.xilinx.com/univ/).

In this chapter, we describe some basic aspects of VHDL and a few simple circuit design cases. In-depth coverage on VHDL and its wide applications can be found in the many books readily available in most book stores and at online book shops.

5.2 VHDL Programming Environment

There are four types (or styles) of VHDL programs commonly seen in practice: structural, functional, behavioral, and hierarchical. A structural VHDL program describes the exact interconnection of basic and elementary components in a circuit design in the form of a *netlist*. Typical components used in structural VHDL programs are logic gates, such as NAND, NOR, XOR, and so on. A functional VHDL description describes signal relations with Boolean functions, without specifying the types of logic gates to be used in implementations. A behavioral VHDL program describes the functions of a circuit on the basis of how the circuit is expected to work and how data is processed, especially at the *register transfer level* (RTL). RTL descriptions mimic the data movements between registers regardless of the details of logic and circuit implementations. RTL concepts will be covered in some detail in Chapter 9. A hierarchical VHDL program entails the use of circuit blocks and their interconnections in a circuit. Hierarchical VHDL programs are commonly used for large systems designed with circuit-building blocks. Typical circuit-building blocks are various arithmetic circuits, such as adders, counters, shifters, and so on, as well as memory blocks, bus structures, and other control circuits.

To facilitate writing and documenting complex VHDL programs, comments lines, starting with the symbol "--" and ending with end-of-line breaks, are often used. Keywords (or reserved words) are often seen in VHDL program lines that end with a semicolon (;). Once a VHDL program is prepared, it is imported to a compiler that processes VHDL program lines and prepares various circuit design data and extracts circuit characteristics for applications such as syntax check, schematic display, circuit simulation, and timing analysis, and so on.

5.3   Structural VHDL

In a structural VHDL program, there are typically two sections: definition and netlist. In the definition section, use of specific libraries and interface of a circuit are identified. In the netlist section, use of components as well as their interconnections is defined and specified.

Figure 5.1 shows a sample VHDL program for a 2-to-4 decoder. The code line numbers on the right-hand side are added for the ease of discussion (i.e., -- 1, or -- 2, etc.). **Bold**-faced words are keywords (or reserved words) in VHDL. The first two lines are comments as the

```
-- A structural VHDL description of                        -- 01
-- a 2-to-4 decoder using INVERTER and NAND gates          -- 01
library ieee, lcdf_vhdl;                                   -- 03
use ieee.std_logic_1164.all, lcdf_vhdl.func_prims.all      -- 04
entity decoder_2to4 is                                     -- 05
   port(En, X0, X1: in std_logic;                          -- 06
        Y0, Y1, Y2, Y3: out std_logic);                    -- 07
end decoder_2to4;                                          -- 08
                                                           -- 09
architecture structural_2to4decoder of decoder_2to4 is     -- 10
   component INV1                                          -- 11
      port(in1: in std_logic;                              -- 12
           out1: out std_logic);                           -- 13
   end component;                                          -- 14
                                                           -- 15
   component NAND3                                         -- 16
      port(in1, in2, in3: in std_logic;                    -- 17
           out1: out std_logic);                           -- 18
   end component;                                          -- 19
                                                           -- 20
   signal X0_b, X1_b: std_logic;                           -- 21
                                                           -- 22
   begin                                                   -- 23
      G1: INV1 port map(in1 => X0, out1 => X0_b);          -- 24
      G2: INV1 port map(in1 => X1, out1 => X1_b);          -- 25
      G3: NAND3 port map(in1 => X0_b, in2 => X1_b,         -- 26
                         in3 => En, out1 => Y0);           -- 27
      G4: NAND3 port map(in1 => X0, in2 => X1_b,           -- 28
                         in3 => En, out1 => Y1);           -- 29
      G5: NAND3 port map(in1 => X0_b, in2 => X1,           -- 30
                         in3 => En, out1 => Y2);           -- 31
      G6: NAND3 port map(in1 => X0, in2 => X1,             -- 32
                         in3 => En, out1 => Y3);           -- 33
end structural_2to4decoder;                                -- 34
-- End of the VHDL description                             -- 35
```

FIGURE 5.1
A structural VHDL description of a 2-to-4 decoder.

heading of the program. In line 3, the keyword **library**, specifies the external libraries—collections of precompiled VHDL code, such as ieee and lcdf_vhdl—to be referenced (or used) in the program. In the next line (line 4), keywords **use** and **all** indicate that all of the two identified packages, std_logic_1164 in ieee and func_prims in lcdf_vhdl, are referenced and used in the program. Keywords **entity** and **is** in line 5 declare the unit to be a 2-to-4 decoder, while the *port* statement in lines 6 and 7 specifies the input and output interfaces of the 2-to-4 decoder, with signals En, X0, X1 declared as inputs and Y0, Y1, Y2, and Y3 as outputs with signal type of std_logic (for standard logic). Declaration of a signal type specifies the values that may appear on the signals. The most common values in the std_logic set are the usual binary values 0, 1, X (for unknown), and U (for uninitialized). An *entity* statement is ended with an *end* statement, as in line 8 where the **end** is a keyword and the phrase next to it must be the same as the one next to its **entity** keyword.

Next, beginning with the *architecture* statement in line 10, is the netlist section describing the internal circuit implementation of the defined **entity** of 2to4_decoder. The statement **architecture** structural_2to4decoder **of** 2to4_decoder **is**, paired with the *end* statement in line 34, declares one implementation, named structural_2to4decoder, for the 2-to-4 decoder entity. An entity may have several implementations associated with different names declared right after each architecture keyword.

There are typically three blocks within each architecture section. Usually, the first block declares the gate types to be used as components, while the second declares the set of internal signals and the third describes the instances of gates and interconnections.

In the *component* block, each **component** keyword, ending with **end component** keywords, declares a gate type. For example, line 11 in the 2-to-4 decoder example, as shown in Figure 5.1, declares an inverter called INV1. A *port* statement within a component block declares the interface of the component. Lines 12 and 14, therefore, specify that the input terminal of INV1 is called din and the output terminal is called dout, both of signal type std_logic. Note that the input and output segments in a *port* statement are separated by a semicolon. Similarly, lines 16 to 19 declare a NAND gate called NAND3.

The *signal* block, sometimes with multiple statements starting with the keyword **signal**, declares the names of internal signals along with their types. For example, line 21 declares two internal signals, X0_b and X1_b, with the type std_logic.

The third block starts with the keyword **begin**, as shown in line 23. In this block, each statement specifies the name (or number) of a gate instance, the type, and the interface declared with the keywords **port map**. Each statement from lines 24–33 declares one gate instance. For example, line 24 specifies that a gate instance named G1 is an inverter of INV1 type, with its input terminal (din, as declared in line 13) connected with the signal X0 (as defined in line 6) and its output terminal (dout, as declared in line 13) connected to internal signal X0_b (which is declared in line 21). The symbol ⇒ indicates a physical electric connection between two terminals (such as din => X0) or a terminal and an internal signal (such as dout => X0_b). Similarly, lines 28 and 29 declare that the instance G4 is of type NAND3, with its input terminal in1 connected to X0_b, in2 connected to X1, in3 connected to En, and its output terminal dout connected to Y1.

With VHDL, signals and gate terminals can also be declared in the form of a *vector*. Figure 5.2 shows a structural VHDL description of a 4-to-1 multiplexer, where some of the terminals and internal signals are declared in vectors, as shown in lines 6, 7, 27, and 28. Note the one-to-one mapping between a signal (terminal) in a vector and a single-point signal (or terminal), as illustrated in lines 31–37.

The VHDL statement styles represented by lines 24–33 in Figure 5.1 and by lines 31–37 in Figure 5.2 are different. With the former, the matching of the signal pair is indicated by

```
-- A structural VHDL description of a 4-to-1 multiplexer        -- 01
-- using INVERTER, AND and OR gates                             -- 01
library ieee, lcdf_vhdl;                                        -- 03
use ieee.std_logic_1164.all, lcdf_vhdl.func_prims.all;          -- 04
entity multiplexer_4to1 is                                      -- 05
   port(Sel: in std_logic_vector(0 to 1);                       -- 06
        Din: in std_logic_vector(0 to 3);                       -- 07
        Dout: out std_logic);                                   -- 08
end multiplexer_4to1;                                           -- 09
                                                                -- 10
architecture structural_4to1mux of multiplexer_4to1 is          -- 11
                                                                -- 12
   component INV1                                               -- 13
      port(in1: in std_logic; out1: out std_logic);             -- 14
   end component;                                               -- 15
                                                                -- 16
   component AND3                                               -- 17
      port(in1, in2, in3: in std_logic;                         -- 18
           out1: out std_logic);                                -- 19
   end component;                                               -- 20
                                                                -- 21
   component OR4                                                -- 22
      port(in1, in2, in3, in4: in std_logic;                    -- 23
           out1: out std_logic);                                -- 24
   end component;                                               -- 25
                                                                -- 26
   signal Sel_b: std_logic_vector(0 to 1);                      -- 27
   signal mt: std_logic_vector(0 to 3);                         -- 28
                                                                -- 29
   begin                                                        -- 30
      G1: INV1 port map(Sel(0), Sel_b(0));                      -- 31
      G2: INV1 port map(Sel(1), Sel_b(1));                      -- 32
      G3: AND3 port map(Sel_b(1),Sel_b(0),Din(0),mt(0));        -- 33
      G4: AND3 port map(Sel_b(1),Sel(0),Din(1),mt(1));          -- 34
      G5: AND3 port map(Sel(1),Sel_b(0),Din(2),mt(2));          -- 35
      G6: AND3 port map(Sel(1),Sel(0),Din(3),mt(3));            -- 36
      G7: OR4 port map(mt(0),mt(1),mt(2),mt(3),Dout);           -- 37
   end structural_4to1mux;                                      -- 38
-- End of the VHDL description                                  -- 39
```

FIGURE 5.2
A structural VHDL description of a 4-to-1 multiplexer.

the symbol =>. With the latter, the matching is implied by the relative position of variables in a *port* statement and in a *port map* statement. For example, consider line 31 in Figure 5.2. Since signal Sel(0) is in the first position of the *port map* statement, it is mapped to the first variable in the *port* statement of line 14, where the interface of INV1 is defined.

5.4   Functional VHDL

Statements in VHDL programs can also be written on the basis of functional descriptions. Consider, for example, the VHDL program shown in Figure 5.3, describing a 2-to-4 decoder. There is no *component* statement in this functional VHDL program. Instead, the description of the 2-to-4 decoder is based on the Boolean functions between and of signals. Instead of

```
-- A functional VHDL description of          -- 01
-- a 2-to-4 decoder                          -- 01
library ieee;                                -- 03
use ieee.std_logic_1164.all;                 -- 04
entity decoder_2to4 is                       -- 05
   port(En, X0, X1: in std_logic;            -- 06
        Y0, Y1, Y2, Y3: out std_logic);      -- 07
end decoder_2to4;                            -- 08
                                             -- 09
architecture functional_2to4_decoder of decoder_2to4 is    -- 10
                                             -- 11
   signal X0_b, X1_b: std_logic;             -- 12
                                             -- 13
   begin                                     -- 14
      X0_b <= not X0;                        -- 15
      X1_b <= not X1;                        -- 16
      Y0 <= not (X0_b and X1_b and En);      -- 17
      Y1 <= not (X0 and X1_b and En);        -- 18
      Y2 <= not (X0_b and X1 and En);        -- 19
      Y3 <= not (X0 and X1 and En);          -- 20
end functional_2to4_decoder;                 -- 21
-- End of the VHDL description                -- 22
```

FIGURE 5.3
A functional VHDL description of a 2-to-4 decoder.

describing connections of specific instances of logic gates, statements of logic assignments are used with signals.

For example, line 15 in Figure 5.3 *assigns* "**not** X0" to signal X0_b, where **not** is a Boolean logic operator (and therefore, a keyword in VHDL) and the symbol $<=$ indicates the directional assignment. Similarly, lines 16–20 make functional assignments to output signals Y0, Y1, Y2, and Y3. Most basic Boolean operators, such as AND, OR, NAND, NOR, and XOR, are often used in functional VHDL statements. For example, line 20 in Figure 5.3 describes that output Y3 of the 2-to-4 multiplexer is connected with the output of a NAND function with inputs X0, X1, and En.

Typical conditional statements can also be used in VHDL. For example, Figure 5.4 illustrates a functional VHDL description of a 4-to-1 multiplexer using conditional *when-else* statements, in lines 13–17. The use of *with-select* statement in the VHDL description, as shown in Figure 5.5, demonstrates the simplification of *when-else* statements.

There are many constraints in VHDL. One of them is that outputs defined in the **port** statement of an entity cannot be used in the source-side of assignment statements in the entity's **architecture** section. Instead, internal signals feeding the defined entity outputs should be used. In most cases, built-in program syntax checking functions are provided with VHDL design and synthesis software packages to verify the correctness and to identify syntax errors. Another constraint is that the **library**, **use**, and **entity** statements go together as a group, meaning for each entity statement there must be **library** and **use** statements preceding it.

5.5 Behavioral VHDL

Designs of large and complex digital systems often can be described at the register-transfer-level (RTL) with *behavioral* specifications. RTL descriptions show RTL data movements

```
-- A functional VHDL description for a 4-to-1 multiplexer      -- 01
library ieee;                                                 -- 01
use ieee.std_logic_1164.all;                                  -- 03
entity multiplexer_4to1 is                                    -- 04
   port(Sel: in std_logic_vector(1 downto 0);                 -- 05
        Din: in std_logic_vector(0 to 3);                     -- 06
        Dout: out std_logic);                                 -- 07
end multiplexer_4to1;                                         -- 08
                                                              -- 09
architecture functional_4to1_mux of multiplexer_4to1 is       -- 10
                                                              -- 11
   begin                                                      -- 12
      Dout <= Din(0) when Sel = "00" else                     -- 13
              Din(1) when Sel = "01" else                     -- 14
              Din(2) when Sel = "10" else                     -- 15
              Din(3) when Sel = "11" else                     -- 16
              'X';                                            -- 17
end functional_4to1_mux;                                      -- 18
-- End of the VHDL description                                -- 19
```

FIGURE 5.4
A functional VHDL description of a 4-to-1 multiplexer using *when-else* statement.

```
-- A functional VHDL description for a 4-to-1 multiplexer      -- 01
library ieee;                                                 -- 01
use ieee.std_logic_1164.all;                                  -- 03
entity multiplexer_4to1 is                                    -- 04
   port(Sel: in std_logic_vector(1 downto 0);                 -- 05
        Din: in std_logic_vector(0 to 3);                     -- 06
        Dout: out std_logic);                                 -- 07
end multiplexer_4to1;                                         -- 08
                                                              -- 09
architecture functional_4to1_mux_ws of multiplexer_4to1 is    -- 10
                                                              -- 11
   begin                                                      -- 12
      with Sel select                                         -- 13
         Dout <= Din(0) when "00",                            -- 14
                 Din(1) when "01",                            -- 15
                 Din(2) when "10",                            -- 16
                 Din(3) when "11",                            -- 17
                 'X' when others;                             -- 18
end functional_4to1_mux_ws;                                   -- 19
-- End of the VHDL description                                -- 20
```

FIGURE 5.5
A functional VHDL description of a 4-to-1 multiplexer using *with-select* statement.

regardless of the details of logic and circuit implementations. Typically, behavioral VHDL descriptions of digital systems are more compact than most structural and functional descriptions, due to the absence of implementation details.

A simple behavioral VHDL description of a 4-bit adder is shown in Figure 5.6, in which the bit-cascade operator & is used to augment data bits. In the example, '0' & A in line 17 represents a signal vector with the following five elements: '0' A(3) A(2) A(1) A(0). Similarly, ''0000'' & Cin represents a signal vector containing the following five elements: '0' '0' '0' '0' Cin. It is important to note that the resulting vectors of '0' & A and A & '0' are not the same, with the latter being a 5-element vector A(3) A(2) A(1) A(0) '0'. The purpose of using the bit-cascade & operator in

```
-- A behavioral VHDL description of a 4-bit adder        -- 01
library ieee;                                            -- 03
use ieee.std_logic_1164.all;                             -- 04
use ieee.std_logic_unsigned.all;                         -- 05
entity adder_4bit is                                     -- 06
   port(A, B: in std_logic_vector(3 downto 0);           -- 07
           Cin: in std_logic;                            -- 08
           Sum: out std_logic_vector(3 downto 0);        -- 09
           Cout: out std_logic);                         -- 10
end adder_4bit;                                          -- 11
                                                         -- 12
architecture behavioral_4bit_adder of adder_4bit is      -- 13
                                                         -- 14
   signal Si: std_logic_vector(4 downto 0);              -- 15
                                                         -- 16
      begin                                              -- 17
         Si <= ('0' & A) + ('0' & B) + ("0000" & Cin);  -- 18
         Cout <= Si(4);                                  -- 19
         Sum <= Si(3 downto 0);                          -- 20
end behavioral_4bit_adder;                               -- 21
-- End of the VHDL description
```

FIGURE 5.6
A behavioral VHDL description of a 4-bit adder.

line 17 in the example is to *align* the data variables properly, so that all involved data are of 5-element vectors. The symbol + in line 17 represents the arithmetic operator for binary addition.

Writing behavioral VHDL design descriptions requires fundamental understanding of RTL and digital system architectures.

5.6 Hierarchical VHDL

For large digital systems designed around the concepts of building blocks, hierarchical design representation is of critical importance. Figure 5.7 shows a hierarchical VHDL description of a 4-bit ripple-carry adder. In this example, multiple **entity** and **architecture** sections, describing building blocks at different integration levels, are used. The first entity in the example is defined as a half adder; the second is a full adder, in which two half adders are used; and the third is a 4-bit adder, in which four full adders are used.

Also note that, in this example, both the structural and functional VHDL styles are used. The first entity, the half adder, is described in functional VHDL, while the second and third entities, the full adder and the 4-bit ripple-carry adder, respectively, are described in structural VHDL. Here, the building blocks are the half and full adders used in the 4-bit adder description, while the 4-bit adder itself is a building block in larger circuits. In order to make different pieces (or building blocks) of VHDL descriptions working together correctly, the interfaces of the building blocks, defined by the **port** statements in the **entity** sections of VHDL building blocks, must be correctly matched with the **port map** statements in the **architecture** sections, allowing the implementation details defined within the **architecture** sections to be isolated from each other. This is very much like the *capsulation* principles practiced in software system development.

```
-- A hierarchical VHDL description of a 4-bit adder       -- 01
library ieee;                                             -- 03
use ieee.std_logic_1164.all;                              -- 04
entity half_adder is                                      -- 05
    port(A, B: in std_logic; Sum, Cout: out std_logic);  -- 06
end half_adder;                                           -- 07
                                                          -- 08
architecture functional_adder of half_adder is           -- 09
  begin                                                   -- 10
      Sum <= A xor B;                                     -- 11
      Cout <= A and B;                                    -- 12
end functional_adder;                                     -- 13
                                                          -- 14
library ieee;                                             -- 15
use ieee.std_logic_1164.all;                              -- 16
entity full_adder is                                      -- 17
    port(A, B, Cin: in std_logic; Sum, Cout: out std_logic); -- 18
end full_adder;                                           -- 19
                                                          -- 20
architecture structural_functional_adder of full_adder is -- 21
                                                          -- 22
    component half_adder                                  -- 23
        port(A, B: in std_logic; Sum, Cout: out std_logic); -- 24
    end component;                                        -- 25
                                                          -- 26
    signal half_sum, half_carry, tc: std_logic;          -- 27
                                                          -- 28
    begin                                                 -- 29
       HA1: half_adder port map(A, B, half_sum, half_carry); -- 30
       HA2: half_adder port map(half_sum, Cin, Sum, tc); -- 31
       Cout <= tc or half_carry;                          -- 32
end structural_functional_adder;                          -- 33
                                                          -- 34
library ieee;                                             -- 35
use ieee.std_logic_1164.all;                              -- 36
entity adder_4bit is                                      -- 37
    port(A, B: in std_logic_vector(3 downto 0);          -- 38
         Cin: in std_logic;                               -- 39
         Sum: out std_logic_vector(3 downto 0);          -- 40
         Cout: out std_logic);                            -- 41
end adder_4bit;                                           -- 42
                                                          -- 43
architecture structural_4bit_adder of adder_4bit is       -- 44
                                                          -- 45
    component full_adder                                  -- 46
      port(A, B, Cin: in std_logic;                       -- 47
           Sum, Cout: out std_logic);                     -- 48
    end component;
```

FIGURE 5.7
A hierarchical VHDL description of a 4-bit ripple-carry adder (part 1).

5.7 Logic Circuit Synthesis with Xilinx WebPACK ISE Project Navigator

In this section, we demonstrate step-by-step how to use the Xilinx WebPACK ISE Project Navigator to synthesize a logic circuit based on a VHDL program. The Project Navigator is a design platform where we manage design projects, sharing design modules and libraries,

```
      signal C: std_logic_vector(4 downto 0);                  -- 49
                                                                -- 50
                                                                -- 51
      begin                                                     -- 52
         BIT0: full_adder                                       -- 53
            port map(A(0), B(0), C(0), Sum(0), C(1));           -- 54
         BIT1: full_adder                                       -- 55
            port map(A(1), B(1), C(1), Sum(1), C(2));           -- 56
         BIT2: full_adder                                       -- 57
            port map(A(2), B(2), C(2), Sum(2), C(3));           -- 58
         BIT3: full_adder                                       -- 59
            port map(A(3), B(3), C(3), Sum(3), C(4));           -- 60
         C(0) <= Cin;                                           -- 61
         Cout <= C(4);                                          -- 62
      end structural_4bit_adder;                                -- 63
      -- End of the VHDL description                            -- 64
```

FIGURE 5.8
A hierarchical VHDL description of a 4-bit ripple-carry adder (part 2).

and maintain design data files. The first step is to activate the Xilinx ISE design software by double-clicking the Project Navigator icon

on the desktop of a computer. This will bring up a project window as shown in Figure 5.9. This window has four panes:

1. The **Sources in Project** (Source, in short) pane shows the organization of the source files that makes up a design.
2. The **Processes for Current Source** (Process, in short) pane lists various operations that can be applied to a given object in the Source pane.
3. The Log pane displays various messages from a current process.
4. The larger pane on the right is called the **Editor** pane, where VHDL code is typed in.

We start a design project by creating a new project directory by selecting File → New Project in the pull-down menu, as shown in Figure 5.10. This will pop up a New Project window (Figure 5.11) where the location of our project files, project name, the target device for the design, and the tools used to synthesize logic from source files are entered.

To set the family of FPGA devices for this design in the New Project window, click in the Value field corresponding to the Device Family in the Project Device Option section and under Property Name. Select entry Spartan2 (or other suitable entry) from the pull-down menu that appears, as shown in Figure 5.12.

Then click in the Value field for Device and select the xc2s100 (or other suitable) entry. Device xc2s100 is used in the popular XSA-100 experimental board on which small designs can be synthesized and tested. Similarly, select tq144 as the entry for Package. Finally, select

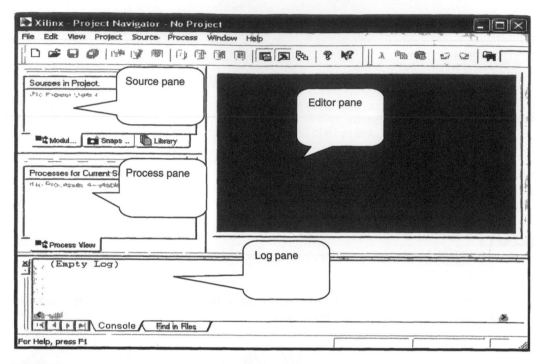

FIGURE 5.9
Xilinx WebPACK ISE Project Navigator window.

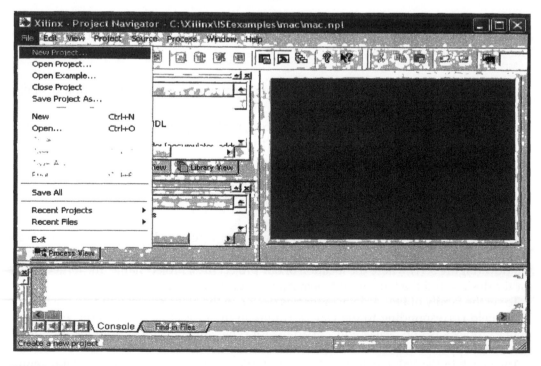

FIGURE 5.10
To bring up a New Project window.

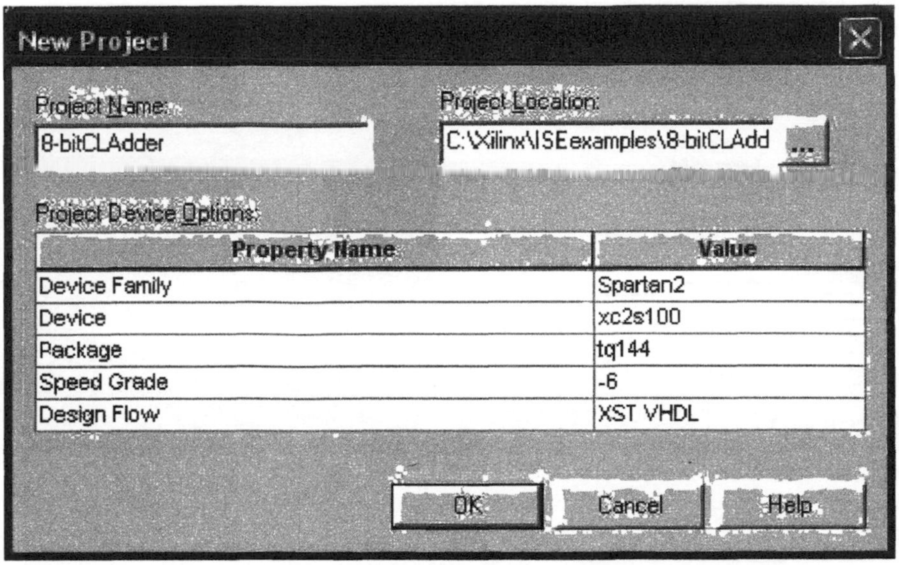

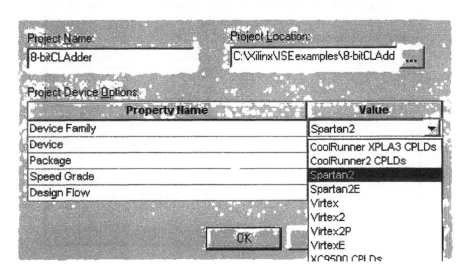

FIGURE 5.11
The New Project window.

FIGURE 5.12
Selecting FPGA device.

XST VHDL for Design Flow. This enables the Xilinx VHDL synthesizer. Once the fields are set, click on OK in the New Project window to close it.

The next step is to import a VHDL program to the new project. This is accomplished by selecting Project → New Source on the menu toolbar's pull-down menu, as shown in Figure 5.13. A New Source window pops up, in which the VHDL Module is selected for this exercise, the file name is typed in and its location is specified. Then, click on Next and the **Define VHDL Source** window appears, in which the property of inputs and outputs are declared. Click on Next in the window and the definition information is summarized in a pop-up window. Click Finish on the summary window to complete this step. These two windows are illustrated in Figure 5.14.

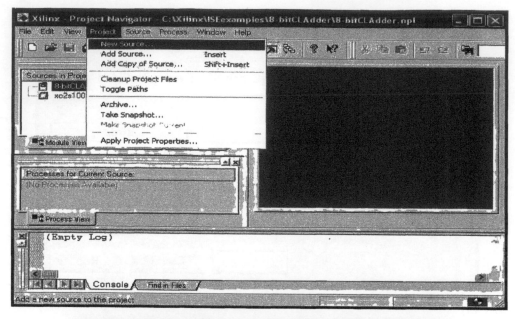

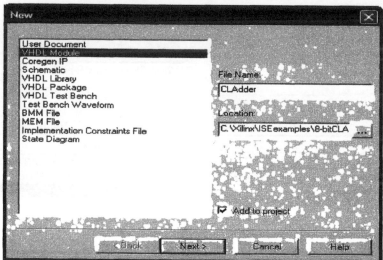

FIGURE 5.13
Importing a VHDL program.

Now, the Editor pane in the Project Navigator window displays a VHDL program frame, where various VHDL code components can be either edited directly in the pane or with cut-and-paste. It is advised to click on the Save button 📳 often to prevent accidental loss of design data, which is shown in Figure 5.15.

It is also recommended to use the Check Syntax (under the Synthesis in the Processes pane) process to verify the correctness of a VHDL program. This syntax check can be performed at any time (by double-click on Check Syntax) while editing and composing a VHDL program—a practice known to be time saving. Messages from a Check Syntax process are displayed in the Log pane, and usually will pinpoint syntax or other errors in the VHDL code. It is important to note that it is a good practice to save the VHDL program before running Check Syntax process. Otherwise, the program being checked

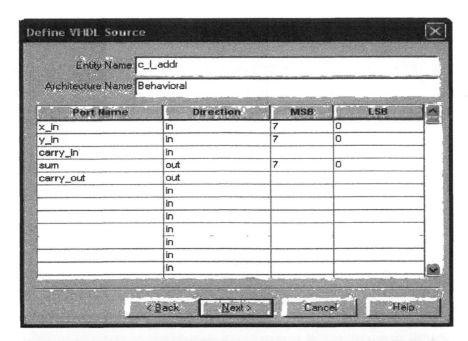

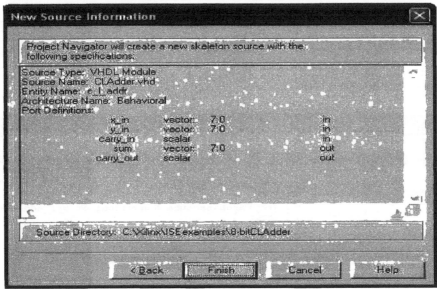

FIGURE 5.14
Pop-up windows for define VHDL source and summary information.

may be different from the one shown in the Editor pane, causing unnecessary confusion. A check mark ✔ appears next to Check Syntax if there is no syntax error. A warning mark ⚠ sometimes appears to indicate noncritical but nevertheless important issues that may become potential problems. Warning messages must be analyzed to understand and determine the causes and implications.

When a VHDL program is composed, the next step is to synthesize the logic circuit. To start the synthesis process, double-click on Synthesis in the Processes pane, as shown in Figure 5.16, with process messages displayed in the Log pane. Often, a VHDL program is

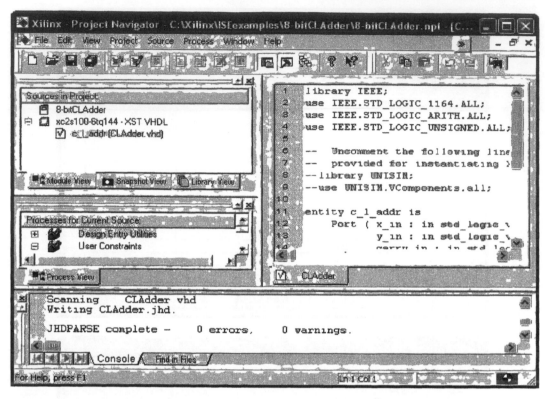

FIGURE 5.15
Editing VHDL program and save often.

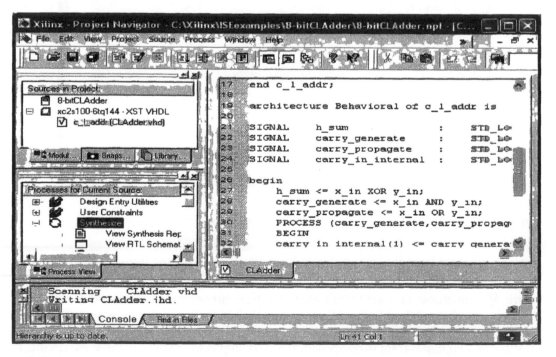

FIGURE 5.16
Synthesize logic circuit.

```
Release 5.1i - xst F.23
Copyright (c) 1995-2002 Xilinx, Inc.  All rights reserved.
--> Parameter TMPDIR set to __projnav
CPU : 0.00 / 2.80 s | Elapsed : 0.00 / 2.00 s

--> Parameter xsthdpdir set to ./xst
CPU : 0.00 / 2.80 s | Elapsed : 0.00 / 2.00 s

--> Reading design: c_1_addr.prj

TABLE OF CONTENTS
    1) Synthesis Options Summary
    2) HDL Compilation
    3) HDL Analysis
    4) HDL Synthesis
       4.1) HDL Synthesis Report
    5) Low Level Synthesis
    6) Final Report
       6.1) Device utilization summary
       6.2) TIMING REPORT


=========================================================================
*                       Synthesis Options Summary                       *
=========================================================================
---- Source Parameters
Input File Name                   : c_1_addr.prj
Input Format                      : VHDL
Ignore Synthesis Constraint File  : NO

---- Target Parameters
```

FIGURE 5.17
A typical synthesis process message in the Log pane.

modified to correct issues being raised during synthesis. In these cases, process messages often provide useful clues as to what the problems are. Double-clicking on View Synthesis Report brings up a message window, as shown in Figure 5.17, which provides valuable information and statistics.

Finally, double-clicking on View RTL Schematic under Synthesis in the Processes pane brings up a hierarchical schematic description of a synthesized circuit, as shown in Figure 5.18. The schematic view describes a graphical connection between the various components. This schematic view is hierarchical because components are displayed as "blocks" at various levels. To view detailed circuits, first select a component by single-clicking the block and then clicking on the down-arrow (as shown) in the tool bar on the Schematic View window. Right click on the top-level schematic symbol of a block to view various options for the schematic viewer. Select Push Into the Selected Instance in the pull-down menu (Figure 5.19) to see a detailed schematic as shown in Figure 5.20.

To display a detailed schematic diagram of a module, select the part in the RTL Design Hierarchy pane and then click 🔲 on the tool bar, as illustrated in Figure 5.21.

5.8 Simulation of Timing Characteristics

Simulation is commonly used to verify logic circuit designs. It begins with preparing a testbench by selecting Project → New Source and then, in the pop-up New Source window, selecting VHDL Test Bench and entering a file name for the testbench and its location,

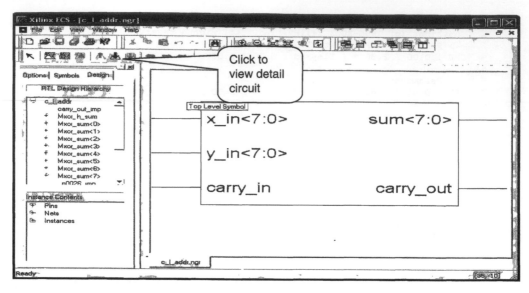

FIGURE 5.18
Schematic view of logic circuit.

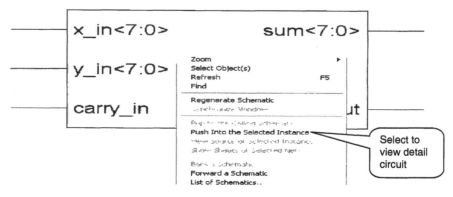

FIGURE 5.19
Pull-down menu for schematic display options.

as shown in Figure 5.22. Then click on Next. In the next window, select lab and click on Next and then click on Finish. Then add test codes in the blank space of the testbench file and click 🖫 to save the testbench.

To run simulation, first select the testbench file in the Source pane. Then double-click Simulate Behavioral VHDL in the Process pane, as shown in Figure 5.23. The ModelSim Simulator will run automatically. If there is no error in the testbench file, several pop-up windows will appear to show the simulation results, as shown in Figures 5.24 and 5.25.

5.9 Logic Circuit Implementation with FPGA Device

There are several steps involved in implementing a logic circuit onto an FPGA. These steps involve translating a design, mapping circuit blocks with programmable resources

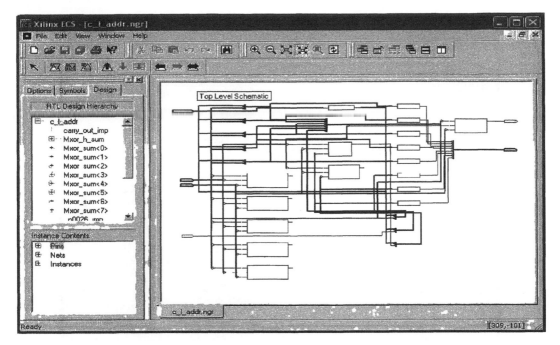

FIGURE 5.20
Display of detailed circuit schematic diagram.

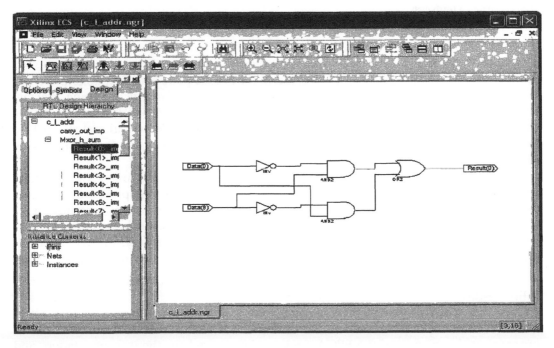

FIGURE 5.21
View schematic of selected design module.

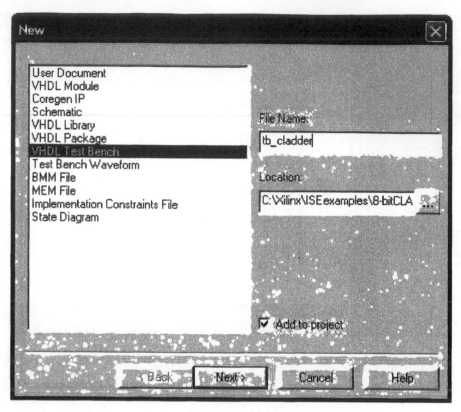

FIGURE 5.22
Start preparing a testbench.

on an FPGA device, placing individual components, and routing interconnects. To start this process, select the xc2s100-6tq 144 object in the Source pane and then double-click on Implement Design in the Process pane, which runs the floor-planning, mapping, placing, and routing in an automated sequence. A successful implementation is indicated by the ✔ mark next to Implement Design. To view the individual implementation steps, click on the + sign next to Implement Design to view the list of subprocesses. Clicking on the + sign of a subprocess expands to the individual steps.

The Translate process converts the netlist output by the synthesizer into a Xilinx-specific format and annotates it with any design constraints specified. The Map process decomposes the netlist and rearranges it so that it fits nicely into the circuitry elements contained in the selected FPGA device (Figure 5.26). The Place & Route process assigns the mapped elements to specific locations in the FPGA and sets the switches to route signals in between. If the Implement Design process failed, a ✖ mark would appear next to the subprocesses where errors occurred. A ⚠ mark indicates successful completion with warnings in subprocesses.

Once the circuit implementation is completed, it is a good practice to observe a few specifics about the implementation. For example, it is useful to know how much of the programmable resource of the FPGA device is being used for the implementation, and which and how many of the available pins are assigned as inputs and outputs, and so on. To start viewing the implementation reports, double-click on Place & Route Report and Pad Report in the Process pane, as shown in Figure 5.27.

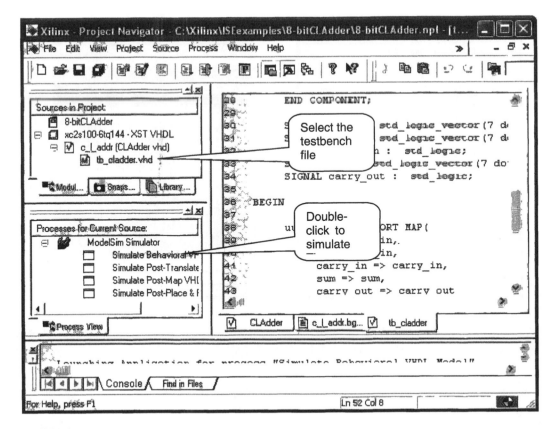

FIGURE 5.23
To run simulation with a selected testbench.

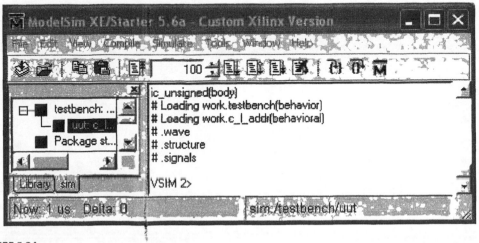

FIGURE 5.24
Window showing simulation results.

A graphic depiction of the FPGAs complex logic blocks (CLBs) and pins usages is also available by highlighting the project design in the Source pane and then double-clicking View/Edit Placed Design (FloorPlanner) in the Process pane, as shown in Figure 5.28. This brings up a FloorPlanner window with three panes (Figure 5.29): the Design Hierarchy

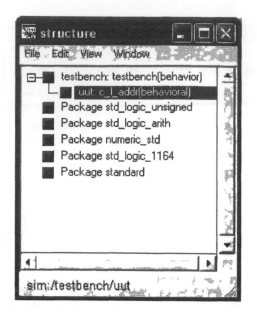

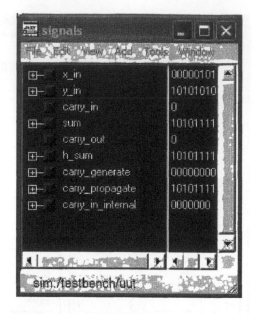

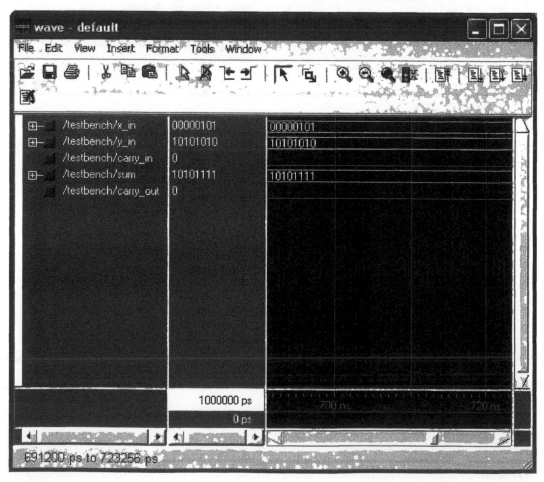

FIGURE 5.25
Pop-up windows showing simulation results.

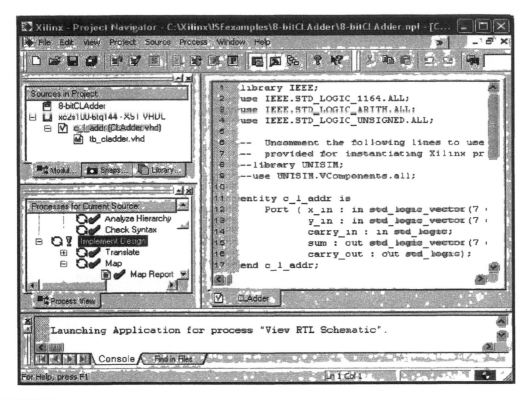

FIGURE 5.26
Implementing logic circuit with FPGA.

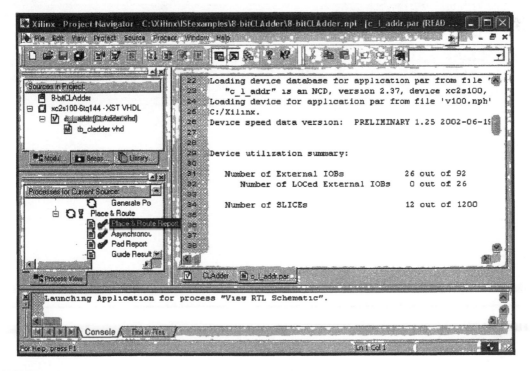

FIGURE 5.27
View implementation statistic reports.

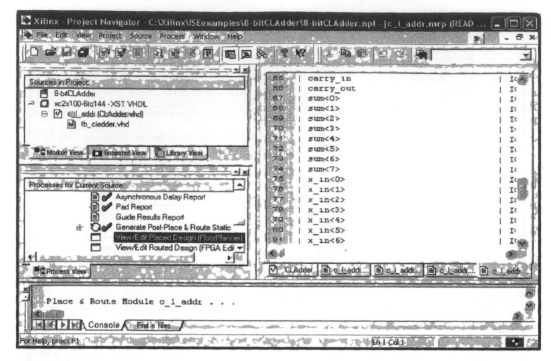

FIGURE 5.28
View a graphic depiction of implementation.

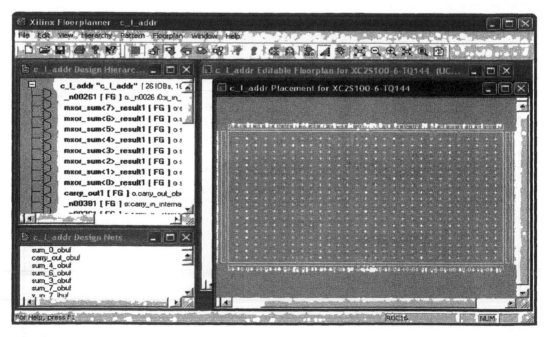

FIGURE 5.29
FloorPlanner window.

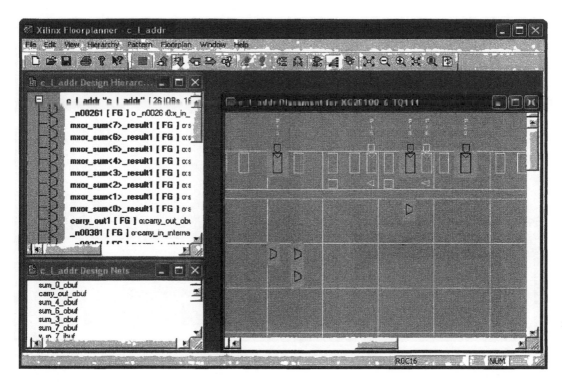

FIGURE 5.30
Enlarged view of used CLBs on the FPGA device.

pane lists the design's I/Os and LUTs assigned to various CLBs; the Design Nets pane lists the various signals; and the Placement pane shows a 30×20 array of slices in the FPGA device, with I/O pins around the periphery.

The CLBs used by the circuit of the project design are highlighted in light green and sometimes clustered near the edge of the CLB array. To enlarge this region of array, click on the 🔍 button and select around the highlighted CLBs in the Placement pane. The enlarged view of the CLBs used by the circuit will appear as shown in Figure 5.30. Clicking on a CLB shows the inputs and outputs of the CLB, as shown in Figure 5.31. Clicking on an input pin shows the CLBs using the input signal, as illustrated in Figure 5.32.

By default, the Placement pane only shows FPGA resources that are used by the project design. To see all the logic resources in each CLB, click on Edit → Preferences . . . and check all the boxes, as shown in Figure 5.33.

Now the Placement pane shows all LUTs, flip-flops, carry-chains and tristate buffers included in each CLB, as shown in Figure 5.34. Viewing the placement of the circuit elements after the Place & Route process can give insights into the resource usage of certain VHDL language constructs. In addition to viewing the placement of the design, the FloorPlanner can be used to rearrange and optimize the placement.

Once the logic circuit is synthesized and mapped onto the FPGA device with correct pin assignments, it is ready to generate the bitstream that is used to program an actual FPGA chip. To start the programmer, highlight the project design object in the Source pane and double-click on Generate Programming File in the Process pane, as shown in Figure 5.35. Within a few seconds, a 🖋 mark appears next to Generate Programming File, indicating that a file detailing the bitstream generation is created, with the name of the project design as the file name (Figure 5.36).

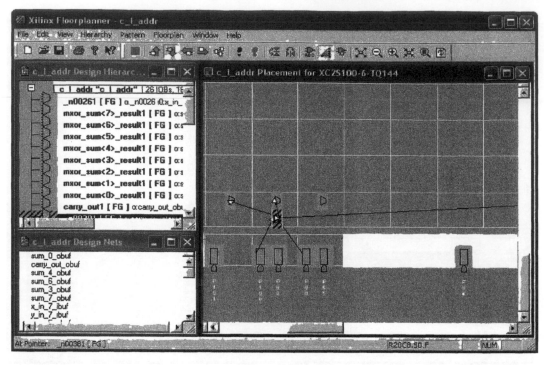

FIGURE 5.31
I/O view of a used CLB.

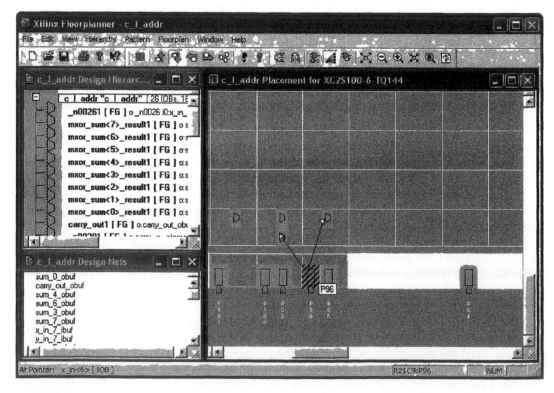

FIGURE 5.32
View of CLBs using a selected input.

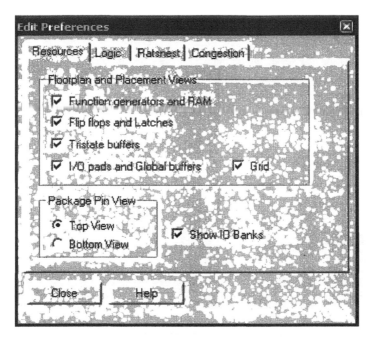

FIGURE 5.33
Editing placement view options.

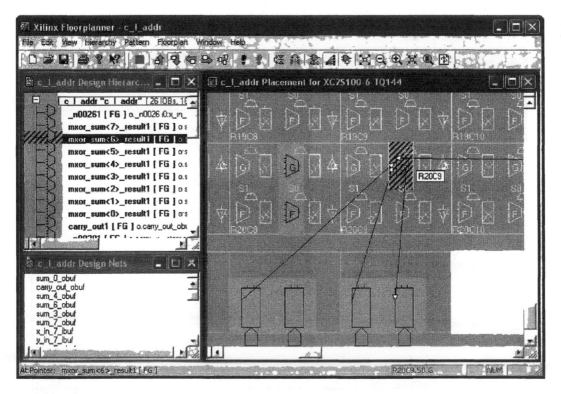

FIGURE 5.34
Full view of CLBs.

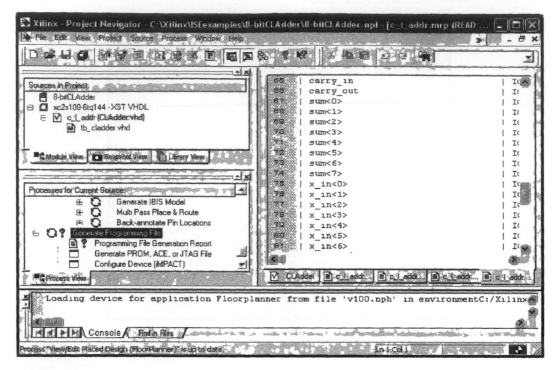

FIGURE 5.35
Generating programming file.

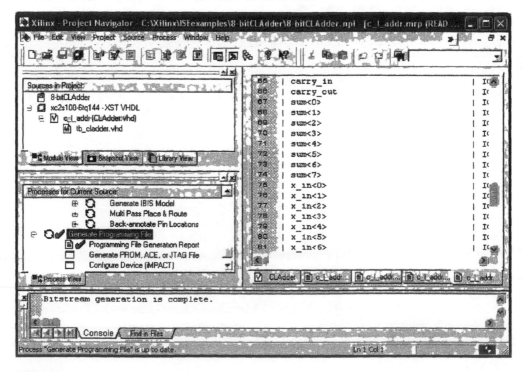

FIGURE 5.36
Completion of generating programming file.

Once the bitstream file is generated, the next step is to download it into the FPGA chip on the Xilinx XSA board (shown in Figure 5.37). The XSA board is powered with a 9 V DC power supply and is connected to a PCs parallel port with a standard 25-wire cable. The GXSLOAD software (supplied with the purchase of the board) is used to program the XSA board (Figure 5.38).

FIGURE 5.37
The Xilinx XSA-100 board.

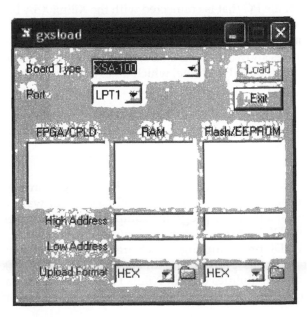

FIGURE 5.38
XSA-100 programming setup window.

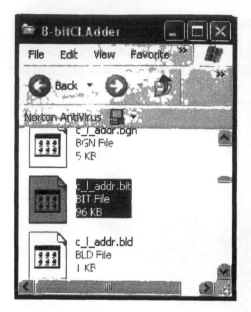 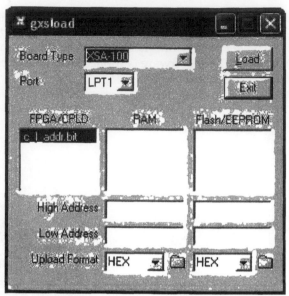

FIGURE 5.39
Select a programming file and start download.

To start the programming download action, double-click the

icon on the desktop of the PC that is connected with the Xilinx XSA board, and a programming setup window appears. Then, open a folder window that shows the bitstream file and simply drag the file (in this case the c_l_addr.bit file) into the GXSLOAD window's FPGA/CPLD section, as shown in Figure 5.39. Next, click on the Load button to start the programming of the XSA board with the, which takes a few seconds.

5.10 Summary

In this chapter, we have given an introduction to VHDL programming and the logic circuit synthesis process using the Xilinx WebPACK ISE Design software and the FPGA platform. The simple and small VHDL program examples are aimed at helping beginners. For advanced practice of VHDL programming, students are encouraged to explore further with in-depth books on VHDL programming, which are readily available in most bookstores.

The logic circuit synthesis process illustrated with Xilinx XSA-100 platform demonstrates one of the many capabilities that the Xilinx FPGA devices are capable of. The online Help provides many helpful directions that can assist students exploring other advanced applications.

Bibliography

Parnell, K. and Mehta, N., *Programmable Logic Design—Quick Start Hand Book*. Xilinx, Inc., 2003.

Thomas, D.T. and Moorby, P.R., *The Verilog® Hardware Description Language*, 5th edn. Kluwer Academic Publishers, 2002.

Yalamanchili, S., *Introductory VHDL from Simulation to Synthesis*. Prentice-Hall, New Jersey, 2001.

Problems

1. Design a 4-bit carry-select adder by first composing a structural VHDL program, compile and simulate the design, and then synthesize the logic circuit. The report should include: the VHDL program, the captured logic circuit schematic diagram, the simulation waveform chart, the synthesis process report, and the FPGA CLB usage map; all produced by using Xilinx WebPACK ISE design software.

2. Write a behavioral VHDL program for the above 4-bit carry-select adder, and work through the same synthesis process.

3. Design an 8-bit ALU by first composing a hierarchical VHDL program, compile and simulate the design, and then synthesize the logic circuit. The report should include: the VHDL program, the captured logic circuit schematic diagram, the simulation waveform chart, the synthesis process report, and the FPGA CLB usage map; all produced by using Xilinx WebPACK ISE design software.

4. Write a behavioral VHDL program for the above 8-bit ALU design, and work through the same synthesis process.

5. Design a traffic-control circuit that controls the traffic lights in a street intersection, by first composing a structural VHDL program, compile and simulate the design, and then synthesize the logic circuit. The report should include: the VHDL program, the captured logic circuit schematic diagram, the simulation waveform chart, the synthesis process report, and the FPGA CLB usage map; all produced by using Xilinx WebPACK ISE design software.

6. Write a behavioral VHDL program for the above traffic-control circuit and work through the same synthesis process.

6

Design of Modular Combinatorial Components

6.1 Introduction

A *combinatorial* logic circuit refers to a combination of various logic devices such that each of its outputs is a function of one or more of the inputs. In the absence of any feedback as shown in Figure 6.1, its outputs are functions of only the present values of the inputs. In reality, however, there always exists a finite delay between the application of the inputs and the corresponding appearances of the outputs. This finite delay accounts for the worst-case propagation delay encountered by the signals going through the various logic elements that are present along the circuit path. In particular, propagation delay associated with the most critical circuit path determines the duration the circuit would take to yield its outputs. In comparison, if the circuit output is also a function of the past values of the input variables, it is referred to as a *sequential* circuit. Such logic circuits are easy to identify as they include one or more feedback paths between the output and the input.

In previous chapters, we have studied various tools to design combinatorial logic circuits. The main design objective typically has been to construct a logic circuit utilizing a minimum number of logic gates and gate inputs from the behavioral specification of the circuit. The design process leading to the realization of a combinatorial logic circuit consists of the following design steps:

1. The given logic problem is analyzed and, if necessary, decomposed into a set of smaller but nontrivial subfunctions.
2. A truth table is developed describing the functional relationship between the inputs and the desired outputs.
3. The output functions are simplified using an appropriate minimization scheme.
4. The logical expressions for the circuit outputs are retooled, if necessary, to suit the needs of the available small-scale integrated (SSI), medium-scale integrated (MSI), or very large-scale integrated (VLSI) components.
5. The constructed logic circuits are assembled and tested to see if they yield the desired outputs for all possible combinations of the inputs.

As part of Step 4, the designer will often consider various practical limitations such as the number of logic gates, interconnections, gate inputs, fan-out requirements, and

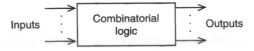

FIGURE 6.1
A combinatorial circuit module.

propagation delay. In the not too distant past the only option available to the designer was to assemble the entire logic circuit with SSI circuits. However, many of the more complex logic circuits are available now in the form of integrated circuits (ICs). The only limitation to the use of either the MSI or the large-scale integrated (LSI) is that we may not locate a logic device that exactly meets the specifications. It is thus typical for a designer to consider one of two choices. Either use available subfunction-level devices to construct the desired function-level device or one or more nearly-matching available standard devices and suitably combine them with other SSI or MSI chips so that the combination functions as desired.

In designing more involved digital systems, the to-be-designed high-level function is decomposed into lower-level subfunctions, each of which is nonabstract. The decomposition process may be repeated again until the design has been reduced to a set of functions or modules, each of which is well-defined and can be readily realized using available ICs. Each of the modules is then designed, implemented, and tested individually. Finally, these lower-level modules are interconnected to complete the high-level system. Often the designer is faced with the question of whether or not to use an already designed module or to develop a new one for one or more of the modules. Many important modules have already been developed and are often available as standard functions. These include devices such as decoders, encoders, adders, subtracters, and error-control logic. In this chapter, we introduce some of these more significant combinatorial components as well as their applications in combinatorial logic design. These components as well as custom application-specific ICs, make up the architecture of all processors. These are popularly used in interconnection, data transformation, and data conversion applications.

The combinatorial interconnect components typically includes multiplexers, demultiplexers, and buses that are used to connect the arithmetic and storage sections of a processor. Data transformation components perform logic operations, arithmetic operations, comparison operations (such as less than, equal to, and greater than), and bit manipulation operations (shift, rotation, etc.). Data conversion components, on the other hand, are used for conversion between different codes. We have already addressed the interconnect components in Chapter 4. We shall consider in this chapter mostly data transformation and data conversion components. Each of their designs will be studied since their understanding is vital in designing higher-level functions. They all play an important part in the development of more advanced digital systems. While experience and intuition are very important, there is no substitute for understanding exactly what these devices can do.

6.2 Special-Purpose Decoders and Encoders

In Chapter 4, we introduced the basic concepts pertaining to the decoders and demultiplexers. In case of the decoders, in particular, we could select any one of up to 2^n outputs by means n address inputs. An n-to-2^n line decoder may be treated simply as a minterm generator. We have also shown that a decoder can be reorganized for it to function like a demultiplexer, that is, as an inverse multiplexer. A typical demultiplexer receives input from a single line and directs it to only one of 2^n possible outputs. The particular destination output is determined by means of n selectors. Herein, we shall be considering again the decoders but now in the context of special-purpose code converters. In digital systems, one finds it necessary to convert one code to another. Often a decoder is an integral part of this conversion process. Typically, it is a specially organized combinatorial logic circuit that translates a code to a more useful or meaningful form. In this section we shall look at a few of these decoding functions.

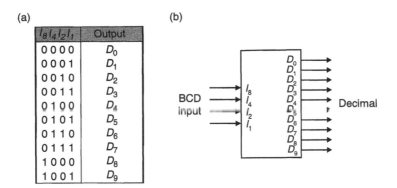

FIGURE 6.2
BCD-to-decimal decoder: (a) truth table and (b) block diagram.

One of the more frequently used decoders is a BCD-to-decimal decoder. Examples of other such decoders that may be used to convert data coded in one scheme into another format include XS3 to decimal, binary to XS3, and so on. The truth table listing the various combinations of the inputs and their corresponding outputs is shown in Figure 6.2a. Figure 6.2b shows a functional block diagram of this decoder. This decoder is similar to a 4-to-16 line decoder but with only ten outputs. To design such a decoder, we can construct a K-map for each of these ten functions and then obtain minimum output functions. Each of these K-maps includes a single minterm and six don't cares. The don't cares 10 through 15 refer to those minterms that do not correspond to valid binary coded decimal (BCD) codes. The set of ten logic equations are given by

$$D_0 = \overline{I_8}\,\overline{I_4}\,\overline{I_2}\,\overline{I_1} \quad D_1 = \overline{I_8}\,\overline{I_4}\,\overline{I_2}I_1$$
$$D_2 = \overline{I_4}I_2\overline{I_1} \quad D_3 = \overline{I_4}I_2I_1$$
$$D_4 = I_4\overline{I_2}\,\overline{I_1} \quad D_5 = I_4\overline{I_2}I_1$$
$$D_6 = I_4I_2\overline{I_1} \quad D_7 = I_4I_2I_1$$
$$D_8 = I_8\overline{I_1} \quad D_9 = I_8I_1$$

Fortunately, since the binary and corresponding BCD codes are identical to digit values 0 through 9, we can realize a BCD-to-decimal decoder also using a 4-to-16 line decoder. In that case, we need to simply ignore its last six outputs D_{10} through D_{15}.

A common decoding application often is the conversion of encoded data to a format particularly suitable for driving a numeric display. Typically such display device is a 7-segment light-emitting diode (LED) display such as the one already introduced in Example 4.17. Predefined combinations of the LED segments can be used to display numeric digits as well as other symbols. Figure 6.3 shows two such LED arrangements. An LED is turned on to emit light when the voltage at its anode is driven higher than the voltage at its cathode. However, to minimize the number of control signals, either the anodes are connected to a common anode as shown in Figure 6.3a, or all cathodes are connected to a common cathode as shown in Figure 6.3b. As an example, LED segments a, f, g, c, and d have to be turned on to display the digit 5. Typically, there are two choices for representing a 1: either f and e or b and c.

Consider the design of a BCD-to-7-segment decoder. This particular decoder is expected to accept BCD numbers as inputs and provide proper outputs to drive a 7-segment LED device. The corresponding logic circuit is expected to decode BCD bits into readable digits. Typically, in instruments such as a pocket calculator, for example, we encounter three

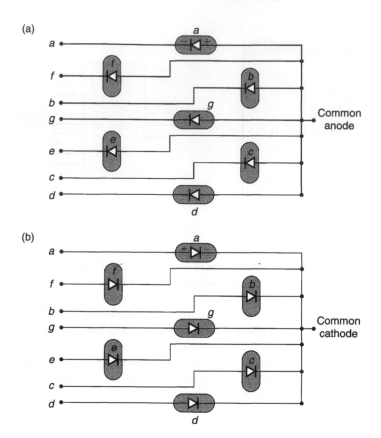

FIGURE 6.3
7-Segment LED configurations: (a) common anode and (b) common cathode.

(a)

DCBA	a	b	c	d	e	f	g
0000	1	1	1	1	1	1	0
0001	0	1	1	0	0	0	0
0010	1	1	0	1	1	0	1
0011	1	1	1	1	0	0	1
0100	0	1	1	0	0	1	1
0101	1	0	1	1	0	1	1
0110	0	0	1	1	1	1	1
0111	1	1	1	0	0	0	0
1000	1	1	1	1	1	1	1
1001	1	1	1	0	0	1	1

(b)

FIGURE 6.4
A 7-segment LED decoder: (a) truth table and (b) block diagram.

different types of data conversion. The decimal inputs are first mapped to their equivalent binary form by means of BCD-to-binary converter modules. The binary data is then manipulated and processed and the resultant binary output is converted back to its equivalent BCD form and usually displayed finally by means of 7-segment LED devices. The truth table for transforming BCD codes to their equivalent LED 7-segments is obtained as shown in Figure 6.4.

Boolean expressions corresponding to each LED segment may now be determined from the truth table. Since there are fewer numbers of 0s in the output column, we may prefer to derive NOR–NOR expressions from the table as follows:

$$a(D,C,B,A) = \overline{(D+C+B+\bar{A}) + (D+\bar{C}+B+A) + (D+\bar{C}+\bar{B}+A)}$$

$$b(D,C,B,A) = \overline{(D+\bar{C}+B+\bar{A}) + (D+\bar{C}+\bar{B}+A)}$$

$$c(D,C,B,A) = \overline{D+C+\bar{B}+A}$$

$$d(D,C,B,A) = \overline{(D+C+B+\bar{A}) + (D+\bar{C}+B+A) + (D+\bar{C}+\bar{B}+\bar{A}) + (\bar{D}+C+B+\bar{A})}$$

$$e(D,C,B,A) = \overline{(D+C+B+\bar{A}) + (D+C+\bar{B}+\bar{A}) + (D+\bar{C}+B+A) + (D+\bar{C}+B+\bar{A})}$$
$$\overline{+(D+\bar{C}+\bar{B}+\bar{A}) + (\bar{D}+C+B+\bar{A})}$$

$$f(D,C,B,A) = \overline{(D+C+B+\bar{A}) + (D+C+\bar{B}+A) + (D+C+\bar{B}+\bar{A}) + (D+\bar{C}+\bar{B}+\bar{A})}$$

$$g(D,C,B,A) = \overline{(D+C+B+A) + (D+C+B+\bar{A}) + (D+\bar{C}+\bar{B}+\bar{A})}$$

Alternatively, we can construct K-maps for these seven functions and accordingly minimize the logical functions for circuit implementation.

Example 6.1
Using 7-segment LED display devices, design a 4-digit display module for displaying up to four BCD digits. The module must not display any leading string of 0s.

Solution
For designing this display module, we need four BCD-to-7-segment LED decoders shown in Figure 6.4b. However, in order to make sure that leading 0s are not displayed, we may have to expand the scope of the BCD-to-7-segment decoder and accordingly modify it. In short, the circuit of Figure 6.4b will need added feature to suppress the display of leading 0s. Detection of whether or not there are leading 0s can be approached as follows: We first determine if the most significant digit is a 0. If it is, then the next significant digit is tested, and so on. As soon as the least significant 0 of the string of the leading 0s has been detected, the logic circuit may not need to search the other digit positions. Accordingly, one needs to redesign the module of Figure 6.4b for the *i*-th digit so that we could test if it is a zero or not. Figure 6.5 shows the block diagram of this to-be-designed module

A high *SI* input causes the particular decoder module as shown in Figure 6.5 to be searched for the presence of a 0. A high *SO* output, on the other hand, implies that the next decoder module still needs to be searched for the presence of a 0. This feature may be accomplished by adding a modifier logic circuit such as that shown in Figure 6.6 to the already designed logic circuit of Figure 6.4b. Its output is a 1 only when $DCBA = 0000$.

When the BCD digit is a 0 and the *SI* input is high, the output of circuit shown in Figure 6.6 must be able to disable that particular decoder circuit and suppress the corresponding 7-segment display device. The *SO* output resulting from the logic circuit causes the next digit to be searched for further search for 0s, if there are any. However, if the BCD digit is not a 0, the particular 7-segment display device is not disabled and further search is discontinued. The resulting BCD-to-7-segment decoder equations are obtained so as to handle not only those D, C, B, A values for which it is a 0 but also SO.

$$a(D, C, B, A) \Rightarrow 1, 4, 6, SO$$
$$b(D, C, B, A) \Rightarrow 5, 6, SO$$
$$c(D, C, B, A) \Rightarrow 2, SO$$
$$d(D, C, B, A) \Rightarrow 1, 4, 7, 9, SO$$
$$e(D, C, B, A) \Rightarrow 1, 3, 4, 5, 7, 9, SO$$
$$f(D, C, B, A) \Rightarrow 1, 2, 3, 7, SO$$
$$g(D, C, B, A) \Rightarrow 0, 1, 7, SO$$

The overall 4-digit display circuit is organized as shown by the logic circuit of Figure 6.7.

An *encoder* is a combinatorial logic module that accepts a digit on its inputs and converts it to a unique coded output. In fact, an encoder performs a function that is the inverse of a decoder. Typically, if an encoder module has n inputs, the number of unique outputs is equal to or greater than $\log_2 n$. Consider, for example, the design of a decimal-to-BCD encoder such as the one we typically encounter on the input side of a hand-held calculator. Such an encoder may have ten inputs, one for each decimal digit, and four binary outputs to represent the corresponding BCD numbers. The logical expressions for the encoder outputs

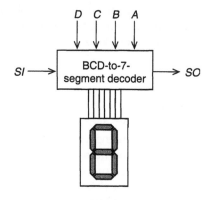

FIGURE 6.5
Block diagram for Example 6.1.

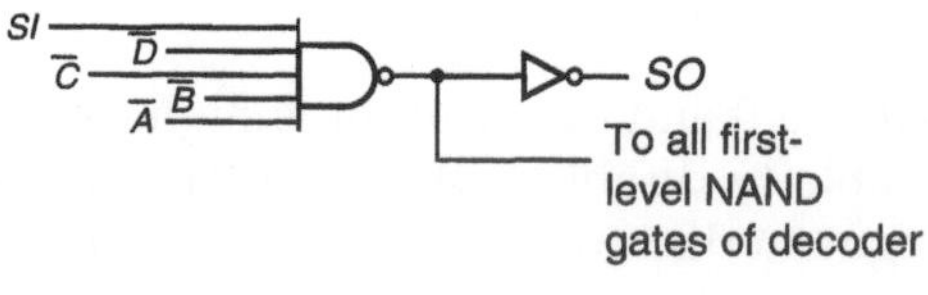

FIGURE 6.6
A modifier logic circuit for Example 6.1.

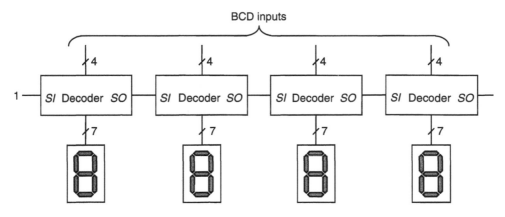

FIGURE 6.7
4-digit display circuit for Example 6.1.

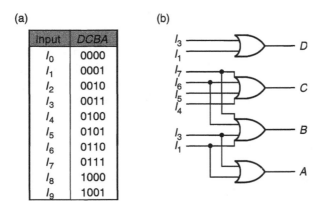

FIGURE 6.8
A decimal-to-BCD encoder: (a) truth table and (b) logic circuit.

may be readily obtained from the truth table shown in Figure 6.8a as

$$D = I_8 + I_9$$
$$C = I_4 + I_5 + I_6 + I_7$$
$$B = I_2 + I_3 + I_6 + I_7$$
$$A = I_1 + I_3 + I_5 + I_7 + I_9$$

The resulting OR-based logic circuit is shown in Figure 6.8b. The input I_0 is just not required in this logic circuit.

In the next example, we shall consider a particular type of encoder referred to as a *priority encoder*. Oftentimes in many practical situations, we want to identify an event and then assign and transmit a code based on some priority. Such events are typically assigned a priority on the basis of their relative function or role in the overall system. Typically, the system takes on a specific set of tasks or actions in response to such prioritization.

(a)　　　　　　　　　　　　(b)　　　　　　　　　　　　(c)

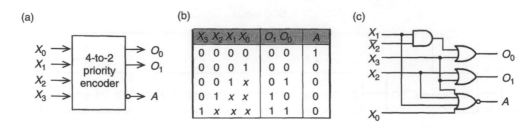

X_3	X_2	X_1	X_0	O_1	O_0	A
0	0	0	0	0	0	1
0	0	0	1	0	0	0
0	0	1	x	0	1	0
0	1	x	x	1	0	0
1	x	x	x	1	1	0

FIGURE 6.9
Priority encoder for Example 6.2.

Example 6.2

Design a 4-input priority encoder wherein the input D_i, for example, has priority over the input Dj if $i > j$. The encoder produces a binary output code corresponding to the input having the highest priority.

Solution

To design this logic circuit, the highest priority is assigned to the input that has the numerical value as its subscript. The next highest priority is assigned to the one that has the next highest numerical value as its subscript, and so on. The outputs O_1 and O_0 correspond to the particular input having the highest priority. An additional output A is turned on to indicate that one or more inputs are active. The block diagram for such a device is shown in Figure 6.9a. The corresponding truth table is obtained as shown in Figure 6.9b where we have used "x" to denote the presence of a don't care. With $X_3 X_2 X_1 X_0 = 01\text{-}\text{-}$, for example, the output is 10 since X_2 is a 1 and X_3 is a 0. A K-map may be drawn up now with these truth table entries to obtain the logical equations for the outputs as follows:

$$O_1 = X_2 + X_3$$

$$O_0 = X_3 + X_1 \overline{X_2}\,\overline{X_3} = X_3 + X_1 \overline{X_2}$$

$$A = \overline{X_3 + X_2 + X_1 + X_0}$$

We see that the two output functions O_1 and O_0 are both independent of X_0. The 4-input priority encoder circuit is obtained as shown in Figure 6.9c.

A priority encoder output is very much like the complement of a decoder output since the priority encoder when connected to the outputs of a decoder would produce an identity function. For example, if the logic circuit of Figure 6.9 were to be connected to the output of a 2-to-4 line decoder, then the output of the priority encoder will be the same as the input to the decoder. However, if a 2-to-4 line decoder is connected to the outputs of the logic circuit shown in Figure 6.9, the resulting logic circuit will not produce an identity function. This is because the priority encoders are designed by disregarding all the 1s present at less significant bit positions.

As it is true in the case of the decoders, smaller-order priority encoders can be suitably cascaded to form larger-order priority encoders. If we were to use 2-to-1 line priority encoders, for example, for designing an n-input priority encoder ($n > 2$), the inputs are all grouped into pairs and then each pair is introduced as inputs to a 2-to-1 line priority encoder. This circuit level includes a total of $n/2$ priority encoders. This circuit level includes a total of $n/2$ priority encoders. One of these $n/2$ encoder outputs is expected to be selected by an 1-of-$(n/2)$ multiplexer to be the least significant address bit O_0. If n is a power of 2, this section can be constructed using $n/4$ of 1-of-2 multiplexers in the second level, $n/8$ of the same in third level, and so on. In case of the design of an 8-to-3 line priority encoder, however, we would not have to go beyond the third level since by that time we have already selected one from $n/2$ encoders.

To determine the next more significant address bit O_1, we next encode $n/2$ outputs from the first level of the priority encoders that indicates that at least one of the inputs (of the first level of the priority encoders) is a 1. In case of the logic circuit of Figure 6.9, for example, the output A plays that role. For encoding the first level A outputs at the second level, they are first grouped in pairs and then processed using $n/4$ of 2-to-1 priority encoders. These in turn will produce $n/4$ A outputs for the third level wherefrom the next more significant address bit O_2 is determined. In the second level, a 1-of-$(n/4)$ multiplexer is used, for example, in the logic circuit of a 8-to-3 line priority encoder as shown in Figure 6.10a. This encoding process is repeated until when at a level only one A output is available. Each 1-of-2 multiplexers selects one of the outputs from a group of two priority encoders, using their A outputs as the selector input for the multiplexer in question. Figure 6.10b shows a variation of the 8-to-3

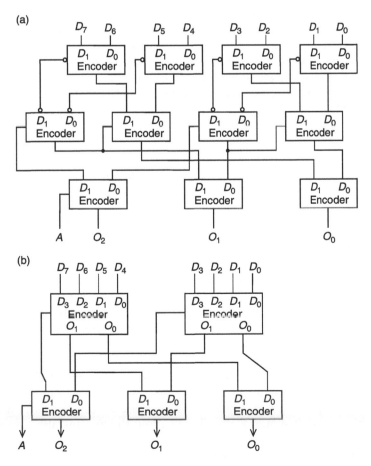

FIGURE 6.10
An 8-to-3 line priority encoder using (a) 2-to-1 line priority encoders and (b) 4-to-2 line priority encoders.

line priority encoder using 4-to-2 line priority encoders in the first level. In comparison, this latter logic circuit is a lot faster as it involves one less level of subfunction.

6.3   Code Converters

In Chapter 1, we have already encountered a number of different binary codes that may be used in digital systems. Consequently, it often becomes necessary to transfer coded data from one subsystem to another as well as convert one form of binary code to another. Converter circuit may thus play an important role in digital systems. Many of these converters use combinatorial logic while there are some that may use sequential logic. In this section, in particular, we consider only the ones that are based strictly on combinatorial logic.

The design of a converter circuit could follow the steps same as those used for any other combinatorial logic circuits. Fortunately, we may not have to always take the typical design route. There are many codes that have been defined so as to have distinct bit patterns in them. The pattern may not be obvious always, but this may become apparent when one scans the truth table entries of such functions. In case a bit pattern has been identified, the designer may not have to take the design process through K-map or Q-M tabular reduction steps for minimizing logical functions. In fact, for some of these functions, the traditional process may not always lend itself to finding better logical circuits. Nature of the bit patterns often dictates the degree of simplification. The corresponding process or steps that may not at all involve traditional scheme for determining the output functions varies from problem to problem. Such design process is thus referred to as heuristic. The following examples illustrate application of a variety of heuristic techniques for designing code converters. We shall consider a few of these conversions: Gray-to-binary, binary-to-Gray, binary-to-BCD, and BCD-to-binary.

Example 6.3
Design a logic circuit that converts a 5-bit Gray code into its binary equivalent.

Solution
By definition, the Gray code is a reflected binary code. The Gray code representation for any one number differs from that of the next number in only one bit position. Two consequent Gray codes are thus said to be one unit distance apart. The 5-bit Gray code and their corresponding binary numbers are listed in the truth table of Figure 6.11.

The obvious next step in the combinatorial design process is to construct five 5-variable K-maps for determining B_4, B_3, B_2, B_1, and B_0 in terms of G_4, G_3, G_2, G_1, and G_0. However, we may pause for a moment before taking that route.

Upon examination of the truth table, we find that G_4 and B_4 are identical. In addition, B_3 is a 1 when either G_4 or G_3 is a 1. Likewise, B_1 is a 1 when either any one or three of the set $\{G_4, G_3, G_2,$ and $G_1\}$ are 1; and, finally, B_0 is a 1 when either one or any three or all five of the set of five Gray code inputs are 1. These observed relationships can help eliminate the need for going through the K-map minimization scheme, although as a general rule, K-maps will always give a correct, if not minimum, solution. In this particular combinatorial design, it is possible to avoid the minimization step altogether since we have in our memory bank the characteristics of various logic elements that could easily map each of the aforementioned observations to a logic circuit. In each of these observations, we note that the output is expected to be high when odd number of inputs are high. We recall that an exclusive-OR function (XOR) logic gate yields a high output when there are odd number of inputs that are high. Accordingly, the role of XOR logic in the design of this converter circuit is rather obvious. The binary outputs of the converter module are readily obtained as

$$B_4 = G_4$$
$$B_3 = G_3 \oplus G_4$$
$$B_2 = G_2 \oplus G_3 \oplus G_4$$
$$B_1 = G_1 \oplus G_2 \oplus G_3 \oplus G_4$$
$$B_0 = G_0 \oplus G_1 \oplus G_2 \oplus G_3 \oplus G_4$$

We also note that each one of the more significant binary output bits can be used to realize its next less significant binary output since

$$B_n = G_n \oplus B_{n+1}$$

The resulting logic circuit thus consists of only four XOR logic gates as shown in Figure 6.12. The alternative to this would have involved using four different XOR logic gates with number of inputs ranging respectively from two to five.

$G_4G_3G_2G_1G_0$	$B_4B_3B_2B_1B_0$
0 0 0 0 0	0 0 0 0 0
0 0 0 0 1	0 0 0 0 1
0 0 0 1 1	0 0 0 1 0
0 0 0 1 0	0 0 0 1 1
0 0 1 1 0	0 0 1 0 0
0 0 1 1 1	0 0 1 0 1
0 0 1 0 1	0 0 1 1 0
0 0 1 0 0	0 0 1 1 1
0 1 1 0 0	0 1 0 0 0
0 1 1 0 1	0 1 0 0 1
0 1 1 1 1	0 1 0 1 0
0 1 1 1 0	0 1 0 1 1
0 1 0 1 0	0 1 1 0 0
0 1 0 1 1	0 1 1 0 1
0 1 0 0 1	0 1 1 1 0
0 1 0 0 0	0 1 1 1 1
1 1 0 0 0	1 0 0 0 0
1 1 0 0 1	1 0 0 0 1
1 1 0 1 1	1 0 0 1 0
1 1 0 1 0	1 0 0 1 1
1 1 1 1 0	1 0 1 0 0
1 1 1 1 1	1 0 1 0 1
1 1 1 0 1	1 0 1 1 0
1 1 1 0 0	1 0 1 1 1
1 0 1 0 0	1 1 0 0 0
1 0 1 0 1	1 1 0 0 1
1 0 1 1 1	1 1 0 1 0
1 0 1 1 0	1 1 0 1 1
1 0 0 1 0	1 1 1 0 0
1 0 0 1 1	1 1 1 0 1
1 0 0 0 1	1 1 1 1 0
1 0 0 0 0	1 1 1 1 1

FIGURE 6.11
Truth table for Example 6.3.

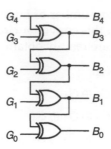

FIGURE 6.12
Logic circuit for Example 6.3.

The 5-bit Gray-to-binary code converter module can be used as the basis for designing an arbitrary n-bit Gray-to-binary code converter when $n > 5$. To design an n-bit converter, the number of 5-bit converter modules that will need to be cascaded is $I([\{n - 5\}/5] + 1)$ where $I(x)$ is the integer value of x.

Example 6.4

Using a 4-to-16 line decoder, show the design of a logic circuit that converts 4-bit binary input to its equivalent 4-bit Gray code.

Solution

For this problem, we can make use of the truth table entries shown in Figure 6.11. To suit our current need, we have to just treat its inputs as the new outputs and its outputs as the new inputs. We find that

$$G_3 = \sum m(8,9,10,11,12,13,14,15) = B_3$$

$$G_2 = \sum m(4,5,6,7,8,9,10,11)$$

$$G_1 = \sum m(2,3,4,5,10,11,12,13)$$

$$G_0 = \sum m(1,2,5,6,9,10,13,14)$$

A 4-to-16 line decoder can be employed to design the converter circuit. The inputs to the decoder follow directly from the equations listed above. The converter logic circuit is obtained as shown in Figure 6.13.

Example 6.5

Design an n-bit converter for determining Gray code equivalent of binary numbers.

Solution

The only initial difficulty with this problem is that the number of inputs is not limited. However, we have already seen in Example 6.3 that Gray codes and their corresponding binary equivalents are pattern driven except that in this current case one needs to determine Gray codes in terms of the binary inputs. By scanning the truth table entries of Example 6.11, however, one may solve for G_0 through G_4 in terms of B_0 through B_4.

We notice that G_0 is a 1 only when either B_0 or B_1 is a 1. Likewise, G_1 is a 1 when either B_1 or B_2 is a 1; G_2 is a 1 when either B_2 or B_3 is a 1, and so on; with the exception that the most significant Gray bit code $G_{n-1} = B_{n-1}$. In this case, we can use again the property of XOR logic. Accordingly, we obtain

$$G_{n-1} = B_{n-1}$$

$$\cdots \quad \cdots$$

$$G_3 = B_3 \oplus B_4$$

$$G_2 = B_2 \oplus B_3$$

$$G_1 = B_1 \oplus B_2$$

$$G_0 = B_0 \oplus B_1$$

The resulting n-bit binary-to-Gray converter logic circuit is obtained as shown in Figure 6.14. This module is different from that of the Gray-to-binary converter in that the logic circuit of Figure 6.14 is fast since none of its outputs is a function of any other outputs.

Before we go through the design of a binary-to-BCD conversion module in the next example, we consider studying the interrelationship between a binary number and its equivalent BCD number. Using Equation 1.4, the equivalent decimal value D_{10} of an n-bit

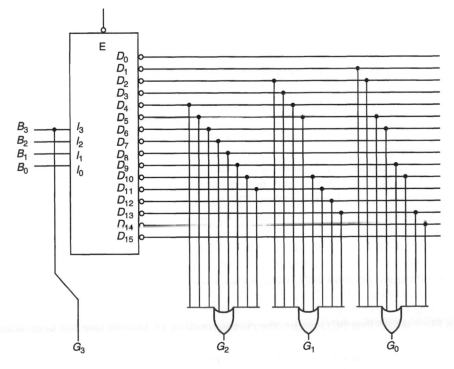

FIGURE 6.13
Logic circuit for Example 6.4.

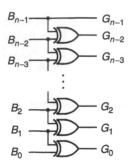

FIGURE 6.14
Gray-to-binary converter for Example 6.5.

BCD	Shifted left BCD	Corrected BCD
0000	0000 0000	0000 0000
0001	0000 0010	0000 0010
0010	0000 0100	0000 0100
0011	0000 0110	0000 0110
0100	0000 1000	0000 1000
0101	0000 1010	0001 0000
0110	0000 1100	0001 0010
0111	0000 1110	0001 0100
1000	0001 0000	0001 0110
1001	0001 0010	0001 1000

FIGURE 6.15
Shifting of the BCD numbers.

binary number is given by

$$D_{10} = a_{n-1}2^{n-1} + a_{n-2}2^{n-2} + a_{n-3}2^{n-3} + \cdots + a_3 2^3 + a_2 2^2 + a_1 2^1 + a_0 2^0$$
$$= ((\ldots (((a_{n-1})2 + a_{n-2})2 + a_{n-3})2 + \cdots + a_2)2 + a_1)2 + a_0$$

where each of the coefficients, a_j, is either a 0 or a 1. Accordingly, the equivalent decimal number can be obtained from the binary coefficients provided the binary fraction $(0.a_{n-1}a_{n-2}a_{n-3}\ldots a_2a_1a_0)_2$ is shifted to the left n times as follows: Typically, a binary shift-left operation is equivalent to multiplying the unshifted binary number by 2. As binary digits are shifted to the left bit by bit, the first group of four consecutive bits (beginning with the binary point to the left) may not necessarily be the same as a BCD number. If the shift were to cause a number larger than 1001 to appear, the number will cease to be a valid BCD code. However, with proper correction, the shifted bits can be made to represent a decimal digit.

To understand the scope of correction when one shifts binary digits to derive its BCD equivalent, consider the following explanation. A binary shift-left within the string of four bits (representing a decimal digit) doubles the value of the original number. However, a bit passing from the most significant of the four bits representing a decimal digit to the least significant of the next four bits increases its value from 8 to 10 (rather than from 8 to 16). The table in Figure 6.15 lists BCD digits, the corresponding BCD digit that has been shifted left one bit. It also lists the correct BCD version of the shifted bits. For example, 0111 (equivalent to 7 in BCD) when shifted to the left one bit yields 1110 (equivalent to 14 in binary). But in order to represent 14 in BCD, the corrected form should rather be 0001 0100 instead of 0000 1100.

In the truth table shown in Figure 6.15, we see that for BCD values less than or equal to 0100, a single left-shift always yields the correct BCD number. Considering the fact that six of the 4-bit binary numbers (i.e., 1010 through 1111) are undefined in BCD, the correct version of the shifted BCD number is obtained by adding 0110 to the shifted bits. Alternatively, however, 0011 can be added first to the given number (before shift) and then the resultant binary sum is shifted to the left one bit. Thus, a BCD number can be realized from its corresponding binary fraction by repeatedly performing n shift-left operations. In case the preshifted number is greater than 0100, we may first add 0011 to it and then shift the binary sum to the left one bit. The device yet to be designed in Example 6.7 can be used to add 0011 to the 4-bit binary input if the four bits represent a value greater than 0100.

To understand the function of a modular binary-to-BCD converter in converting binary digits of more than five bits, an example involving a 7-bit binary number will be considered next in Example 6.6. The steps employed in this example will lead us to an algorithm that can be used then for the conversion of any n-bit binary number. The design approach, in particular, that we shall be considering will correct each group of 4-bits of binary numbers before they are shifted to the left. Since combinatorial logic design does not involve any feedback or memory, a shift-left can be realized by hard-wiring multiple binary-to-BCD converter modules. For each shift operation, a separate set of converter modules would be necessary.

Example 6.6
Convert 1101011_2 to its BCD equivalent number.

Solution

Step 1. *Setup*: We begin with the corresponding binary fraction $(0.1101011)_2$ that will need to be shifted to the left seven bits. This follows from the understanding that $(0.1101011)_2 \times 2^7$ may be used to determine the desired BCD equivalent number.

Step 2. *Shift with no correction*: When $(0.1101011)_2$ is shifted left, we first get $(1.1110110)_2$ and then $(11.1101100)_2$. Both 1 and 11 on the left of the binary point are less than $(100)_2$. Therefore, no correction is required for the first two left-shift operations. Note that the shift-left operations just performed brought in bits from the right of the binary point to the left for them to be included in the BCD portion.

Step 3. *Third left-shift*: The third left-shift, in this case, will cause the number to be $(110.1011000)_2$ and we see for the first time that the number $(110)_2$ on the left of the binary point is greater than $(100)_2$. Accordingly, we shall have to correct this integer before we shift it again to the left. If we do not correct it before the shift, the shifted number becomes an invalid BCD number.

Step 4. *Correct and then shift*: For the fourth shift-left, the sequence of operations will involve the following transformations:

110.1011000×2^4	after the third shift
1001.1011000×2^4	with correction before the fourth shift
10011.0110000×2^3	after the fourth shift

The resulting number $(10011.0110000) \times 2^3$, which can be rewritten as $(0001\ 0011.0110) \times 2^3$ is a hybrid number where the integer is in BCD and the fraction is in binary. The decimal value of this hybrid number is $\{13 + (6/8)\} \times 2^3$.

Step 5. *Fifth Shift*: Following the fourth shift, the four bits (0011) on the immediate left of the binary point has turned out to be less than 0100. Thus the fifth shift requires no correction. The fifth shift results in (0010 0110.1100000) $\times 2^2$.

Step 6. *Sixth Shift*: Now, the four bits (i.e., 0110) on the immediate left of the binary point is greater than 0100. Accordingly, we correct the integer before the next shift-left operation.

The transformations are as follows:

0010 0110.1100000 $\times 2^2$ after the fifth shift
0010 1001.1100000 $\times 2^2$ with correction before the sixth shift
0101 0011.1000000 $\times 2^1$ after the sixth shift

The resultant hybrid number is $\{53 + (1/2)\} \times 2^1$.

Step 7. *Seventh Shift*: Since 0011 is less than 0100, no correction is needed in the lower significant BCD digit. However, the more significant BCD digit (i.e., 0101) needs correction. The resulting transformations are

0101 0011.1000000 $\times 2^1$ after the sixth shift
1000 0011.1000000 $\times 2^1$ with correction before the seventh shift
0001 0000 00111 $\times 2^0$ after the seventh shift

At the end of the seventh shift, the process is now complete since the resultant number is now devoid of any fractional part. The final result amounts to 107 in BCD.

It is interesting to note that the intermediate hybrid results such as $\{13 + (6/8)\} \times 2^3$ obtained at the end of Step 4 and $\{53 + (1/2) \times 2^1\}$ obtained at the end of Step 6 are all equal to 107_{10}.

Let us re-examine the bit manipulation pattern in Example 6.6. Conversion of an *n*-bit binary number to its BCD equivalent can be performed using multiple 4-bit converter modules. In designing this converter module, we need to be cognizant of the fact that bits may not have to change position provided the converter modules are moved to the right one bit at a time to effect shift-left operations. The ability to perform such a hardwired shift-left operation is important since we are planning now to use only combinatorial components (as opposed to using sequential circuits that are yet to be discussed) for this data conversion. Furthermore, test pertaining to whether or not the 4-bit number is greater than 0100 when the number is being corrected and shifted to the left, becomes nontrivial only when the third more significant digit is a 1. Provided that such a converter module is available (yet to be designed later in Example 6.7), the *n*-bit *binary-to-BCD conversion algorithm* can be summarized as follows:

1. Insert a 0 to the left of the most significant bit of the given binary number. This 0 and the three next significant bits of the binary number are fed as input to a 4-bit binary-to-BCD converter module.

2. Feed the three less significant processed bits (i.e., from the 4-bit output of the binary-to-BCD converter module) and the most significant unprocessed bit (i.e.,

$B_4 B_3 B_2 B_1 B_0$	$D_7 D_6 D_5 D_4 D_3 D_2 D_1 D_0$
0 0 0 0 0	0 0 0 0 0 0 0 0
0 0 0 0 1	0 0 0 0 0 0 0 1
0 0 0 1 0	0 0 0 0 0 0 1 0
0 0 0 1 1	0 0 0 0 0 0 1 1
0 0 1 0 0	0 0 0 0 0 1 0 0
0 0 1 0 1	0 0 0 0 0 1 0 1
0 0 1 1 0	0 0 0 0 0 1 1 0
0 0 1 1 1	0 0 0 0 0 1 1 1
0 1 0 0 0	0 0 0 0 1 0 0 0
0 1 0 0 1	0 0 0 0 1 0 0 1
0 1 0 1 0	0 0 0 1 0 0 0 0
0 1 0 1 1	0 0 0 1 0 0 0 1
0 1 1 0 0	0 0 0 1 0 0 1 0
0 1 1 0 1	0 0 0 1 0 0 1 1
0 1 1 1 0	0 0 0 1 0 1 0 0
0 1 1 1 1	0 0 0 1 0 1 0 1
1 0 0 0 0	0 0 0 1 0 1 1 0
1 0 0 0 1	0 0 0 1 0 1 1 1
1 0 0 1 0	0 0 0 1 1 0 0 0
1 0 0 1 1	0 0 0 1 1 0 0 1

FIGURE 6.16
Truth table for Example 6.7.

the bit from the next less significant position of the given binary number) together as inputs to another binary-to-BCD converter module. On the other hand, the most significant processed bit (MSPB), that is, the MSB of the 4-bit output, remains unprocessed during this step since at this time there is only a single such bit. We have seen in Example 6.6 that the more significant bits remain uncorrected for the first two shift-left operations. But then starting with the next shift-left (i.e., as soon as we have had three MSPBs), a correction is again warranted.

3. Repeat Step 2 by feeding three less significant bits from the output of the converter module and the next most significant unprocessed bit to a next-level binary-to-BCD converter. This process is repeated until all except the least significant of the given binary digits have been processed. The least significant bit need not be processed since it requires no correction.

4. Insert a 0, each time we obtain a string of three MSPBs, before the three MSPBs and feed these four bits to a next-level binary-to-BCD converter module. Accordingly, we process the MSPBs only once after we have had three shift-left operations. Once a converter module has been added to a level, the number of converter modules remains unchanged until three MSPBs have been accumulated at which time another converter module is added to the next level.

Example 6.7
Design a 4-bit combinatorial logic module that may be used in the conversion of an n-bit binary number to its BCD equivalent.

Solution
The truth table for the binary-to-BCD conversion is obtained as shown in Figure 6.16. However, we consider here 5-bit inputs instead of 4-bit inputs since the least significant bit (LSB) of the 5-bit input number need not be fed through this to-be-designed module. This is evident since we see that $B_0 = D_0$. On the other hand, binary inputs larger than 10011 need not be considered in this truth table since for binary inputs larger than 10011, D_5 becomes a 1.

In the table of Figure 6.16, we see that the first ten values of $B_4B_3B_2B_1B_0$ (i.e., 00000 through 01001) need no correction. However, for the rest of the $B_4B_3B_2B_1B_0$ values, the corrected BCD equivalent $D_4D_3D_2D_1D_0$ is obtained by adding 0011 to the corresponding $B_4B_3B_2B_1$ values. Also, $B_0 = D_0$. Accordingly, it follows that

$$(B_4B_3B_2B_1), \quad \text{if } 0000 \leq B_4B_3B_2B_1 \leq 0100$$
$$D_4D_3D_2D_1 = (B_4B_3B_2B_1 + 0011), \text{ otherwise}$$

The to-be-designed binary-to-BCD converter can be used to convert five input bits: the four bits fed to it and the one immediately to the right of those fed as inputs. This least significant single bit input appears unchanged in the output. The typical way to design this conversion module will involve obtaining reduced Boolean expressions for D_4, D_3, D_2, and D_1 from their corresponding K-maps shown in Figure 6.17. For this particular case in question, $B_4B_3B_2B_1$ values 1010 through 1111 may be considered as don't cares. Accordingly, in the K-maps of Figure 6.17, the corresponding minterms appear as don't care. When minterms are grouped properly, equations for the circuit outputs are obtained as

$$D_4 = B_3B_1 + B_3B_2 + B_4$$
$$D_3 = B_4B_1 + B_3\bar{B}_2\bar{B}_1$$
$$D_2 = B_2B_1 + B_4\bar{B}_1 + \bar{B}_3B_2$$
$$D_1 = \bar{B}_4\bar{B}_3B_1 + B_3B_2\bar{B}_1 + B_4\bar{B}_1$$

The resulting logic circuit for the conversion module is obtained as shown in Figure 6.18.

The 4-bit binary-to-BCD modules can be used to implement logic circuit for converting any n-bit binary number to its BCD equivalent. The rules for cascading such 4-bit converter modules have been identified already in this section. Figure 6.19 shows, for example, the

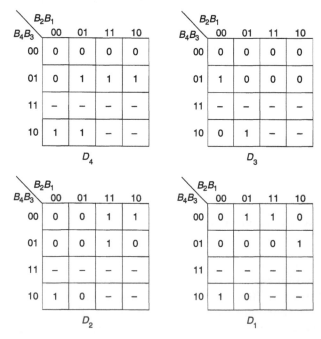

FIGURE 6.17
K-maps for a 4-bit binary-to-BCD converter module.

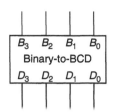

FIGURE 6.18
4-bit binary-to-BCD module.

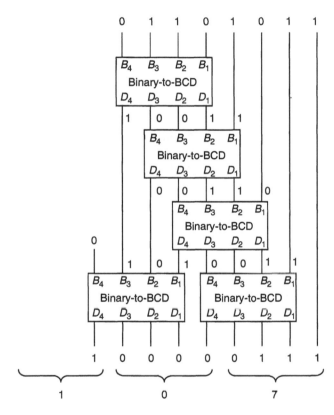

FIGURE 6.19
Logic circuit for converting a 7-bit number to its BCD equivalent.

logic circuit for converting a 7-bit binary number such as 1101011 (as in Example 6.6) to its BCD equivalent. For ease of understanding, the binary values are listed next to both inputs and outputs of each of the converter modules.

Example 6.8
Use binary-to-BCD converter modules to convert a 12-bit number to its BCD equivalent. Show the data transformation for the input 101101110110_2.

Solution
The logic circuit for converting a 12-bit number into its BCD equivalent follows directly from the n-bit binary-to-BCD conversion algorithm. The resulting logic circuit configuration is obtained as shown in Figure 6.20. After every three levels, that is, as soon as three MSPBs have been generated, the number of modules in the next level increases by one. The output of the logic network for the given binary input

becomes 0010 1001 0011 0100, that is, 2934_{BCD}, as expected. We can check the validity of the result by noting $101101110110_2 = 2^{11} + 2^9 + 2^8 + 2^6 + 2^5 + 2^4 + 2^2 + 2^1 = (2048 + 512 + 256 + 64 + 32 + 16 + 4 + 2)_{10} = 2934_{10}$.

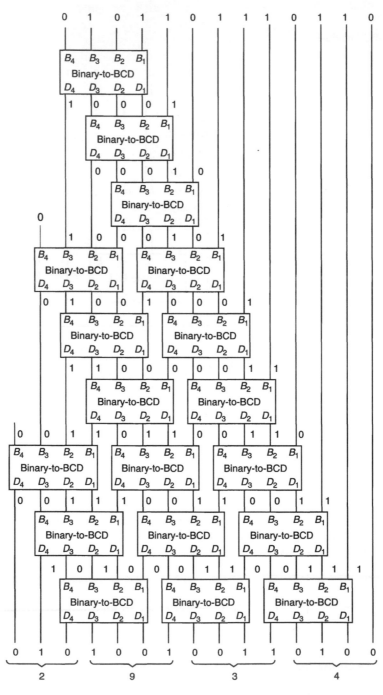

FIGURE 6.20

Logic circuit for Example 6.8.

Example 6.9
Design a 4-bit BCD-to-binary conversion module.

Solution
In Example 6.7, we have already seen that the least significant bits of a binary number and its equivalent BCD number are identical. Accordingly, in designing a BCD-to-binary converter module, we shall consider only the next more significant four bits for each of the given BCD numbers.

The design of a BCD-to-binary converter module will involve undoing the repeatedly performed shift and correction process that was used in defining the binary-to-BCD converter module. Consequently, we now need to consider shift-right (instead of shift-left) operation. The necessary correction is achieved here by subtracting (instead of adding) 0011. The function outputs that may be used for describing this conversion module can be described as,

$$(D_4 D_3 D_2 D_1), \quad \text{if } (D_4 D_3 D_2 D_1) < 1000$$

$$B_4 B_3 B_2 B_1 = (D_4 D_3 D_2 D_1 - 0011), \quad \text{otherwise}$$

The corresponding truth table for the converter module is obtained as shown in Figure 6.21.

For six different inputs, the corresponding output bits are listed as don't cares. These inputs are not expected to exist since that would otherwise imply the given BCD digit to be valid. For example, a 0110 will never occur after a shift-right operation since that would imply the presence of an unlikely BCD number, either 1100 or 1101, before shift operation. The truth table of Figure 6.21 also indicates that the BCD digits equal to or less than 0100 require no correction. However, for BCD digits 8 through 12, 0011 needs to be subtracted from the input to achieve correction. By minimizing K-maps corresponding to $B_4, B_3, B_2,$ and B_1, we may obtain the following logical equations for the converter outputs:

$$B_4 = D_3 D_4 + D_1 D_2 D_3$$

$$B_3 = D_2 \bar{D}_4 + D_1 \bar{D}_2 D_4 + \bar{D}_1 \bar{D}_3 D_4$$

$$B_2 = D_2 \bar{D}_4 + D_1 \bar{D}_2 D_4 + \bar{D}_1 D_2$$

$$B_1 = D_1 \bar{D}_4 + \bar{D}_1 D_4 = D_1 \oplus D_4$$

The logic circuit for the BCD-to-binary converter module may now be obtained as shown in Figure 6.22.

Example 6.10
Use the module designed in Example 6.9 as the basis for converting 326_{10} (given in BCD) to its binary equivalent.

Solution
We note that the input $326_{10} = (0011\ 0010\ 0110)_{\text{BCD}}$. Its conversion to its binary equivalent should follow a process that is an inverse of the scheme used earlier in

binary-to-BCD conversion. The input data as it gets changed in each of the steps are described as follows:

> **Step 1.** *Shift-Right and Correction*: The binary version of the given BCD number (i.e., 0011 0010 0110) is shifted to the right by one bit. Accordingly, the least significant integer bit moves to the right of the decimal point. Here, the bits appearing on the right side of the radix point are in binary and those appearing on the left side are in BCD. The shift-right operation first results in an uncorrected BCD value. The BCD-to-binary converter modules are used then to correct these uncorrected results (when the number is greater than or equal to 1000) before the next shift-right operation as follows:
>
> 0001 1001 0011.0 $\times$ 2^1 uncorrected BCD
> 0001 0110 0011.0 $\times$ 2^1 corrected BCD
>
> Herein, the middle digit 1001 needs correction since it is greater than 1000. The resulting mixed number is given by $(163) \times 2 = 326_{BCD}$.

> **Step 2.** *Shift-Right and Correction*: After the next shift-right operation, the middle digit again needs correction as follows:
>
> 0000 1011 0001.10 $\times$ 2^2 uncorrected BCD
> 0000 1000 0001.10 $\times$ 2^2 corrected BCD
>
> The resulting mixed number still is $\{81 + (1/2)\} \times 2^2 = 326_{BCD}$.

> **Step 3.** *Three Right Shifts*: The next three shift-right operations require no digit correction since none of the digits is larger than 1000. The three shift-right operations result in
>
> $$0000\ 0100\ 0000.110 \times 2^3$$
> $$0000\ 0010\ 0000.0110 \times 2^4$$
> $$0000\ 0001\ 0000.00110 \times 2^5$$

> **Step 4.** *Shift-Right and Correction*: The next right shift operation results in a least significant BCD digit that needs correction. The transformations lead to
>
> 0000 0000 1000.000110 $\times$ 2^6 uncorrected BCD
> 0000 0000 0101.000110 $\times$ 2^6 corrected BCD
>
> The resulting mixed number $\{5 + (6/64)\} \times 2^6 = 326_{BCD}$.

> **Step 5.** *Three Right Shifts*: The next three shift-right operations involve no correction. The resultant value $(0000\ 0000\ 0000.101000110) \times 2^9 = 101000110_2$. We can verify the validity of this answer since $101000110_2 = 2^8 + 2^6 + 2^2 + 2^1 = 256 + 64 + 4 + 2 = 326_{10} = (0011\ 0010\ 0110)_{BCD}$.

The resulting logic circuit using BCD-to-binary converter modules is obtained as shown in Figure 6.23. Notice that the least significant bit of the BCD number is left unprocessed in this logic circuit.

D_4	D_3	D_2	D_1	B_4	B_3	B_2	B_1
0	0	0	0	0	0	0	0
0	0	0	1	0	0	0	1
0	0	1	0	0	0	1	0
0	0	1	1	0	0	1	1
0	1	0	0	0	1	0	0
0	1	0	1	-	-	-	-
0	1	1	0				
0	1	1	1	-	-	-	-
1	0	0	0	0	1	0	1
1	0	0	1	0	1	1	0
1	0	1	0	0	1	1	1
1	0	1	1	1	0	0	0
1	1	0	0	1	0	0	1
1	1	0	1	-	-	-	-
1	1	1	0	-	-	-	-
1	1	1	1	-	-	-	-

FIGURE 6.21
Truth table for Example 6.9.

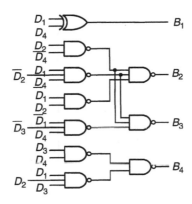

FIGURE 6.22
BCD-to-binary converter module: (a) block diagram and (b) logic circuit.

Just as we had summarized earlier the *n*-bit binary-to-BCD conversion algorithm, the *n*-bit *BCD-to-binary conversion algorithm* can have the following steps:

1. Feed all binary bits (in fours) of the BCD number except for the least significant bit into 4-bit BCD-to-binary converter modules. If at any level the left-most converter were to have three or less number of nonzero bits as its input, the left-most converter module is eliminated from the logic circuit. Typically, when the most significant output bit of the left-most converter module is a 0, it need not be further processed in the next level of converter modules.

2. Reassign all processed binary bits except for the least significant processed bit are fed in fours into the next level of 4-bit BCD-to-binary converter modules.

3. Continue reassignment and processing of bits until the MSB (processed or unprocessed) from a prior level has been included as the most significant input to a 4-bit BCD-to-binary converter module.

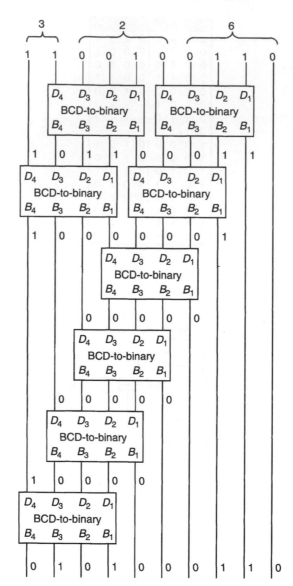

FIGURE 6.23
A BCD-to-binary converter circuit.

6.4   Error-Detecting and Error-Correcting Circuits

We have considered a number of code converters, for example, in the earlier sections that can be used in both computers and communication systems. There is always a possibility, however small, that a random-noise pulse in computers and communications equipment will change a 0 to a 1 or a 1 to a 0. Oftentimes, it becomes difficult to detect if and when these errors occur. In case of a binary number, an error would always yield another valid binary number. When working with BCD numbers, however, the occurrence of an error may be detected since the error may cause the BCD number to take on values from one of the six 4-bit combinations, 1010 through 1111.

In Chapter 1, we considered error detection coding principle such that a single-bit error always converts the valid code word into an invalid code word. One easier way is to attach an extra bit called a *parity* bit to the data bits. Before transmitting a code, the number of 1s are counted and the parity bit is set to make the number of 1s either odd or even as per choice. At the receiving end, the receiver can test the incoming bit stream and, accordingly, detect whether or not an error has occurred during transmission. If even parity, for example, has been used but an odd number of 1s has been received, it would imply that an error has occurred. This scheme, however, would fail when there are even number of errors. Thus, for situations where the probability of multiple errors is high, a more sophisticated coding scheme must be used.

To generate the parity bit and then to test the parity status of a given code word, it is necessary to determine whether an odd or even number of 1s is present on the receiver end. As defined in Chapter 2, an XOR (XNOR) logic gate functions in such a way that the output of an even (odd) number of 1s is always a 0, and the output of an odd (even) number of 1s is always a 1. Logic gates such as XOR and exclusive-NOR function (XNOR) are thus ideal candidates for designing both parity generator and error detector. The logic circuit of Figure 6.24, for example, generates an even parity bit for the BCD input on the transmitter side and then on the receiving side tests the parity status to determine whether or not an error has occurred during transmission.

The parity generator circuit examines the four data lines $B_8, B_4, B_2,$ and B_1 and, accordingly, generates a parity bit P (given by $B_8 \oplus B_4 \oplus B_2 \oplus B_1$) so that the resulting code word (five bits in all) has even parity. The parity detection circuit, on the other hand, determines if an error has occurred or not on the basis of the parity status (given by $B_8 \oplus B_4 \oplus B_2 \oplus B_1 \oplus P$) of the code word. A high output indicates the occurrence of an error during transmission. The logic circuit of Figure 6.24 uses cascaded 2-input XOR logic gates instead of 5- and 4-input XOR logic gates. This logic circuit can be modified to handle odd parity as well in which case the final XOR logic gates of both the parity generator and parity detector circuits are replaced with XNOR logic gates. In case one is interested in fully parallel operation, however, a 2-level AND-OR logic or its equivalent will need to be employed. In general, the SOP from an n-variable parity function will have 2^{n-1} product terms, each of which is a minterm. Thus, for example, for an 8-variable binary number, one would have needed 128 8-input AND gates and one 128-input OR gate! Obviously, these requirements are difficult if not impossible to realize using standard AND and OR gates. One reasonable alternative may be to use programmable logic function devices such as those considered in Chapter 4. Otherwise, one will have to settle for a pseudo-parallel scheme such as that shown in Figure 4.20b for realizing the parity bit.

As described in Chapter 1, several other schemes are also available for coding decimal digits. The most common of these are the 2-out-of-5 code and 2-out-of-7 (i.e., biquinary)

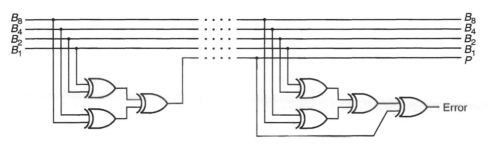

FIGURE 6.24
Logic circuit for even parity generation and error detection.

Decimal	Odd parity	Even parity	2-out-of-5	Biquinary
0	00001	00000	01001	0100001
1	00010	00011	00011	0100010
2	00100	00101	00101	0100100
3	00111	00110	00110	0101000
4	01000	01001	01010	0110000
5	01011	01010	01100	1000001
6	01101	01100	10001	1000010
7	01110	01111	10010	1000100
8	10000	10001	10100	1001000
9	10011	10010	11000	1010000

FIGURE 6.25
Error-detecting codes.

code. These codes (having positional weights) are listed along with parity-coded BCD in Figure 6.25. In 2-out-of-5 code, however, only decimal 0 is unweighted. In both these m-out-of-n codes, there are m 1's and $(n - m)$ 0's. Each code word in an m-out-of-n code is at least two bit changes away from all other code words. An occurrence of a single error will thus convert a valid code word into an invalid code word. Consequently, these codes can be used to detect single errors. For example, consider the 2-out-of-5 code word 01100 that got changed to 01110. Since there are three 1's in 01110, we would know immediately that an error has occurred. However, we wouldn't know for sure if the original code word had been either 00110, or 01100, or 01010.

While the parity-based coding scheme shown in Figure 6.25 can determine only whether or not an error has occurred, the *Hamming code* is particularly useful since it has the added capability to pinpoint and correct the single errors. This is because each code word in the Hamming code requires changes in at least three bits to transform itself to another valid code word. At least two errors will have to occur before there can be confusion in error detection. The Hamming code uses multiple parity bits and interleaves them at specific bit locations of the resulting code word. This positioning scheme allows for the detector circuit to pinpoint the error and, consequently, it becomes possible to regenerate the undistorted string of input bits. The *parity generation algorithm* for identifying the multiple parity bits is summarized as follows:

1. The number of parity bits p needed to code an m-input number, is the smallest integer value of p that satisfies $2^p \geq m + p + 1$.

2. The parity bits are positioned respectively at bit locations 1, 2, 4, 8, 16, and so on (i.e., bit positions corresponding to powers of 2), of the code word. The m-bits of the given number, on the other hand, are positioned respectively at remaining bit locations, that is, 3, 5, 6, 7, 9, 10, and so on.

3. Each parity bit covers only a select few of the bits of the code word. To determine the particular bits of the code word that are covered by a particular parity bit, every bit position of the code word is expressed in binary. A parity bit having a 1 in location x of its binary representation covers all those bit positions of the code word (and itself) that have a 1 in location x of their binary representations.

Figure 6.26 shows a block diagram summarizing the implementation of Hamming code-based parity generation, error detection, and error correction scheme. The m-bits of the input are encoded with p parity bits before transmission. The resulting code word is transmitted through medium where, for example, an additive noise can change either a 0 to a 1 or a 1 to a 0. The received code word is tested with an error detector circuit to see if any error has

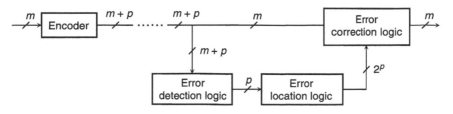

FIGURE 6.26
A Hamming code-based transmission system.

occurred. The error locator circuit (it actually is a decoder) then locates the exact position of error, if any, and the bit in error is corrected thereafter by an error corrector circuit.

The error detection and correction algorithm consists of the following steps:

1. Check each of the parity bits P_n and the correspondingly covered bits for which it provides parity.
2. Assign a 0 to the n-th test result C_n if the test indicates preservation of an assumed parity (as originally determined on the transmitter side) and a 1 to the test result for a failed test.
3. Identify the error bit location from the binary number $C_n \ldots C_2 C_1$ by cascading the test results.
4. Complement the value of the bit in error.

It is rather easy to design the logic circuits of the various modules of a Hamming code-based transmission system shown in Figure 6.26. The module for generating multiple parity bits can be realized by using the principle that has already been used in the circuit of Figure 6.24 for generating single parity bit. The design for modules of a Hamming code-based system is based on the number of input bits. For illustration, consider transmitting a 5-bit input. When $m = 5$, $p = 4$. Therefore, the parity generating logic circuit will have to generate four parity bits: P_1, P_2, P_3, and P_4. We know that XOR (XNOR) logic gates are ideal for detecting the presence of an odd (even) number of 1s. The parity bits, for example, can be generated readily using following logic equations:

$$P_1 = M_1 \oplus M_2 \oplus M_4 \oplus M_5$$
$$P_2 = M_1 \oplus M_3 \oplus M_4$$
$$P_3 = M_2 \oplus M_3 \oplus M_4$$
$$P_4 = M_5$$

where M_j is the j-th input bit. The resulting logic circuit for even parity assumption is obtained as shown in Figure 6.27. Correspondingly, for an odd parity assumption only the final XOR logic gates of all stages need to be replaced with XNOR logic gates and M_5 needs to be simply complemented to generate P_4.

Design of an error detection circuit for verifying the parity status also follows the same principle as the one that was used in the circuit of Figure 6.24. The parity of each parity bit and its correspondingly covered data bits are tested together. The logical equations for the

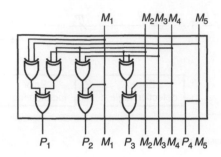

FIGURE 6.27
Parity (even) generator circuit capable of covering 5-bit number.

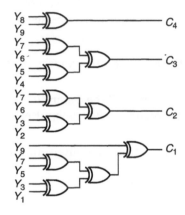

FIGURE 6.28
Error detection circuit for a 9-bit data transmission.

9-input parity-testing circuit are described by

$$C_1 = Y_1 \oplus Y_3 \oplus Y_5 \oplus Y_7 \oplus Y_9$$

$$C_2 = Y_2 \oplus Y_3 \oplus Y_6 \oplus Y_7$$

$$C_3 = Y_4 \oplus Y_5 \oplus Y_6 \oplus Y_7$$

$$C_4 = Y_8 \oplus Y_9$$

where $Y_1, Y_2, \ldots, Y_8$, and Y_9 are the bits received (after transmission) corresponding to $P_1, P_2, M_1, P_3, M_2, M_3, M_4, P_4$, and M_5 output of the parity generator circuit. The parity detection circuit is realized also using XOR logic gates and is shown in Figure 6.28. Should an error occur, the binary value of $C_4C_3C_2C_1$ pinpoints the bit in error. The test results are thus used then to correct the bit in error. Since our sole interest is in recovering corrected data bits, we make use of XOR logic gates here as a programmable inverter. The logical equations for the corrected bits are thus obtained as follows:

$$D_1 = (\overline{C}_4 \cdot \overline{C}_3 \cdot C_2 \cdot C_1) \oplus Y_3$$

$$D_2 = (\overline{C}_4 \cdot C_3 \cdot \overline{C}_2 \cdot C_1) \oplus Y_5$$

$$D_3 = (\overline{C}_4 \cdot C_3 \cdot C_2 \cdot \overline{C}_1) \oplus Y_6$$

$$D_4 = (\overline{C}_4 \cdot C_3 \cdot C_2 \cdot C_1) \oplus Y_7$$

$$D_5 = (C_4 \cdot \overline{C}_3 \cdot \overline{C}_2 \cdot C_1) \oplus Y_9$$

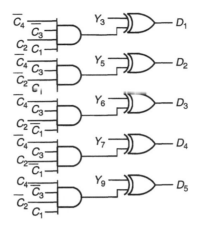

FIGURE 6.29
Error location and correction circuit for a 9-bit data transmission.

The bit in error is corrected by complementing the bit in error. All of the other bits are left unaltered. For example, when $C_4C_3C_2C_1 = 0110$, the bit in error is located at bit position 6. Thus when $\overline{C_4}C_3C_2\overline{C_1}$ is a 1, the bit Y_6 is complemented to give D_3. But if instead $\overline{C_4}C_3C_2\overline{C_1}$ is a 0, the bit Y_6 remains unchanged. The corresponding error-correcting logic circuit is obtained as shown in Figure 6.29.

6.5 Binary Arithmetic

The ability to add binary numbers is significant since most other arithmetic operations can be realized by manipulating addition operations. We have studied already in Chapter 1 that subtraction operation $A - B$ is same as performing the addition operation $A + (-B)$. Operation such as multiplication can be realized by repetitively performing addition operations. The equivalent of division operation, on the other hand, is realized by having repeatedly performed subtraction operations each of which in turn is again a function of addition operation.

At the least significant bit position, addition involves only two operands, the *addend* and *augend*. The addition operation at all other bit positions, however, requires a third input (the *carry-in*) generated by the addition operation performed at the next less significant bit position. Like any other combinatorial logic design, an n-bit adder can be realized also in several different ways, each of the resulting logic circuits differing in its speed and cost characteristics. Because of the difficulty in assigning cost figures (since cost estimation process is rather dynamic and often a function of the various technologies involved) to a particular design, we shall consider here multiple designs whenever possible. Estimation of speed is not a trivial matter either. However, a 2-level logic circuit would obviously prove to be the fastest of all combinatorial logic circuits. Such a logic network, however, would require a large number of both logic gates and gate inputs. To add two n-bit numbers, for example, we may need up to 2^{2n} NAND gates of $2n + 1$ inputs and one NAND gate of 2^{2n} inputs. Thus, even for a small n, this indeed is a very demanding design. For $n = 4$, for example, we may need 9-input NAND gates (256 of them) and one 256-input NAND gate. Not only the number of logic gates and gate inputs are large but they all have serious fan-in and fan-out implications. Thus an attempt at identifying a 2-level logic circuit seems a staggering problem. The alternative to this very expensive design, however, is rather straightforward in which each pair of input operand bits are treated separately. It follows directly from our observation of the mechanics of an n-bit add operation.

Irrespective of the number of bits in the operands, the process of adding augend and addend is same at all bit positions except at the least significant bit position. Thus the addition process may be implemented one bit at a time, starting with the least significant digit. A parallel n-bit addition circuit, therefore, can be designed by first designing a total of n single-bit addition circuits. Then in order to allow for n-bit addition, these n single-bit adders will need to be cascaded together. These single-bit adders at all bit positions except the least significant bit position will also need to be full adders (i.e., adders with three inputs: addend, augend, and carry-in). At the least significant bit position, however, a single-bit full adder may be used but then its carry-in is fixed at 0. Or, we may use simply a half adder (i.e., adder with two inputs: addend and augend at that bit location). At each of the remaining bit positions, carry-in input of the single-bit full adder is fed with the carry-out of the single-bit full adder located at the next less significant bit position.

This particular design approach demonstrates the fact that the design of a not-so-manageable logic function is managed by, first, designing smaller manageable subfunction units. Finally, these subfunction units are cascaded together in accordance with the understanding of how they may generate the truth table pertaining to the original function.

In this section, we first consider the design of a half adder as an introduction to the design of adder circuits in general. In principle, an n-bit adder circuit can be designed using $n-1$ full adders and one half adder. A *half adder* (HA for short) module is a 2-input, 2-output combinatorial logic circuit used for adding two input bits (addend and augend) and no carry-in. The truth table for the two inputs, A_0 and B_0, and the two outputs, S_0 (sum) and C_0 (carry-out), are obtained as shown in Figure 6.30. When used within the context of an n-bit adder, it is used at its least significant bit position. The logical equations for the sum S_0 and carry-out C_0 are as follows:

$$S_0 = A_0 \oplus B_0 \tag{6.1}$$

$$C_0 = A_0 B_0 \tag{6.2}$$

Equations 6.1 and 6.2 can be implemented in many different ways. Figure 6.31 shows implementation schemes using (a) an AND and an XOR gate, (b) NOR gates, (c) NAND gates, and (d) multiplexers.

A *full adder* (FA for short) is a 3-input, 2-output logic circuit that adds two binary digits, A_i and B_i, and a carry-in C_{i-1} from bit position $(I-1)$. The block diagram and the corresponding truth table for a full adder are shown in Figures 6.32a and b. The K-maps for the sum bit,

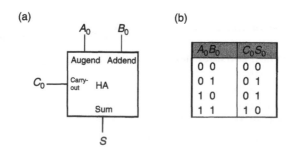

FIGURE 6.30
A half adder module: (a) block diagram and (b) truth table.

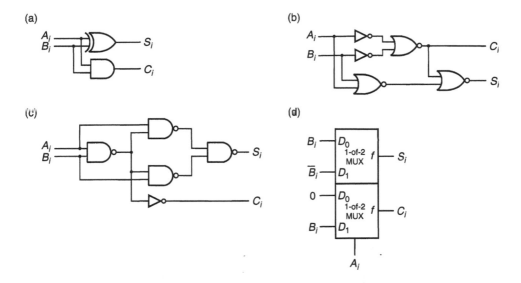

FIGURE 6.31
A half adder logic circuit using (a) AND and XOR gates, (b) NOR gates, (c) NAND gates, and (d) multiplexers.

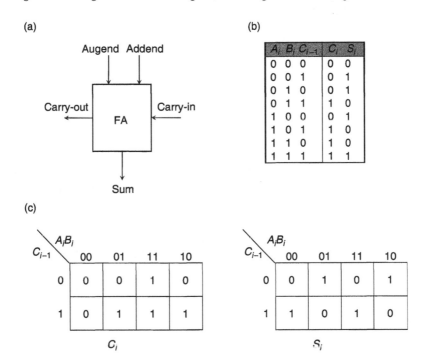

FIGURE 6.32
Full adder: (a) block diagram, (b) truth table, and (c) K-maps.

S_i, and the carry-out, C_i, are then constructed from the truth table and obtained as shown in Figure 6.32c. The logical equations for the sum and carry-out bits as obtained from their corresponding K-maps are given by,

$$S_i(A_i, B_i, C_{i-1}) = \Sigma m(1, 2, 4, 7) = A_i \oplus B_i \oplus C_{i-1} \tag{6.3}$$

(a)

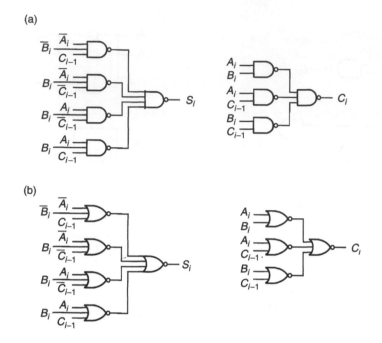

(b)

FIGURE 6.33
Full adder using (a) only NAND gates and (b) only NOR gates.

and

$$C_i(A_i, B_i, C_{i-1}) = A_iB_i + B_iC_{i-1} + C_{i-1}A_i \tag{6.4}$$

Just as in the case of a half adder, a full adder may also be implemented in several different ways. In order to obtain the fastest possible logic circuit using two levels of logic gates, we may have to undo Equation 6.3 to obtain its canonical form. The NAND- and NOR-only versions of the full adder logic circuit may then be obtained as shown in Figure 6.33. Each of these logic circuits requires 12 logic gates (six with two gate inputs, five with three gate inputs, and one with four gate inputs) for a total of 31 gate inputs. However, if we were to instead realize Equation 6.3, it can be also realized using only NAND gates, since a 2-input XOR logic operation can be generated using four 2-input NAND gates.

Alternatively, Equation 6.4 may be rewritten by making use of the bridging technique. The entries in the K-map for carry-out suggests that it can be bridged with an XOR function to yield (see Chapter 4, Problem 37)

$$C_i = (A_i \oplus B_i)C_{i-1} + A_iB_i \tag{6.5}$$

It is interesting to note that in accordance with Equations 6.1 and 6.2, a half adder can be used to realize both $A_i \oplus B_i$ and A_iB_i. Consequently, we can rewrite Equations 6.3 and 6.5 as follows:

$$S_i = S_{i(HA)} \oplus C_{i-1} \tag{6.6}$$

$$C_i = S_{i(HA)} \cdot C_{i-1} + C_{i(HA)} \tag{6.7}$$

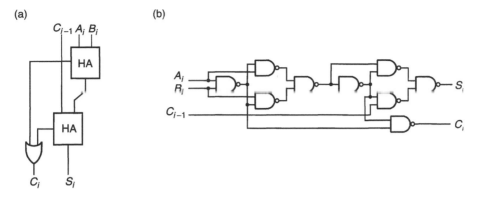

FIGURE 6.34
Full adder using (a) half adders and (b) multilevel NAND gates.

where $S_{i(HA)}$ and $C_{i(HA)}$ are the sum and carry-out of a half adder. This half adder takes in A_i and B_i as its inputs. Equation 6.7 can be realized by feeding the carry-in and the half adder sum output to a 2-input XOR gate. In principle, however, half adder circuits are all able to yield XOR logic. Accordingly, if we were to feed the sum output of a half adder along with the carry-in into a second half adder, the final sum output of the second half adder can be treated as equivalent to the full adder sum given by Equation 6.7.

Fortunately, the carry-out of this second half adder is given by $(A_i \oplus B_i)C_{i-1}$ while that of the first half adder is given by A_iB_i. These two carry-out outputs are same as the two product terms of Equation 6.5. Consequently, if we take an OR of the carry-out outputs of these half adders, the full adder carry-out, C_i, as given by Equation 5.6 can be generated. Figure 6.34a shows the resulting full adder logic circuit that uses two half adders and an OR gate. The full adder circuit may be realized also using either only NAND gates or only NOR gates. The NAND-only full adder circuit is obtained as shown in Figure 6.34b. This latter implementation of the full adder requires only nine 2-input NAND gates and a total of eighteen gate inputs. By comparing with the earlier NAND-only full adder circuit shown in Figure 6.33, this latter design led to a logic circuit of three fewer gates and thirteen fewer gate inputs. Thus, the bridging concept, in this case, has allowed us to design a less expensive logic circuit than would be necessary had we not made use of bridging technique. However, this cost improvement in the design comes at a cost since there are more NAND levels through which the input signals have to propagate. One must be aware of the fact that an AND–OR–INVERT (AOI) logic gate can be used instead of a NAND–NAND format for improving the speed. Note, however, that an AOI gate in reality is a set of NAND logic gates followed by an internally wired-AND gate. This latter wired-AND gate contributes to no additional propagation delay.

Since the design of full adder is of primary significance in digital computers, a large effort has gone into the problem of generating its most economical and fastest realizations. The logic circuit that is the most optimized is a function often of the technology used. An interesting form of full adder design follows directly from a particular repackaging of the K-maps shown in Figure 6.32c. As shown in Figure 6.35a, the K-map of $(A_i + B_i + C_{i-1})$ and that of the sum are seen to have five cell entries common. By intersecting (i.e., taking an AND of) this map and a map of $\overline{C_i}$, we obtain a K-map as shown also in Figure 6.35a. This latter K-map is not quite the same as that of the sum bit S_i shown in Figure 6.32c. However, by combining (i.e., by taking an OR of) a map of $A_iB_iC_{i-1}$ and this latter K-map of Figure 6.35a, we may obtain a K-map for the sum. Accordingly, we can express the sum

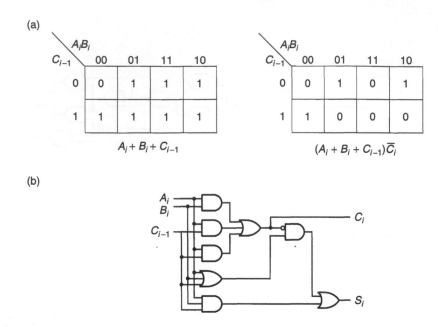

FIGURE 6.35
An alternate full adder design: (a) relevant K-maps and (b) logic circuit.

bit as

$$S_i(A_i, B_i, C_{i-1}) = \Sigma m(1, 2, 4, 7)$$
$$= A_i \oplus B_i \oplus C_{i-1}$$
$$= (A_i + B_i + C_{i-1})\overline{C_i} + A_i B_i C_{i-1} \qquad (6.8)$$

The full adder logic circuit equivalent to Equation 6.8 is obtained as shown in Figure 6.35b. Here, the sum is a function of the carry-out.

The half and full adder modules are commercially available in the form of MSI circuits. A designer can use such modules to readily design a multibit adder circuit. Commercially available 4-bit full adders often use four single-bit full adders in a serial format. Such 4-bit full adder circuits can be used to add up to two 4-bit quantities, $A_3A_2A_1A_0$ and $B_3B_2B_1B_0$. Correspondingly, the adder module will yield a 4-bit sum, $S_3S_2S_1S_0$, and a carry-out resulting from the addition of A_3, B_3, and C_2. A serially cascaded logic circuit of n full adders, as shown in Figure 6.36, may serve as an n-bit adder. Each full adder represents a particular bit position. At the least significant bit position, however, we can use a half adder since a carry-in from a yet less significant bit position does not exist. Because of the serial format of the logic circuit shown in Figure 6.36, the propagation delay of an n-bit ripple adder is same as the product of the total number of single-bit full adders and propagation delay for a single-bit full adder. Total propagation delay is thus $n\Delta$ where Δ is the propagation delay (from carry-in to carry-out) of a single-bit full adder. This propagation delay is cumulative since the carry into a multibit ripple adder will have to propagate through all of the single-bit full adders before the final carry-out output can be generated. This is why such adders are referred to as a *ripple adders*. It should be pointed out that when n is too large, a multi-input ripple adder logic circuit may not provide much advantage.

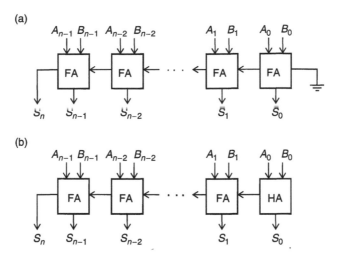

FIGURE 6.36
An n-bit adder circuit using (a) n full adders and (b) $n - 1$ full adders and one half adder.

Example 6.11
Implement a 2-bit ripple adder using a programmable logic array (PLA) that has four input lines, three output lines and two bidirectional lines.

Solution
From Equations 6.3 and 6.4, we obtain for the i-th bit:

$$S_i(A_i, B_i, C_{i-1}) = \Sigma m(1, 2, 4, 7)$$
$$= \overline{A_i}\,\overline{B_i}C_{i-1} + \overline{A_iB_i\overline{C_{i-1}}} + A_i\overline{B_i}\,\overline{C_{i-1}} + A_iB_iC_{i-1}$$

and

$$C_i(A_i, B_i, C_{i-1}) = A_iB_i + B_iC_{i-1} + C_{i-1}A_i$$

In a ripple adder scheme, the carry-out output C_0 of the single-bit full adder at any bit is fed to the single-bit full adder located at the next higher bit as its carry-in input. Each of the sum bits will require a 4-input OR operation and each of the carry bits will require a 3-input OR operation. The sum-related OR logic will need to process four 3-input product terms (in its canonical form) as its inputs while the carry-related OR logic will need to process three 2-input product terms (in their reduced form). In this 2-bit addition, C_1 needs to be fed back to the AND array to generate S_0 and C_0 and likewise C_0 needs to be fed back to generate S_1 and C_1.

A 2-bit adder requires five inputs (two 2-bit data and a carry-in) but the given PLA has only four input lines. To implement this logic circuit, therefore, one of the bidirectional lines is used as an input (conveniently chosen to feed C_{-1}). The intermediate carry C_0, on the other hand, is fed to the AND array through the other bidirectional line. Figure 6.37 shows the corresponding PLA implementation. The tristate driver of the line connected to C_{-1} is disabled since the AND term corresponding to it is a 0. This is accomplished by having included all the inputs and their complements as inputs to this AND term.

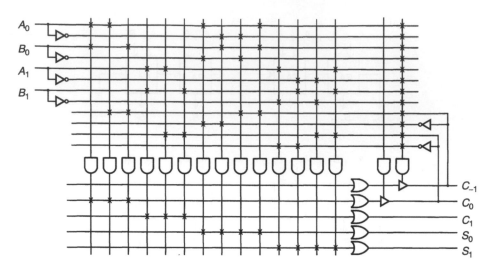

FIGURE 6.37
A 2-bit ripple adder.

As has been pointed out already, multiplication of multibit binary numbers can be realized using binary multibit addition operations. It follows from the idea that may be used to implement multidigit multiplication of decimal numbers. The multiplication of two 4-bit numbers, $A = A_3A_2A_1A_0$ and $B = B_3B_2B_1B_0$, for example, can be obtained as shown below:

$$
\begin{array}{cccccccc}
 & & & & A_3 & A_2 & A_1 & A_0 \\
 & & & & B_3 & B_2 & B_1 & B_0 \\
 & & & & A_3B_0 & A_2B_0 & A_1B_0 & A_0B_0 \\
 & & & A_3B_1 & A_2B_1 & A_1B_1 & A_0B_1 & \\
 & & A_3B_2 & A_2B_2 & A_1B_2 & A_0B_2 & & \\
 & A_3B_3 & A_2B_3 & A_1B_3 & A_0B_3 & & & \\
P_7 & P_6 & P_5 & P_4 & P_3 & P_2 & P_1 & P_0 \\
\end{array}
$$

By generalizing this multiplication operation, we may conclude that an additional diagonal of product terms will be required for each additional bit of the multiplicand A. Likewise, an additional row of product terms will be required for each additional bit of the multiplier B. Each row, called a partial product, is formed by a bit-by-bit multiplication of multiplicand and a particular multiplier bit. In general, when an n-bit number is multiplied with an m-bit number, we get $m + n$ product bits.

In binary representation, the product term A_iB_j (denoted as p_{ij}, for simplicity) is equivalent to realizing an AND of A_i and B_j. Accordingly, we can design a combinatorial multiplication circuit by first realizing the product terms A_iB_j and then use full adders to add such AND outputs column wise. But that would mean that we need to first figure out how to add more than three single-bit numbers column wise. In the next example, accordingly, we consider the mechanics of adding multiple single-bit numbers column wise using multiple single-bit full adders. Alternatively, for example, one could begin to add the partial product of a row by having it shifted to the right one bit before it is added to the next partial product set of in the next row as shown in Figure 6.38. This logic circuit uses three 4-bit ripple adders to successively add the four 4-bit partial products. In later chapters, however, we shall be considering both sequential devices and sequential design

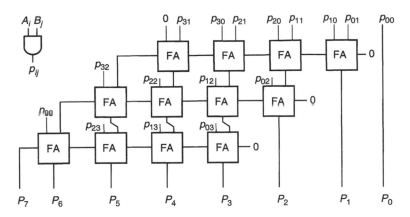

FIGURE 6.38
A 4 × 4 binary multiplier using three 4-bit adder.

techniques. We shall show then as to how a multiplier may be designed using a repetitive multiplication and addition algorithm.

Example 6.12
Design a logic circuit that may be used to add a column of seven single-bit numbers.

Solution
If the seven binary bits a_0, a_1, a_2, a_3, a_4, and a_6 were to be all 1, then the sum would be 111. This suggests that the to-be-designed adder circuit will need to accommodate seven inputs and three outputs. One approach to designing this logic circuit is to construct a truth table and then the corresponding K-maps. We may then determine the three outputs, S_2, S_1, and S_0 in terms of the seven inputs. However, one may also use already designed single-bit full adders to add numbers. But since each single-bit full adder can add up to three single bits, we may use a second full adder to add either the sum and two other single bits or add up to only three other single bits. This process may be repeated until we have added all of the bits. An optimized version of the circuit using cascaded single-bit full adders can be determined from an understanding of the mechanics of addition as follows: We may first use two single-bit full adders each used to add a set of three single-bits.

$$
\begin{array}{cc}
a_0 & a_3 \\
a_1 & a_4 \\
a_2 & a_5 \\
\hline
c_1 d_1 & c_2 d_2
\end{array}
$$

The two resultant sum bits can then be added to a_6 using a third single-bit full adder. The sum output of this third single-bit full adder then is equivalent to the least significant sum bit S_0. The three carry-out output of the three single-bit full adders may be added then using a fourth single-bit full adder to generate the other two sum bits $S_2 S_1$ as follows:

$$
\begin{array}{cc}
d_1 & c_1 \\
d_2 & c_2 \\
a_6 & c_3 \\
\hline
c_3 S_0 & S_2 S_1
\end{array}
$$

The resultant logic circuit thus requires four single-bit full adders and is obtained as shown in Figure 6.39.

However, if we instead add the sum of the first single-bit full adder and two of the other single bits, and so on, then the design follows the following addition steps:

$$
\begin{array}{ccc}
a_0 & a_3 & a_5 \\
a_1 & a_4 & a_6 \\
a_2 & d_1 & d_2 \\
\hline
c_1 d_1 & c_2 d_2 & c_3 S_0
\end{array}
$$

Finally the three carry-outs may be added to produce the other two sum outputs, $S_2 S_1$.

$$
\begin{array}{c}
c_1 \\
c_2 \\
c_3 \\
\hline
S_2 S_1
\end{array}
$$

The resultant logic circuit is obtained as shown in Figure 6.40. It is same as that shown in Figure 6.39 in terms of the number of single-bit full adders. However the circuit of Figure 6.40 is slower since the signals need to propagate through one additional level of single-bit full adder.

Now that we know how to add single-bit numbers column wise, the 4-bit by 4-bit multiplication problem that we introduced earlier may be solved. The sixteen product terms are realized first using sixteen 2-input AND gates and then these product terms are added using twelve single-bit full adders as shown in the logic circuit of Figure 6.41. An additional diagonal of single-bit full adders will be required for each additional bit in the multiplicand A. An additional row of single-bit full adders will be needed for each additional bit in the multiplier B. For an n-bit A and an m-bit B, the resulting multiplication logic circuit will require mn AND gates and $(n-1)m$ single-bit full adders. The overall multiplication time of the logic circuit of Figure 6.41 depends on the choice of full adders involved. Such a multiplication circuit is typically referred to as an *array multiplier*. The term array is used to describe the multiplier because it is organized as an array structure. The propagation times of more significance include that required for the signals to propagate from (a) any input to the sum output, and (b) any input to the carry-out output.

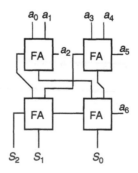

FIGURE 6.39
Logic circuit for Example 6.12.

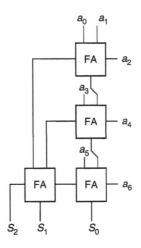

FIGURE 6.40
Alternate logic circuit for Example 6.12.

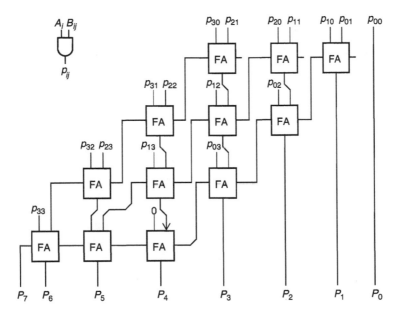

FIGURE 6.41
A 4-bit by 4-bit multiplier module using single-bit full adders.

In Example 6.13, we next consider another application of adders in digital systems. Here though we use ripple adders both for cost advantage and simplicity. This example demonstrates how a system that is designed to handle n-bit numbers can also perform operations on $2n$-bit numbers. Such an operation may be referred to as a *multiple-precision operation*. Such feat is often accomplished by partitioning the function into several subfunctions. As will be shown later, each of the subfunctions in turn could use predesigned modules as well.

Example 6.13
Design an 8-bit by 8-bit multiplication logic circuit using 4-bit by 4-bit multiplier PROM modules. Each of these modules is functionally equivalent to the logic circuit shown in Figure 6.41.

Solution

To realize 8-bit by 8-bit multiplication, we could use a single large PROM with sixteen inputs (eight multiplier bits and eight multiplicand bits) and sixteen outputs. Such a PROM is programmed with values listed in the corresponding multiplication table. This table will have 2^{16} combinations of the two 8-bit inputs. However, rather than relying on this huge PROM, we could use 4-bit by 4-bit multiplier PROM modules. These PROM modules are programmed with 2^8 sets of input combinations listed in a multiplication table. To design using these PROMs, it would be necessary that we partition the overall multiplication function into subfunctions that can accommodate only up to 4-bit operands. For the case of this 4-bit by 4-bit multiplier PROM, the two to-be-multiplied 4-bit numbers make up its 8-bit address (see Chapter 4, Problem 27 for a similar PROM).

Consider two 8-bit numbers, X_1 and X_2, each consisting of a sequence of the least significant four bits and a sequence of the most significant four bits. The numerical value of each of the 8-bit numbers can then be described by

$$X_i = 2^4(M_i) + L_1$$

where M_i is shifted to the left 4 bits before being added to L_1. By taking an approach such as this, we can consider the partitioning of an 8-bit operand into two 4-bit operands. The next challenge is to figure out the exact mapping scheme to relate an 8-bit by 8-bit product in terms of the various 4-bit by 4-bit component products. In other words, we need to determine how and which 4-bit by 4-bit product can be generated and combined to produce the desired 16-bit product. The numerical value of the 16-bit product, P, is given by

$$P = [2^4(M_1) + L_1] + [2^4(M_2) + L_2]$$

$$= 2^8(M_1M_2) + 2^4(M_1L_2 + M_2L_1) + (L_1L_2)$$

In other words, the 16-bit product may be obtained by using four 4-bit by 4-bit multiplier PROM modules. The four 4-bit by 4-bit multiplier PROM modules (i.e., PROM1 through PROM4) would respectively have (i) M_1 and M_2, (ii) M_1 and L_2, (iii) M_2 and L_1, and (iv) L_1 and L_2 as their respective 4-bit inputs. Each of the PROMs is designed/programmed to produce an 8-bit product of the two 4-bit inputs. The outputs of two of the 8-bit PROM modules (from (i) and (iii)) have the same positional value of significance.

The 16-bit product of two 8-bit numbers is thus realized by shifting PROM1 product bits 8 bits to the left, and the sum of the part (ii) and part (iii) product bits 4 bits to the left and then adding the two to the part (iv) product bits. An 8-bit PROM adder module can be used to obtain the sum of part (ii) product bits and part (iii) product bits as shown in Figure 6.42. The final sum and all the necessary shift left-operations are obtained using a 12-bit PROM adder. The least significant four bits (i.e., PROM4 outputs) are then concatenated to the right of this latter 12-bit sum to yield the desired 16-bit product. Alternatively, five 4-bit adders may also be employed instead of the two PROM adders. In such a case, two such adders are cascaded for generating the first 8-bit sum, and the other three are cascaded to yield the second set of 12-bit sum. In the logic circuit shown in Figure 6.42, the necessary shift-left operations are implicit in the way the inputs are fed to the two adder modules.

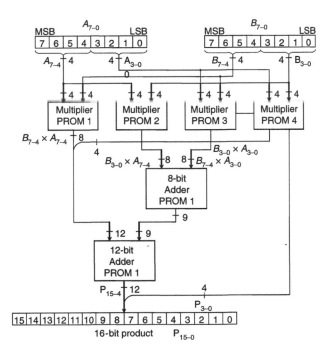

FIGURE 6.42
Logic circuit for Example 6.13.

A single PROM for multiplying two 8-bit numbers would have required 1,048,576 bits, that is, $2^{16} \times 16$ bits. Each multiplier PROM modules that we have considered in the circuit of Figure 6.42 had a memory size of only 2048 bits. Thus the memory size of the multiplier PROMs together amounts to 8192 bits. Except for the cost incurred in having two multibit adders, the logic circuit of Figure 6.42 would probably be less expensive. However, it is expected to be slower since it needs to perform the two additions. As always, speed and cost trade-off decision is very common in most design-related decisions.

Example 6.14

Design a binary-to-BCD converter module using off-the-shelf adder modules.

Solution

From Example 6.7, the outputs D_4, D_3, D_2, and D_1 are repeated here again:

$$D_4D_3D_2D_1 = \begin{cases} (B_4B_3B_2B_1), & \text{if } 0000 \leq B_4B_3B_2B_1 \leq 0100 \\ (B_4B_3B_2B_1 + 0011), & \text{otherwise} \end{cases}$$

This implies that whenever $B_4B_3B_2B_1 > 0100$, 0011 needs to be added to the 4-bit input ($B_4B_3B_2B_1$) to obtain the corrected output. From K-map formed of all input combinations greater than 0100, we obtain the corresponding logical function to be $\overline{B_4} + B_3B_1 + B_3B_2$. Accordingly, we want to add 0011 to $B_4B_3B_2B_1$ only when $\overline{B_4} + B_3B_1 + B_3B_2$ is a 1. Otherwise, $B_4B_3B_2B_1$ is left unaltered which is the same as that obtained by adding 0000 to $B_4B_3B_2B_1$. Thus, the logic function $\overline{B_4} + B_3B_1 + B_3B_2$ is fed directly the two least significant addend inputs of a 4-bit adder while the other 4-bit input (i.e., $B_4B_3B_2B_1$) is fed as its augend. The 4-bit sum obtained from this 4-bit adder unit, as shown in Figure 6.43, would then yield the desired BCD output.

The multibit adder circuit shown in Figure 6.36 can be used to add two n-bit binary numbers. Thus if we have to add more than two n-bit binary numbers, more than one level of ripple adder circuits may be needed. Figure 6.44, for example, shows a logic circuit for adding three n-bit numbers, P, Q, and R. However, the circuit of Figure 6.44 is not necessarily the only solution for adding such numbers. A valid solution can be found also by saving carries and then adding them separately. Such an addition is often referred to as *carry-save addition*. This scheme inhibits carry propagation by storing them in memory devices. Then during the last addition, the carries are allowed to propagate just as in a ripple adder. Figure 6.45 shows, for example, a logic circuit for adding three 3-bit numbers, $P_2P_1P_0$, $Q_2Q_1Q_0$, and $R_2R_1R_0$ using this latter scheme. When the circuits of Figures 6.44 and 6.45 are compared, we see that the latter circuit requires one less full adder.

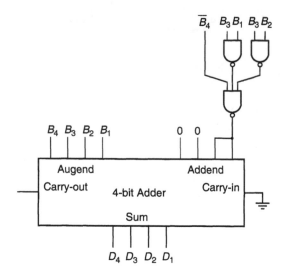

FIGURE 6.43
Binary-to-BCD converter module for Example 6.14.

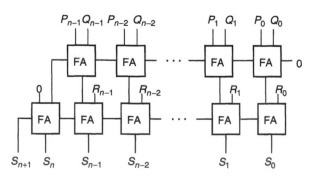

FIGURE 6.44
A layered ripple adder for adding three n-bit binary numbers.

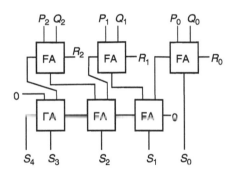

FIGURE 6.45
A carry-save adder for adding three 3-bit binary numbers.

Example 6.15
Using carry-save algorithm, obtain the sum of decimal numbers 7, 8, 9, 5, 11, and 12.

Solution

Step 1

$7 \Rightarrow 0111$		$5 \Rightarrow 0101$
$8 \Rightarrow 1000$		$11 \Rightarrow 1011$
$9 \Rightarrow 1001$		$12 \Rightarrow 1100$

$$0110 \quad \text{sum}_1 \qquad\qquad 0010 \quad \text{sum}_2$$
$$1001 \quad \text{carry}_1 \qquad\qquad 1101 \quad \text{carry}_2$$

Step 2

$$\text{sum}_1 \Rightarrow 0110$$
$$\text{carry}_1 \Rightarrow 1001$$
$$\text{sum}_2 \Rightarrow 0010$$

$$10110 \quad \text{sum}_3$$
$$00010 \quad \text{carry}_3$$

Step 3

$$\text{carry}_2 \Rightarrow 1101$$
$$\text{sum}_3 \Rightarrow 10110$$
$$\text{carry}_3 \Rightarrow 00010$$

$$001000 \quad \text{sum}_4$$
$$010110 \quad \text{carry}_4$$

Step 4. Now Sum$_4$ and Carry$_4$ may be added using the normal ripple addition scheme as follows:

$$\text{sum}_4 \Rightarrow 001000$$
$$\text{carry}_4 \Rightarrow 010110$$

$$0110100 \quad \text{final sum}$$

As expected, $(0110100)_2 = 52_{10}$.

6.6   Binary Subtraction

There are many occasions when one will be in need of a logic circuit capable of subtracting numbers. Just as we have done so with adder circuits, we can also design a full-blown subtracter circuit by attempting to first design a half subtracter. A half subtracter does not have a borrow input while a full subtracter does. This will require that we construct a truth table similar to that of Figure 6.30b. The output equations for (difference and borrow-out) of a half subtracter circuit as derived from such a truth table are respectively given by

$$D_i(X_i, Y_i) = X_i \oplus Y_I \tag{6.9}$$

$$B_i(X_i, Y_i) = \overline{X_i}Y_i \tag{6.10}$$

where Y_i is the subtrahend bit that is being subtracted from the minuend bit X_i. Not surprisingly, the logic equations for the half subtracter have almost the same form as those of a half adder. We can follow an identical process for designing a full subtracter. Accordingly, K-maps are formed for both the difference and the borrow-out outputs. Then by grouping the K-map entries, we obtain respectively the difference and the borrow-out as given by

$$D_i(X_i, Y_i) = X_i \oplus Y_i \oplus B_{i-1} \tag{6.11}$$

$$B_i(X_i, Y_i) = \overline{X_i}Y_i + B_{i-1}(\overline{X_i} + Y_i) \tag{6.12}$$

where B_{i-1} is the borrow-in at the next less significant bit position. The single-bit full subtracter modules so designed can be cascaded to form a multibit ripple subtracter. It too will be slow since the borrow-bit needs to propagate through all of the single-bit full subtracter modules.

We have already seen that this latter set of equations is similar to that of the full adder. In fact, between a subtracter and an adder the equation format for difference and sum are exactly alike. The borrow-out and carry-out differ in that X_i appears complemented during subtraction. This similarity implies though that subtraction can be realized using an adder but by complementing one of the inputs. In particular, we have seen in Chapter 1 that subtraction can be realized by adding complement of the subtrahend to the minuend. Consequently, rather than designing a subtracter from scratch, a multibit subtracter can be realized using complement arithmetic technique and a multibit adder. In any event, this scheme will have to accommodate the introduction of complemented subtrahend.

To meet the requirement of our present design, certain modification is necessary for obtaining complement of the subtrahend. If each bit of the n-bit subtrahend is complemented, the corresponding 1's complement of the number is obtained. The corresponding 2's complement number, on the other hand, may be realized next by adding a 1 to the LSB of this 1's complement number. The most straightforward subtracter design will involve an n-bit adder. Each of the subtrahend bits is first introduced to a NOT gate. The corresponding NOT output and the minuend bits are fed as inputs to the adder. Addition of a 1 at the LSB is rather easy since in that case we shall have to feed a 1 through the carry-in input (at the least significant bit) of the corresponding n-bit ripple adder. It has already been pointed out in Chapter 4 that a 2-input XOR logic gate can be used as a programmable NOT

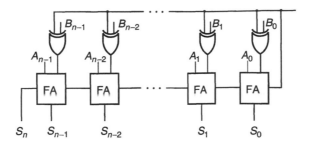

FIGURE 6.46
A n-bit adder/subtracter module.

gate since

$$0 \oplus Y = Y \tag{6.13}$$

$$1 \oplus Y = \bar{Y} \tag{6.14}$$

In the first case where the other input is a 0, the XOR output is the same as the input but when the other input is a 1, the output is a complement of the given input. It may be recalled that we have used XOR logic gates in a similar situation in Chapter 4 but for programming polarities of PLA outputs. This property of XOR logic opens up a desirable opportunity. It is more convenient to construct a dual-purpose functional unit such that it can perform either addition or subtraction operation on demand. Consider having a 2-input XOR gate with one input tied to either a subtrahend bit (in case of subtraction) or an augend bit (in case of addition). The other XOR input is tied to a select line, E. With E set to a 1 (in case of subtraction), complement of the corresponding subtrahend bit appears at the output. On the other hand, when E is a 0 (in the case of addition), the output is same as the input augend bit. Thus, the unique nature of XOR function provides the possibility of having both an adder and a subtracter.

Figure 6.46 shows an n-bit adder/subtracter logic circuit. The enable input, E, allows the module to add or to subtract by taking the 1's complement and adding a 1 to the LSB. E is fed as an input to the least significant full adder as its carry-in input. When $E = 0$, the logic circuit adds and when $E = 1$, the circuit complements B and then adds that to A with a carry-in of 1. The latter operation is identical to subtracting B from A. It must be noted that this programmable NOT function can be designed also using 1-of-2 multiplexers. One of the multiplexer inputs is tied to the variable itself and the other input is fed with its complement.

Example 6.16
Design a combinatorial logic circuit for obtaining 2's complement of an n-bit binary number.

Solution
By comparing any arbitrary multibit number and its 2's complement, we notice that both of the LSB's are identical. However, if the LSB a_0 is a 1, then the next more significant bit, a_1 appears inverted in the 2's complement number. A 2-input XOR gate with a_0 and a_1 as its inputs can be used to test whether or not the LSB is a 1. If and when the LSB is a 1, the other bit is complemented. However, if the LSB is a 0, the next higher significant bit remains unchanged. Similarly, the next higher significant

bit a_2 is complemented if either a_0 or a_1 is a 1. This can be accomplished first by taking an OR of a_0 and a_1 and then feeding the OR output and a_2 to a 2-input XOR gate. Following this process, therefore, the test (at bit n) as to whether or not any one of the lesser significant bits is a 1 can be realized by performing the OR operation $a_{n-1} + \ldots + a_3 + a_2 + a_1 + a_0$.

It is apparent, therefore, that n 2-input XOR logic gates and n different OR logic gates will be needed to obtain the 2's complement of an n-bit number. The OR gates perform tests as to whether or not a bit needs to be inverted while the XOR gates complement the bits when test results dictate so. The resulting logic circuit is obtained as shown in Figure 6.47. Note that this logic circuit has two inherent problems: both processing time and number of inputs feeding the OR gate increase with increasing n.

An arithmetic operation such as addition and/or subtraction may end up generating a value outside the range of acceptable binary numbers. Consequently, it is often necessary to flag an overflow so that the result is not processed any further. The overflow condition is indicated by either the most significant operand bits and the most significant sum bit or the carry-in and the carry-out of the most significant bit position. Designs of such overflow circuits have already been explored in Example 2.1.

6.7   High-Speed Addition

The ripple adders, ripple subtracters, and dual-purpose adder/subtracter modules are all victims of nonnegligible propagation delay. All these combinatorial logic circuits are iterative in nature since the same general set of logic is used to process input data at all bit positions. In general, these iterative networks exhibit properties analogous to those of sequential circuits. For now, however, we shall postpone discussion of such sequential circuits until in Chapter 10.

A single-bit full adder circuit can be designed so that it has no more than two levels of logic gates. Since circuit speed is a function of the number of logic levels, this 2-level full adder circuit is expected to be the fastest. The speed of an n-bit module, however, may not be too desirable since its propagation delay will have to account for the input signals to ripple through all of the single-bit adder circuits. For such devices the maximum delay is directly proportional to the number of FA units. An n-bit ripple adder implies that the

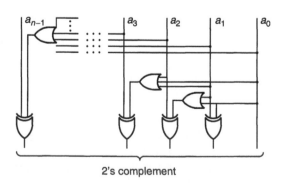

2's complement

FIGURE 6.47
Logic circuit for Example 6.16.

signals may have to pass through at least $2n$ levels of logic gates. Thus increasing adder speed beyond that obtainable using a ripple adder will depend on further reduction of logic gate levels through which carry bits must propagate. The speed also depends on the available technology as that determines the minimum delay for logic gates. Typically, the number of logic gates needs to be increased significantly so as to reduce the overall propagation delay. Circuit designers are thus confronted with a situation again where cost and speed may not be simultaneously enhanced.

To increase speed, it is not necessary that we consider only one bit position at a time. We could design an adder module that can add two 2-bit numbers, for example, and a carry-in. The output of such a logic circuit will include a single carry-out from the more significant bit position and two sum bits. A quick investigation of the resulting logic circuit will convince us that this process of adding two m-bit numbers (where $n \geq m > 2$) dramatically increases functional complexity of the unit thus making the addition process prohibitive.

There are two particular methods to speed up addition operations: *carry-skip* and *carry look-ahead* (CLA). To design these high-speed addition circuits, one needs to consider Figure 6.48 that lists four input combination cases for which a carry-out is generated. The carry-out is the same as the carry-in as long as one of the other two inputs is a 1. Also, the carry-out is always a 1 independent of carry-in when both the other inputs are 1s, and 0 if both are 0. Correspondingly, we define two useful functions: the carry-propagate, P_i, and the carry-generate, G_i

$$P_i = A_i \oplus B_i \tag{6.15}$$

$$G_i = A_i B_i \tag{6.16}$$

where A_i and B_i are the addend and augend bits, respectively, of the i-th full adder.

Consider the gate-level logic circuit of a single-bit full adder shown in Figure 6.49. It can be seen that the carry-propagate line P_i given by $A_i \oplus B_i$ determines if the carry will merely propagate through the single-bit adder. On the basis of such a rationale, a 2-bit carry-skip logic circuit can be designed as shown in Figure 6.50. It includes an AND of two successive propagate functions, P_i and P_{i+1}. Each of the bit i logic circuit as shown in Figure 6.50 is assumed to take in augend, addend, and carry-in bits as its inputs and generate carry and propagate outputs. When $P_i = P_{i+1} = 1$, the carry needs to simply propagate around

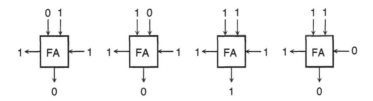

FIGURE 6.48
The input combination cases that lead to a carry-out.

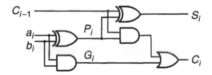

FIGURE 6.49
A single-bit full adder showing carry propagation.

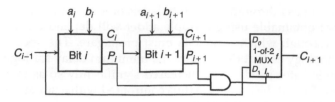

FIGURE 6.50
A 2-bit carry-skip adder.

rather than going through logic circuits corresponding to the two pairs of input bits. C_{i-1} is connected to one of the multiplexer inputs so that it can propagate unhindered should $P_iP_{i+1} = 1$. In that special case, the carry propagates through only the multiplexer rather than through additional logic circuit. A variety of lengths and combinations of carry skips are possible.

   The carry-propagate and carry-generate functions for all bit positions can be realized in parallel. Accordingly, if an adder circuit were to use only propagate and generate functions as its inputs, the resulting logic circuit may provide us with a faster alternative. For designing a CLA adder, therefore, we rewrite the full adder outputs in terms of the carry-propagate and carry-generate functions as

$$S_i = P_i \oplus C_{i-1} \tag{6.17}$$
$$C_i = G_i + P_iC_{i-1} \tag{6.18}$$

For a 4-bit CLA adder, in particular, the four carries may be obtained as follows:

$$C_0 = G_0 + P_0C_{-1} \tag{6.19}$$
$$C_1 = G_1 + P_1C_0$$
$$= G_1 + P_1G_0 + P_1P_0C_{-1} \tag{6.20}$$
$$C_2 = G_2 + P_2C_1$$
$$= G_2 + P_2G_1 + P_2P_1G_0 + P_2P_1P_0C_{-1} \tag{6.21}$$

and

$$C_3 = G_3 + P_3C_2$$
$$= G_3 + P_3G_2 + P_3P_2G_1 + P_3P_2P_1G_0 + P_3P_2P_1P_0C_{-1} \tag{6.22}$$

while the four sum bits are

$$S_0 = P_0 \oplus C_{-1} \tag{6.23}$$
$$S_1 = P_1 \oplus C_0 \tag{6.24}$$
$$S_2 = P_2 \oplus C_1 \tag{6.25}$$

and

$$S_3 = P_3 \oplus C_2 \tag{6.26}$$

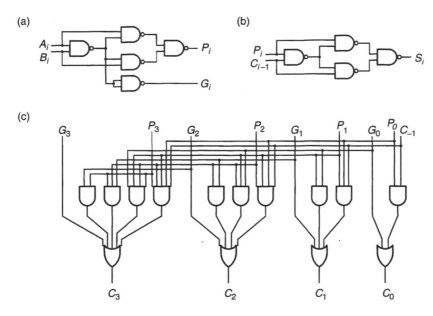

FIGURE 6.51
(a) A propagate-generate unit, (b) a sum unit, and (c) the 4-bit CLA generator.

Equation for each of the carry bits is determined (by repeated substitutions) in terms of the carry-in C_{-1} to the least significant bit positions. Within each bit position, if G_i is a 1, a carry is generated independent of the carry input C_{-1}. Likewise a carry input propagates from the input to the output if for that bit position P_i is a 1.

Equations 6.19 through 6.22 show that each of the carry-out bits depends only on the initial carry, C_{-1}, and the corresponding propagate and generate functions. In other words, if one were to realize a logic circuit corresponding to Equations 6.19 through 6.22, no carry-out bit depends on all of the less significant bit carries. The output C_3, for example, can be implemented with just a two-level logic circuit thus contributing to a high-speed 4-bit adder. The internal hardware for each of the modules of a 4-bit CLA adder is obtained as shown in Figure 6.51. First, the carry-propagate and carry-generate functions are generated using the circuit of Figure 6.51a. Next, the intermediate carries are obtained from these propagate and generate functions using a CLA generator as shown in Figure 6.51b. Finally, the sum bits are computed. Each propagate-generate unit requires five NAND gates, and each sum unit requires four NAND gates. The n-bit carry section, on the other hand, requires a total of $(n^2 + 5n)/2$ NAND gates. A 4-bit CLA adder, therefore, requires a total of 54 NAND gates and involves eight units of NAND delay. In comparison, addition using a 4-bit ripple adder requires 12 units of NAND delay. The 4-bit CLA, therefore, cuts down the time factor by about one-third. Similarly, a 64-bit CLA adder requires almost five times as many NAND gates as a 64-bit ripple adder, but it reduces the propagation delay by a factor of 17. It follows, therefore, that the CLA adder provides a faster addition alternative for large n.

The logic diagram of a 4-bit CLA adder is shown in Figure 6.52. Equations 6.19 through 6.22 suggest that the number of AND logic as well as the number of literals in each AND term increases by one with each successive carry equation. This poses a problem since it may not be possible to consider n too large. To solve this increase in the number of literals, a logic gate may have to be implemented with multiple levels of logic gates. However, that will slow down the addition process that we are trying to avoid using this CLA scheme. The nature of this limitation is compounded by the fact that the carry-in, C_{-1}, must drive a total of $n + 1$ gates. Assuming that an AND–OR logic can be easily realized using a NAND–NAND logic,

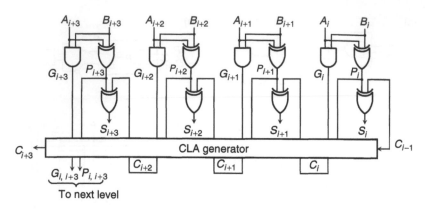

FIGURE 6.52
A 4-bit CLA adder.

we note that for an n-bit CLA adder, the sum and the propagate-generate subunits together require $9n$ 2-input NAND gates at each bit position. The CLA generator requires $(2n + 1)$ 2-input NAND gates and $(n + 3 - m)m$-input NAND gates for $3 \leq m \leq n + 1$. In addition, propagate functions are subjected to a fan-out requirement on the order of $(n + 1)^2/4$. In summary, a CLA adder for too large n also becomes prohibitive because of fan-in and fan-out limitations. The logic circuit complexity makes these implementations rather impractical for large n. As a consequence, therefore, the digital designers limit themselves to designing only 4-bit CLA adders.

Example 6.17

Use the schemes learned in Chapter 4 to realize a performance-optimized NAND-based logic circuit for C_3 for use in CLA adders.

Solution

Equation 6.22 expresses C_3 as

$$C_3 = G_3 + P_3C_2$$
$$= G_3 + P_3G_2 + P_3P_2G_1 + P_3P_2P_1G_0 + P_3P_2P_1P_0C_{-1}$$

The typical AND–OR implementation of this logical equation results in the logic circuit shown in Figure 6.53a. Assuming that we need to limit ourselves to using only 3-input AND and OR gates, we can rewrite Equation 6.22 by decomposing it as

$$C_3 = G_3 + P_3G_2 + \{(P_3P_2G_1 + (P_3P_2P_1)G_0 + (P_3P_2P_1)P_0G_0\}$$

The corresponding logic circuit is obtained as shown in Figure 6.53b. Its NAND-equivalent logic circuit can be obtained then as shown in Figure 6.53c. However, we could also rewrite Equation 6.22 in a fashion so that the speed factor can be improved by reducing the number of gate levels. It follows that

$$C_3 = (G_3 + P_3G_2 + P_3P_2G_1) + (P_3P_2P_1)G_0 + (P_3P_2P_1)P_0G_0$$

The resultant performance-optimized AND–OR logic circuit and its equivalent NAND-only logic circuit is thus obtained as that shown in Figures 6.53d and e respectively.

Typically, commercially available 4-bit adders perform an internal CLA. It can be argued that by connecting multiple 4-bit CLA adders, we could significantly improve the speed of larger adders. The most significant carry-out (as given by Equation 6.22) of a 4-bit CLA can be expressed also as

$$C_3 = G_{0,3} + P_{0,3}C_{-1} \tag{6.27}$$

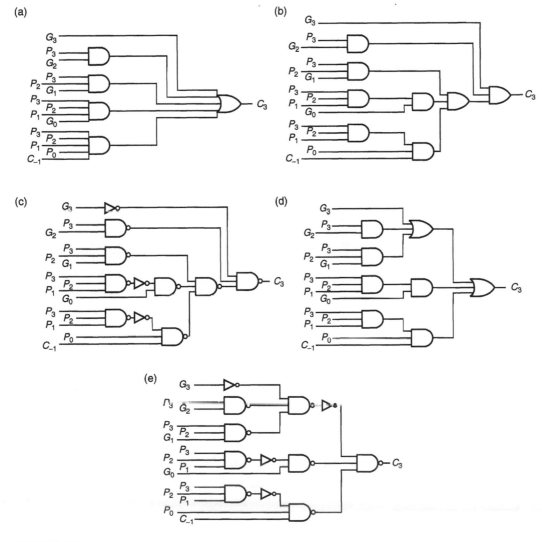

FIGURE 6.53
A performance-optimized circuit for generating C_3.

such that

$$G_{0,3} = G_3 + P_3 G_2 + P_3 P_2 G_1 + P_3 P_2 P_1 G_0 \tag{6.28}$$

$$P_{0,3} = P_3 P_2 P_1 P_0 \tag{6.29}$$

While the carries within each of the 4-bit CLA adders are produced at high speeds, the carries between the successive 4-bit CLA adder units just ripple through. This type of an adder is referred to as a *single-level CLA adder*. For designing a 16-bit adder, for example, the form of Equation 6.27 can be used to generate the carry-out outputs that propagate from one 4-bit CLA to the next as follows:

$$C_3 = G_{0,3} + P_{0,3} C_{-1} \tag{6.30a}$$

$$C_7 = G_{4,7} + P_{4,7} C_3 \tag{6.30b}$$

$$C_{11} = G_{8,11} + P_{8,11} C_7 \tag{6.30c}$$

$$C_{15} = G_{12,15} + P_{12,15} C_{11} \tag{6.30d}$$

Such a 16-bit single-level CLA adder consisting of four 4-bit CLA adders is obtained as shown in Figure 6.54.

We note though that Equations 6.30a through d have the same form as Equation 6.18. Thus it is possible to generate C_3, C_7, C_{11}, and C_{15} outputs using another level of CLA generator as shown in the logic circuit of Figure 6.55. In this latter design, the carries do not ripple between the 4-bit CLA adders; instead the second level CLA circuit generates these carries for use in the first-level CLA generators.

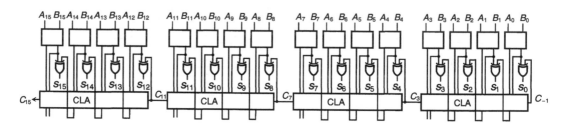

FIGURE 6.54
A single-level CLA adder for adding 16-bit numbers.

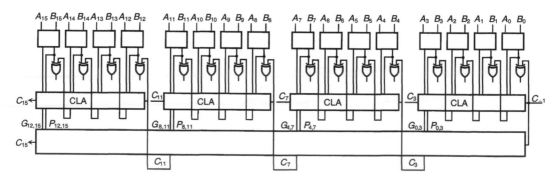

FIGURE 6.55
A two-level CLA generator for adding two 16-bit numbers.

CLA adder with more than two levels can be constructed by cascading two levels of 16-bit CLA adders. We note that C_{15} can be derived same following the argument that was used in Equation 6.27 as

$$C_{15} = G_{0,15} + P_{0,15}C_{-1} \qquad (6.31)$$

The 16-bit carry-generate functions $G_{0,15}$ and the carry-propagate functions $P_{0,15}$ have the same form as those listed in Equations 6.28 and 6.29 and as such can be used for generating higher-order carries. In general, an n-bit addition is performed using $\text{Int}(\log_4 n + 1)$ levels of CLA generators. A 16-bit adder with two levels of CLA generators is nearly 30% faster than that using a single level of CLA generator which itself is four times faster than a ripple adder. Gate delay in a CLA adder grows logarithmically with the number of bits, while that in the case of ripple adders grows linearly.

6.8 BCD Arithmetic

There are computers and calculators that operate on the inputs that are decimal rather than binary. Typically, decimal numbers are introduced using BCD code. In comparison, a BCD-based digital system requires more memory to store given information as it is associated with a less efficient coding. Fortunately, the final arithmetic result of BCD operations performed in these systems do not have to be decoded before displaying it as decimal digits. We have already encountered digital logic circuits that can be used for converting binary numbers into their BCD equivalents and BCD into their binary equivalents. A general procedure for designing BCD adders, for example, would be to first convert the BCD numbers into binary, then add the resulting binary numbers using binary additive scheme, and then reconvert the binary sum back into its BCD equivalent. This design approach requires both postcorrection and preprocessing circuits and, therefore, can be rather wasteful and unnecessarily costly.

We shall consider an alternative approach for BCD arithmetic so that we may be able to avoid the back-and-forth conversion steps. The to-be-designed BCD adder will need to have nine inputs and five outputs. Of the nine inputs, four inputs are designated for the BCD augend, another four for the BCD addend, and the ninth input for a carry-in from the next less significant digit position. Likewise, of the five outputs, one is for the carry-out to the next more significant digit position and the remaining four inputs are for the BCD sum. The truth table corresponding to such a BCD adder will include $2^9 = 512$ different input combinations, many of which will lead to don't-care conditions. The two obvious design choices for such a BCD adder circuit will involve using a PLD or minimize the multioutput logic functions using Q M technique and then generate the corresponding logic circuit using logic gates. In either case, the logic circuit will be rather involved for it to be pursued any further.

The easiest route to designing a BCD adder is to base its operation on a 4-bit FA module. However, BCD arithmetic using such binary adders is expected to be complicated since some of the sums or differences may turn out to be invalid in BCD code. When two BCD numbers are added using a 4-bit binary adder, it is possible to obtain 20 different sums of which only ten 4-bit numbers are valid. Six of the other 4-bit numbers are undefined in BCD and the other four are in fact 5-bit numbers. Figure 6.56 shows a truth table listing these 20 different sums, $C_3'S_2'S_2'S_1'S_0'$. The corresponding output column lists the desired BCD output $C_3S_2S_2S_1S_0$. While ten of the entries (the first ten) require no correction, the other ten entries need to be corrected in order for them to be interpreted as a valid BCD

$C_3'S_3'S_2'S_1'S_0'$	$C_3S_3S_2S_1S_0$
0 0 0 0 0	0 0 0 0 0
0 0 0 0 1	0 0 0 0 1
0 0 0 1 0	0 0 0 1 0
0 0 0 1 1	0 0 0 1 1
0 0 1 0 0	0 0 1 0 0
0 0 1 0 1	0 0 1 0 1
0 0 1 1 0	0 0 1 1 0
0 0 1 1 1	0 0 1 1 1
0 1 0 0 0	0 1 0 0 0
0 1 0 0 1	0 1 0 0 1
0 1 0 1 0	1 0 0 0 0
0 1 0 1 1	1 0 0 0 1
0 1 1 0 0	1 0 0 1 0
0 1 1 0 1	1 0 0 1 1
0 1 1 1 0	1 0 1 0 0
0 1 1 1 1	1 0 1 0 1
1 0 0 0 0	1 0 1 1 0
1 0 0 0 1	1 0 1 1 1
1 0 0 1 0	1 1 0 0 0
1 0 0 1 1	1 1 0 0 1

FIGURE 6.56
Outputs of a single-digit BCD adder that uses a 4-bit binary adder.

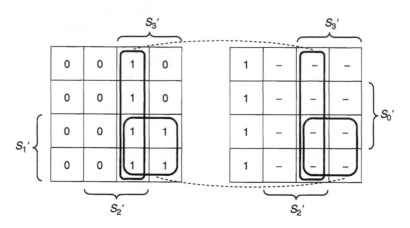

FIGURE 6.57
K-map for the BCD carry-out.

sum. The largest possible uncorrected sum is 10011, corresponding to the sum of two BCD 9s and a carry-in of 1. The binary sums greater than 01001, therefore, are all undefined and are in need of correction. If we assume only valid BCD numbers as inputs, then the 5-bit binary sums corresponding to 10100 through 11111 can be treated simply as don't cares.

As shown in the truth table of Figure 6.56, when the binary sum of two BCD digits exceeds 01001, a carry is generated and the sum will need to be corrected by means of a correction circuit. We see that the corrected sum can be obtained from the uncorrected sum simply by adding to it 00110. We also note that the BCD carry-out C_3 becomes a 1 each time we correct the binary sum. The BCD adder design thus requires the circuit to take two actions: decide when the correction is warranted and if needed correct the sum. The condition for generating a BCD carry-out can be found readily by minimizing C_3 using the K-map shown in Figure 6.57. This K-map is constructed using the entries listed in Figure 6.56. The desired

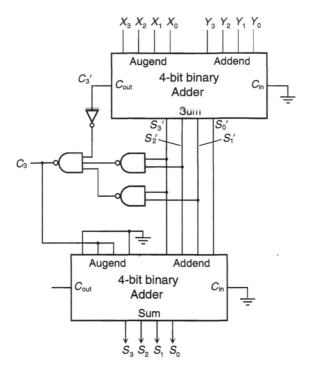

FIGURE 6.58
A single-digit BCD adder using 4-bit binary adders.

BCD carry-out is given by

$$C_3 = C_3' + S_1'S_3' + S_2'S_3' \tag{6.32}$$

A high C_3 indicates that a correction is warranted. To correct the binary sum, we may use a second 4-bit adder to add 0110.

A single-digit BCD adder may be constructed as shown by the logic circuit of Figure 6.58. The sum bits of the first 4-bit binary adder are introduced at the augend inputs of the second 4-bit binary adder. When the corrected carry-out, C_3, is a 0, the uncorrected sum is expected to remain unchanged. This is guaranteed by making sure that 0000 has been fed at the addend input of this second 4-bit binary adder. When C_3 is a 1, however, the uncorrected sum is added to 0110 (fed through the addend input). The corrected sum at the output of this second 4-bit binary adder. To assure proper correction, therefore, C_3 is fed directly to the two addend bits in the middle while the other two addend bits are grounded.

Should the designer be concerned with propagation delay, it is possible to use three single-bit full adders (or two single-bit full adders and one half adder) instead of a 4-bit binary adder for the correction circuit. This improvement is possible since we note that least significant bit S_0 needs no correction. A slightly improved version of this BCD adder is obtained as shown in Figure 6.59.

We next consider the design of a BCD subtracter. Subtraction process is more complex than addition since there is this distinct possibility of obtaining a negative difference. From Chapter 1, we know though that there are at least two ways to represent negative BCD numbers: either by 10's complement or by 9's complement. The technique using 10's complement would result in a logic circuit similar to that obtained for binary subtraction as shown in Figure 6.46. The scheme using 9's complement, however, will require a mandatory end-round-carry into the system. Such BCD subtracter that uses 9's complement

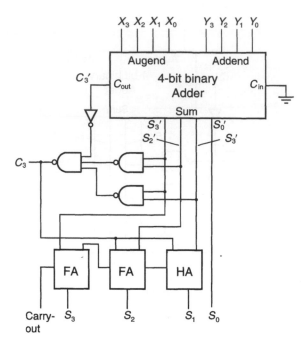

FIGURE 6.59
A single-digit BCD adder using smaller correction circuit.

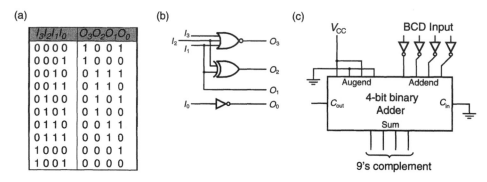

FIGURE 6.60
Realizing 9's complement: (a) truth table, (b) using combinatorial logic circuit, and (c) using 4-bit binary adder.

numbers can be designed using a BCD adder and a 9's complement logic circuit. The truth table for obtaining 9's complement is obtained as shown in Figure 6.60a.

Using the table of Figure 6.60a, we obtain the 9's complement outputs as

$$O_3 = \overline{I_3 + I_2 + I_1} \qquad \qquad (6.33a)$$

$$O_2 = I_1 \oplus I_2 \qquad \qquad (6.33b)$$

$$O_1 = I_1 \qquad \qquad (6.33c)$$

$$O_0 = \bar{I}_0 \qquad \qquad (6.33d)$$

The resulting BCD 9's complement logic circuit can be readily obtained as shown in Figure 6.60b. Figure 6.60c shows a variation of this logic circuit but it uses a 4-bit binary

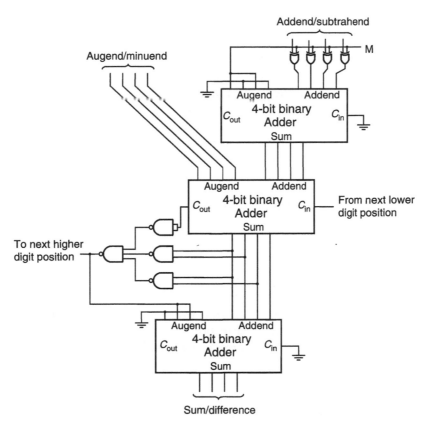

FIGURE 6.61
A single-digit BCD adder/subtracter.

adder. This 4-bit binary adder first takes the 1's complement of the BCD digit and then adds to it 1010 (the 2's complement of six, 0110). The circuit of Figure 6.60c is more preferable since then the 9's complement circuit can be used to construct a programmable BCD adder/subtracter. Each of the inverters of Figure 6.60c can be replaced using a 2-input XOR gate. The resulting single-digit BCD adder/subtracter is obtained then as shown in Figure 6.61. Each decade unit consists of three 4-bit binary adders. The first 4-bit adder, functions as the backbone for taking the 9's complement, the second functions as an adder and the third realizes postcorrection. When M is a 0, the unit performs addition with the complement circuit simply adding 0000 to the given BCD number. On the other hand, when M is a 1, the subtrahend is complemented and is then added to the minuend to achieve the equivalent of a subtraction. When several such single-digit BCD adder/subtracters are cascaded to make a multidigit BCD adder/subtracter, the carry-out output is fed as an input to the carry-in of the next more significant single-digit BCD adder/subtracter. The carry-out of the most significant digit BCD adder/subtracter is fed as a carry-in to the least significant digit BCD adder/subtracter. This particular data routing guarantees the accommodation of an end-around-carry.

Example 6.19
Show the following arithmetic operations: (a) $48+33$, (b) $33-48$, and (c) $48-33$ using XS3 code.

Solution

We have already seen in Section 1.7 that the binary sum of two given XS3 numbers will need to be postprocessed further for obtaining from it the correct XS3 sum. While determining the binary sum using a 4-bit binary adder, carry is allowed to go from its current digit position to the next higher significant digit position. Accordingly, each of the incorrect result digits is postprocessed as follows: When a carry has been produced, a 0011 is added to the incorrect result. On the other hand, a 0011 is subtracted from the incorrect result digit if a carry were not produced at that digit position. This latter subtraction can be accomplished by adding 1101 (the 2's complement of 0011) and ignoring any carry if so produced by any of the digit positions. During this final correction phase, however, the carry-out of a digit position is restricted from going to the next higher significant digit position. The operations may be carried out as follows:

(a)

48_{10}	0111	1011	XS3
$+33_{10}$	0110	0110	XS3
81	1110	0001	incorrect result (with no carry-out)
	+1101	0011	correction
	1011	0100	correct result (with a carry-out)

(b) Since $48_{10} = 0111\ 1011_2$, $-48_{10} = 1000\ 0100_2$. Therefore, the subtraction in question can be carried out as

33_{10}	0110	0110	XS3
-48_{10}	1000	0100	9's complement of XS3
-15_{10}	1110	1010	incorrect result (with no carry-out)
	1101	1101	correction
	1011	0111	result (with a carry-out)
	0100	1000	correct result

(c) Since $33_{10} = 0110\ 0110_2$, $-33_{10} = 1001\ 1001_2$. Therefore, the addition can be carried out as

48_{10}	0111	1011	XS3
-33_{10}	1001	1001	9's complement of XS3
15_{10}	0001	0100	incorrect result (with carry-out)
		1	end-around carry
	0011	0011	correction
	0100	1000	correct result (with no carry-out)

In part (c), that the result is positive is indicated by the carry resulting from the most significant bit when adding the 9's complement number. In such a case, a 1 (only at the least significant digit position) and the necessary correction bits are added to the incorrect result.

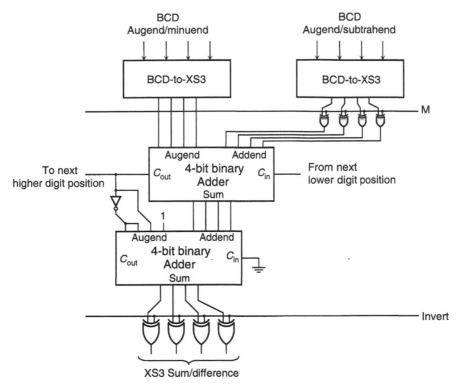

FIGURE 6.62
A single-digit BCD adder/subtracter that uses XS3 code for its arithmetic operations.

The 9's complement of an XS3 number is equivalent to taking its 1's complement. Consequently, 9's complement of an XS3 number is obtained by passing each of the component bits through a level of either NOT gates or controllable 2-input XOR logic gates. Figure 6.62 shows a BCD adder/subtracter circuit that first converts the BCD digit to its equivalent XS3 number. The XS3 numbers (one of these two numbers appears complemented during subtraction) are then added using a binary adder. The carry-out output determines the follow-on correction to be employed at that digit position. The output of the first level binary adder is corrected by adding to it either 0011 or 1101, depending on whether or not a carry has been generated out of that digit position. The INVERT input is activated when there is no end-around-carry (for example, as that shown in Example 6.19c). With INVERT = 1, the output is complemented yielding the final result. If one wanted to obtain instead the result in BCD form, each of the sum digits will need to be converted next using an XS3-to-BCD converter.

6.9 Comparators

A binary *comparator* is a logic circuit that determines the relative magnitude of two strings (often of equal length) of binary numbers. The logic circuit after having treated two *n*-bit numbers (*A* and *B*) conveys one of three decisions: *A* is greater than *B*, *A* is less than *B*, or *A* is equal to *B*. All three of these output conditions (f_1, f_2, and f_3 respectively) are of utmost importance in many digital systems. Each of these three relational decisions has a complement as well. The complement of "greater than" is "less than or equal to", that of "less than" is "greater than or equal to", and that of "equal to" is "not equal to."

When both A and B are only single-bit numbers, the three relational outputs are obtained from their corresponding K-maps as follows:

$$f_1 = \overline{A_0}B_0 \qquad\qquad (6.34\text{a})$$

$$f_2 = A_0\overline{B_0} \qquad\qquad (6.34\text{b})$$

$$f_3 = \overline{A_0 \oplus B_0} \qquad\qquad (6.34\text{c})$$

In the case where the two given numbers are 2-bits in size, the outputs turn out to be

$$f_1 = f(A_1, A_0, B_1, B_0) = \Sigma m(1,2,3,6,7,11) \qquad\qquad (6.35\text{a})$$

$$f_2 = f(A_1, A_0, B_1, B_0) = \Sigma m(4,8,9,12,13,14) \qquad\qquad (6.35\text{b})$$

$$f_3 = f(A_1, A_0, B_1, B_0) = \Sigma m(0,5,10,15) \qquad\qquad (6.35\text{c})$$

While Equations 6.35a and 6.35b are reducible a little, Equation 6.35c is simply not reducible. If one wants to design an n-bit binary comparator with $n > 2$, the traditional process of design will result in unwieldy consequences. Fortunately, we have a choice to explore an iterative design for the comparator. This is possible since particular aspects of decision making process are identical when one scans the given pair of n-bit numbers during comparison.

Example 6.20
Design a stand-alone PLD circuit that compares two 4-bit numbers.

Solution
For the two numbers $A = (A_3A_2A_1A_0)$ and $B = (B_3B_2B_1B_0)$, the three comparator outputs are as follows:

$$f_1 = \overline{A_3}B_3 + (A_3 \otimes B_3)\overline{A_2}B_2 + (A_3 \otimes B_3)(A_2 \otimes B_2)\overline{A_1}B_1$$
$$+ (A_3 \otimes B_3)(A_2 \otimes B_2)(A_1 \otimes B_1)\overline{A_0}B_0$$

$$f_2 = A_3\overline{B_3} + (A_3 \otimes B_3)A_2\overline{B_2} + (A_3 \otimes B_3)(A_2 \otimes B_2)A_1\overline{B_1}$$
$$+ (A_3 \otimes B_3)(A_2 \otimes B_2)(A_1 \otimes B_1)A_0\overline{B_0}$$

$$f_3 = (A_3 \otimes B_3)(A_2 \otimes B_2)(A_1 \otimes B_1)(A_0 \otimes B_0)$$

where $A_j \otimes B_j \equiv C_j = A_jB_j + \overline{A_j}\,\overline{B_j}$. When expressed in its canonical form, therefore, the first two outputs will have 15 product terms and the third output will include sixteen product terms. Most PALs, for example, may not be able to accommodate that many product terms. Consequently, we may first obtain C_j and then use it to obtain the output functions as follows:

$$f_1 = \overline{A_3}B_3 + C_3\overline{A_2}B_2 + C_3C_2\overline{A_1}B_1 + C_3C_2C_1\overline{A_0}B_0$$

$$f_2 = A_3\overline{B_3} + C_3A_2\overline{B_2} + C_3C_2A_1\overline{B_1} + C_3C_2C_1A_0\overline{B_0}$$

$$f_3 = C_3C_2C_1C_0$$

The corresponding output equations now have at most four product terms. A PAL capable of accommodating eight inputs (for the two 4-bit numbers) and seven outputs (for the four Cs and three fs) can be used now to realize this 4-bit comparator.

In order to be able to use a meaningful iterative cell, we shall consider designing 4-bit comparators. Assume further that its design incorporates the decision making thought process that one encounters when examining two strings of numbers from the right to the left. Such 4-bit comparators will need to be suitably cascaded to perform comparison of n-bit numbers where $n > 4$. If the two 4-bit numbers (inputs to the j-th comparator cell) under consideration are equal, for example, decision pertaining to the two strings of $4j$ bits processed so far will be the same as that determined by the $(j\text{-}1)$-th comparator cell. In other words, the three decision functions of a comparator take on the same values as that obtained at the next less significant comparator provided the latest two 4-bit numbers were found to be equal. On the other hand, if the value of the 4-bits of A were to be determined to be less (or greater) than that of the 4-bits of B, then the decision functions of the j-th comparator cell will supersede all previously obtained comparator decisions on its right. The comparator under consideration then makes the decision that the string of $4j$ bits of A processed so far is less (or greater) in value than the corresponding $4j$ bits of B.

In general, iterative cells must convey decisions pertaining to the number just examined and feed those decisions outputs to the next higher significant iterative cell as its inputs. Accordingly, if we were to design a 4-bit comparator as the basic iterative cell, each such cell will have eleven inputs (corresponding to two 4-bit numbers and three decision outputs from the less significant cell). The three particular comparator inputs that take decision-out information (f_1, f_2, f_3) from the next less significant comparator will be referred to as the decision-in inputs (D_1, D_2, and D_3 respectively). The $D_1D_2D_3$ input of the least significant comparator is set to 001 implying that the bits on the right of the least significant comparator are equal. The logic circuit of Example 6.20 may not be usable here since it has made no room for accepting any decision-in inputs.

Figure 6.63b shows the truth table for a 4-bit binary comparator cell. The derivation of the corresponding decision-out functions is left as an end-of-chapter exercise (Problem 6.42). Figure 6.64 shows the logic circuit of a 16-bit comparator that has four 4-bit comparators cascaded together. The three decision-out outputs of the most significant comparator yields the final result.

6.10 Combinational Circuit Design Using VHDL

In this section we illustrate the steps for combinational circuit design using VHDL with several circuit design examples discussed in this chapter early. Consider the design of a 5-bit Gray-to-binary converter discussed in Example 6.3. One of the possible implementations,

(a)

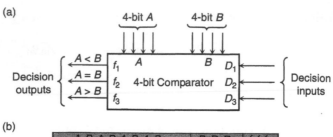

(b)

$A_3B_3A_2B_2A_1B_1A_0B_0$	$D_1D_2D_3$	$f_1f_2f_3$
$A_3>B_3$	- - -	1 0 0
$A_3<B_3$	- - -	0 1 0
$A_3=B_3\ A_2>B_2$	- - -	1 0 0
$A_3=B_3\ A_2<B_2$	- - -	0 1 0
$A_3=B_3\ A_2=B_2\ A_1>B_1$	- - -	1 0 0
$A_3=B_3\ A_2=B_2\ A_1<B_1$	- - -	0 1 0
$A_3=B_3\ A_2=B_2\ A_1=B_1\ A_0>B_0$	- - -	1 0 0
$A_3=B_3\ A_2=B_2\ A_1=B_1\ A_0<B_0$	- - -	0 1 0
$A_3=B_3\ A_2=B_2\ A_1=B_1\ A_0=B_0$	1 0 0	1 0 0
$A_3=B_3\ A_2=B_2\ A_1=B_1\ A_0=B_0$	0 1 0	0 1 0
$A_3=B_3\ A_2=B_2\ A_1=B_1\ A_0=B_0$	0 0 1	0 0 1

FIGURE 6.63

A 4-bit comparator cell: (a) logic module and (b) truth table.

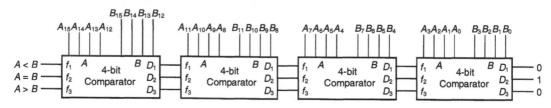

FIGURE 6.64

A 16-bit comparator.

as discussed in Section 6.3, uses four XOR gates. A functional VHDL description for this implementation is shown in Figure 6.65. This implementation has a shortcoming in that input–output pairs have different logic levels, therefore, with unequal delay, which may be of concern in digital systems. An alternative structural VHDL implementation is shown in Figure 6.66, where logic levels between any input–output pair are identical. Figure 6.67 illustrates a functional VHDL description of the 5-bit Gray-to-binary converter. Accordingly, a functional VHDL description of a 5-bit binary-to-Gray converter is shown in Figure 6.68.

Example 6.7 discusses n-bit binary-to-BCD conversion. A 4-bit binary-to-BCD converter circuit is illustrated in Figure 6.18. To describe this circuit in VHDL, a structural description and a functional description are shown in Figures 6.69 and 6.70, respectively. A functional VHDL description of a 4-bit BCD-to-binary converter is shown in Figure 6.71.

6.11 Arithmetic Logic Unit

The previous sections have identified many situations where we may need to employ multibit arithmetic operations. However, it may as well be necessary to perform bit-by-bit

```
-- A structural VHDL description of                        -- 01
-- a 5-bit Gray-to-binary converter                        -- 02
library ieee, lcdf_vhdl;                                   -- 03
use ieee.std_logic_1164.all, lcdf_vhdl.func_prims.all;     -- 04
entity Gray2Binary_5bit is                                 -- 05
   port(G0, G1, G2, G3, G4: in std_logic;                  -- 06
        B0, B1, B2, B3, B4: out std_logic);                -- 07
end Gray2Binary_5bit;                                      -- 08
                                                           -- 09
architecture structural_G2B of Gray2Binary_5bit is        -- 10
                                                           -- 11
   component XOR 2                                         -- 12
      port(din1, din2: in std_logic;                       -- 13
           dout: out std_logic);                          -- 14
   end component;                                          -- 15
                                                           -- 16
   signal  B3_t, B2_t, B1_t: std_logic;                    -- 17
                                                           -- 18
   begin                                                   -- 19
      B4 <= G4;                                            -- 20
      B3 <= B3_t;                                          -- 21
      B2 <= B2_t;                                          -- 22
      B1 <= B1_t;                                          -- 23
      g3: XOR2 port map(G4, G3, B3_t);                     -- 24
      g2: XOR2 port map(B3_t, G2, B2_t);                   -- 25
      g1: XOR2 port map(B2_t, G1, B1);                     -- 26
      g0: XOR2 port map(B1_t, G0, B0);                     -- 27
end structural_G2B;                                        -- 28
-- End of the VHDL description                             -- 29
```

FIGURE 6.65
A structural VHDL description of a 5-bit gray-to-binary converter using four 2-input XOR gates.

logic of multibit operands. Accordingly, a multifunction logic circuit that can operate on groups of bits is found to be very useful to the digital designers. Familiar combinatorial techniques may be employed to construct such a multifunction circuit, referred to as an *arithmetic logic unit* (ALU). Its different arithmetic and logic operations are activated by suitably supplying the corresponding address bits to this functional unit.

The various logic outputs may be most easily realized by routing the inputs to a series of logic gates that perform those logic functions and then feed the gate outputs to a MUX. The MUX selectors are used then to select one from among these logic outputs. The size of the MUX used will depend on the number of logic operations that will have to be performed. For an n-bit ALU, however, one needs to employ n such iterative cells. Commercially available ALU often operates on two 4-bit numbers to yield up to 32 arithmetic and 16 logic operations selected by a combination of five selector inputs, and a carry-in input. In this section, we consider the design of a relatively less complex ALU to demonstrate the process involving some typical functions.

Consider an ALU that will realize eight functions: four arithmetic operations (add, subtract, increment, and decrement) and four logic operations (AND, OR, XOR, and NOT). Since there are eight operations, three selectors, as shown in Figure 6.72a, will need to be employed for designing the ALU. The operands all go through the same arithmetic or logic transformations and consequently the same selector values may be used in each of these n single-bit ALU modules. The overall logic circuit can be represented thus by an interconnected group of single-bit ALU modules. For the most part, a logic operation at any one bit position is independent of the logic outputs at other bit positions. On the other hand, the four arithmetic operations are realized using a single-bit full adder at each of the bit positions. The exact arithmetic operation is selected by determining what exact operands

```
-- A structural VHDL description of                          -- 01
-- a 5-bit Gray-to-binary converter                          -- 02
library ieee, lcdf_vhdl;                                     -- 03
use ieee.std_logic_1164.all, lcdf_vhdl.func_prims.all;       -- 04
entity Gray2Binary_5bit is                                   -- 05
   port(G0, G1, G2, G3, G4: in std_logic;                    -- 06
        B0, B1, B2, B3, B4: out std_logic);                  -- 07
end Gray2Binary_5bit;                                        -- 08
                                                             -- 09
architecture structural_G2B of Gray2Binary_5bit is          -- 10
                                                             -- 11
   component XOR2                                            -- 12
      port(din1, din2: in std_logic;                         -- 13
           dout: out std_logic);                            -- 14
   end component;                                            -- 15
                                                             -- 16
   component XOR3                                            -- 17
      port(din1, din2, din3: in std_logic;                   -- 18
           dout: out std_logic);                            -- 19
   end component;                                            -- 20
                                                             -- 21
   component XOR4                                            -- 22
      port(din1, din2, din3, din4: in std_logic;             -- 23
           dout: out std_logic);                            -- 24
   end component;                                            -- 25
                                                             -- 26
   component XOR5                                            -- 27
      port(din1, din2, din3, din4, din5: in std_logic;       -- 28
           dout: out std_logic);                            -- 29
   end component;                                            -- 30
                                                             -- 31
   component AND2                                            -- 32
      port(din1, din2: in std_logic;                         -- 33
           dout: out std_logic);                            -- 34
   end component;                                            -- 35
                                                             -- 36
   begin                                                     -- 37
      g4: AND2 port map(G4, G4, B4);                         -- 38
      g3: XOR2 port map(G4, G3, B3);                         -- 39
      g2: XOR3 port map(G4, G3, G2, B2);                     -- 40
      g1: XOR4 port map(G4, G3, G2, G1, B1);                 -- 41
      g0: XOR5 port map(G4, G3, G2, G1, G0, B0);             -- 42
end structural_G2B;                                          -- 43
-- End of the VHDL description                               -- 44
```

FIGURE 6.66
An alternative structural VHDL description of a 5-bit gray-to-binary converter with equal input–output delay.

need to be fed to this single-bit full adder. The output of the adder and that of the logic unit can then be multiplexed further. The selector input (based on its value) to this 1-of-2 multiplexer determines whether or not the desired operation is an arithmetic or a logic operation.

Figure 6.72b shows the logic block diagram for the j-th single-bit ALU while in Figure 6.72c this same single-bit ALU is further decomposed to show its two subcells: logic unit (LU) and arithmetic unit (AU). The top-level selector S_2 selects output from either the LU or AU. Note that the output of arithmetic operations at any one bit often depends on that at the less significant bit positions. From Figure 6.72a, we find that A is a common operand for all of the arithmetic operations. Consequently, A_i can be fed directly to one of the full adder inputs. The carry-in input at bit i is supplied by the carry-out output of the full adder at bit i-1. We need to consider next the status of the carry-in input and that of the other input of the AU full adder. For "add" operation, B_i is fed as the other input of the full

```
-- A functional VHDL description of            -- 01
-- a 5-bit Gray-to-binary converter            -- 02
library ieee;                                  -- 03
use ieee.std_logic_1164.all;                   -- 04
entity Gray2Binary_5bit is                     -- 05
    port(G0, G1, G2, G3, G4: in std_logic;     -- 06
         B0, B1, B2, B3, B4: out std_logic);   -- 07
end Gray2Binary_5bit;                          -- 08
                                               -- 09
architecture functional_G2B of Gray2Binary_5bit is   -- 10
                                               -- 11
    begin                                      -- 12
        B4 <= G4;                              -- 13
        B3 <= G3 xor G4;                       -- 14
        B2 <= G2 xor G3 xor G4;                -- 15
        B1 <= G1 xor G2 xor G3 xor G4;         -- 16
        B0 <= G0 xor G1 xor G2 xor G3 xor G4;  -- 17
end functional_G2B;                            -- 18
-- End of the VHDL description                 -- 19
```

FIGURE 6.67
A functional VHDL description of a 5-bit gray-to-binary converter.

```
-- A functional VHDL description of            -- 01
-- a 5-bit binary-to-Gray converter            -- 02
library ieee;                                  -- 03
use ieee.std_logic_1164.all;                   -- 04
entity Binary2Gray_5bit is                     -- 05
    port(B0, B1, B2, B3, B4: in std_logic;     -- 06
         G0, G1, G2, G3, G4: out std_logic);   -- 07
end Binary2Gray_5bit;                          -- 08
                                               -- 09
architecture functional_B2G of Binary2Gray_5bit is   -- 10
                                               -- 11
    begin                                      -- 12
        G4 <= B4;                              -- 13
        G3 <= B3 xor B4;                       -- 14
        G2 <= B2 xor B3;                       -- 15
        G1 <= B1 xor B2;                       -- 16
        G0 <= B0 xor B1;                       -- 17
end functional_B2G;                            -- 18
-- End of the VHDL description                 -- 19
```

FIGURE 6.68
A functional VHDL description of a 5-bit binary-to-Gray converter.

adder. However, since "subtraction" operation involves 2's complement representation, 1's complement of B_i is fed directly to the full adder while at the same time carry-in at the least significant bit position is set to a 1.

For the "increment" operation a 1 should be fed through the carry-in input of the full adder at the least significant bit position. During increment, however, a 0 (instead of B_i) should be fed through the other full adder input. "Decrement", on the other hand involves adding A_i to the 2's complement of 1. This is accomplished by setting the least significant carry bit to a 1 and by feeding a 1 to each of this other full adder inputs. Figure 6.73a shows the corresponding n-bit ALU where a carry-generate (CG) unit placed on the right of the least significant full adder supplies the necessary carry-in input. The subfunction YG shown in Figure 6.73b supplies either B_i or its variations, as required, to the full adder.

```
-- A structural VHDL description of                          -- 01
-- a 4-bit binary-to-BCD converter                           -- 02
library ieee, lcdf_vhdl;                                     -- 03
use ieee.std_logic_1164.all, lcdf_vhdl.func_prims.all;       -- 04
entity Binary2BCD_4bit is                                    -- 05
   port(B1, B2, B3, B4: in std_logic;                        -- 06
        D1, D2, D3, D4: out std_logic);                      -- 07
end Binary2BCD_4bit;                                         -- 08
                                                             -- 09
architecture structural_B2BCD_4bit of Binary2BCD_4bit is     -- 10
                                                             -- 11
   component OR3                                             -- 12
      port(in1, in2, in3: in std_logic;                      -- 13
           out1: out std_logic);                             -- 14
   end component;                                            -- 15
                                                             -- 16
   component OR2                                             -- 17
      port(in1, in2: in std_logic;                           -- 18
           out1: out std_logic);                             -- 19
   end component;                                            -- 20
                                                             -- 21
   component INV1                                            -- 22
      port(in1: in std_logic; out1: out std_logic);          -- 24
   end component;                                            -- 25
                                                             -- 26
   component AND3                                            -- 27
      port(in1, in2, in3: in std_logic;                      -- 28
           out1: out std_logic);                             -- 29
   end component;                                            -- 30
                                                             -- 31
   component AND2                                            -- 32
      port(in1, in2: in std_logic;                           -- 33
           out1: out std_logic);                             -- 34
   end component;                                            -- 35
                                                             -- 36
   signal  a31, a32, a41, a32b1b, a21, a41b: std_logic;       -- 37
   signal  a3b2, a4b3b1, a321b, b1b, b2b: std_logic;          -- 38
   signal  b3b, b4b: std_logic;                              -- 39
                                                             -- 40
   begin                                                     -- 41
      g0: INV1 port map(B1, b1b);                            -- 42
      g1: INV1 port map(B2, b2b);                            -- 43
      g2: INV1 port map(B3, b3b);                            -- 44
      g3: INV1 port map(B4, b4b);                            -- 45
      g4: AND2 port map(B3, B1, a31);                        -- 46
      g5: AND2 port map(B3, B2, a32);                        -- 47
      g6: AND2 port map(B4, B1, a41);                        -- 48
      g7: AND2 port map(B2, B1, a21);                        -- 49
      g8: AND2 port map(B2, b1b, a41b);                      -- 50
      g9: AND2 port map(b3b, B2, a3b2);                      -- 51
      g10: AND3 port map(B3, b2b, b1b, a32b1b);              -- 52
      g11: AND3 port map(b4b, b3b, B1, a4b3b1);              -- 53
      g12: AND3 port map(B3, B2, b1b, a321b);                -- 54
      g13: OR3 port map(a31, a32, B4, D4);                   -- 55
      g14: OR2 port map(a41, a32b1b, D3);                    -- 56
      g15: OR3 port map(a21, a41b, a3b2, D2);                -- 57
      g16: OR3 port map(a4b3b1, a321b, a41b, D1);            -- 58
                                                             -- 59
end structural_B2BCD_4bit;                                   -- 60
-- End of the VHDL description                               -- 61
```

FIGURE 6.69
A structural VHDL description of a 4-bit binary-to-BCD converter.

```
-- A functional VHDL description of           -- 01
-- a 4-bit binary-to-BCD converter            -- 02
library ieee;                                 -- 03
use ieee.std_logic_1164.all;                  -- 04
entity Binary2BCD_4bit is                     -- 05
   port(B1, B2, B3, B4: in std_logic;         -- 06
        D1, D2, D3, D4: out std_logic);       -- 07
end Binary2BCD_4bit;                          -- 08
                                              -- 09
architecture functional_B2BCD of Binary2BCD_4bit is  -- 10
                                              -- 11
   signal B1_b, B2_b, B3_b, B4_b: std_logic;  -- 12
                                              -- 13
   begin                                      -- 14
      B1_b <= not B1;                         -- 15
      B2_b <= not B2;                         -- 16
      B3_b <= not B3;                         -- 17
      B4_b <= not B4;                         -- 18
      D1 <= ((B4_b and B3_b and B1) or        -- 19
             (B3 and B2 and B1_b) or (B4 and B1_b));  . -- 20
      D2 <= ((B2 and B1) or (B4 and B1_b) or  -- 21
             (B3_b and B2));                  -- 22
      D3 <= ((B4 and B1) or (B3 and B2_b and B1));  -- 23
      D4 <= ((B3 and B1) or (B3 and B2) or B4);  -- 24
end functional_B2BCD;                         -- 25
-- End of the VHDL description                -- 26
```

FIGURE 6.70
A functional VHDL description of a 4-bit binary-to-BCD converter.

```
-- A functional VHDL description of           -- 01
-- a 4-bit BCD-to-binary converter            -- 02
library ieee;                                 -- 03
use ieee.std_logic_1164.all;                  -- 04
entity BCD2Binary_4bit is                     -- 05
   port(D1, D2, D3, D4: in std_logic;         -- 06
        B1, B2, B3, B4: out std_logic);       -- 07
end BCD2Binary_4bit;                          -- 08
                                              -- 09
architecture functional_BCD2B of BCD2Binary_4bit is  -- 10
                                              -- 11
   signal D1_b, D2_b, D3_b, D4_b: std_logic;  -- 12
                                              -- 13
   begin                                      -- 14
      D1_b <= not D1;                         -- 15
      D2_b <= not D2;                         -- 16
      D3_b <= not D3;                         -- 17
      D4_b <= not D4;                         -- 18
      B1 <= D1 xor D4;                        -- 19
      B2 <= ((D2 and D4_b) or (D1 and D2_b and D4) or  -- 20
             (D1_b and D2));                  -- 21
      B3 <= ((D2 and D4_b) or (D1 and D2_b and D4) or  -- 22
             (D1_b and D3_b and D4));         -- 23
      B4 <= ((D3 and D4) or (D1 and D2 and D3));  -- 24
end functional_BCD2B;                         -- 25
-- End of the VHDL description                -- 26
```

FIGURE 6.71
A functional VHDL description of a 4-bit BCD-to-binary converter.

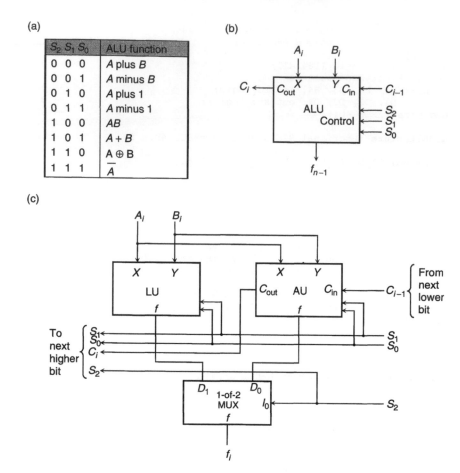

FIGURE 6.72
A single-bit arithmetic logic unit: (a) truth table, (b) block diagram, and (c) its subfunctions.

The truth table of Figure 6.72a can be used now to derive the LU output while that of Figure 6.74 can be used to obtain CG and YG outputs. These three outputs are given by

$$f_j = \overline{S_1}A_jB_j + S_0A_j\overline{B_j} + S_0\overline{A_j}B_j + S_1\overline{S_0}\,\overline{A_j} \tag{6.36a}$$

$$C_{-1} = S_1 \oplus S_0 \tag{6.36b}$$

$$YG_j = (\overline{S_j}B_j) \oplus S_0 \tag{6.36c}$$

The corresponding LU, CG, and YG subfunctions for the desired ALU are obtained as shown in Figure 6.75. Note that in the logic circuit of Figure 6.75, LU has been derived instead using a 1-of-4 multiplexer. Equation 6.36a can be realized also using NAND–NAND logic circuit. This may be desirable if we want to improve the overall circuit speed.

We have shown that AU, for example, can perform only arithmetic operations. Figure 6.76a, however, shows an AU subfunction module that in addition to performing pure arithmetic operations can also perform mixed arithmetic/logic operations. Figure 6.76b shows the corresponding function truth table.

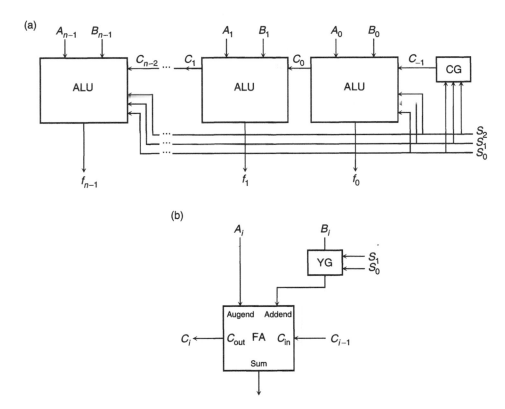

FIGURE 6.73
(a) An *n*-bit ALU showing the CG unit and (b) a single-bit AU.

Function	$S_1 S_0$	CG	YG
Add	0 0	0	B_j
Subtract	0 1	1	$\overline{B_j}$
Increment	1 0	1	0
Decrement	1 1	0	1

FIGURE 6.74
CG and YG outputs.

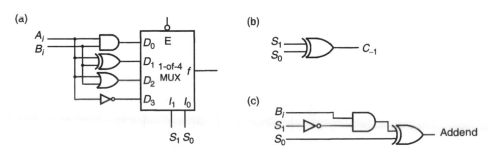

FIGURE 6.75
Logic circuit for the subfunctions: (a) LU, (b) CG, and (c) YG.

(a)

$S_3S_2S_1S_0$	Function ($C_{-1} = 0$)	Function ($C_{-1} = 1$)
1 0 0 0	A plus B	A plus B plus 1
1 0 0 1	A plus $\overline{B}$	A plus $\overline{B}$ plus 1
1 0 1 0	A	A plus 1
1 0 1 1	A minus 1	A
1 1 0 0	A plus AB	A plus AB plus 1
1 1 0 1	A plus $A\overline{B}$	A plus $A\overline{B}$ plus 1
1 1 1 0	A plus (A + B)	A plus (A + B) plus 1
1 1 1 1	A plus (A + $\overline{B}$)	A plus (A + $\overline{B}$) plus 1

(b)

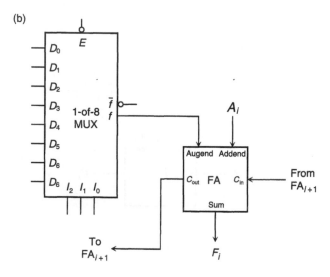

FIGURE 6.76
An AU: (a) truth table and (b) logic diagram.

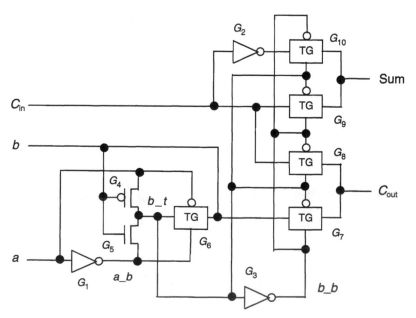

FIGURE 6.77
An optimized TG-based full adder.

```
-- A hierarchical VHDL description of a TG-based adder    -- 01
library ieee;                                             -- 02
use ieee.std_logic_1164.all;                              -- 03
entity PchT is                                            -- 04
   port(in1, gc: in std_logic; out1: out std_logic);     -- 05
end PchT;                                                 -- 06
                                                          -- 07
architecture functional_PchT of PchT is                  -- 08
                                                          -- 09
   begin                                                  -- 10
      out1 <= in1 when gc = '0' else                      -- 11
              'X';                                         -- 12
end functional_PchT;                                      -- 13
                                                          -- 14
library ieee;                                             -- 15
use ieee.std_logic_1164.all;                              -- 16
entity NchT is                                            -- 17
   port(in1, gc: in std_logic; out1: out std_logic);     -- 18
end NchT;                                                 -- 19
                                                          -- 20
architecture functional_NchT of NchT is                  -- 21
                                                          -- 22
   begin                                                  -- 23
      out1 <= in1 when gc = '1' else                      -- 24
              'X';                                         -- 25
end functional_NchT;                                      -- 26
                                                          -- 27
library ieee;                                             -- 28
use ieee.std_logic_1164.all;                              -- 29
entity INV is                                             -- 30
   port(in1: in std_logic; out1: out std_logic);         -- 31
end INV;                                                  -- 32
                                                          -- 33
architecture functional_INV of INV is                    -- 34
                                                          -- 35
   begin                                                  -- 36
      out1 <= not in1;                                    -- 37
end functional_INV;                                       -- 38
                                                          -- 39
library ieee;                                             -- 40
use ieee.std_logic_1164.all;                              -- 41
entity TG is                                              -- 42
   port(in1, gc, gc_b: in std_logic; out1: out std_logic); -- 43
end TG;                                                   -- 44
                                                          -- 45
architecture functional_TG of TG is                      -- 46
                                                          -- 47
   begin                                                  -- 48
```

FIGURE 6.78
A hierarchical VHDL description of TG-based full adder (part 1).

6.12 ALU Design Using VHDL

In this section we illustrate the design of ALU circuits in VHDL descriptions. Figure 6.72a shows a truth table of the operation selection circuit of an ALU unit that realizes eight operations.

Figure 6.77 shows the optimized adder circuit discussed in Chapter 4.11. It is shown with component instances (individual gates, in this case) named as *Gn* and internal signals labeled accordingly. A hierarchical VHDL description for this TG-based full adder design

```
        out1 <= in1 when (gc = '1' and gc_b = '0' else          -- 49
                  'X';                                           -- 50
    end functional_TG;                                           -- 51
                                                                 -- 52
    library ieee;                                                -- 53
    use ieee.std_logic_1164.all;                                 -- 54
    entity DOT2 is                                               -- 55
       port(in1, in2: in std_logic; out1: out std_logic);       -- 56
    end DOT2;                                                    -- 57
                                                                 -- 58
    architecture functional_DOT2 of DOT2 is                      -- 59
                                                                 -- 60
       begin                                                     -- 61
          out1 <= in1 when (in2 = 'X' or in1 = in2) else         -- 62
                  in2 when (in1 = 'X' or in1 = in2) else          -- 63
                  'X';                                           -- 64
    end functional_DOT2;                                         -- 65
                                                                 -- 66
    library ieee;                                                -- 67
    use ieee.std_logic_1164.all;                                 -- 68
    entity DOT3 is                                               -- 69
       port(in1, in2, in3: in std_logic; out1: out std_logic);   -- 70
    end DOT3;                                                    -- 71
                                                                 -- 72
    architecture functional_DOT3 of DOT3 is                      -- 73
                                                                 -- 74
       begin                                                     -- 75
          out1 <= in1 when (in1 = in2 and in1 = in3) else        -- 76
                  in1 when (in2 = 'X' and in2 = in3) else          -- 77
                  in1 when (in2 = 'X' and in1 = in3) else          -- 78
                  in1 when (in3 = 'X' and in1 = in3) else          -- 79
                  in2 when (in1 = 'X' and in1 = in3) else          -- 80
                  in2 when (in1 = 'X' and in2 = in3) else          -- 81
                  in3 when (in1 = 'X' and in1 = in2) else          -- 82
                  'X';                                           -- 83
    end functional_DOT3;                                         -- 84
                                                                 -- 85
    library ieee;                                                -- 86
    use ieee.std_logic_1164.all;                                 -- 87
    entity TG_adder is                                           -- 88
       port(a, b, Cin: in std_logic; Sum, Cout: out std_logic);  -- 89
    end TG_adder;                                                -- 90
                                                                 -- 91
    architecture structural_TG_adder of TG_adder is              -- 92
                                                                 -- 93
       component INV                                             -- 94
         port(in1: in std_logic; out1: out std_logic);          -- 95
       end component;                                            -- 96
```

FIGURE 6.79
A hierarchical VHDL description of TG-based full adder (part 2).

is shown in Figure 6.78 (which is continued onto Figures 6.79 and 6.80). This full adder can be used as part of the ALU circuit to perform the four arithmetic operations listed in Figure 6.72a. A functional VHDL description of a 4-bit ripple-carry adder is shown in Figure 6.81.

In the hierarchical VHDL description of this TG-based full adder, the first **entity-architecture** pair defines a p-channel transistor device (named PchT), while the second pair defines an n-channel transistor (named NchT), the third pair an inverter (named INV) and the fourth pair a transmission gate (named TG). The fifth **entity-architecture** pair describes the TG-based full adder structurally.

```
                                                        -- 97
   component PchT                                       -- 98
     port(in1, gc: in std_logic; out1: out std_logic); -- 99
   end component;                                       --100
                                                        --101
   component NchT                                       --102
     port(in1, gc: in std_logic; out1: out std_logic); --103
   end component;                                       --104
                                                        --105
   component TG                                         --106
     port(in1, gc, gc_b: in std_logic; out1: out std_logic);--107
   end component;                                       --108
   component DOT2                                       --109
     port(in1, in2: in std_logic; out1: out std_logic);--110
   end component;                                       --111
                                                        --112
   component DOT3                                       --113
     port(in1, in2, in3: in std_logic; out1: out std_logic);--114
   end component;                                       --115
                                                        --116
   signal a_b, b_b, b_t, b_t4, b_t5, Cin_b: std_logic; --117
   signal b_tt, dt11, dt12, dt21, dt22: std_logic;     --118
                                                        --119
   begin                                                --120
     G1: INV port map(a, a_b);                          --121
     G2: INV port map(Cin, Cin_b);                      --122
     G3: INV port map(b_t, b_b);                        --123
     G4: PchT port map(a, b, b_t4);                     --124
     G5: NchT port map(a_b, b, b_t5);                   --125
     G6: TG port map(b, a_b, a, b_tt);                  --126
     G7: TG port map(b, b_b, b_t, dt11);                --127
     G8: TG port map(Cin, b_t, b_b, dt12);              --128
     G9: TG port map(Cin, b_b, b_t, dt21);              --129
     G10: TG port map(Cin_b, b_t, b_t, dt22);           --130
     G11: DOT2 port map(dt11, dt12, Cout);              --131
     G12: DOT2 port map(dt21, dt22, Sum);               --132
     G13: DOT3 port map(b_t4, b_t5, b_tt, b_t);         --133
   end structural_TG_adder;                             --134
-- End of the VHDL description                          --135
```

FIGURE 6.80
A hierarchical VHDL description of TG-based full adder (part 3).

```
-- A functional VHDL description of                     -- 01
-- a 4-bit ripple-carry adder                           -- 02
library ieee;                                           -- 03
use ieee.std_logic_1164.all;                            -- 04
entity rc_adder_4bit is                                 -- 05
   port(A, B: in std_logic_vector(3 downto 0);          -- 06
        Cin: in std_logic;                              -- 07
        Sum: out std_logic_vector(3 downto 0);          -- 08
        Cout: out std_logic);                           -- 09
end rc_adder_4bit;                                      -- 10
                                                        -- 11
architecture functional_4rca of rc_adder_4bit is       -- 12
                                                        -- 13
   signal Si: std_logic_vector(4 downto 0);             -- 14
                                                        -- 13
   begin                                                -- 14
     Si <= (('0' & A) or ('0' & B) or ("0000" & Cin));  -- 15
     Cout <= Si(4);                                     -- 16
     Sum <= Si(3 downto 0);                             -- 17
end functional_4rca;                                    -- 18
-- End of the VHDL description                          -- 19
```

FIGURE 6.81
A functional VHDL description of A 4-bit ripple-carry adder.

```
-- A structural VHDL description of a one-stage Logic Unit  -- 01
library ieee, lcdf_vhdl;                                    -- 02
use ieee.std_logic_1164.all; lcdf_vhdl.func_prims.all;      -- 03
entity logic_unit_1bit is                                   -- 04
   port(dinA, dinB: in std_logic;                           -- 05
          Sel: in std_logic_vector(1 downto 0);             -- 06
          lu_out: out std_logic);                           -- 07
end logic_unit_1bit;                                        -- 08
                                                            -- 09
architecture structural_1bitLU of logic_unit_1bit is        -- 10
                                                            -- 11
   component INV1                                           -- 12
     port(in1: in std_logic; out1: out std_logic);          -- 13
   end component;                                           -- 14
                                                            -- 15
   component AND2                                           -- 16
     port(in1, in2: in std_logic; out1: out std_logic);     -- 17
   end component;                                           -- 18
                                                            -- 19
   component OR2                                            -- 20
     port(in1, in2: in std_logic; out1: out std_logic);     -- 21
   end component;                                           -- 22
                                                            -- 23
   component XOR2                                           -- 24
     port(in1, in2: in std_logic; out1: out std_logic);     -- 25
   end component;                                           -- 26
                                                            -- 27
   component MUX4                                           -- 28
     port(Din: in std_logic_vector(3 downto 0);             -- 29
           Sel: in std_logic_vector(1 downto 0);            -- 30
           Dout: out std_logic);                            -- 31
   end component;                                           -- 32
                                                            -- 33
   signal Dtmp: std_logic_vector(3 downto 0);               -- 34
                                                            -- 35
   begin                                                    -- 36
     G1: AND2 port map(dinA, dinB, Dtmp(0));                -- 37
     G2: OR2 port map(dinA, dinB, Dtmp(1));                 -- 38
     G3: XOR2 port map(dinA, dinB, Dtmp(2));                -- 39
     G4: INV1 port map(dinA, Dtmp(3));                      -- 40
     G5: MUX4 port map(Dtmp, Sel, lu_out);                  -- 41
end structural_1bitLU;                                      -- 42
-- End of the VHDL description                              -- 43
```

FIGURE 6.82
A structural VHDL description of a single-stage logic unit.

The last four operations specified in Figure 6.72a are logic operations performed by a LU. A multiplexer-based single-stage (or one bit) LU circuit design is shown in Figure 6.75 and a structural VHDL description for this single-stage LU is shown in Figure 6.82. A functional VHDL description of the single-stage LU is shown in Figure 6.83.

6.13 Summary

We have demonstrated how a problem of a high level of abstraction can be decomposed into more concrete lower-level subfunctions. Each of the subfunctions is then defined in terms of its truth table, and correspondingly its minimized Boolean form is identified using either observation or a suitable minimization technique. After all minimized functions have been identified, they are constructed and then tested. VHDL is used in this chapter to design

```
-- A functional VHDL description of a 1-bit logic unit    -- 01
library ieee;                                             -- 02
use ieee.std_logic_1164.all;                              -- 03
entity logic_unit_1bit is                                 -- 04
   port(Sel: in std_logic_vector(1 downto 0);             -- 05
        dinA, dinB: in std_logic;                         -- 06
        lu_out: out std_logic);                           -- 07
end logic_unit_1bit;                                      -- 08
                                                          -- 09
architecture functional_1bitLU of logic_unit_1bit is      -- 10
                                                          -- 11
   signal dAND, dOR, dXOR, dINV: std_logic;               -- 12
                                                          -- 13
   begin                                                  -- 14
      dAND <= dinA and dinB;                              -- 15
      dOR <= dinA or dinB;                                -- 16
      dXOR <= dinA xor dinB;                              -- 17
      dINV <= not dinA;                                   -- 18
      lu_out <= dAND when Sel = "00" else                 -- 19
                dOR when Sel = "01" else                  -- 20
                dXOR when Sel = "10" else                 -- 21
                dINV when Sel = "11" else                 -- 22
                'X';                                      -- 23
   end functional_1bitLU;                                 -- 24
```

FIGURE 6.83
A functional VHDL description of a single-stage logic unit.

combinatorial circuit, in general, and an arithmetic logic unit, in particular. Finally, such smaller and manageable subfunction modules are often appropriately cascaded together to realize an *n*-bit system. Examples of such *n*-bit circuit includes decoders, code converters, encoders, binary and nonbinary adders/subtracters, high-speed arithmetic circuit, comparators, and arithmetic logic unit. Some of these devices are used in the later chapters for developing more complex logic concepts.

Bibliography

Bartee, T.C., *Digital Computer Fundamentals*. New York, NY. McGraw-Hill, 1985.

Bywater, R.E.H., *Hardware/Software Design of Digital Systems*. Englewood Cliffs, NJ. Prentice-Hall International, 1981.

Comer, D.J., *Digital Logic and State Machine Design*. 3rd edn. New York, NY, Oxford University Press, 1994.

Floyd, T., *Digital Fundamentals*. 8th edn. Englewood Cliffs, NJ. Prentice Hall, 2003.

Hamming, R.W., *Coding and Information Theory*. Englewood Cliffs, NJ. Prentice Hall, 1980.

Hwang, K., *Computer Arithmetic*, New York, NY. Wiley, 1979.

Johnson, E.J. and Karim, M.A., *Digital Design: A Pragmatic Approach*. Boston, MA. PWS-Kent Publishing, 1987.

Karim, M.A. and Awwal, A.A.S., *Optical Computing: An Introduction*. New York, NY. John Wiley & Sons, 1992.

Katz, R.H., *Contemporary Logic Design*. Boston, MA. Addison Wesley, 1993.

Mowle, F.J.A., *A Systematic Approach to Digital Logic Design*. Reading, MA. Addison-Wesley, 1976.

Nelson, V.P.P., Nagle, H.T., and Carroll, B.D., *Digital Logic Circuit Analysis and Design*. Englewood Cliffs, NJ. Prentice Hall, 1995.

Rajeraman, V. and Radhakrishnan, T., *Introduction to Digital Computer Design*. Englewood Cliffs, NJ. Prentice-Hall, 1983.

Wakerley, J., *Error Detecting Codes, Self-Checking Circuits and Applications*. New York, NY. North-Holland, 1978.

Waser, S. and Flynn, M.J., *Introduction to Arithmetic for Digital Systems Designers*. New York, NY. Holt, Rinehart & Winston, 1982.

Problems

1. Use a 3-to-8 line decoder to realize $f(A, B, C) = \sum m(0, 1, 3, 5)$.

2. Design a controllable, dual-purpose, 4-bit converter that converts binary to Gray and Gray to binary.

3. Obtain the circuit for a 16-input, 4-output priority encoder.

4. Use 4-bit binary-to-BCD converter modules to construct a logic circuit for converting input binary numbers of size: (a) 15-bit, (b) 20-bit, and (c) 25-bit.

5. Use 4-bit BCD-to-binary converter modules to construct a logic circuit for converting BCD numbers (each 4-bit in length) of size: (a) 3 digit, (b) 4 digit, and (c) 5 digit.

6. Shift the binary number 1101101 to the left bit by bit as follows: After each left shift, subtract 0101 from the processed number that is greater than or equal to 0101 before the next shift is made and a 1 is set up as a carry into the next higher-order digit position. Show that the resulting number is a BCD equivalent of 1101101. If one were to design a binary-to-BCD converter module based on the scheme just described, what is the implication on the overall logic circuit that will be needed for converting an n-bit binary number to its BCD equivalent? Compare this with that already designed in Example 5.8.

7. Shift the BCD number 0101 0011 to the right bit by bit as follows: If a 1 is shifted right out of some digit position, then 0101 is added to the next lower digit position. Show that the resulting number is binary equivalent of 0101 0011. If one were to design a BCD-to-binary converter module based on the scheme just described, what is the implication on the overall logic circuit that will be needed for converting an n-digit BCD to its binary equivalent?

8. Analyze the behavior of a 4-input 4-output logic circuit given by the following Boolean functions:

$$F_1 = BC \qquad F_2 = B\bar{C} + A\bar{B}$$
$$F_3 = A \oplus C \qquad F_4 = D$$

9. Obtain a carry-save logic circuit for adding m n-bit numbers. Compare this circuit with that based on layered ripple addition scheme in terms of both circuit complexity and speed.

10. (a) Find the binary equivalent of the Gray code number 101101, (b) Find the Gray equivalent of the binary number 101101.

11. Using only a 4-bit binary adder, design decimal code converters for the following conversions:
    (a) 2-4-2-1 to 8-4-2-1
    (b) 8-4-2-1 to XS3
    (b) XS3 to BCD
    (c) XS6 to XS3
    (d) BCD to XS3

8-4-2-1	5-2-1-1	8-4-$\bar{2}$-$\bar{1}$	2-4-2-1
0000	0000	0000	0000
0001	0001	0111	0001
0010	0011	0110	0010
0011	0101	0101	0011
0100	0111	0100	0100
0101	1000	1011	1011
0110	1010	1010	1100
0111	1100	1001	1101
1000	1000	1000	1110
1001	1111	1111	1111

FIGURE 6.P1

12. Using appropriate sized decoders, design (a) an XS3-to-BCD code converter and (b) a BCD-to-XS3 code converter.

13. Besides XS3, there are a few more self-complementing codes such as $5-2-1-1$, $8-4-\bar{2}-\bar{1}$, and $2-4-2-1$ codes as shown in Figure 6.P1. Design decimal code converters for converting 8-4-2-1 codes to (a) $5-2-2-1$, (b) $8-4-\bar{2}-\bar{1}$, and (c) $2-4-2-1$ codes.

14. Design a minimized binary-to-decimal converter whose four inputs are active-high BCD-encoded numbers and whose ten outputs are active-low.

15. Design a serial-to-parallel converter circuit that routes a long sequence of binary digits into four different output lines as specified by external control signals.

16. Design a BCD adder circuit that first converts 2-digit decimal numbers into binary, then adds the resulting binary numbers, and finally converts the binary sum to its equivalent BCD sum.

17. The Hamming-coded message received is 1001111001011. Obtain the corresponding correction circuit assuming that odd parity was used in encoding the message. Determine if the message has any error and the correct message.

18. Design an encoder for four input signals if one and only one is active at any moment in time.

19. Design an encoder for four input signals that outputs a zero code unless one and only one input line is active.

20. Obtain a single-bit full adder using (a) a 3-to-8 line decoder and NAND gates, and (b) 1-of-4 multiplexers, (c) only NOR gates.

21. Use decoders to design (a) a full adder and (b) a full subtracter.

22. Design a 2-bit adder circuit using two-level NAND gates for each output.

23. Show that the design of a 2-bit adder (shown in Problem 22) is effectively a cascade of a half adder and a full adder.

24. Using only single-bit half adders, obtain a logic circuit that can add three single-bit numbers.

25. Design a 3-bit full adder using the carry look-ahead concept.

26. Design a 4-bit full adder using a ROM.

27. Design a four-bit CLA circuit where the propagate function is defined as $P_i = A_i + B_i$ instead of $P_i = A_i \oplus B_i$. How does the corresponding logic design differ from the one discussed in Section 6.7?

28. From the n-bit CLA equations (when $n > 13$), show that the maximum fan-out is dependent on variable $P_{(n-2)/2}$ and is equal to $\{[(n+1)^2/4] + 2\}$ for odd n. For

even n, show that the maximum fan-out is dependent on both $P_{(n/2)-1}$ and $P_{(n/2)}$ and is equal to $\{[n(n+2)]/4 + 2\}$.

29. Use bridging to implement a full subtracter using XOR gates.

30. Design a decimal full adder/subtracter assuming 2-4-2-1 code (as listed in the truth table of Figure 6.P1) for representing the inputs.

31. Design a special-purpose unit using single-bit full adders for adding (a) 12 single-bit numbers and (b) 19 single-bit numbers.

32. Design a half subtracter circuit using (a) only NOR gates and (b) only MUXs.

33. Design a 1-bit full subtracter module using only NOR gates and then obtain a 4-bit full subtracter using these modules.

34. Design a full subtracter module using half subtracter modules.

35. Design a borrow-look-ahead 4-bit subtracter module and show how it could be cascaded to perform 16-bit subtraction.

36. Design a circuit for dividing a 4-bit number by a 2-bit number.

37. Design a logic circuit that multiplies an input decimal digit (expressed in BCD) by 5. The output must appear in BCD form. Show that the outputs can be obtained from the input lines without using any logic gates.

38. Obtain the most minimal logic circuit that squares a 3-bit number.

39. Design a multiplier/adder circuit to realize $f = (A$ times $B)$ plus C plus D, where $A, B, C,$ and D are BCD inputs.

40. Design a $n \times n$ BCD multiplier using the multiplier/adder unit obtained in Problem 39. Design a combinatorial logic circuit capable of comparing two 8-bit binary integers (without sign bits) X and Y. The output Z should be a 1 whenever $X \geq Y$.

41. Design a logic unit that will perform the following combinations of operations: (a) NAND, NOR, transfer, and invert operations; (b) XOR and XNOR; and (c) AND, OR, and NOT.

42. Obtain and discuss the logic circuit implementation of a 4-bit comparator as described in Figure 6.63b.

43. Design an n-bit binary comparator circuit to test if an n-bit number A is equal to, larger than, or smaller than an n-bit number B as follows: It should have one half comparator and $n - 1$ full comparator modules cascaded together. Each of the modules generates two outputs: G_n and L_n, such that (i) $G_n = 0$ and $L_n = 1$ if $A_n < B_n$ (ii) $G_n = 1$ and $L_n = 0$ if $A_n > B_n$, and (iii) $G_n = G_{n-1}$ and $L_n = L_{n-1}$ for a full comparator and $G_n = L_n = 0$ for a half comparator if $A_n = B_n$. Describe each of the modules and the overall n-bit comparator.

44. Implement a parallel scheme for 16-bit comparison circuit using the comparator modules (developed in Problem 43) as follows: At the first level only two bits are to be compared. Then the results of these 2-bit comparisons are compared at the next level, yielding 4-bit comparisons, which are then compared on the next level, yielding 8-bit comparisons, and so on. Compare the serial implementation (as in Problem 43) and this parallel implementation in terms of both cost and speed.

45. Design an n-bit comparator where each of the modules is different from that used in Problem 43 in that the comparison process begins by scanning the string of bits from the left to the right.

46. Design the serial and parallel versions of a comparator to compare the following type of binary numbers: (a) sign and magnitude and (b) 2's complement.

47. Design an 8-bit barrel left rotator.

48. Design an 8-bit left and right shifter.

49. Design an 8-bit left and right shifter/rotator.

50. Show how an ALU can be used to (a) subtract 1 from a given number and (b) add 1 to a given number.

51. Design the logic extender (LE) and arithmetic extender (AE) subfunctions for an ALU that is organized as shown in Figure 6.P2 but follows the truth table of Figure 6.72a.

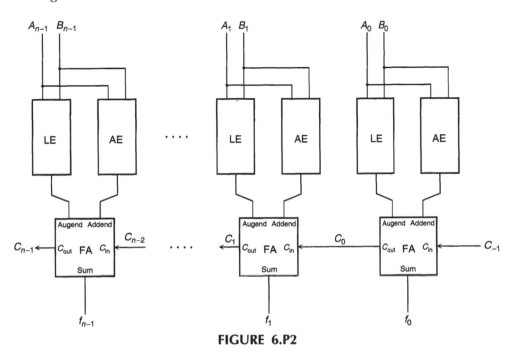

FIGURE 6.P2

52. Describe each of the logic circuits in Figure 6.P3 in terms of their functional characteristics and possible applications.

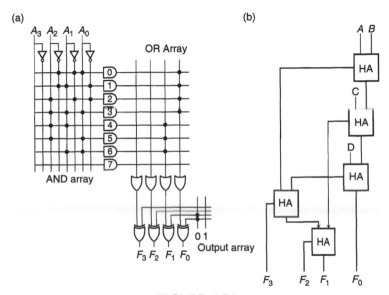

FIGURE 6.P3

7

Sequential Logic Elements

7.1 Introduction

The values of future steady-state outputs of a purely combinatorial logic circuit depend only on its current inputs. In previous chapters, we have seen numerous examples of such logic circuits. The time between the introduction of input signals and the generation of the corresponding steady-state output signals is characterized as the inherent propagation delay of the circuit. In comparison, in *sequential circuits* the future values of the outputs are dependent on both the present and past values of the inputs. There are many digital applications where the signals are first interpreted by the system and then necessary outputs are generated in accordance with the sequence in which the input signals are received. Such systems require logic circuits that respond to the past history of the inputs.

In general, sequential circuits have built-in feedback paths in them. Accordingly, one or more of the outputs just generated may contribute to the future values of the same or a different set of outputs. Sequential circuits typically store information and are used widely in digital systems. Counters and registers are typical examples of the sequential circuits. Properly designed counters, for example, can be used to count the number of days. Every 24 h, the circuit needs to conclude that a day has just passed and accordingly increment the latest day count by 1. Thus, this circuit must be able to store information, remember what was the value of the day count 24 h ago, and then should be able to update the value of the day count.

An integral part of a sequential circuit is the memory unit. The memory elements are also referred to as bistable electronic circuits; that is, they exist indefinitely in one of two binary states. Binary data are stored in a memory element by transitioning it into the 1 state to store a 1 and the 0 state to store a 0. The one or more inputs feeding the memory circuits are known as *excitation inputs* since they excite the circuit to reach a newer steady state. The two commonly used sequential memory elements are latches and flip-flops. Typically, the flip-flop circuit output indicates the latest state of the memory element. There are a number of different flip-flops available, each differing from one another in the number of inputs they have and in the manner in which its binary state is affected by the inputs. The changes in the values of the outputs of flip-flops often are a direct consequence of the frequency with which the circuit inputs change their values. However, there exists a special class of sequential memory device, known as a *monostable multivibrator*, whose output is often independent of the changes or rate of changes in the inputs. In this chapter, we introduce the characteristics of the various flip-flops.

7.2 Latches

A *latch* is a bistable memory unit whose state is determined by its excitation inputs. It is the fundamental building block of a flip-flop. The latch is basically a combinatorial circuit that has one or more of its outputs fed back as its inputs. If an input signal drives the latch output to a 1, the memory element is called a *set latch*. On the other hand, it is referred to as a *reset latch* when the excitation inputs force the output to a 0. If the device has both set and reset excitation inputs, then the memory element is referred to as a *set–reset latch*. A *flip-flop*, on the other hand, differs from a latch in that it has a triggering signal called clock. The clock signal triggers a command that allows the flip-flop to change its state in accordance with its excitation inputs. While a latch changes its state immediately in accordance with its excitation inputs, a flip-flop waits for its clock signal before it changes its state.

Figure 7.1 shows timing diagrams pertaining to the operation of latch and flip-flop. The latch closely follows all of its inputs continuously and changes its outputs independent of a clocking signal. Although latches can be useful for storing binary values, they are not practical for use in many of the sequential circuits. However, we shall study latches to better understand their behavior and to design a latch from scratch.

Let us now consider what might happen if we were to introduce a feedback path between the output and some input. The output of a 2-input OR gate, for example, is a 0 as long as the two inputs are 0. If its output were to be connected to one of the inputs as shown in Figure 7.2a, the gate output would continue to remain 0. Now if a 1 were to be introduced at input S as shown in Figure 7.2b, the circuit output becomes a 1. After that event even if we were to switch S back to a 0 as shown in Figure 7.2c, the circuit output will remain permanently latched to a 1. This memory cell is thus called a *set latch*.

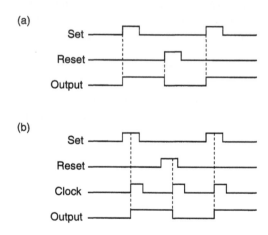

FIGURE 7.1
Representative timing diagrams of (a) latch and (b) flip-flop.

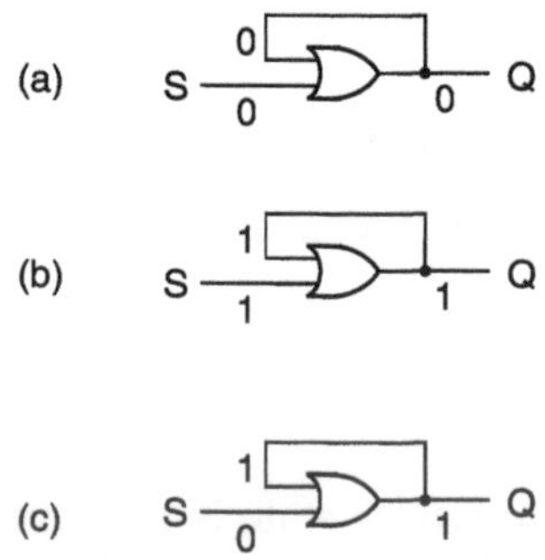

FIGURE 7.2
A 2-input OR gate with feedback when (a) Q (low) is connected to the input with $S = 0$, (b) S is switched to a 1, and (c) S is switched back to a 0 with $Q = 1$.

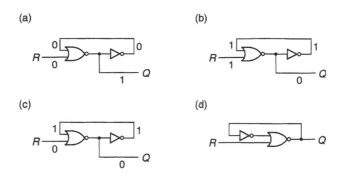

FIGURE 7.3
A 2-input NOR gate and a follow-on NOT gate with feedback (a) $Q = 1$ with 0 inputs, (b) R is switched to a 1, (c) R is switched back to a 0, and (d) a redrawn version of the reset latch.

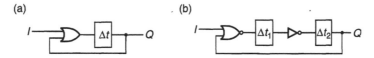

FIGURE 7.4
Latch for storing logic 1 (a) using OR gate and (b) using a NOR gate and a NOT gate.

An equivalent set latch may be obtained by replacing the OR gate shown in Figure 7.2 with a 2-input NOR gate and a follow-on NOT gate as shown in Figure 7.3a. Note that, in Figure 7.3a, we have relabeled the circuit. The NOR gate output is labeled Q and the free input is renamed as R. With the two inputs tied to a 0, Q is a 1. The NOT gate assures that the two inputs continue to maintain its 0 state which in turn maintains a stable output. If we were to now feed a 1 through the R input as shown in Figure 7.3b, Q will become a 0. After that no matter what is introduced at input R, the circuit shown in Figure 7.3c remains latched on to a 0. This particular circuit is referred to as the *reset latch*. Figure 7.3d shows a redrawn alternative for the reset latch.

Neither of the two circuits in Figures 7.2 and 7.3 has yet considered fully the impact of propagation delays. Consider a latch whose output goes high as soon as or nearly after its input goes high. Furthermore, the output remains high thereafter no matter what happens to the input. Considering that Δt accounts for the propagation delay of the gate used, the output Q can be expressed in terms of the input I using a time-dependent equation:

$$Q(t + \Delta t) = Q(t) + I(t) \tag{7.1}$$

so that $Q(t) = 0$ when $t = 0$. In Equation 7.1, t is used to denote time to emphasize the fact that both input and output are functions of time. After a time Δt, the next value of Q is determined by the present values of both Q and I. The latch circuit corresponding to Equation 7.1 is obtained as shown in Figure 7.4a while an alternative form of the same is obtained as shown in Figure 7.4b, where Δt_1 and Δt_2 are the propagation delays respectively of the NOR and NOT gates. For this latter sequential circuit, however, the output is given by

$$Q(t + \Delta t_1 + \Delta t_2) = Q(t) + I(t) \tag{7.2}$$

To make sure that $Q(0) = 0$, it will be necessary to open the feedback route and introduce a 0 either at the OR input of the circuit shown in Figure 7.4a or at the NOR input of the circuit shown in Figure 7.4b.

Consider the logic circuit of Figure 7.5a, which can be obtained from that of Figure 7.4a by introducing a 2-input AND logic gate at the output of the OR gate. The propagation delay

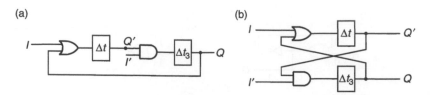

FIGURE 7.5
(a) An AND–OR latch and (b) A modified drawing of AND–OR latch.

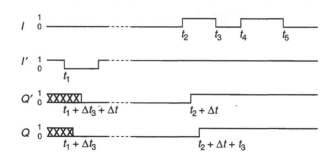

FIGURE 7.6
A timing diagram pertaining to the AND–OR latch of Figure 7.5b.

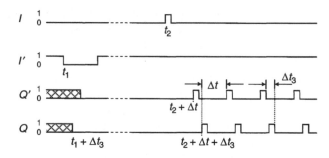

FIGURE 7.7
Response of an AND–OR latch to a narrow input pulse.

associated with the AND gate is given by Δt_3. The two circuits (shown in Figures 7.4a and 7.5a) are logically identical to one another. The circuit in Figure 7.5a has an added advantage. By feeding a 0 at I', the output Q can be initialized to a 0. This capability allows the circuit of Figure 7.5a to have an output that will remain 0 unless and until the input I becomes a 1. The logic circuit of Figure 7.5a can be redrawn now to give a cross-coupled form as that shown in Figure 7.5b.

The response pattern of the latch in question can be envisioned by studying its timing diagram shown in Figure 7.6. The shaded region of the timing diagram corresponds to the initial phase when the signal is either a 0 or a 1. Following the first $0 \rightarrow 1$ transition of the input, that is, I, the output Q latches on to a 1. Even though the input I has been subsequently switched back and forth several times, the output Q continues to be a 1 until I' switches back to a 0.

The latch shown in Figure 7.5b functions in accordance to the timing diagram as long as the pulse width, $t_3 - t_2$ or $t_5 - t_4$, is larger than the sum of the propagation delays, $\Delta t + \Delta t_3$. With a noise pulse of duration less than the sum of the propagation delays, Q becomes oscillatory. Figure 7.7 shows an example of such an oscillatory output.

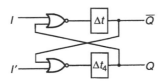

FIGURE 7.8
A NOR–NOR latch.

R(t)	S(t)	Q(t)	Q(t + Δt)
0	0	0	0
		1	1
0	1	0	0
		1	0
1	0	0	1
		1	1
1	1	0	-
		1	-

FIGURE 7.9
Truth table for a basic latch.

The AND–OR latch shown in Figure 7.5 has been derived from that in Figure 7.4a. However, if we had instead chosen to modify the circuit of Figure 7.4b, the NOT logic gate can be replaced by a 2-input NOR gate as shown in Figure 7.8. The outputs Q and $\bar{Q}$ of the resulting NOR–NOR circuit are complements of each other as long as $I' = 0$. Thus, when $I' = 0$, the circuits of Figures 7.4b and 7.8 are identical in their logical behavior. In addition, the output Q can be forced to have a value of 0 at $t = 0$ by making sure that I' is a 1. Such a feature of latches or flip-flops turns out to be very useful in the design of sequential machines.

The latching concept explored in this section will be used next to design flip-flops. When the values stored (the *current state*) in the flip-flops change, we say that the sequential circuit has undergone a transition to its *next state*. Typically, a flip-flop has two outputs, called Q and $\bar{Q}$, that are complements of each other. The truth table shown in Figure 7.9 lists the input conditions of one such basic latch, which lead to the desired outputs. The variable t denotes time while Δt is the time span between a change in the input and a possible change in the output. Here, the time span Δt time is equivalent to the lumped delay of the latch circuit in question. The two inputs, S (set) and R (reset), are used to determine the next state $Q(t + \Delta t)$ on the basis of the current state of the output $Q(t)$. With $R = 0$ and $S = 1$, the output is turned on if not already on. When $R = 1$ and $S = 0$, the output is turned off if not already off. When both S and R are 0, no output change occurs. The to-be-designed latch circuit, however, manifests an undesirable condition when both inputs become a 1. With $S = R = 1$, the two outputs, Q and $\bar{Q}$, cease to be complements. In addition, the next state of the latch becomes unpredictable if the inputs were to be returned to a 0. Consequently, the input condition $S = R = 1$ is forbidden.

The desired latch circuit may be determined readily from the K-map corresponding to the truth table of Figure 7.9. This K-map is obtained as shown in Figure 7.10a. The equation for $Q(t + \Delta t)$ is given as follows:

$$Q(t + \Delta t) = S(t) + \overline{R(t)}Q(t)$$
$$= (S(t) \bullet \overline{R(t)}Q(t))' \tag{7.3}$$

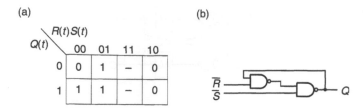

FIGURE 7.10
A set–reset latch (a) K-map and (b) logic circuit.

Q(t) R(t)		Q(t + Δt)
0	0	S(t)
0	1	0
1	0	1
1	1	0

FIGURE 7.11
Revised truth table of a set–reset latch.

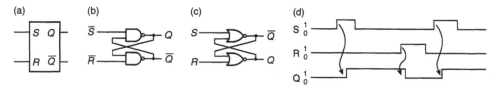

FIGURE 7.12
Set–reset latch (a) symbol, (b) NAND-only logic circuit, (c) NOR-only logic circuit, and (d) timing diagram.

This equation, known also as *next-state equation*, states that after a duration equivalent to Δt, the next state of Q is determined by the current values of Q, R, and S at time t. The corresponding logic circuit is obtained as shown in Figure 7.10b. If R and S values were to change at time t, a new value of Q will result Δt time later. Typically, Δt accounts for the total gate delay of the two NAND gates. Traditionally, the output of NAND gate next to the reset input is referred to as $\bar{Q}$ output since the output of the other NAND is labeled as Q and the outputs of these two NAND gates are complements of each other.

One may obtain the truth table of Figure 7.11 by rewriting the one in Figure 7.9. Each of the don't-care input conditions of this set–reset latch is assumed to be a 0. The equation for $Q(t + \Delta t)$ may be rewritten (from the corresponding MEV K-map) then as follows:

$$\begin{aligned} Q(t + \Delta t) &= \bar{Q}(t)\bar{R}(t)S(t) + Q(t)\bar{R}(t) \\ &= \bar{R}(t)[Q(t) + \bar{Q}(t)S(t)] \\ &= \bar{R}(t)[Q(t) + S(t)] \\ &= \overline{R(t) + [\overline{Q(t) + S(t)}]} \end{aligned} \tag{7.4}$$

This NOR form of latch may now be implemented by grouping the 0s of the K-map of Figure 7.10a. The corresponding latch circuits are obtained as shown in Figure 7.12. Figure 7.12d shows a representative timing diagram of the set–reset latch. The logic circuit of Figure 7.12c is identical to that shown in Figure 7.8. The two latch outputs are complements of each other. However, the forbidden state occurs when $S = R = 1$. As long as both S and R inputs are set at 1, Q, and $\bar{Q}$ are forced to take the same value simultaneously, thus violating the basic complementary nature of the latch outputs.

FIGURE 7.13
Logic circuit for Example 7.1.

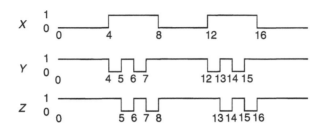

FIGURE 7.14
Timing diagram for Example 7.1.

Example 7.1
Obtain the timing diagram for the logic circuit shown in Figure 7.13 for a square wave input with period 8 units.

Solution
The Boolean equations for Y and Z are respectively given by

$$Y(t) = \overline{X(t)Z(t)}$$

$$Z(t + \Delta t) = \overline{X(t)Z(t)} = \bar{X}(t) + \bar{Z}(t)$$

Accordingly, the timing diagram is obtained as shown in Figure 7.14. When X changes from a 1 to a 0, the output Z is forced to be a 1 after a delay of Δt time. A later transition of X when it changes from a 1 to a 0, the output Z oscillates back and forth with a period of $2\Delta t$.

The set–reset latch described by Equations 7.3 and 7.4 has its time delay lumped together. However, there is another form for the delay model known as the *distributed gate delay model*. Consider, for example, the NAND latch shown in Figure 7.12b where the top and bottom gates are assumed to have gate delays t_1 and t_2 respectively. The corresponding next-state equations are then given by

$$Q(t + t_1) = \overline{\bar{S}(t) \cdot \bar{Q}(t)} = S(t) + Q(t) \tag{7.5}$$

$$Q(t + t_2) = \overline{\bar{R}(t) \cdot Q(t)} = R(t) + \bar{Q}(t) \tag{7.6}$$

Equations 7.5 and 7.6 now may be cascaded to give,

$$\bar{Q}(t + t_1 + t_2) = \overline{\bar{R}(t + t_1)Q(t + t_1)}$$

$$= \overline{\bar{R}(t + t_1)[S(t) + Q(t)]} \tag{7.7}$$

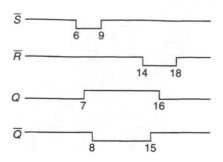

FIGURE 7.15
NAND-latch timing diagram.

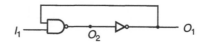

FIGURE 7.16
Latch for Example 7.2.

Taking the complement of Equation 7.7, we may obtain for the next state,

$$Q(t + t_1 + t_2) = \bar{R}(t + t_1)[S(t) + Q(t)] \tag{7.8}$$

For simplicity, we may assume that the two gates have equal gate delays when Equation 7.8 reduces to

$$Q(t + \Delta t) = \bar{R}\left(t + \frac{\Delta t}{2}\right)[S(t) + Q(t)] \tag{7.9}$$

Figure 7.15 shows the timing characteristics of a NAND latch under the assumption of the distributed gate delay model. For simplicity, we assume further that both of the logic gates have gate delays of 1 unit and that Q and $\bar{Q}$ at time $t = 0$ are, respectively, 0 and 1. Although we may conclude that the latch functions as it was intended, however, under certain input conditions the latch may contribute to unforeseen problems. Example 7.2 illustrates the nature of the impact that one of these input conditions may have on the latch outputs.

Example 7.2
For the latch shown in Figure 7.16, trace the timing diagram when the input I_1 changes from a 1 to a 0 for a period smaller than the total gate delay.

Solution
For this logic circuit, we have assumed the gate delays for the NAND and NOT gates to be respectively 3 and 2 units. The timing diagram as obtained in Figure 7.17 shows that the output becomes oscillatory in nature.

Assuming that gate delays are very small (order of a nanosecond), a simple low-pass filter may be placed at the input to eliminate these high-frequency noises. The statistical chance of occurrence of such oscillatory response is very low but one should be aware of this problem.

Switches are normally used to feed user inputs to a digital system. The corresponding make or break operation generally appears as an instantaneous event to the human users.

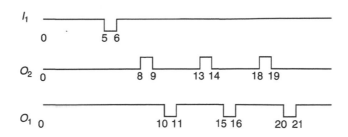

FIGURE 7.17
Timing diagram for Example 7.2.

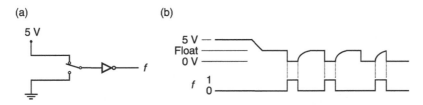

FIGURE 7.18
Bouncing of mechanical switches (a) single-pole double-throw switch and (b) timing diagram for the $1 \rightarrow 0$ transition.

But in actuality when a switch is transferred from one setting to another it goes through multiple number of mechanical bouncing. This in turn causes changes in logic levels, such as those shown in Figure 7.18, which are easily discernible by high-speed digital circuits. Figure 7.18a shows a typical single-pole double-throw switch that may be used to generate logic input. It has a "break-before-make" behavior because of which the terminal often "floats" during the switch depression phase. The logic gate following the switch interprets the floating input to be a 1. A typical switch may bounce for 10–25 ms, a rather long duration compared to the switching speeds of typical digital gates. Such contact bouncing may pose a problem if the switch, for example, is used to count or indicate an event. A binary counter that has as its input signal a switch of the type just described, may end up counting several times each time the switch is activated. Thus it will be necessary to provide a circuit to debounce the switch so as to be able to provide only a single signal change for switch activation. Incidentally, we have devised latches already just for that purpose.

A way to accomplish debouncing is by using a NAND-based set–reset latch, for example, and a pair of pull-up resistors along with the single-pole double-throw switch as shown in Figure 7.19a. Because of the very nature of the switch, both $\bar{S}$ and $\bar{R}$ may not both be zero simultaneously. With the switch in its upper position, Q becomes a 1 and remains so even though $\bar{S}\bar{R}$ bounces back and forth between 01 and 11 before finally settling to 01. Again as the switch is moved to its lower position, Q becomes a 0 and remains so even though $\bar{S}\bar{R}$ bounces back and forth between 10 and 11 until finally settling to 10. As shown by the timing diagram of Figure 7.19b, Q is not directly affected by the mechanical bouncing action.

Switch debouncing is one of the more important applications of a latch. The outputs of these devices, as we have already seen, appear at arbitrary instants independent of external control inputs. Next, we shall introduce latch that normally samples inputs and changes its outputs only at instants as determined by an additional input. It is desirable often to use a gate signal to inhibit state changes especially when, for example, S and R inputs of a set–reset latch are changing. When both S and R inputs are stable, this gate input may

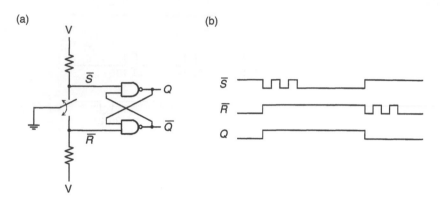

FIGURE 7.19
A bounce-free switch (a) logic circuit and (b) timing diagram.

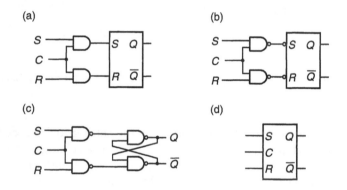

FIGURE 7.20
A gated SR latch (a) using a NOR latch (b) using a NAND latch (c) NAND-only logic diagram, and (d) logic symbol.

be activated such that the set–reset latch may then respond to the newer values of S and R inputs. Such a latch is commonly referred to as a *gated SR latch* since the gate signal functions as a gate through which the signals S and R propagate.

Figure 7.20a shows a gated set–reset latch formed by including two AND gates external to the set–reset latch. When $C = 0$, the other inputs to the gated set–reset latch are disabled and, consequently, the set–reset latch on account of having both of its inputs equal to 0 moves on to its hold (i.e., no change) state. When $C = 1$, the gated set–reset latch functions exactly as a set–reset latch in accordance to the truth table of Figure 7.9. When the AND logic gates of the gated SR latch shown in Figure 7.20a are replaced with NAND logic gates as shown in Figure 7.20b, one finds an opportunity to introduce NAND-only logic circuit for the set–reset latch. The resulting gated set–reset latch is shown in Figure 7.20c.

In spite of many advantages the gated SR latch still has a serious limitation as illustrated in Figure 7.21. When the gate input goes high, the flip-flop responds according to the S and R inputs. The flowchart of Figure 7.21a shows the desired latch operation and the consequence of having a gate pulse that is too long. If and when Q output changes before the termination of the gate input, then the input conditions SR change again, and this leads to possible follow-on changes in Q. Consequently, Q may be in an indeterminate state when C is undergoes a $1 \rightarrow 0$ transition. The latch output takes on a value that corresponds to the last possible set of latch inputs. Such a possibility exists as long as Δt is less than the duration for which gate signal $C = 1$. It is desirable that state change occurs only once during the

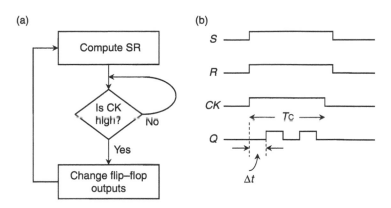

FIGURE 7.21
A gated *SR* latch (a) flowchart and (b) timing diagram.

period when the gate input is a 1. This may be accomplished easily by maintaining a clock width much smaller than the total delay but not too small that prevents predictable change from occurring, as demonstrated in Example 7.2.

7.3 Set–Reset Flip-Flop

In the previous section we have investigated the logic circuits for set–reset latch. Typically, these set–reset latches are not usable for reliable synchronous operation. In a gated *SR* latch, for example, when *C* is activated, the other inputs are gated directly to the latch output. A change in these other inputs readily causes change in the latch output. Such changes will continue to occur as long as *C* continues to remain at 1. In a typical sequential circuit, at least one of the combinatorial outputs is fed as input to memory elements whose outputs may be used in turn to drive the combinatorial part of the circuit. When *C* is a 1, the gated latch if it were to be used as a memory element, functions as a combinatorial logic circuit. Thus, we may have in a sequential circuit two cascaded combinatorial logic circuits feeding each other and contributing to both oscillations and unstable transient behavior.

For dependable operation of the gated latches, transient pulses will need to be prevented from appearing at its inputs. This problem of the gated latch, in general, is resolved by including a *clock* signal to restrict the duration during which the gated latch may change its state. The clock pulses can be periodic or a set of random pulses. Almost always, however, they are periodic. This particular version of the gated latch is referred to as a *flip-flop*. The purpose of the clock input is to force the flip-flop to remain in its rest (or hold) state when changes are allowed to occur on the set and reset inputs. Clock input is set to logic 1 once the inputs have settled. The gated *SR* latch shown in Figure 7.20 can be treated as an *SR* flip-flop as long as we can assure that the clock signal is being fed through its gate input *C*. In order to operate flip-flops effectively, the following conditions must be met:

1. The flip-flop inputs should be allowed to change only when the clock input is a 0.
2. The clock input should be long enough so that the flip-flop outputs may reach their steady states.
3. The condition $S = R = 1$ must not be allowed to occur when the clock input is a 1. $S(t)R(t)$ is thus set equal to 0.

(a)

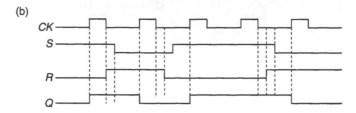

$C(t)$	$S(t)$	$R(t)$	Mode	$Q(t+\Delta t)$	$\overline{Q}(t+\Delta t)$
0	-	-	Hold	$Q(t)$	$\overline{Q(t)}$
1	0	0	Hold	$Q(t)$	$\overline{Q(t)}$
1	0	1	Reset	0	1
1	1	0	Set	1	0
1	1	1	Illegal	–	–

(b)

FIGURE 7.22
The *SR* flip-flop (a) truth table and (b) timing diagram.

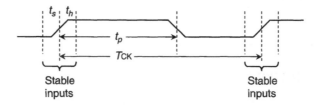

FIGURE 7.23
Flip-flop timing characteristics for determining clock frequency.

The circuit action occurs only when the clock input is high. When the clock input is low, the flip-flop outputs remain unchanged. However, the inputs S and R may be high simultaneously when the clock is absent since the flip-flop is then fully inhibited. The overall characteristics of the SR flip-flop are illustrated by the truth table of Figure 7.22. Note in the timing diagram shown in Figure 7.23 that it is necessary to consider the logic circuit only at the instant when CK changes from low to high to see if the output changes. The next-state equation of a gated SR latch is given by,

$$Q(t+\Delta t) = S(t)C(t) + \overline{R(t)}Q(t) + \overline{C(t)}Q(t) \qquad (7.10)$$

Consider now the operational characteristics of the flip-flops. Figure 7.23 shows some of these specifications, of which setup and hold times are the most important ones. The setup time, t_s, is the time necessary for the input data to stabilize before the triggering edge of the clock. Its value is extremely critical since it manifests itself either by ignoring actions or by resulting in partial transient outputs, commonly referred to as *partial set* and *partial reset* outputs. Consequently, it is possible to begin a set or reset mode, causing the output to change and withdraw back to its initial state. In some cases the output might become metastable wherein the flip-flop is neither set nor reset. On the other hand, the *hold time*, t_h, is the time it takes for the data to remain stabilized beyond the triggering edge of the clock. This is also a critical parameter in determining the correct behavior of a flip-flop.

The clock frequency for a flip-flop is determined from a knowledge of setup time; hold time; FF propagation delay, t_p; and propagation delay of the next-state decoder, t_{NS}. The clock frequency, f_{CK}, may not exceed the worst-case condition given by

$$f_{CK} = \frac{1}{T_{CK}} \leq \frac{1}{t_s + t_p + t_{NS}}$$

(7.11)

Equation 7.11 serves as the flip-flop limiting condition.

Example 7.3

Determine $Q(t + \Delta t)$ as a function of the inputs and $Q(t)$ for the logic circuit shown in Figure 7.24.

Solution

From Equation 7.10, the next-state equation follows as

$$Q(t + \Delta t) = S(t) + \bar{R}(t)Q(t)$$

where $C = 1$. In this logic circuit $S(t) = \overline{X(t) + Y(t)} = \bar{X}(t) \cdot \bar{Y}(t)$ and $R(t) = Y(t)$. Note also that $S(t)R(t) = \bar{X}(t) \cdot \bar{Y}(t) \cdot Y(t) = 0$ which is desirable for SR flip-flop operation. Consequently, we may obtain

$$Q(t + \Delta t) = \bar{X}(t)\bar{Y}(t) + \bar{Y}(t)Q(t)$$

Figure 7.25 lists all possible combinations of the flip-flop inputs and the corresponding flip-flop outputs.

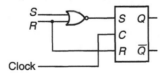

Clock

FIGURE 7.24
Logic circuit for Example 7.3.

X(t)	Y(t)	Q(t)	Q(t+Δt)
0	0	0	1
0	0	1	1
0	1	0	0
0	1	1	0
1	0	0	0
1	0	1	1
1	1	0	0
1	1	1	0

FIGURE 7.25
Listing for Example 7.3.

(a)   (b)   (c)

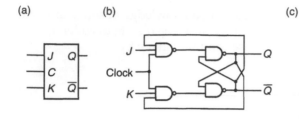

C(t)	J(t)	K(t)	Mode	Q(t+Δt)	Q̄(t+Δt)
0	-	-	Hold	Q(t)	Q̄(t)
1	0	0	Hold	Q(t)	Q̄(t)
1	0	1	Reset	0	1
1	1	0	Set	1	0
1	1	1	Toggle	Q̄(t)	Q(t)

FIGURE 7.26
A *JK* flip-flop (a) block diagram, (b) logic circuit, and (c) truth table.

7.4   *JK* Flip-Flop

In earlier sections, we have noticed that the set–reset latch, gated *SR* latch, as well as *SR* flip-flop have an indeterminate state. When using an *SR* flip-flop, for example, one needs to be very careful about the flip-flop inputs. This limitation of the *SR* flip-flop, however, can be removed by modifying it. This modified *SR* flip-flop, as shown in Figure 7.26, is referred to as the *JK* flip-flop. This modification involves feeding the outputs of the flip-flop back as the inputs of the *SR* flip-flop in a cross-coupled format. The resulting block diagram and its functional behavior are shown in Figures 7.26b and c.

When the inputs *J* and *K* are both 1, the outputs of the first-level NAND gates may not simultaneously become 0. With $Q = 0$, the lower of the first-level NAND gates outputs a 1, and when $Q = 1$, the upper of the first-level NAND gates outputs a 1. Consequently, the input restriction otherwise associated with *SR* FF is eliminated by default. The cross-coupled feedback provides for an additional switching mode, called *toggle*. From the truth table of Figure 7.26c, we may obtain the next-state equation as follows:

$$Q(t + \Delta t) = J(t)\bar{Q}(t) + \bar{K}(t)Q(t) \tag{7.12}$$

When $J = 1$ and $K = 0$, the flip-flop gives out an output of 1 if not already 1. Similarly when $J = 0$ and $K = 1$, the flip-flop is reset to 0 if not already reset. Finally, when $J = K = 1$, the FF output is complemented (toggled), and when $J = K = 0$, no change takes place. Thus except for the toggle state, the *SR* and *JK* flip-flops are equivalent.

7.5   Master–Slave Flip-Flop

One way to eliminate unstable behavior of latches is to employ two gated latches in a master–slave configuration as follows: This type of a flip-flop is formed of two cascaded subunits: the master gated latch and the slave gated latch. This device depends not on the synchronous clocking of both subunits, but rather on their alternate turn-on and turn-off sequences. This out-of-phase antisynchronization is realized by placing a NOT logic gate on the clock line between the master and slave subunits. This scheme introduced into the FF circuitry eliminates the requirement for limiting the clock width to a particular value as determined by the circuit gate delays.

The logic circuit of an *SR* master–slave flip-flop is shown in Figure 7.27. It consists of a master flip-flop, a slave flip-flop, and an inverter for achieving out-of-phase clocking of the

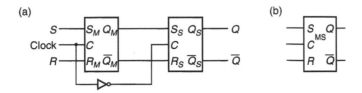

FIGURE 7.27

Master–slave *SR* flip-flop (a) logic circuit and (b) logic symbol.

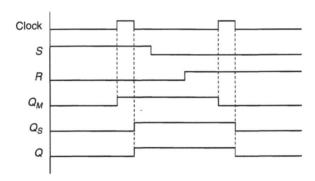

FIGURE 7.28

Timing diagram of a master–slave *SR* flip-flop.

two units. With a NOT logic gate present as shown in Figure 7.27, the master flip-flop is enabled and the slave flip-flop is disabled when clock is high. When clock is low, on the other hand, the master flip-flop is disabled and the slave flip-flop is enabled. The master–slave flip-flop functions as follows: for all inputs of S and R, except when $S = R$, $Q_M = S$ and $\bar{Q}_M = R$ when clock is high. During this phase, the slave flip-flop remains disabled. Then when the clock input goes to a 0, $Q_S = Q_M$, $\bar{Q}_S = \bar{Q}_M$, and the master flip-flop is disabled. Now that the clock input to the master flip-flop is a 0, one can safely allow the combinatorial inputs S and R to change so that the proper next state of the master flip-flop can be realized. The clock should remain low until the two inputs have reached their respective steady states. When the clock is turned high again, the master flip-flop is first turned on followed by the slave flip-flop.

Since the master–slave principle utilizes both $1 \rightarrow 0$ and $0 \rightarrow 1$ transitions of the clock input, this type of a flip-flop is said to be *pulse-triggered*. The next state of the output is generated only after the clock input has completed a $1 \rightarrow 0$ transition. In general, the pulse-triggered flip-flops require both a rising and falling edge on the clock. Figure 7.28 shows a timing diagram illustrating the sequence of operations that takes place in a master–slave *SR* flip-flop. The master–slave configuration, in particular, introduces a buffering mechanism to eliminate unstable transient conditions otherwise present in a gated latch. The overall master–slave outputs shown in Figure 7.27 change at the negative edge of the clock input. However, it is possible to have master–slave flip-flop whose outputs change at the positive edge of the clock input. We can also design SR master–slave configuration where the next state is generated only after a $0 \rightarrow 1$ clock transition.

The master–slave cascading may be realized also in the case of other flip-flops by introducing an inverter between the two gated latches. Figure 7.29 shows the logic diagram of a master–slave *JK* flip-flop. This is slightly different from that of a master–slave *SR* flip-flop in that the outputs of the last-level NAND logic gates are introduced as inputs (in a cross-coupled format) to the first-level NAND logic gates. It was shown in Figure 7.21, that

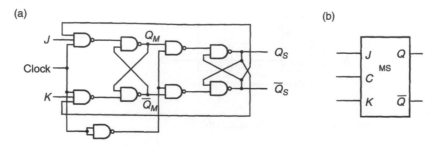

FIGURE 7.29
A master–slave *JK* flip-flop (a) logic circuit and (b) logic symbol.

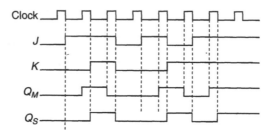

FIGURE 7.30
Timing diagram for a master–slave *JK* flip-flop.

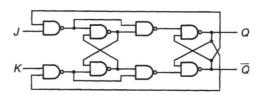

FIGURE 7.31
Logic circuit of a nonisolated master–slave *JK* flip-flop.

Q and $\bar{Q}$ may change several times during a clock pulse yielding an undesirable flip-flop condition. Presence of similar clock input would still cause the master outputs, Q_M and $\bar{Q}_M$, to change; but the slave outputs, Q_S and $\bar{Q}_S$, would not change because the inverted clock pulse will have disabled the slave gated latch. In case of a *JK* flip-flop, for example, the values for *J* and *K* are still determined by the preclock values of Q_S and $\bar{Q}_S$. When the clock input feeding the master gated latch goes low, the clock input to the slave gated latch becomes high, transferring Q_M and $\bar{Q}_M$ to the slave gated latch. Accordingly, the problem of clock pulses that are too wide is fully eliminated by using master–slave configuration. A representative timing diagram for a master–slave *JK* flip-flop is shown in Figure 7.30.

An additional modification can be made to the logic circuit of Figure 7.29a by removing the NOT isolator. But to continue with the master–slave working scheme, the output of the first-level NAND gates are fed as the input of the corresponding third-level NAND gates (instead of the clock input feeding these NAND gates). Such a nonisolated master–slave *JK* flip-flop is shown in Figure 7.31. Both nonisolated and isolated master–slave flip-flops respond correctly to the *J* and *K* values taking on new values when the clock is low. These two types of master–slave flip-flops respond correctly to input changes from a 0 to a 1 after the clocking line has been set high provided that these two inputs remain unchanged for enough duration before the clock becomes a 0. However, an input changing from a 1 to a 0 after the clock has become high causes both of these flip-flops to malfunction. The impact is immediate in the nonisolated case and at the trailing edge in the case of the isolated master–slave flip-flop.

A master–slave JK flip-flop has its own problem. The master flip-flop is vulnerable during the period when the clock is high. The flip flop may be set or reset by appropriate changes of the input. When $Q_S = 0$, $\bar{Q}_S = 1$, and the clock input is high and while still high, the input J becomes high, Q_M is set, and consequently Q_S *catches* a 1 on the trailing edge of the clock input. Again when $Q_S = 1$, $\bar{Q}_S = 0$, and K becomes a 1 after the clock has already become high, Q_M is reset, and Q_S *catches* a 0 on the trailing edge of the clock input. It is important that no such input changes gain entry into the flip-flop. These situations are commonly referred to as the *1s and 0s catching* problems.

The timing diagram of Figure 7.32 shows 1s and 0s catching in addition to the normal behavior of a JK flip-flop. The output of the slave flip-flop, Q_S, always follows the output of the master flip-flop, Q_M, each time the clock input makes a $1 \rightarrow 0$ transition. However, one must be careful in that the master–slave flip-flop may contribute also to improper circuit outputs whenever the inputs are relatively unstable before the transition $0 \rightarrow 1$ of the clock.

To partially eliminate the 1s and 0s catching problems, a data lock-out feature can be added to the master–slave flip-flops. With data lock-out, the master–slave flip-flop will be subjected to both setup and hold time requirements for the inputs as shown in Figure 7.33. Accordingly, in a master–slave flip-flop equipped with a data lock-out feature, the flip-flop inputs will have to be stable for a period starting from t_s (before the $0 \rightarrow 1$ transition of clock) through t_h (after the $0 \rightarrow 1$ transition of the clock input). In the case of an ordinary master–slave flip-flop, the inputs have to be stable for a relatively longer time given by $t_s + t_p + t_h$. In general, flip-flops are driven by a symmetrical square wave signal whose maximum frequency limit is determined by the time required to transfer data from flip-flop

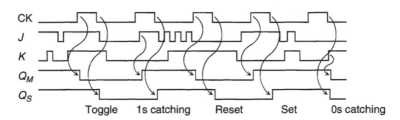

FIGURE 7.32
Timing diagram of a master–slave JK flip-flop showing 1s and 0s catchings.

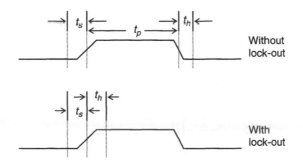

FIGURE 7.33
Setup and hold time characteristics for the master–slave flip-flop.

input to flip-flop output. The time needed for the input combinatorial logic to be stabilized must be included in this calculation.

Often a flip-flop is equipped with two additional inputs for control: preset and clear. The *preset* ($\overline{PR}$) and *clear* ($\overline{CLR}$) inputs allow initializing the flip-flop respectively to either a set ($Q = 1$) or a reset ($Q = 0$) condition. Unless mentioned otherwise, both preset and clear inputs are typically considered to be active when low. In general, preset and clear inputs are indicated by vertical inverted inputs respectively at the top and bottom of the corresponding flip-flop logic symbol. These two inputs allow one to either set or reset the flip-flop without the necessity of a clock pulse.

Figure 7.34 shows how $\overline{PR}$ and $\overline{CLR}$ input can be added to a conventional set–reset flip-flop. When $\overline{PR} = 0$, Q will make a transition to 1 and $\bar{Q}$ will make a transition to 0 irrespective of the other flip-flop inputs. Similarly when $\overline{CLR} = 0$, Q will make a transition to 0 and $\bar{Q}$ will make a transition to 1 regardless of the other flip-flop inputs. In like manner, in the case of master–slave flip-flops, the $\overline{PR}$ and $\overline{CLR}$ inputs are introduced to both master and slave sections. Figures 7.35 and 7.36 respectively show the circuits of a master–slave SR flip-flop and a master–slave JK flip-flop equipped with two asynchronous inputs. Since both $\overline{PR}$ and $\overline{CLR}$ inputs are unclocked inputs, there exists a distinct possibility of logical hazards that may cause false flip-flop outputs to occur. Consequently, these two inputs should be used only when $CK = 0$ as this would eliminate the clocked inputs from becoming dominant.

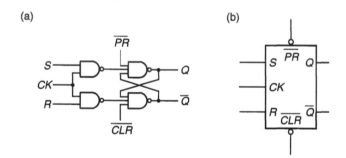

FIGURE 7.34
A set–reset flip-flop with preset and clear inputs.

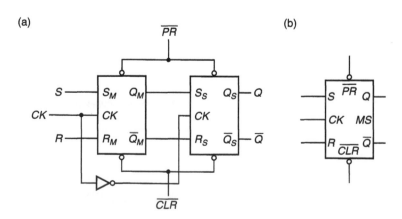

FIGURE 7.35
A master–slave set–reset flip-flop with asynchronous controls (a) circuit and (b) symbol.

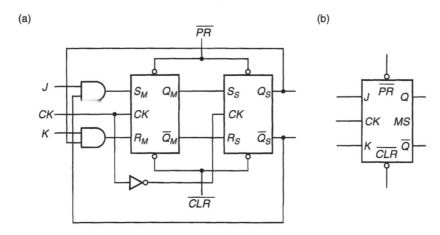

FIGURE 7.36
A master–slave *JK* flip-flop with asynchronous controls (a) circuit and (b) symbol.

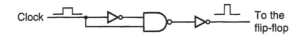

FIGURE 7.37
A pulse-narrowing circuit.

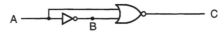

FIGURE 7.38
A pulse-narrowing circuit for Example 7.4.

7.6 Edge-Triggered Flip-Flops

Another way to eliminate the problem caused by a clock pulse width that is too long is to design the flip-flops in a way so as to respond to only transitions of the clock, either $1 \rightarrow 0$ or $0 \rightarrow 1$. Such flip-flops are referred to as *edge-triggered* flip-flops. One of the means often involves using a pulse-narrowing circuit, as shown in Figure 7.37. One of the inputs to the NAND gate is delayed since it has to first go through a NOT gate. The NAND output thus often becomes spike-like at the very beginning of the clock pulse. This narrow pulse may be used as the clock input of the flip-flop eliminating the necessity for narrow clock pulses. Depending on the parameters of the gates that are being used in the circuit of Figure 7.37, this pulse could be too narrow to trigger an FF. In such an event more than one, but only an odd number of, cascaded NOT gates could be used in place of the first NOT gate. Again the number of NOT gates should not be too large, otherwise a clock input that is too wide might result.

Example 7.4
Consider that each of the gates of Figure 7.38 have the same delay of one arbitrary unit. Plot the output waveform for an input waveform that remains high for a period of five arbitrary units.

322   *Digital Design: Basic Concepts and Principles*

Solution

The timing diagram is obtained as shown in Figure 7.39. The logic circuit of Figure 7.38 locates the trailing edge of the input pulse. It is different than the logic circuit of Figure 7.37 that locates instead the leading edge of an input pulse.

A pulse narrowing circuit such as that in Figures 7.37 and 7.38 is not very reliable compared to other alternatives such as master–slave (pulse-triggered) and edge-triggered flip-flops. We have already considered in the previous section master–slave configuration that introduces a buffering mechanism to eliminate unstable transient conditions. An edge-triggered flip-flop neither uses a pulse-narrowing circuit nor a master–slave configuration. For example, Figures 7.40a, b show the logic circuits respectively of a positive edge-triggered SR flip-flop and a positive edge-triggered JK flip-flop. For simplicity, the exact design of these logic circuits is avoided here. These sequential circuits form a distinct class called fundamental mode circuit. Chapter 11 covers the methodology involved in the design of such sequestial circuits.

Figure 7.41 shows the logic symbols for the different edge-triggered flip-flops. In order to activate a positive edge-triggered flip-flop, it is necessary that the clock makes a $0 \rightarrow 1$ transition. In a similar manner, a negative edge-triggered flip-flop is activated with a $1 \rightarrow 0$ transition of the clock input. But with an additional NOT gate in series with the clock input, the positive and negative edge-triggered flip-flops may be activated respectively with the $1 \rightarrow 0$ and $0 \rightarrow 1$ transitions of the clock.

The timing characteristics of edge-triggered flip-flops are shown in Figure 7.42. Both the setup time t_s and hold time t_h may be measured with respect to $0 \rightarrow 1$ and $1 \rightarrow 0$

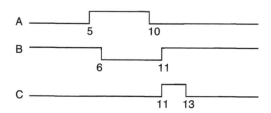

FIGURE 7.39
Timing diagram for Example 7.4.

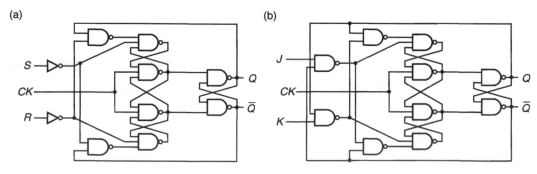

FIGURE 7.40
Circuits of positive edge-triggered (a) SR flip-flop and (b) JK flip-flop.

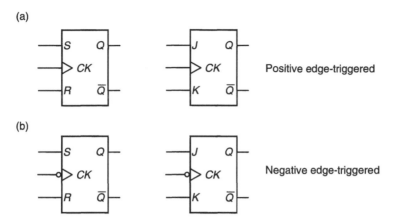

FIGURE 7.41
Logic symbols for edge-triggered flip-flops (a) positive edge-triggered and (b) negative edge-triggered.

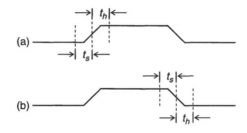

FIGURE 7.42
Timing characteristics of edge-triggered flip-flops (a) positive edge-triggered and (b) negative edge-triggered.

transition respectively for the positive edge-triggered and negative edge-triggered flip-flops. With an additional NOT gate in series with the CK input, the positive and negative edge-triggered flip-flops can be activated with a clock that makes $1 \rightarrow 0$ and $0 \rightarrow 1$ transitions respectively. The flip-flop inputs must be stable during the setup time and they can be changed only after the passage of hold time so as not to affect the flip-flop output. The output of the edge-triggered flip-flop follows in accordance to the inputs and appears after a period equivalent to the sum of the propagation delays of the component logic gates.

In the last section, we have shown how preset and clear inputs are introduced to the master–slave flip-flops. Likewise, it is also possible to include asynchronous preset and clear inputs to an edge-triggered flip-flop. Figure 7.43 shows, for example, a negative edge-triggered JK flip-flop that is equipped with such asynchronous inputs.

Example 7.5
Obtain the timing diagram for the sequential circuit shown in Figure 7.44 assuming that both of the flip-flops are initially in reset state.

Solution
The timing diagram is readily obtained as shown in Figure 7.45 by making use of the truth table of a JK flip-flop. Within two clock periods, Q_1 is reset and Q_2 is set.

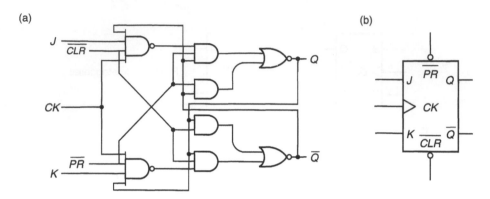

FIGURE 7.43
Negative edge-triggered *JK* flip-flop with asynchronous controls (a) logic circuit and (b) logic symbol.

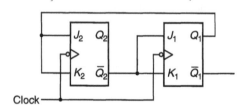

FIGURE 7.44
Negative edge-triggered *JK* flip-flop with asynchronous controls (a) logic circuit and (b) logic symbol.

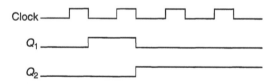

FIGURE 7.45
Timing diagram for Example 7.5.

7.7   Delay and Trigger Flip-Flops

There are two other types of flip-flops that are commonly used: the *delay* and the *toggle* flip-flops. Unlike *SR* and *JK* flip-flops, these two flip-flops have one input in addition to the clock, clear and preset. Both of these single-input flip-flops can be realized from *JK* flip-flops by externally manipulating its traditional inputs.

One of the most frequent digital operations is storing data. The stored bits are then manipulated in accordance to the predefined arithmetic and logic operations. The resulting bits are then stored again. In other words, we need memory elements that continues to remember what has been fed to it. The *delay* (*D*) flip-flop performs this function. The output of this flip-flop follows the input whenever a clock pulse is 1 and holds the value the input had when the clock changes to 0. The logic diagram and the truth table for a *D* flip-flop are shown in Figures 7.46a, b. A comparison of this truth table with that for the *JK* flip-flop (in Figure 7.26) reveals that a *D* flip-flop is realizable from a *JK* FF by making $K = \bar{J}$ and using *J* as the *D* input, as illustrated in Figure 7.46c. The next-state equation of the *D* flip-flop is

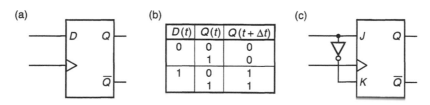

FIGURE 7.46
D flip-flop (a) logic symbol, (b) truth table, and (c) logic circuit.

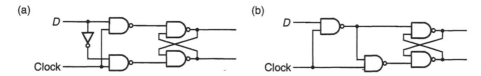

FIGURE 7.47
Realization of D flip-flops using (a) a SR flip-flop and an inverter and (b) a modified NAND-based latch.

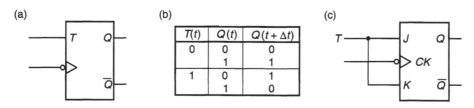

FIGURE 7.48
T flip-flop (a) logic symbol, (b) truth table, and (c) logic circuit.

given by

$$Q(t + \Delta t) = D(t) \tag{7.13}$$

If the restriction so placed on the clock input of a SR flip-flop poses no problem, the SR flip-flop can also be used to design a D flip-flop. The resulting circuit is shown in Figure 7.47a. Figure 7.47b shows a slight variation of the circuit of Figure 7.47a where advantage is taken of the special properties of NAND gates to eliminate one gate and still retain the characteristics of a D flip-flop.

The *toggle* (T) flip-flop has an input that triggers the output to change each time a pulse occurs at the input. The output remains unchanged as long as $T = 0$. The logic diagram and the truth table for a T flip-flop are shown in Figures 7.48a and b. It should be noted that the JK flip-flop has already this mode available. The JK flip-flop can be reorganized for realizing a T flip-flop, as shown in Figure 7.48c. With both the T input and the clock input high, the flip-flop output undergoes a transition. Its next-state equation, therefore, is given by:

$$Q(t + \Delta t) = Q(t) \oplus T(t) \tag{7.14}$$

An alternate version of the T flip-flop involves a single-input device. Both J and K inputs of a JK flip-flop are tied to a 1 to realize this unclocked T flip-flop. The input data are typically introduced at the clock input. The corresponding logic circuit for the unclocked T FF is shown in Figure 7.49. The unclocked T flip-flops are sometimes useful but they are typically not available. While it is derived from a JK flip-flop, it can also be obtained using

(a)   (b)

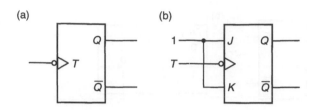

FIGURE 7.49
Unclocked T flip-flop (a) logic symbol and (b) logic circuit.

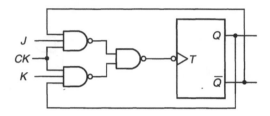

FIGURE 7.50
A JK flip-flop using an unclocked T flip-flop.

a D flip-flop. It is interesting to note that we can modify an unclocked T flip-flop to obtain a JK flip-flop from it. Figure 7.50 shows the logic circuit that converts an unclocked T flip-flop to a JK flip-flop.

Example 7.6
Determine the characteristics of the logic circuit shown in Figure 7.51.

Solution
The next-state equation for this circuit may be expressed as

$$Q(t + \Delta t) = D(t) = Y(t)\bar{Q}(t) + \bar{X}(t)Q(t)$$

This is very much similar to Equation 7.12 given by

$$Q(t + \Delta t) = J(t)\bar{Q}(t) + \bar{K}(t)Q(t)$$

describing the next state of a JK flip-flop. Accordingly, the logic circuit shown in Figure 7.51 describes a JK flip-flop wherein Y is treated as the J input and X as the K input.

7.8   Monostable Flip-Flop

We introduce in this section monostable flip-flop, a specialized logic circuit, used in the implementation of certain sequential logic circuits. The *monostable* flip-flop, also known as a *one-shot*, is an edge-triggered device used for producing a single output pulse of a duration that is independent of the input frequency. Typically, they produce a single output pulse of

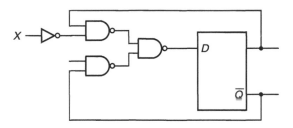

FIGURE 7.51
Logic circuit for Example 7.6.

(a)

(b)

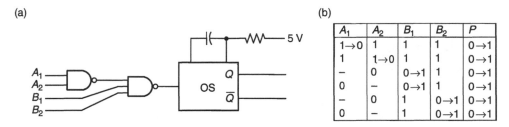

A_1	A_2	B_1	B_2	P
1→0	1	1	1	0→1
1	1→0	1	1	0→1
–	0	0→1	1	0→1
0	–	0→1	1	0→1
–	0	1	0→1	0→1
0	–	1	0→1	0→1

FIGURE 7.52
Retriggerable one-shot (a) logic diagram and (b) truth table.

specified width that is initiated by an input trigger signal changing from a 0 to a 1 and/or from a 1 to a 0. The output returns to its quiescent state after a specified time. The pulse duration is determined often by parameters of the resistor-capacitor network external to the one-shot. One-shot with a variable pulse duration is referred to as also delay units. While in a one-shot the output returns from a 1 to a 0, the output of a flip-flop remains either a 1 or a 0 when the clock is low.

One-shots can be either nonretriggerable or retriggerable. The nonretriggerable one-shot ignores the second of two successive trigger pulses if the two are separated in time by an amount less than the width of the output pulse generated by a single trigger. The retriggerable one-shot, on the other hand, is activated by the second trigger pulse. It results in an output pulse of width approximately $\tau + \Delta t$ where τ is the separation between successive trigger pulses and Δt is the width of the output pulse generated by a single trigger. By applying a succession of trigger pulses separated in time by $\tau < \Delta t$, one can maintain the output of a retriggerable one-shot indefinitely. It must be understood that one-shot should be used only when no other alternative is available. Circuits with a number of one-shots are rather difficult to troubleshoot. Furthermore, one-shots can be falsely triggered by signal noise.

A one-shot may be designed using basic logic gates and a resistor–capacitor network. However, it is more convenient to use an IC one-shot because they are widely available and relatively inexpensive. Figure 7.52 shows the logic diagram and the truth table of a retriggerable one-shot. The four inputs, A_1, A_2, A_3, and A_4, feed signals to two logic gates whose output in turn serves as the input P of the one-shot. The availability of these four inputs provides added flexibility in the operation of a one-shot.

The capacitor, C, and the resistor, R, are external to the IC one-shot and are used to determine the duration Δt of the output pulse. Adjustable resistors and/or capacitors may be used to trim the output pulse to the desired width. In general, Δt is given by

$$\Delta t = f(R, C) \tag{7.15}$$

where $f(R,C)$ is a function of the combination of resistance and capacitance. Typically, the manufacturer provides the exact numerical relationship or curves, providing the output pulse width as a function of both resistance and capacitance. To obtain the minimum width output pulse, no external capacitor is required. However, there will always be some stray capacitance existing between the terminals even in the absence of an external capacitor. The triggering conditions, as listed in Figure 7.52b, cause P to change from a 0 to a 1. A nonretriggerable one-shot may be realized from a retriggerable one-shot by feeding its Q output as one of the NAND gate inputs, say, A_2, while the other input A_1 serves as the triggering input. The remaining two inputs, B_1 and B_2, are both tied to a 1.

7.9   Design of Sequential Elements Using VHDL

In this section we illustrate several VHDL descriptions for sequential circuit elements. Latches, as discussed in Section 7.2, can be implemented with a pair of looped inverters separated by a transmission gate (TG). Figure 7.53 shows the circuit diagram of such an implementation, which is perhaps the simplest for D-type latch. Figure 7.54 shows a functional VHDL description of this D latch, where an *if-then-else* conditional statement is used. A more complex conditional section can also include *elsif* statements as needed, as will be demonstrated later. As with the case in many other programming languages, these conditional statements may also be nested within each other.

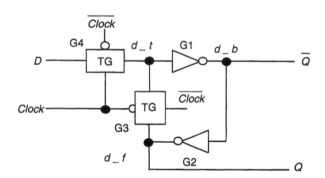

FIGURE 7.53
Inverter-based D latch implementation.

```
-- A functional VHDL description of a D-latch              -- 01
library ieee;                                              -- 02
use ieee.std_logic_1164.all;                               -- 03
entity TG is                                               -- 04
   port(din, clk, clk_b: in std_logic; dout: out std_logic); -- 05
end TG;                                                    -- 06
                                                           -- 07
architecture functional_TG of TG is                       -- 08
   begin                                                   -- 09
      dout <= din when (clk = "1" and clk_b = "0") else    -- 10
              'X';                                          -- 11
end functional_TG;                                         -- 12
                                                           -- 13
-- End of the VHDL description                             -- 14
```

FIGURE 7.54
A functional VHDL description of a D-latch.

```
-- A hierarchical VHDL description of a D-latch          -- 01
library ieee, lcdf_vhdl;                                 -- 02
use ieee.std_logic_1164.all, lcdf_vhdl.func_prims.all;   -- 03
entity D_latch is                                        -- 04
   port (D, clk, clk_b: in std_logic; Q, Q_b: out std_logic); -- 05
end D_latch;                                             -- 06
                                                         -- 07
architecture hierarchial_D_latch of D_latch is          -- 08
                                                         -- 09
   component NOT1                                        -- 10
     port(din: in std_logic; dout: out std_logic);      -- 11
   end component;                                        -- 12
                                                         -- 13
   component TG                                          -- 14
     port(din, clk, clk_b: in std_logic;                -- 15
          dout: out std_logic);                         -- 16
   end component;                                        -- 17
                                                         -- 18
   signal d_t: std_logic;                               -- 19
                                                         -- 20
   begin                                                 -- 21
     G1: NOT1 port map(d_t, Q_b);                        -- 22
     G2: NOT1 port map(Q_b, Q);                          -- 23
     G3: TG port map(Q, clk_b, clk, d_t);                -- 24
     G4: TG port map(D, clk, clk_b, d_t);                -- 25
end hierarchical_D_latch;                                -- 26
-- End of the VHDL description                           -- 27
```

FIGURE 7.55
A hierarchical VHDL description of a D-latch.

Figure 7.55 illustrates a hierarchical VHDL of a D-type latch, where an entity called TG is defined functionally within, while an inverter component called NOT1 and defined externally in *func_prims* is referenced. Regardless being defined internally or externally, interfaces of components referenced by an architecture section must be defined within. This is clearly illustrated by lines 22–29 in Figure 7.55. Figures 7.56 and 7.57, in which the concept of *process* is employed, illustrate descriptions of positive- and negative-edge-triggered D flip-flops, respectively, with asynchronous set and reset.

The VHDL *process* is often used to describe complex digital systems embedded with many sequential components. The keyword **process** may be preceded by an optional process name followed by a colon and is followed by a list of process-control input signals included in parentheses. This is illustrated by lines 13 in Figures 7.56 and 7.57, as well as line 12 in Figure 7.58. The description of a process must be included by a pair of **begin** and **end process** statements. A rule-of-thumb in deciding process-control signals is that input signals to be used in conditional statements within a process are typically identified as the process-control signals. This is illustrated in Figures 7.56 and 7.57 with input signals clk, Res, and Set. Figure 7.58 shows a VHDL process description of a positive edge-triggered RS-type flip-flop.

7.10 Sequential Circuits

The general model of a sequential circuit is shown in Figure 7.59. It has associated with it the inputs, represented by the m-tuples $(X_1 \ldots X_m)$, the outputs, represented by the n-tuples $(Z_1 \ldots Z_n)$, the present state and next state, represented respectively by the r-tuples $(q_1 \ldots q_r)$

```
-- A VHDL process description of a positive edge      -- 01
-- trigged asynchronous-set-reset D-FlipFlop          -- 02
library ieee;                                         -- 03
use ieee.std_logic_1164.all;                          -- 04
entity D_flipflopis                                   -- 05
   port(D,clk, Res, Set: in std_logic;               -- 06
        Q,Qb: out std_logic);                         -- 07
end D_flipflop;                                       -- 08
                                                      -- 09
architecture functional_D_FF of D_flipflop is         -- 10
                                                      -- 11
   begin                                              -- 12
   D_FF: process (clk, Res, Set)                      -- 13
      begin                                           -- 14
        if(Res = '1')then                             -- 15
           Q <= '0';                                  -- 16
           Qb <= '1';                                 -- 17
        elsif(Set= '1')then                           -- 18
           Q <='1';                                   -- 19
           Qb <='0';                                  -- 20
        elsif(clk'event and clk='1')then              -- 21
           Q <= D;                                    -- 22
           Qb <= not D;                               -- 23
        end if;                                       -- 24
    end process;                                      -- 25
end functional_D_FF;                                  -- 26
-- End of the VHDL description                        -- 27
```

FIGURE 7.56
A VHDL process description of a positive edge-triggered D flip-flop with asynchronous set and reset.

```
-- A VHDL process description of a negative edge      -- 01
-- trigged asynchronous-set-reset D-FlipFlop          -- 02
library ieee;                                         -- 03
use ieee.std_logic_1164.all;                          -- 04
entity D_flipflop is                                  -- 05
   port(D, clk, Res, Set: in std_logic;              -- 06
        Q, Qb:out std_logic);                         -- 07
end D_flipflop;                                       -- 08
                                                      -- 09
architecture functional_D_FF of D_flipflop is         -- 10
                                                      -- 11
   begin                                              -- 12
   D_FF: process (clk, Res, Set)                      -- 13
      begin                                           -- 14
        if (Res ='1') then                            -- 15
           Q <='0';                                   -- 16
           Qb <='1';                                  -- 17
        elsif (Set='1') then                          -- 18
           Q <='1';                                   -- 19
           Qb <='0';                                  -- 20
        elsif (clk'event and clk='0')then             -- 21
           Q <= D;                                    -- 22
           Qb <= not D;                               -- 23
        end if;                                       -- 24
    end process;                                      -- 25
end functional_D_FF;                                  -- 26
-- End of the VHDL description                        -- 27
```

FIGURE 7.57
A VHDL process description of a negative edge-triggered D flip-flop with asynchronous set and reset.

```
-- A VHDL process description of a positive edge        -- 01
-- trigged RS-FlipFlop                                  -- 02
library ieee;                                           -- 03
use ieee.std_logic_1164. all;                           -- 04
entity RS_flipflop is                                   -- 05
   port(R,S, clk: in std_logic;Q, Qb:out std_logic);    -- 06
end RS_flipflop;                                         -- 07
                                                         -- 08
architecture functional_RS_FF of RS_flipflop is         -- 09
                                                         -- 10
   begin                                                -- 11
   RS_FF: process(clk, R, S)                            -- 12
     begin                                              -- 13
       if (clk'event and clk = '1') then                -- 14
         if (R ='1'and S = '0') then                    -- 15
            Q <= '0';                                   -- 16
            Qb <= '1';                                  -- 17
         elsif (R = '0' and S = '1') then               -- 18
            Q <= '1';                                   -- 19
            Qb <= '0';                                  -- 20
         elsif(R = '1'and S = '1') then                 -- 21
            Q <= 'X';                                   -- 22
            Qb <= 'X';                                  -- 23
         end if;                                        -- 24
       end if;                                          -- 25
     end process;                                       -- 26
end                                                      -- 27
-- End of the VHDL description                           -- 28
```

FIGURE 7.58
A VHDL process description of a positive edge-triggered RS flip-flop.

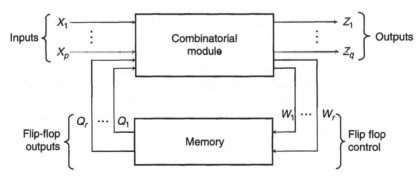

FIGURE 7.59
Model for a generalized sequential circuit.

and ($Q_1 \ldots Q_r$), and a clock input (in the case of synchronous circuit). The memory circuits may be of several types: flip-flops, magnetic devices, delay lines, and many others. The combinatorial module of the sequential circuit generates flip-flop control variables for the correct changes to occur on the basis of the input values and the current flip-flop states. In addition the combinatorial module also generates the correct outputs. Thus, the present inputs (including the clock in the case of synchronous sequential circuits) and the present state information stored in the circuit's memory is used to generate the present outputs as well as the next state of the sequential circuits.

An alternative means of representing the characteristics of a sequential circuit involves what is known as a state transition diagram. This diagram consists of one labeled node (represented by a circle) for each state and a set of labeled transition paths leaving each

$Q(t)$	$Q(t+\Delta t)$	$S(t)R(t)$	$J(t)K(t)$	$T(t)$	$D(t)$
0	0	0-	0-	0	0
0	1	10	1-	1	1
1	0	01	-1	1	0
1	1	-0	-0	0	1

FIGURE 7.60
Flip-flop control characteristics.

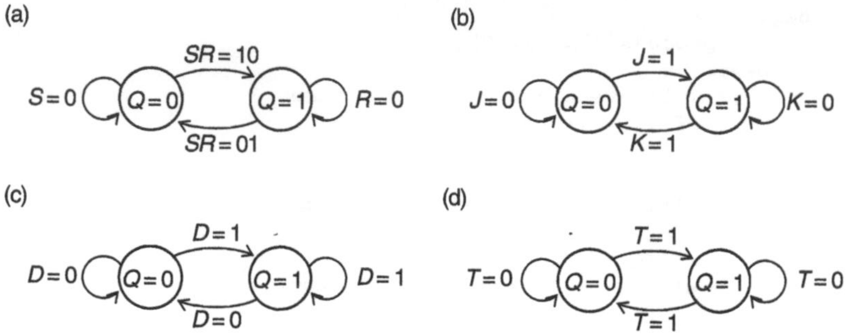

FIGURE 7.61
Transition diagrams: (a) SR flip-flop, (b) JK flip-flop, (c) D flip-flop, and (d) T flip-flop.

state and terminating at the next state. The label of the transition paths lists the input condition(s) responsible for the transition in question. As an option, however, one may even list the output(s) resulting from such transition. The state transitions between $Q = 0$ and $Q = 1$ for each of the four flip-flops are shown in the Figure 7.60 where the conditions for transitions are indicated next to the transition lines. Figure 7.61 shows the corresponding state transition diagrams. These flip-flop characteristics will be recalled time and again for designing complex sequential circuits.

One of the advantages of the sequential systems is that it may offer a cost-effective solution since the same circuit can be used repetitively. Some examples of such applications are in the design of multibit adder/subtracter, multibit comparator, and multibit code converters. As was seen in Chapter 6, we need typically n different combinatorial modules to realize an n-bit parallel operation. However, the price we pay for using the same logic circuit repetitively is often in the speed of the operation.

Consider, for example, an n-bit adder. This addition operation is referred to as parallel since the two operands are simultaneously available to the adder, and after adequate delay, the resultant sum bits become available simultaneously. By making intelligent choice of sequential circuit, it is also possible to implement addition serially. Figure 7.62 shows such a sequential circuit that can perform serial addition. The process may begin by feeding A_0 and B_0 to the single-bit full adder. The resultant sum is stored in a memory device (often in a register made up of several cascaded flip-flops), and the carry-out is stored in a flip-flop and then fed back into the single-bit full adder as carry-in input for the next cycle of addition. This transmission addition storage process is repeated until all of the bits have been processed. The final sum would consist of the last carry-out (the most significant bit) and the remaining sum bits stored in the register. The characteristics of the registers and how the inputs are fed sequentially to the adder will be considered in Chapters 9 and 10. For the time being it is enough to understand that the registers in question are capable moving stored bits from left to the right.

Sequential systems are not devoid of problems. Some of these problems often are very critical to the functioning of the system and thus must be considered very carefully. Consider the

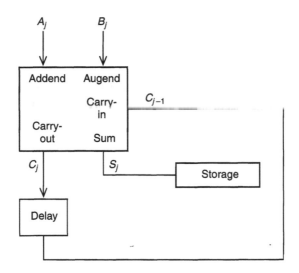

FIGURE 7.62
A serial addition circuit.

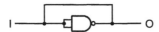

FIGURE 7.63
An indeterminate logic circuit.

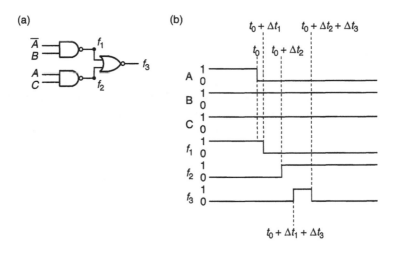

FIGURE 7.64
(a) Combinatorial circuit and (b) its corresponding timing diagram.

logic circuit shown in Figure 7.63. When the input is a 0, the output becomes a 1. Since the output in turn is fed back as a input, the values of the output disagree with that of the input resulting in an indeterminate feedback system. Such problems are avoided by eliminating the conditions of continuous oscillations.

To comprehend another frequently encountered problem, consider next the combinatorial logic circuit shown in Figure 7.64a where the gates numbered 1, 2, and 3, respectively, have gate delays of Δt_1, Δt_2, and Δt_3. We arbitrarily chose that at time t_0 all three inputs are 1 and at time $t > t_0$ the input A, for example, changes from a 1 to a 0. Figure 7.64b shows the resulting timing diagram provided that $\Delta t_1 < \Delta t_2$. It is apparent that the combinatorial circuit output results in a transient error pulse for a period given by $\Delta t_2 - \Delta t_1$. This type of an output is a consequence of the hazard problem discussed earlier in Chapter 2. Such error pulse is small but may not be considered negligible. If the circuit output were to be introduced as the clock, preset, or clear inputs to a flip-flop, the sequential circuit could malfunction. Fortunately, errors will not occur until the pulse width exceeds the time necessary to trigger the flip-flop. The designer must be aware of these problems and take proper precautions either during the design or during the operation of a sequential system or both.

The sequential circuits shown in Figure 7.59 are of two types. They are classified according to the characteristics of their inputs and memory module.

Synchronous sequential circuits: Synchronous sequential circuits involve flip-flop action that occurs in synchronization with the clock input. The input variables that control the flip-flop states may change while the clock is low. For reliable operation, all transients due to the previous clock input are assumed to have disappeared before the beginning of the next clock. The synchronous sequential circuits account for the overwhelming majority of sequential circuits.

Asynchronous sequential circuits: This is the general class of sequential circuit where clock input is absent. On the basis of the type of inputs, however, the asynchronous sequential circuits can be further divided into two classes: pulse- and level-mode. In the case of *pulse-mode* asynchronous circuits, the input variables can have only mutually exclusive pulses. No two input pulses are allowed to remain high simultaneously. The *fundamental-mode* (or *level-mode*) circuits involve level inputs and asynchronous memory devices. The memory changes state whenever an input variable logic level changes.

7.11 Summary

In this chapter, we have explored the design and working principles of latches, flip-flops, and monostable multivibrators with emphasis on their various practical limitations. Finally, we have introduced the idea of using such sequential devices in useful sequential circuits. The design and the characteristics of these sequential systems are explored in detail in the next four chapters.

Bibliography

Comer, D.J., *Digital Logic and State Machine Design*. 3rd edn. New York, NY. Oxford University Press, 1994.

Floyd, T., *Digital Fundamentals*. 8th edn. Englewood Cliffs, NJ. Prentice-Hall, 2003.

Hwang, K., *Computer Arithmetic*, New York, NY. Wiley, 1979.

Johnson, E.J. and Karim, M.A., *Digital Design: A Pragmatic Approach*. Boston, MA. PWS-Kent Publishing, 1987.

Karim, M.A. and Awwal, A.A.S., *Optical Computing: An Introduction*. New York, NY. John Wiley & Sons, 1992.

Katz, R.H., *Contemporary Logic Design*. Boston, MA. Addison Wesley, 1993.

Kline, R.M., *Structured Digital Design Including MSI/LSI Components and Microprocessors*. Englewood Cliffs, NJ. Prentice-Hall, 1983.

Mowle, F.J.A., *A Systematic Approach to Digital Logic Design*. Reading, MA. Addison-Wesley, 1976.

Nelson, V.P.P., Nagle, H.T., and Carroll, B.D., *Digital Logic Circuit Analysis and Design*. Englewood Cliffs, NJ. Prentice Hall, 1995.

Waser, S. and Flynn, M.J., *Introduction to Arithmetic for Digital Systems Designers*. New York, NY. Holt, Rinehart & Winston, 1982.

Problems

1. Show a timing diagram for the given logic circuit shown in Figure 7.P1 assuming that $Q_1(0)Q_2(0) = 00$.

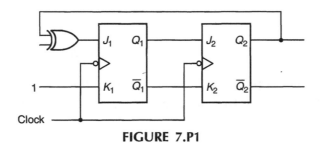

FIGURE 7.P1

2. What sequence should repeat for the sequential circuit of Figure 7.P2 for the following initial inputs $Q_3Q_2Q_1$: (a) 001 and (b) 100.

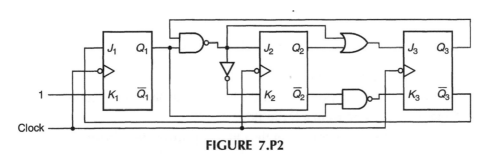

FIGURE 7.P2

3. Find the output and state sequences for the circuit of Figure 7.P3 if the initial state is $Q = 0$ and the input sequence x is (a) 10011011000 and (b) 10110111001.

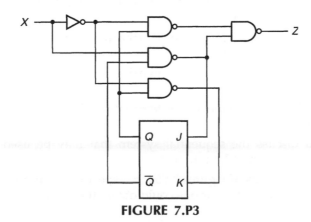

FIGURE 7.P3

4. Design a synchronous sequential circuit that satisfies the following next-state equations:

$$Q_1(t + \Delta t) = A(t)Q_1(t) + \overline{A(t)}Q_2(t)$$
$$Q_2(t + \Delta t) = A(t)\overline{Q_1(t)} + \overline{Q_2(t)}$$

using (a) SR flip-flops, (b) JK flip-flops, (c) D flip-flops, and (d) T flip-flops.

5. Design a synchronous sequential circuit that satisfies the following next-state equations:

$$Q_1(t + \Delta t) = A(t)Q_1(t) + Q_2(t)$$
$$Q_2(t + \Delta t) = \overline{A(t)}Q_2(t) + \overline{Q_1(t)}$$

using (a) SR flip-flops, (b) JK flip-flops, (c) D flip-flops, and (d) T flip-flops.

6. A sequential circuit has two inputs, X and Y, one output, Z, and two JK flip-flops such that

$$J_1 = XQ_2 + \overline{YQ_2}$$
$$J_2 = X\overline{Q_1}$$
$$K_2 = X\overline{Y} + Q_1$$
$$K_1 = X\overline{Y}Q_2$$
$$Z = XYQ_1 + \overline{XY}Q_2$$

What are its next-state equations?

7. Discuss the working principles of the logic circuit shown in Figure 7.49.

8. Obtain a T flip-flop from a D flip-flop.

9. Determine the characteristics of the following sequential circuits:

10. Repeat the problem of Example 7.2 when (a) NAND and NOT gates respectively have a delay of 4 and 3 units, (b) gate delays are lumped together, and (c) NAND and NOT gates have a delay of 4 and 3 units respectively assuming a lumped model.

11. Obtain and discuss the sequential system that may be used for realizing a multidigit BCD-to-binary conversion.

12. Obtain and discuss the sequential system that may be used for realizing a multibit binary-to-BCD conversion.

13. Obtain and discuss the sequential system that may be used for realizing a multibit gray-to-binary conversion.

14. Obtain and discuss the sequential system that may be used for realizing a multibit binary-to-gray conversion.

15. Obtain and discuss the sequential system that may be used for realizing a comparison of two multibit numbers.

16. Show that the circuits of Figures 7.P5a, b respectively represent edge-triggered D and T flip-flops. Show how one could obtain them from the circuits shown in Figure 7.40.

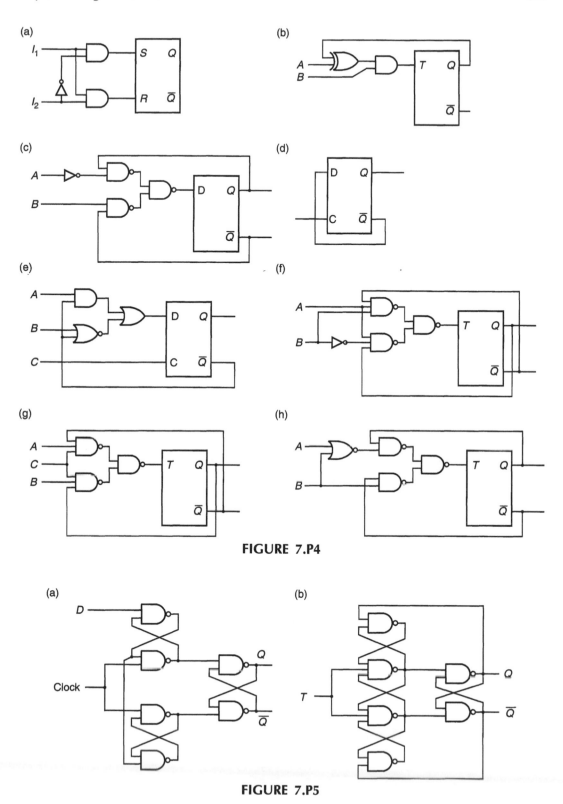

FIGURE 7.P4

FIGURE 7.P5

17. Add asynchronous preset and clear inputs to the edge-triggered D and T flip-flops shown respectively in Figure 7.P5a and b.

18. Consider the serial addition circuit shown in Figure 7.60 where the single-bit full adder receives two external inputs from augend and addend while the carry-in input comes from the output of a D flip-flop that stored the carry-out of the next less significant pairs of bits. Derive state equations for the sum and carry-out outputs.

8

Synchronous Sequential Circuits

8.1 Introduction

In this chapter, synchronous sequential circuits will be explored and analyzed. Our ultimate objective is to understand the process using which one can design such a circuit to meet the application goals that are under consideration. In the next chapter, for example, we shall be using this mechanics to design a few modular sequential components. These circuits employ both combinatorial logic circuit and flip-flops. In particular, all circuit action takes place in synchronization with a periodic sequence of clock. The circuit in response to its inputs will either remain in its present state or transition onto its next state. Since clock pulses regulate all the transitions, glitches that otherwise occur due to the imperfect nature of the logic devices have no effect. This becomes possible since the clock period is chosen so that all glitches due to multiple delay paths end before the flip-flop encounters future changes.

Typically, sequential circuit design process consists of obtaining a state diagram and a state table describing each one of the input sequences, desired outputs, and internal states. Boolean expressions are then derived from the state table by incorporating in it the behavior patterns of flip-flops. In the following sections, we shall explore these design steps along with pertinent examples.

8.2 Formalism

A typical synchronous sequential circuit, as shown in Figure 8.1, has associated with it the inputs, represented by the p-tuples $(X_1 \ldots X_p)$, the outputs, represented by the q-tuples

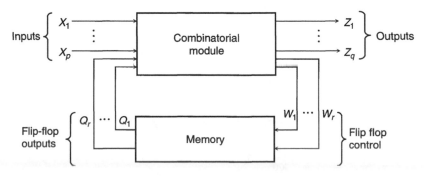

FIGURE 8.1
Model for a synchronous sequential circuit.

$(Z_1 \ldots Z_q)$, the present state and next state, represented respectively by the r-tuples $(q_1 \ldots q_r)$ and $(Q_1 \ldots Q_r)$, and a clock input. For its memory module, the designers are more likely to use either edge-triggered and/or pulse-triggered flip-flops. Its combinatorial module, in addition to generating the correct outputs, computes the flip-flop control variables on the basis of the input values and the flip-flop present states for the correct next state to occur. In summary, the present inputs and the present state information stored in the flip-flops are used to generate the outputs as well as the next state of the sequential circuit.

As discussed, there exists functional interdependency between the inputs, the outputs, the present state, and the next state. Typically, this interdependency is characterized by either a state diagram or a state table. The *state diagram* is a graphical representation of a sequential circuit wherein the states are shown represented by circles and transitions between states are shown by arrows. Few of the basic 2-state state diagrams were already introduced in Figure 7.61.

The state diagram shown in Figure 8.2, for example, represents a synchronous sequential circuit characterized by four states, $A, B, C,$ and D, an input variable, x, and a clock input. The corresponding circuit determines the state it is in and what the current value of x is, and then sets up the flip-flop inputs such that the correct next state results when the clock input occurs. The arrows connecting the states represent the occurrence of clock input, and the variables written alongside the arrow show the input condition that causes that transition to be followed and the corresponding output.

Assuming that the circuit is currently in state A, it will maintain its current state as long as $x = 1$ and the clock is present. This is represented by the transition path originating from A and then returning back to A, and the $x/0$ (i.e., input/output) designation of that transition. On the other hand, if $x = 0$, the circuit enters into state B when the clock occurs. This latter transition is represented by $\bar{x}/0$ designation. Either way, the circuit output remains a 0 when transitioning from state A. Continuing on in like manner, we see that if the circuit is in state B and $x = 0$, the circuit returns back to state A with the next clock but yielding no output (as represented by $\bar{x}/0$). But if $x = 1$, the state C is entered with the next clock and an output is generated. This transition is represented by $x/1$ written next to the transition path. The circuit transitions coincident with the next clock from state C to states A and D respectively when $x = 0$ and $x = 1$ producing no output. Finally, the circuit transitions from state D to state A when $x = 0$ but maintains its current state when $x = 1$. The state diagram must include each of the states of the circuit and all input conditions necessary for entering or

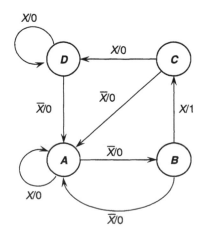

FIGURE 8.2
A 4-state synchronous sequential system.

PS	NS (when x=0), Z	NS (when x=1), Z
A	B,0	A,0
B	A,0	C,1
C	A,0	D,0
D	A,0	D,0

FIGURE 8.3
The state table for the circuit of Figure 8.2.

PS	NS (when x=0), Z	NS (when x=1), Z
A	B,0	A,0
B	A,0	C,1
C	A,0	C,0

FIGURE 8.4
The reduced state table for the circuit of Figure 8.3.

exiting the states. In this case, since there is a single input, two transition paths originate from each of the states. On closer examination, we notice that this sequential circuit is able to detect the occurrence of the sequence 001. Thus if the circuit input x for example, is 1011001000110010, the corresponding output Z will be 0000001000100010.

The relationship between the number of circuit states n and the number of flip-flops m is given by the expression:

$$2^{m-1} \le n \le 2^m \tag{8.1}$$

A 4-state sequential circuit such as that shown in Figure 8.1 thus requires two flip-flops. Likewise sequential circuits having the number of states 5 through 8, for example, require three flip-flops. The outputs of these flip-flops are referred to as the *state variables* and are used to identify which state the circuit is in.

An additional design tool that contains the same information as the state diagram but in tabular form is the *state table*. The state table of the synchronous sequential system shown in the state diagram of Figure 8.2 is obtained as shown in Figure 8.3. The *PS* and *NS* entries respectively refer to the present and next states and Z refers to the circuit output. By examining the entries in the table of Figure 8.3, we see that the last two entries pertaining to the states C and D are identical. This suggests that the states C and D are interchangeable. We refer to this state table as the one having a *redundant* state. In fact, all but one of these states could be eliminated from the state table as long as all mention of those eliminated states is substituted by the designation of the surviving state. The reduced state table is thus obtained as shown in Figure 8.4, where D has been removed and all mention of D has been replaced with C. In this particular case, by eliminating a redundant state, the designer may not gain much since either way the sequential circuit needs two flip-flops. But for example if one could have reduced it further, we would need one less flip-flop. It is thus good practice for the designers to identify and then eliminate redundant states until it cannot be reduced further.

Example 8.1
Obtain the state diagram and the state table for a synchronous sequential machine that outputs a 1 for each sequence of 110. The output is reset only with a subsequent occurrence of 10.

Solution

Figure 8.5 shows the state diagram for this machine. In it, the state A (with output $z = 0$) represents the state of the machine before introducing any input pulse. With an input of 1, the machine proceeds to state B (also with output 0) that is equivalent to have already detected first bit of a possible 110 sequence. If on the other hand the input is 0, the circuit transitions to state C where the machine waits for the first input of 1. The machine remains in state C as long as the input is 0 but transitions to state B as soon as the first 1 has been detected. With an input of 0, the machine at state B falls backward to state C since this implies that the input sequence 10 will not contribute to the desired sequence 110. But with an input of 1, the machine transitions to state D since the occurrence of 11 may lead to a possible sequence of 110. The state D is equivalent to that state of the machine representing the detection of the first two bits of the desired sequence. The machine state D is retained as long as the input continues to be 1. Otherwise, the machine transitions to state E giving an output of 1 since that would imply the detection of the sequence 110.

Once the output has been generated, the machine keeps on generating a string of 1s as long until the occurrence of the sequence 10. Thus, as long as the input received is 0, the machine residing at state E retains its state. However, with an in input of 1, the machine transitions to state F (representing the detection of the first bit of a possible 10 sequence). The circuit remains at state E as long as the input continues to be 1. But with an input of 0, the circuit transitions readily to state C and gives out an output of 0. It is now ready again to detect the next occurrence of 110.

The corresponding state table is obtained as that shown in Figure 8.6.

We see that there is no single state that goes to state A. However, we see that the states A and C are equivalent. This issue of equivalent states will be dealt with in a later section.

Example 8.2

Obtain the state diagram of a sequential machine that converts a binary string of numbers to its 2's complement equivalent.

Solution

The 2's complement of a binary number can be obtained as follows. In a parallel scheme, the bits are all complemented to obtain the 1's complement number first and then a 1 is added to the 1's complement. This scheme would also require a multibit adder. The other method is sequential in nature. The numbers are scanned from the right to the left. The 2's complement is obtained by complementing all bits to the left of the least significant 1 detected in that number.

The state diagram of the machine is obtained as shown in Figure 8.7. We assume that the bits are fed to the circuit in sequence starting with the least significant bit. As long as the input bits are 0, the machine resides in state A. With the detection of first 1, the machine moves to state B. All subsequent bits are complemented. To begin a new 2's complement conversion, the machine needs to be reset back to state A.

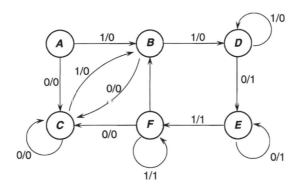

FIGURE 8.5
State diagram for Example 8.1.

PS	NS, z (when x=0)	NS, z (when x=1)
A	C,0	B,0
B	C,0	D,0
C	C,0	B,0
D	E,1	D,0
E	E,1	F,1
F	C,0	F,1

FIGURE 8.6
State table for Example 8.1.

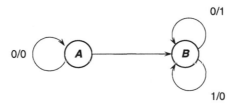

FIGURE 8.7
State diagram for Example 8.2.

8.3 Mealy and Moore Models

Sequential circuits can be classified as either Mealy- or Moore-type. Both Mealy and Moore type circuits are equally applicable to both synchronous and asynchronous circuits while the minimum number of external inputs to any one of these circuits is one. For a synchronous circuit having a single input, the input by default is the system clock. In Mealy model sequential circuit, the outputs are a function of the transition paths (originating state, i.e., present state, and the input) and not just of a transition state. The Mealy sequential circuit is also referred to as *transition-assigned* since the circuit output is associated with the state transitions. In general, the Mealy output and the next state are respectively given by,

$$Z_i = g_i(X_1 \ldots X_m, q_1 \ldots q_r), \quad i = 1, \ldots, n \qquad (8.2a)$$

$$Q_i = h_i(X_1 \ldots X_m, q_1 \ldots q_r), \quad i = 1, \ldots, r \qquad (8.2b)$$

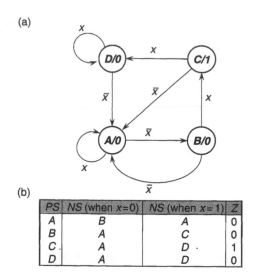

(a)

(b)

PS	NS (when $x=0$)	NS (when $x=1$)	Z
A	B	A	0
B	A	C	0
C	A	D	1
D	A	D	0

FIGURE 8.8
A Moore model (a) state diagram and (b) state table.

where g_i and h_i are Boolean functions. Circuits described by Figures 8.2 through 8.7 are examples of Mealy circuit. Thus when listing the corresponding circuit output in a state table, it is listed alongside with NS. For example, the state B as listed in Figure 8.3 may contribute towards generating an output only when $x = 1$. The NS entered in this case is C. No other transition leads to $Z = 1$.

The Moore output, on the other hand, is associated with only the present state. The Moore output and the corresponding next state are given by,

$$Z_i = g_i(q_1 \ldots q_r), \quad i = 1, \ldots, n \qquad (8.3a)$$

$$Q_i = h_i(X_1 \ldots X_m, q_1 \ldots q_r), \quad i = 1, \ldots, r \qquad (8.3b)$$

The Moore output is given out as soon as the circuit transitions to this particular Moore state. It is immaterial what input caused this circuit to transition into that state. The Moore outputs may change their values only when the states change because of a change of the inputs. Thus in a state diagram representing a Moore sequential circuit, the output is not designated next to the transition path along with the input. It is rather included within the circle representing states. Consider a Moore sequential circuit whose state diagram and state table are shown in Figure 8.8. Figure 8.8a shows the format for a state diagram that includes Moore-type outputs that are circled along with the corresponding present states.

The two state tables of Figure 8.3 and 8.8 are nearly identical. The output has been listed independent of the input. However, we see that in Figure 8.8, the states C and D are not equivalent even though their next states are identical. No state is redundant here because while state C is associated with a high output, the state D is not. Accordingly, it will not be possible to realize a reduced state table from that in Figure 8.8 such as that we derived from the state table in Figure 8.3.

The general forms of the Mealy and Moore circuits are shown in Figure 8.9. These circuits take their names, respectively, from G.H. Mealy and E. F. Moore, two of the most famous pioneers of sequential design. There are many systems that possess both Mealy and Moore outputs; in other words, some outputs may be conditional on both the inputs and the state of the circuit, while others are dependent only on the state of the circuit. In general, the

(a)

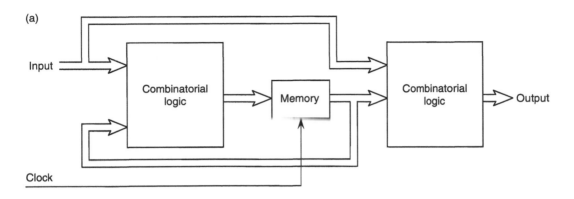

(b)

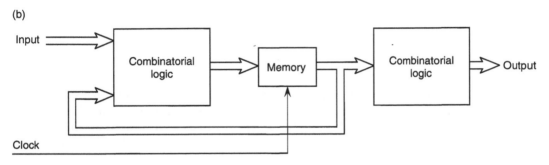

FIGURE 8.9
Synchronous sequential circuits: (a) Mealy model and (b) Moore model.

Mealy outputs are easily convertible to equivalent Moore outputs and vice versa. For now, however, we refrain from that discussion.

Example 8.3

Assuming that the circuits are initially at state A, determine the next state and the outputs when the input sequence $x = 11000101001101110010$ is fed to the circuits respectively represented in Figures 8.3, 8.4, and 8.8.

Solution

The next state and the output are obtained as follows:

$$x = 1\ 1\ 0\ 0\ 0\ 1\ 0\ 1\ 0\ 0\ 1\ 1\ 0\ 1\ 1\ 1\ 0\ 0\ 1\ 0$$
$$NS = A\ A\ B\ A\ B\ C\ A\ A\ B\ A\ A\ A\ B\ C\ D\ D\ A\ B\ C\ A$$
$$Z = 0\ 0\ 0\ 0\ 0\ 1\ 0\ 0\ 0\ 0\ 0\ 0\ 0\ 1\ 0\ 0\ 0\ 0\ 1\ 0$$

for the circuit of Figure 8.3,

$$x = 1\ 1\ 0\ 0\ 0\ 1\ 0\ 1\ 0\ 0\ 1\ 1\ 0\ 1\ 1\ 1\ 0\ 0\ 1\ 0$$
$$NS = A\ A\ B\ A\ B\ C\ A\ A\ B\ A\ A\ A\ B\ C\ C\ C\ A\ B\ C\ A$$
$$Z = 0\ 0\ 0\ 0\ 0\ 1\ 0\ 0\ 0\ 0\ 0\ 0\ 0\ 1\ 0\ 0\ 0\ 0\ 1\ 0$$

for the circuit of Figure 8.4, and

$$x = 1\ 1\ 0\ 0\ 0\ 1\ 0\ 1\ 0\ 0\ 1\ 1\ 0\ 1\ 1\ 1\ 0\ 0\ 1\ 0$$
$$NS = A\ A\ B\ A\ B\ C\ A\ A\ B\ A\ A\ A\ B\ C\ D\ D\ A\ B\ C\ A$$
$$Z = 0\ 0\ 0\ 0\ 1\ 0\ 0\ 0\ 0\ 0\ 0\ 0\ 1\ 0\ 0\ 0\ 1\ 0$$

for the circuit in Figure 8.8.

As expected, we notice that the circuits represented by Figures 8.3 and 8.8 are nearly identical. The next states all change synchronized with the clock. However, Moore output such as that for Figure 8.8 is also synchronized with the clock since this output is a function of only the state and can therefore change only when the state changes. The actual timing diagram for Z for the first two cases will likely be slightly different. This output Z can change any time either the input or the state changes, since it is a function of both.

The distinction between Mealy and Moore outputs is in how these outputs are produced. Their subtle difference becomes more meaningful when we use programmable logic devices (PLDs) for implementing sequential circuits. We have already introduced in Chapter 4 the characteristics of PLDs, however, some additional specifics need to be discussed before emphasizing the consequences of having Mealy and Moore outputs.

There are PLDs, which may offer not only combinatorial but also sequential outputs. The AND–OR array of these special purpose PLDs are quite the same as that in a combinatorial PLD except that these special purpose PLDs have edge-triggered D flip-flops on some of its outputs. These D flip-flops are fed with a common CK input. The outputs not tied to the flip-flops are generally bidirectional in nature thus functioning both as inputs and combinatorial outputs.

Figure 8.10 shows such a special-purpose PLD-based circuit that can be used for designing synchronous sequential circuits. The combinatorial module of the PLD is characterized by t_{pd}, the propagation delay from a primary, bidirectional, or internal feedback input (D flip-flop outputs going through the combinatorial PLD) to the output. The D flip-flops, on the other hand, are characterized by t_{ck}, the propagation delay from $0 \rightarrow 1$ transition of CK to a primary output or feedback input. In designing a synchronous sequential circuit using a PLD, Mealy outputs are derived directly from the AND–OR array. These outputs take on their values at time t_{pd} after the input has taken on a new value. In situations, where the Mealy output is a function of both present state and input, the output becomes stabilized at time $t_{pd} + t_{ck}$ after the $0 \rightarrow 1$ transition of the CK input.

The second kind of Moore output, referred to as Type 2, is generated from the outputs of the D flip-flops. Therefore, these outputs appear t_{ck} time later than the other type of Moore outputs, referred to as Type 1 Moore outputs. Such circuit delay problem is typically overcome by programming an output equation that uses the present states and inputs to predict what the output should be in the next state. This approach is quite analogous to what we employed earlier for generating the look-ahead carries in the *carry-skip* and *carry look-ahead* (CLA) adder (in Chapter 6). This approach of generating Type 2 Moore outputs has the advantage in that the outputs appear only t_{ck} time after the $0 \rightarrow 1$ transition of the CK input. In addition, Type 2 Moore outputs are typically hazard-free. However, as it was the case with the look-ahead carry generation, this approach has its own problem. For example, the to-be-programmed sum-of-product (SOP) expression for z may end up having too many product terms. As a consequence, one may have difficulty in accommodating the output function

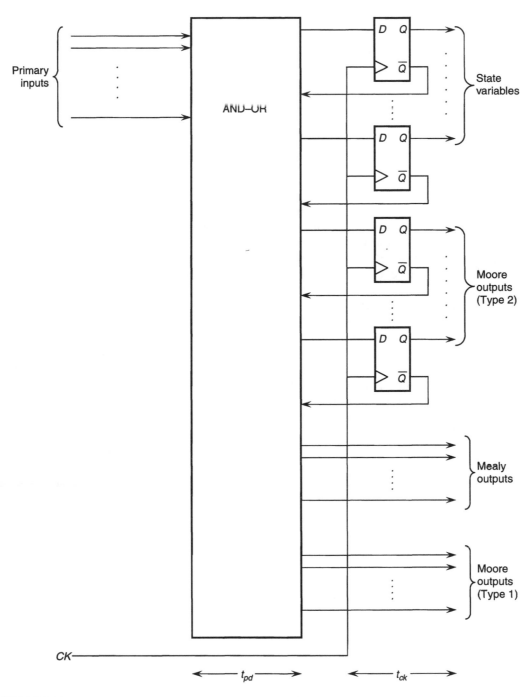

FIGURE 8.10
A PLD organization for synchronous sequential circuits.

with the size of the available PLD. This situation is analogous to that in a multi-input CLA adder where we ended up requiring NAND gates but with large number of inputs.

The Moore outputs are synchronized with clock and thus very well behaved. They do not normally contribute to any glitches. In comparison, in Mealy-type circuit, one must be careful when sampling the output and must allow for the circuit to stabilize after an input change. Noticeably, however, the Mealy model offers an advantage over the Moore model. Mealy outputs are functions of both input and the state. Accordingly, the designer has more flexibility in designing the circuit thus requiring usually fewer number of states and, therefore, fewer flip-flops.

Example 8.4
Obtain the Moore equivalent state table for the Mealy machine described in Figure 8.11.

Solution
As shown in Figure 8.11, the sequential machine is Mealy type since the outputs are not associated with only the states. Outputs are determined by both what state the circuit is in and what inputs have been fed. In comparison, the outputs in a Moore type circuit are associated only with states. To convert a Mealy type circuit to its equivalent Moore type circuit, it will be necessary to associate each of its outputs with a particular state.

It may be seen that the next states, A and C, are associated with two different outputs, 0 and 1. The next states, B and E, are associated with an output of 0 while the state D is associated with an output of 1. Accordingly, states A_0 and A_1 may be introduced to replace the state A and similarly states C_0 and C_1 may be introduced to replace the state C. The states A_0 and C_0 correspond, respectively, to states A and C when the output is a 0. Likewise, the states A_1 and C_1 correspond, respectively, to states A and C when the output is a 1. By introducing only A_0 and A_1, the state table is modified as shown in Figure 8.12. We introduce both of them in the PS column. Next, C_0 and C_1 may be included to obtain the state table as shown in Figure 8.13.

At this time each of the states has only one output associated with itself. Accordingly, one could obtain the equivalent Moore machine as shown in Figure 8.14.

PS	NS, $z(x=0)$	NS, $z(x=1)$
A	B,0	E,0
B	A,1	C,1
C	B,0	C,1
D	C,0	E,0
E	D,1	A,0

FIGURE 8.11
State table for Example 8.4.

PS	NS, $z(x=0)$	NS, $z(x=1)$
A_0	B,0	E,0
A_1	B,0	E,0
B	A_1,1	C,1
C	B,0	C,1
D	C,0	E,0
E	D,1	A_0,0

FIGURE 8.12
Modified state table for Example 8.4.

PS	NS, z(x=0)	NS, z(x=1)
A_0	$B,0$	$E,0$
A_1	$B,0$	$E,0$
B	$A_1,1$	$C_1,1$
C_0	$B,0$	$C_1,1$
C_1	$B,0$	$C_1,1$
D	$C_0,0$	$E,0$
E	$D,1$	$A_0,0$

FIGURE 8.13
Further modified state table for Example 8.4.

PS	NS (when x=0)	NS (when x=1)	Z
A_0	B	E	0
A_1	B	E	1
B	A_1	C_1	0
C_0	B	C_1	0
C_1	B	C_1	1
D	C_0	E	1
E	D	A_0	0

FIGURE 8.14
Moore equivalent state table for Example 8.4.

The Moore machine as obtained in Figure 8.14 is equivalent to the original Mealy machine as given in Figure 8.11 only in the sense that its output appears as pulses. The outputs coincident with A_1 and C_1 are pulses that are high for a full clock period. In the Mealy circuit, however, the pulses are high only as long as the clock is high. In converting this Mealy machine to its equivalent Moore machine, we needed to introduce two extra states. In fact, it is typical for a Moore equivalent machine to have more states than the corresponding Mealy machine. But in this case, since the number of states increased from five to only seven, both of these machines, as will be shown later, can be constructed using three flip-flops.

8.4 Analysis of Sequential Circuits

It is important to analyze sequential circuits before we begin to explore their synthesis or design. Analysis, herein, refers to the process of determining the output response of a given synchronous sequential circuit to a given input sequence. This is accomplished often by determining the corresponding state table or state diagram. Having considered a lumped-delay circuit model, the analysis of sequential circuits can be carried out as follows:

1. Determine the combinatorial logic equation(s) for the flip-flop input(s).
2. Only in case of SR flip-flop, verify that $S \bullet R = 0$. If it is not, the process should be terminated since the circuit may contribute to critical races.
3. Obtain the next state equation(s) for the flip-flop(s) in terms of the combinatorial inputs determined in Step 1.
4. Determine the next states for all possible present states.
5. Use state transitions to obtain the state diagram and, if required, the timing diagram.

(a)

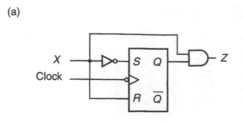

(b)

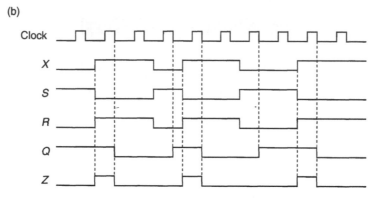

FIGURE 8.15
An *SR* flip-flop based sequential machine (a) the logic circuit and (b) timing diagram.

Consider for example the sequential logic circuit of Figure 8.15a. The corresponding state diagram must have only two states since there is only one flip-flop in the circuit. The corresponding characteristic equations of the logic circuit are as follows:

$$\overline{S} = R = x$$
$$z = xQ$$

The *SR* flip-flop used in this sequential machine is negative edge-triggered. A logic 0 on *x* causes the *SR* flip-flop to set on the negative edge of the clock. The flip-flop remains at logic 1 as long as the input remains a 0. With *x* changing from 0 to 1, the output also changes from 0 to 1. The output *z* remains high until when the flip-flop is reset. A representative timing diagram of the circuit is shown in Figure 8.15b. We may safely conclude that this sequential circuit recognizes the input sequence 01.

Consider next the sequential circuit of Figure 8.16a, which includes *T* flip-flop. Its representative timing diagram shown in Figure 8.16b should convince us that it functions as the sequence detector introduced earlier in Figure 8.15a.

8.5 Equivalent States

A state table is typically a tabular representation of a state diagram, which in turn, is usually derived from description given in words known as the *word statement* of the machine. As we transform the word statement to a state diagram, one or more states that are identical to another state may inadvertently be included. This situation arises quite often in the early

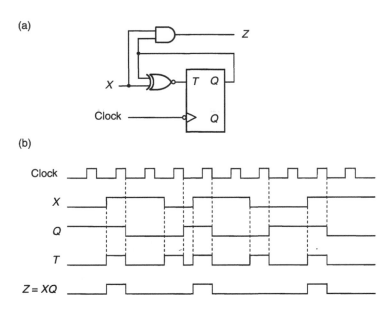

FIGURE 8.16
A T flip-flop based sequential machine (a) the logic circuit and (b) timing diagram.

design phase because of how the designer interprets the word statement of the problem in question. In the state diagram of Figure 8.6, for example, we have already located two states (A and C) that are equivalent. While in this particular case, we were able to identify equivalent states just by visual observation, this often gets complicated in most practical cases. Presence of redundant state or states typically increases the number of total states and may require addition of another flip-flop, making the overall sequential circuit more expensive. In the example of Figure 8.6, we cannot determine from which of the two equivalent states a sequential circuit starts by applying inputs and observing the corresponding outputs. As we shall see later, having redundant states amounts to fewer don't cares or unused states and thus increases the overall complexity of the sequential circuit equations. In addition to cost and complexity issues, routine diagnostic techniques used for failure analysis are based on the assumption that no redundant states exist. Accordingly, a prudent designer always seeks to eliminate redundancies from the state table.

Sometimes identification of redundant states is obvious. We shall seek to exploit a systematic process to reduce a state table to one having the minimum number of states. Every row of a state table representing a present state is associated with its corresponding next states and output. In Mealy circuit, in particular, the output is a function of both input and state while in a Moore circuit, it is only a function of the state. States of a sequential machine are considered *equivalent* if for each input or input combinations they produce the same output and their next states are either identical or equivalent regardless of whether any one of these two states is the initial state. In other words, two states are equivalent, if and only if, for an input,

1. The output produced by one of the states is equal to that produced by another state.
2. The next states are identical.

These two conditions form the basis for all state reduction techniques.

(a)

PS	NS, z (when x=0)	NS,z (when x=1)
A	C,0	B,0
B	C,0	D,0
C	C,0	B,0
D	E,1	D,0
E	E,1	F,1
F	C,0	F,1

(b)

PS	NS, z (when x=0)	NS,z (when x=1)
A	A,0	B,0
B	A,0	D,0
D	E,1	D,0
E	E,1	F,1
F	A,0	F,1

(c)

PS	NS, z (when x=0)	NS,z (when x=1)
B	C,0	D,0
C	C,0	B,0
D	E,1	D,0
E	E,1	F,1
F	C,0	F,1

FIGURE 8.17
State equivalence by visual inspection (a) original state table and (b) and (c) reduced state table.

A *completely specified* machine, in particular, refers to a sequential machine for which all of the next states as well as the outputs considered in its state table are fully specified. There are three techniques for determining equivalent states in a completely specified sequential machine: visual inspection, partitioning, and implication. Each time we determine equivalent states, we eliminate all but one state from the given state table. The scheme for handling incompletely specified state diagram, on the other hand, requires additional care. It is addressed separately in a later section.

Consider the state table of Figure 8.17a (the former Figure 8.6). Since states A and C perform exactly the same function, we can eliminate one of the two to obtain a reduced state table as shown in Figure 8.17b. If we decide to eliminate C, for example, we need to replace then all mention of C with A. However, if we were to eliminate A, no such replacement would have been needed as shown in Figure 8.17c. Both these reduced state tables, however, are equivalent.

Consider next the state table of Figure 8.18a. We notice by visual inspection that if the circuit were to be at either state A or state E, it would retain its original state when $x = 0$. With $x = 1$, on the other hand, both would transition to state F. Hence states A and E are equivalent and thus the reduced state table may be obtained by eliminating E, for example, as shown in Figure 8.18b. The state table example of Figure 8.18c is slightly different than that of Figure 8.18a in that the next states for both A and E are switched. By treating the two states together, we may conclude that the sequential machine does retain its state (and produces an output of 0) when $x = 0$, and transitions to state F (and produces an output of 1) when $x = 1$. In this case, therefore, we may conclude that these two states are equivalent.

Partitioning is a process of identifying equivalent states in a given state diagram by successively considering various combinations of states. The partition P_j is a group of blocks each of which is composed of a group of states. The initial states are all divided amongst these various blocks such that no state is in two or more blocks. On the other hand, each partition accounts for all existing states by including each in one of the blocks.

(a)

PS	NS, z (when x=0)	NS, z (when x=1)
A	A,0	F,1
B	C,0	D,0
C	C,0	B,0
D	E,1	D,0
E	E,0	F,1
F	C,0	F,1

(b)

PS	NS, z (when x=0)	NS, z (when x=1)
A	A,0	F,1
B	C,0	D,0
C	C,0	B,0
D	A,1	D,0
F	C,0	F,1

(c)

PS	NS, z (when x=0)	NS, z (when x=1)
A	E,0	F,1
B	C,0	D,0
C	C,0	B,0
D	E,1	D,0
E	A,0	F,1
F	C,0	F,1

FIGURE 8.18
State equivalence by visual inspection (a) an original state table, (b) reduced state table, and (c) a different original state table that is equivalent to that in Figure 8.18a.

The partitioning scheme of determining redundancies is summarized below:

1. The partition P_1 is determined first such that the states in each of its blocks all produce the same output value for each input value.
2. Next, partition P_j (where $j = 2, 3, 4, \ldots$) is determined so that for each of the states listed in its blocks, the corresponding next states (for each input value) are all listed within a single block of P_{j-1}.
3. Step 2 is repeated until when $P_{j+1} = P_j$. The final partition P_j is representative of the equivalent states.

Consider the state table of Figure 8.19a. Partition P_1 is determined by noting the states that generate the same output for all possible input values. Since the states A, C, E, and F all give the same outputs (1 when $x = 0$, and 0 when $x = 1$) and the states B, D, and H all give the same outputs (0 for both values of x), P_1 is given by,

$$P_1 = (ACEF)(BDH)(G)$$

The case of the last block is self-evident since it has only one state. We need to study the states listed within each of the other two blocks to determine P_2. Between the four states of the first P_1 block, we see that the next states for A and C (i.e., A) and that for E and F (i.e., D) when $x = 0$ lie within the same block, that is, $(ACEF)$ and (BDH) respectively. Similarly we see amongst the states of the second P_1 block, the next states of D and H (i.e., F and C) and that for the next state of B (i.e., B) when $x = 0$ lie within the same P_1 block, that is, $(ACEF)$ and (BDH) respectively. Next we consider the case for $x = 1$. We see that the next states for D and H when $x = 1$ (i.e., C and G respectively) lie respectively in two different

354   *Digital Design: Basic Concepts and Principles*

(a)

PS	NS, z (when x=0)	NS, z (when x=1)
A	A,1	D,0
B	B,0	C,0
C	A,1	D,0
D	F,0	C,0
E	D,1	E,0
F	D,1	F,0
G	G,0	H,1
H	C,0	G,0

(b)

PS	NS, z (when x=0)	NS, z (when x=1)
A	A,1	D,0
B	B,0	A,0
D	E,0	A,0
E	D,1	E,0
G	G,0	H,1
H	A,0	G,0

FIGURE 8.19
Example of partitioning (a) original state table and (b) reduced state table.

P_1 blocks, that is, $(ACEF)$ and (G) respectively. Therefore,

$$P_2 = (AC)(EF)(B)(DH)(G) \quad \text{when } x = 0$$
$$= (AC)(EF)(B)(D)(H)(G) \quad \text{when } x = 1$$

By inspecting A and C of the first P_2 block and E and F of the second P_2 block, we find that they both transition to A when $x = 0$, and to D when $x = 1$. Similarly, both E and F transition to D when $x = 0$ and to E and F respectively when $x = 1$. The next states for each of the P_2 block states for each of the inputs thus lie within the same P_2 blocks. Thus,

$$P_3 = (AC)(EF)(B)(DH)(G) \quad \text{when } x = 0$$
$$= (AC)(EF)(B)(D)(H)(G) \quad \text{when } x = 1$$
$$= P_2$$

We can now obtain the reduced table as in Figure 8.19b by eliminating C (thus replacing each C with A) and F (thus replacing each F with E). The reduced state table now has six states instead of eight as in the original state table.

Example 8.5
Use partitioning to reduce the state table of Figure 8.20.

Solution
The partitions for the state tables are

$$P_1 = (ADEGH)(BCFI)$$
$$P_2 = (ADG)(EH)(BCFI)$$
$$P_3 = (ADG)(EH)(BCF)(I)$$
$$P_4 = P_3$$

The reduced state table is thus obtained by eliminating (i) *D* and *G* (and replacing each mention of either *D* or *G* with *A*), (ii) *H* (by replacing each mention of *H* with *E*), and (iii) *C* and *F* (by replacing each mention of either *C* or *F* with *B*). The reduced state table as shown in Figure 8.21 has four states instead of nine.

We shall see later that had we not reduced the state table, its nine states would have required four flip-flops for the sequential machine. With elimination of redundancies, however, only two flip-flops are needed to cover its four states.

An alternative systematic method for identifying redundant states is accomplished by means of *implication*. In this scheme, an implication table is constructed to compare every possible pairs of states in terms of their next states and output. For an *n*-state state table, the numbers of rows and columns of this table are both equal to $n-1$. This table provides a bookkeeping technique that allows a systematic way to find states that are equivalent. The implication algorithm employed for identifying equivalent states is as follows:

1. The row of the implication table are labeled in order by the symbols used to denote all states except the very first one. The column levels, on the other hand, are denoted by the symbols used to denote all states except the very last one.

2. The entries at each of the cells are determined by comparing the next states of two states that are used to label the cell in question. Nothing is entered in a cell if it corresponds to states that have the same output and the same next states for all input variables. If the equivalency of the two corresponding states is dependent on whether or not the pairs of states P_j are already equivalent, then each of the pairs P_j is entered in the corresponding cell. A cell is marked off with "X" sign, if the outputs of the two states are different for any one or more of the inputs. Such states are referred to as nonequivalent.

PS	NS, z (when x=0)	NS, z (when x=1)
A	D,0	C,1
B	H,1	A,1
C	E,1	D,1
D	D,0	C,1
E	B,0	G,1
F	E,1	D,1
G	A,0	F,1
H	C,0	A,1
I	G,1	E,1

FIGURE 8.20
State table for Example 8.5.

PS	NS, z (when x=0)	NS, z (when x=1)
A	A,0	B,1
B	E,1	A,1
E	B,0	A,1
I	A,1	E,1

FIGURE 8.21
Reduced state table for Example 8.5.

3. The next phase involves marking off as many cells as possible on the basis of the respective cell entries from further consideration. If the pairs of states listed in a cell include at least one pair of states that are nonequivalent (i.e., the cell corresponding to that pair of states has already been crossed out), then that cell in question is eliminated by marking it off.

4. The remaining cell entries are examined during the next iteration to determine if any more cells can be removed from further consideration. The cell having a pair of states S_1 and S_2 should be crossed out if and only if the cell having labels S_1 and S_2 have already been eliminated. This process of elimination is continued until no other cells can be eliminated during an iteration.

5. Redundancy of states is determined by examining the cells of the implication table that did not get marked off. Each unmarked cell indicates that the states that are used as its labels are equivalent.

To fully comprehend this algorithm, we use it to investigate the state table of Figure 8.20. An implication table that locates the possible redundant states of Figure 8.20 is illustrated in Figure 8.22a.

Since there are nine states, the implication table of Figure 8.22a should have eight rows and eight columns. Entries in a cell are made by comparing the next states of those two states that are used as labels of the cell in question. For example, *EH* and *AD* are entered

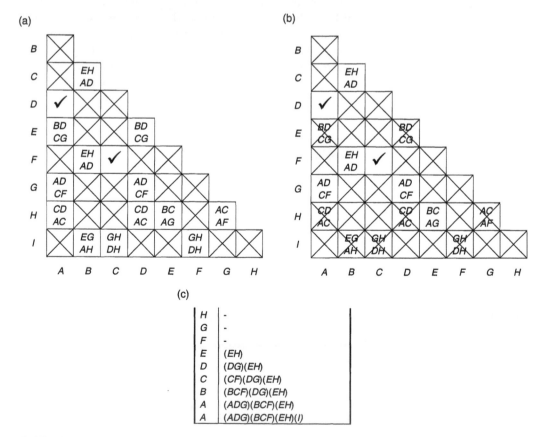

FIGURE 8.22
Reduction by implication (a)–(b) phase 1, (b) phase 2, and (c) equivalence partition table.

in the cell corresponding to B and C. This implies that if E and H were to be equal as well as if A and D were to be equal, B and C will be equivalent. Likewise, both EG and AH are entered in the cell corresponding to B and I since these two states would be equivalent only if both (a) E and G and (b) A and H are equivalent.

For the next phase, consider the cell entries with labels G and H, for example. Since the entries of the cell are AC and AF, we examine both the cells: with labels A and C, and that with labels A and F. Incidentally, both these cells have already been marked off during the previous iteration. Since at least one of the cells has already been marked off, the cell with labels G and H can now be marked off. This process is continued until no more cells can be marked off. The resulting implication table of Figure 8.22b can be used now to construct an equivalence partition table.

An *equivalence partition table*, as shown in Figure 8.22c, is next constructed. Its row labels are the same as all horizontal labels (in the reverse order) of the implication table in Figure 8.22a and b. By scanning the implication table, from right to left, we make the entries of the equivalence partition table. The row and column labels of the surviving cell in Figure 8.22b form an equivalent pair that is noted in the equivalent column identifier in the partition table. Existence of equivalent pairs such as S_1S_2 and S_1S_3 also imply the existence of an equivalent pair S_2S_3. We conclude then that S_1, S_2, and S_3 are all equivalent states. This conclusion is true, in general, in the case of all completely specified state tables.

For each column of the implication table an entry is made in the equivalence partition table provided there is at least one surviving cell in that column. The entries from the earlier row of the equivalence partition table are repeated unless those equivalent states are already included in a newer set of equivalent states. In the example of Figure 8.22b in question, there is no surviving cell in columns H, G, and F. Thus no entries are made in the first three rows of the equivalence partition table as shown in Figure 8.22c. While scanning from right to the left, we note that the cell corresponding to E and H has not been crossed out and, accordingly, the equivalent pair (EH) is listed next to E in the equivalence partition table. Then in the next column, we see that the cell corresponding to D and G has also survived. Accordingly, we list (DG) in the next row of the equivalence partition table. The entry (EH) is repeated here from the previous row. Continuing in this manner, we conclude from the column labeled B that (BCF) represents three equivalent states. While listing the entries next to row B in the equivalence partition table, we list (BCF) along with (DG) and (EH) repeated from the earlier row. Since (CF) is already included in (BCF), we do not repeat (CF) in the row labeled B. In the final step, we find that all the cells labeled I are all checked off in Figure 8.22b, indicating that I is not equivalent to any of the other states. Thus, I is listed all by itself. Our final equivalency result $(BCF)(ADG)(EH)(I)$ is identical to what we have already determined in Example 8.5 but by using partitioning scheme.

Example 8.6
Use implication method to reduce the state table given in Figure 8.23 and then obtain the most reduced state diagram.

Solution
Since there are seven states, the implication table should have only six rows and six columns. The implication table is constructed as shown in Figure 8.24a. The cells in row and column labeled D can all be marked off since the state D is not equivalent to any of the other states. Only this particular Moore state is associated with a high output.

During the next iteration as shown in the implication table of Figure 8.24b, we note that all cells present in the row labeled C and also in the column labeled C got marked off. This implies that the state C is also not equivalent to any other state. In the next phase as shown in Figure 8.24c, all cells that had an entry involving C got eliminated. The follow-on equivalence partition table shown in Figure 8.24d reveals that there are two sets of equivalent states, (AB) and (EFG). The reduced state table and the corresponding state diagram are thus obtained as shown respectively in Figure 8.25a and 8.25b.

H	-
G	-
F	-
E	(EH)
D	(DG)(EH)
C	(CF)(DG)(EH)
B	(BCF)(DG)(EH)
A	(ADG)(BCF)(EH)
A	(ADG)(BCF)(EH)(I)

FIGURE 8.23
State table for Example 8.6.

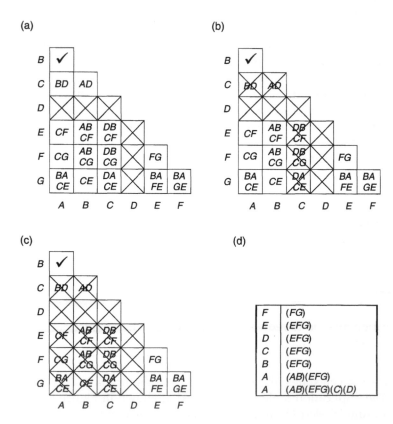

FIGURE 8.24
(a–c) Implication table and (d) equivalence partition table for Example 8.6.

(a)

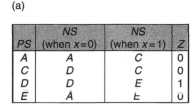

PS	NS (when x=0)	NS (when x=1)	Z
A	A	C	0
C	D	C	0
D	D	E	1
E	A	E	0

(b)

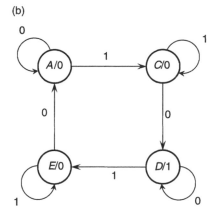

FIGURE 8.25
(a) Reduced state table and (b) state diagram for Example 8.6.

8.6 Incompletely Specified Sequential Circuits

A sequential circuit is said to be *incompletely specified* if its state table contains don't cares either for the next states or for the outputs or for both. Some of these don't cares arise normally in some circuits owing to the fact that a certain set of inputs can ever be applied. In others, the logic circuit function is independent of all those other states. In any event, presence of don't cares often provide opportunities since the resulting logic circuit may be reduced rather significantly. Recall that in the last section, in particular, when we were determining equivalent states from among the present states, the next states as well as the output values were considered in making decision about equivalencies. When dealing with an incompletely specified machine, therefore, determining state equivalencies may pose a rather serious problem.

Consider the state table of Figure 8.26a in which four of the outputs, for example, are unspecified. These don't cares may be assigned either 0 or 1 depending on what lends itself to further simplification. Visual inspection of this state table suggests an obvious strategy for reduction of the number of states. The entries next to states A and C and that next to states E and F are almost the same except for the don't care outputs. Accordingly, if we were to treat the don't care next to the present state A as 0 and that next to present state C as 1, the states A and C can be treated as equivalent states. Similarly if the don't care next to the present state E is treated as 1 and that next to present state F is treated as 0, the states E and F can be treated as equivalent states. The reduced state table shown in Figure 8.26b will thus consist of only four states. But one choice that is not readily obvious might involve assuming the don't cares next to the present states A, C, E, and F treated respectively as 1, 0, 0, and 1. With such a substitution, we may actually end up with more significant simplification. With such an assumption for the unspecified outputs, the states A, D, and F as well as the states B, C, and E are equivalent. The reduced state table obtained thereby is shown in Figure 8.26c. This latter simplification, although is more desirable, was not that obvious from the visual inspection of Figure 8.26a. Accordingly, the case of the incompletely specified sequential machines will need special consideration when determining equivalent states.

While we can assign any value to a don't care, one must be very careful to not assume two different values for the same don't care when trying to establish equivalencies between more states. States S_p and S_q are referred to as *compatible* if and only if for each input sequence

(a)

PS	NS, z (when $x=0$)	NS, z (when $x=1$)
A	A,1	E,-
B	F,0	B,0
C	A,-	E,0
D	A,1	C,1
E	D,0	B,-
F	D,-	B,1

(b)

PS	NS, z (when $x=0$)	NS, z (when $x=1$)
A	E,0	E,0
B	F,0	B,0
D	A,1	A,1
E	D,0	B,1

(c)

PS	NS, z (when $x=0$)	NS, z (when $x=1$)
A	A,1	B,1
B	A,0	B,0

FIGURE 8.26
(a) An incompletely specified state table and (b) and (c) its reduced form.

the same output sequence will be produced when the outputs are specified, whether S_p or S_q is the starting states. A *maximal compatible* is a compatibility class that will cease to be compatible if any state not in the class is included. Two states of an incompletely specified machine are said to be compatible if and only if both of the following conditions are met for all combination of input(s):

1. Outputs, if specified, resulting from the two states are identical and
2. Next states, if specified, for both of the states are compatible

When the two states fail to satisfy these two conditions, they are referred to as *incompatible*.

The existence of state compatible classes (S_1, S_2), (S_1, S_3), and (S_2, S_3) results in the existence of a compatible class (S_1, S_2, S_3). While all these are referred to as compatibles, only (S_1, S_2, S_3) is referred to as maximal compatible provided that we could not form a larger compatible class consisting of these three states and at least one other state. When all of the maximal compatibles are used to obtain a reduced state table, it may not always be the most reduced table. We need to be particularly careful since at times the use of only a subset of the maximal compatibles may result in the most reduced state table. Selection of the best set of maximal compatibles for yielding the most reduced state table must meet the following three conditions:

1. The most reduced state table has the least number of maximal compatibles as its states.
2. The union of the chosen maximal compatibles must contain all the states present in the original state table.
3. The set of chosen maximal compatibles must be closed so that each of the corresponding next states must all be included in a chosen maximal compatible.

Typically, the maximum number of maximal compatibles included in the most reduced table is the minimum of the number of original states and the number of maximal compatibles. It is not desirable that the number of maximal compatibles exceeds the number of states present in the given incompletely specified state table. On the other hand, the lower bound on the number of maximal compatibles for the most reduced state table depends on the set of maximal incompatibles. A group of states is said to be a *maximal incompatible* to which no other incompatible state may be added without destroying the group characteristics. Typically, the lower bound on the number of maximal compatibles is the maximum of the number of states included among the maximal incompatibles.

One may use the implication table introduced earlier to also eliminate redundancies present in an incompletely specified state table. The steps involved herein, however, are somewhat different from that used in the case of fully specified sequential machine. The steps to be followed for obtaining the most reduced state table for an incompletely specified machine involve the following variations:

1. The implication table entries are made exactly as before except that the don't cares are treated as either 1 or 0 depending on which particular choice would lead to the formation of a group of equivalent states.

2. A cell in an implication table is marked off as before on the basis of whether or not the particular cell labeled with states of at least one of the pairs of states (listed within the cell in question as a cell entry) has already been marked off. This step is repeated until no more cells can be marked off.

3. The equivalence partition table is used next to identify the maximal compatibles as well as the maximal incompatibles. One must be careful since the equivalence partition table, unlike in the case of completely specified machines, has to now accommodate a different situation. In the case of incompletely specified state table, the presence of a group of equivalent states (S_1, S_2) and another of a group of equivalent states (S_1, S_3), for example, does not automatically imply the presence of a group of equivalent states (S_1, S_2, S_3) unless there also exists a group of equivalent states (S_2, S_3) or (S_2, S_3, S_j). This restriction avoids of situations where the same don't care variable may have been assumed to take two different binary values in determining either the maximal compatibles or the maximal incompatibles or both.

4. The upper and lower bounds on the number of maximal compatibles to be included in the most reduced state table are determined next. The maximum number of maximal compatibles is the minimum of the two numbers: number of original states and the number of maximal compatibles. The lower bound on the number of maximal compatibles is the maximum of the number of states included among the maximal incompatibles.

5. By trial and error a set of maximal compatibles is determined to obtain the most reduced state table. The next states of the selected maximal compatibles are determined by merging their next states and then listing the appropriate maximal compatibles for these merged next state entries for all of the input conditions.

The reduced state table so obtained can still remain incompletely specified. However, if the number of states in the reduced state table exceeds that in the original state table, the given state table is not reducible any further. The next example illustrates the application of this algorithm.

PS	NS, z (when $x=0$)	NS, z (when $x=1$)
A	A,-	A,-
B	A,-	D,0
C	B,1	B,-
D	A,1	B,-
E	E,-	B,1
F	F,-	C,-

FIGURE 8.27
State table for Example 8.7.

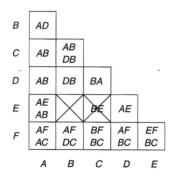

FIGURE 8.28
Implication table for Example 8.7.

Example 8.7
Reduce the incompletely specified state table shown in Figure 8.27.

Solution
The implication table is constructed as shown in Figure 8.28. Since there are six states, the table has five rows and five columns. The entry in the cell labeled *B* and *E* is marked off since these two states are not equivalent. With $x = 1$, these two states yield two different outputs. Accordingly, in the next round the cell (labeled with *C* and *E*) having the entry *BE* is marked off.

The equivalence partition table may now be constructed as shown in Figure 8.29a. We see in this table while make entries next in the row labeled *C*, for example, we could have written the equivalent groups to include (*CDF*), (*DE*), and (*EF*) instead of (*CF*), (*CD*), and (*DEF*). In the same manner, we could have formulated other equivalent groups while completing other rows of the equivalence partition table. The entries obtained using Figure 8.29a thus provide only one particular choice. We note that states *A* and *D* are present in both (*ABCD*) and (*ADEF*). Similarly, the state *F*, for example, is present in three of the groups, (*ADEF*), (*CF*), and (*BF*). The maximal incompatibles are determined in like fashion as shown in Figure 8.29b.

Assuming that we may have up to four maximal compatibles we denote them as

$$S_1 = (ABCD)$$
$$S_2 = (ADEF)$$

(a)

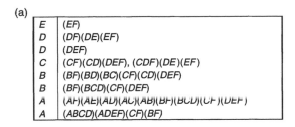

E	(EF)
D	(DF)(DE)(EF)
D	(DEF)
C	(CF)(CD)(DEF), (CDF)(DE)(EF)
B	(BF)(BD)(BC)(CF)(CD)(DEF)
B	(BF)(BCD)(CF)(DEF)
A	(AF)(AE)(AD)(AC)(AB)(BF)(BCD)(CF)(DEF)
A	(ABCD)(ADEF)(CF)(BF)

(b)

E	-
D	-
C	(CE)
B	(BE)(CE)
A	(BE)(CE)

FIGURE 8.29
Identification of (a) maximal compatibles and (b) maximal incompatibles.

$$S_3 = (CF)$$

$$S_4 = (BF)$$

The upper limit of maximal compatibles that may be used in the most reduced state table is thus four (since there are four maximal compatibles but six original states) while the lower limit of maximal compatibles is two (since the largest maximal incompatible class includes just two states). In other words, we could try to reduce the state table of Figure 8.27 to one having between two and four rows.

If we are to merge the states into two states, the obvious choice is to include S_1 and S_2. The corresponding merged state table is obtained as shown in Figure 8.30a. While merging states, we make sure that the merged state will result in a specified output if the output is already specified for at least one of the states. Thus when merging A, B, C, and D, for example, when $x = 0$, the merged next state takes on the output values of 1 since at least one of these four states (in this case both C and D) gives high output when $x = 0$. The reduced table is constructed next as shown in Figure 8.30a by substituting the appropriate maximal compatible for the merged next states. We can settle for this merged state since each one of the merged states, AB, ABD, AEF, and ABC, belong to either S_1 or S_2. The particular maximal compatible $ADEF$ (denoted by S_2), for example, transitions to ABC when $x = 1$. We have no maximal compatible listed as ABC, however, since ABC is included completely in S_1, we list S_1 as the next state for S_2 in the column designated for $x = 1$.

On the other hand, if we had not identified the maximal incompatibles, we would have remained unaware of the lower bound on the number of maximal compatibles. Consequently, if we had settled for all four maximal compatibles as states of the reduced table, the merged state would have been obtained as shown in Figure 8.31a. The corresponding reduced table would have been then obtained as shown in Figure 8.31b. Clearly, the one shown in Figure 8.30b is more desirable than the one shown in Figure 8.31b.

As pointed out earlier, the reduced state table may still be incompletely specified such as that shown in Figure 8.31b. Two of its output values may be treated still as don't cares.

(a)

PS	NS, z (when x=0)	NS, z (when x=1)
S_1	AB,1	ABD,0
S_2	AEF,1	ABC,1

(b)

PS	NS, z (when x=0)	NS, z (when x=1)
S_1	S_1,1	S_1,0
S_2	S_2,1	S_1,1

FIGURE 8.30
State reduction using only two maximal compatibles (a) merging and (b) reduction.

(a)

PS	NS, z (when x=0)	NS, z (when x=1)
ABCD	AB,1	ABD,0
ADEF	AEF,1	ABC,1
CF	BF,1	BC,-
BF	AF,-	CD,0

(b)

PS	NS, z (when x=0)	NS, z (when x=1)
S_1	S_1,1	S_1,0
S_2	S_2,1	S_1,1
S_3	S_1,1	S_1,-
S_4	S_2,-	S_1,0

FIGURE 8.31
State reduction using all four maximal compatibles (a) merging and (b) reduction.

8.7 State Assignments

Up until this time, we have used various English characters to conveniently denote states of the sequential machines. But to assure that it is possible for the circuit to remember which state the circuit is in, one needs to adopt a binary coding scheme for representing these symbolic states of the state table. This process is referred to as *state assignment*. Typically, each bit in the code represents the output of a flip-flop and is called a state variable. The memory of an n-state sequential machine will consist of m flip-flops where m is the smallest integer just greater than or equal to $\log_2(n)$. Thus to represent up to four states, one can use two flip-flops. The states can be assigned then the available binary codes 00, 01, 10, and 11. We could arbitrarily assign 00, for example, to a particular state. Then one of the remaining three codes can be assigned next to the second state and so on. When the symbolic states in a given state table are all replaced with the binary codes, we obtain what we typically refer to as the *transition table*. Any unique assignment is valid; however, it is better to assign codes in such a way that the number of cases where more than one bit in a code change when states change is a minimum. We shall discuss this issue in a later section. Herein, we concentrate more on the synthesis of a sequential machine, efficient or not.

Once the transition table has been constructed and type of flip-flops have already been identified, one could then make use of excitation maps to synthesize the sequential circuit. *Excitation tables/maps* are used to determine what kind of flip-flop inputs (i.e., excitations) need to be provided to the memory so that the combination of memory and combinatorial

(a)

PS	NS (when x=0), Z	NS (when x=1), Z
A	B,0	A,0
B	A,0	C,1
C	A,0	C,0

(b)

$Q_1(t)Q_2(t)$	$Q_1(t+\Delta t)Q_2(t+\Delta t)$, Z, (x=0)	$Q_1(t+\Delta t)Q_2(t+\Delta t)$, Z, (x=1)
00	01,0	00,0
01	00,0	11,1
11	00,0	11,0
10	--,-	--,-

FIGURE 8.32
State assignment example (a) reduced state table and (b) transition table.

$Q(t)$	$Q(t+\Delta t)$	$S(t)R(t)$	$J(t)K(t)$	$T(t)$	$D(t)$
0	0	0-	0-	0	0
0	1	10	1-	1	1
1	0	01	-1	1	0
1	1	-0	-0	0	1

FIGURE 8.33
Flip-flop control characteristics.

components can function as the desired sequential machine. Synthesis, from here on, attempts to simply identify the combinatorial component of the circuit. Consider, for example, the reduced state table of Figure 8.32a, which is the same one that we obtained earlier in Figure 8.4. It has three unique states, and, therefore, at least two flip-flops are needed to synthesize the corresponding logic circuit. Accordingly, it also implies that of the four different codes for Q_1Q_2 (i.e., 00, 01, 10, and 11), only three are used. For this example, if one chooses to assign A = 00, B = 01, and C = 11, the resulting transition table of Figure 8.32b is obtained. Since the assignment 10 is unused, the entries next to 10 state are left as don't cares.

We see that the present state $Q_1Q_2 = 01$, upon receiving the input $x = 0$, transitions to the next state 00 with no change in the output. During this transition, Q_2 changes from a 1 to a 0. But with $x = 1$, only Q_1 changes and contributes to a high output for Z. In either case the binary state changes only one of its bits. The transitions from the state $Q_1Q_2 = 00$ also involve change in either Q_1 or Q_2 but not both. These latter transitions, however, do not result in a high output. However, the present state $Q_1Q_2 = 11$ is different from these two since with $x = 0$, both of the FF bits need to change from 1 to 0. There are situations where such a condition (two flip-flops racing against one another to change their states) could cause problems, as we shall discover later.

In the synthesis phase to determine what particular excitations need to be provided to the flip-flops in the sequential machine, we can begin by assuming to use any type of a flip-flop. Typically, the designer already has a particular flip-flop in mind at the very beginning of the synthesis process. Figure 8.33 shows the excitation conditions for all of the flip-flops that we have considered in Chapter 7 (Figure 7.60). It lists the needed excitations to be fed to the flip-flops to realize any one of the four transitions ($0\rightarrow0, 0\rightarrow1, 1\rightarrow0, 1\rightarrow1$). Although one can use any of the flip-flops, the designers prefer more to use D flip-flops. The reason becomes obvious by studying the table of Figure 8.33. Its next state equation given by $Q(t + \Delta t) = D(t)$ is the simplest of all. Besides, D flip-flops are available more in discrete packages and programmable logic devices. Since the excitation always takes on the value

(a)

Q_1Q_2	D_1D_2, Z $(x=0)$	D_1D_2, Z $(x=1)$
00	01,0	00,0
01	00,0	11,1
11	00,0	11,0
10	--,-	--,-

(b)

Q_1Q_2	D_1D_2 $x=0$	$x=1$	Z $X=0$	$X=1$
00	01	00	0	0
01	00	11	0	1
11	00	11	0	0
10	--	--	-	-

FIGURE 8.34
The excitation table for the state diagram of Figure 8.32b obtained by (a) direct substitution and (b) splitting.

of the desired next state, the very transition table can be used also as the excitation table. In case of other flip-flops, excitation map does not resemble much to the transition table.

Up to this point when considering flip-flops, we have been concerned only with how they respond to various inputs. We shall now utilize them in discussing excitation maps and, accordingly, derive excitation equations for determining the combinatorial component of sequential machines. Consider now the transition table shown in Figure 8.32b. Let us also plan to use D flip-flops first. For example, for the present state 00, the desired next state is 01. The D flip-flops thus must be fed with 0 and 1 (same as the value of the next state) respectively. Figure 8.34a shows the complete excitation table. As expected, the entries for D_1D_2 remains identical to what were already present in the transition table of Figure 8.32b. For clarity, however, this excitation table is split into two separate tables one for the flip-flop excitations and the other for the output as shown in Figure 8.34b.

The rows of these excitation tables of Figure 8.34b are labeled (in order from top to bottom) respectively as 00, 01, 11, and 10, that is, in Gray code order. Fortunately, for the state assignments considered already, these two excitation tables maps are identical to two 3-variable K-maps that we would have constructed respectively for D_1D_2 and Z both in terms of x, Q_1, and Q_2. The excitation tables are thus referred to as the excitation maps. If this were not the case, for example, when the state assignments did not lend themselves to Gray code order, the excitation table would have to be reconstructed so that one can obtain an excitation map fashioned after a K-map. The first excitation map has two entries in each of its cells and, thus, it can be interpreted as two separate K-maps for D_1 and D_2 put together. The second excitation map serves only a single variable, Z. Thus by considering Figure 8.34b as representing three overlapped excitation maps, we obtain:

$$D_1 = Q_2$$
$$D_2 = \overline{Q_1} + Q_2$$
$$Z = \overline{Q_1}Q_2x$$

These Boolean equations can be readily used to design the corresponding sequential machine. Considering the fact that each of the flip-flops has Q and $\overline{Q}$ outputs, the resulting sequential circuit as shown in Figure 8.35 requires two D flip-flops, one 2-input OR gate and a 3-input AND gate.

Now let us explore using JK flip-flops instead to implement the same reduced state table of Figure 8.32a. Since, we are now planning to use JK flip-flops, the JK excitation maps for the sequential machine will have to be constructed using the JK entries of Figure 8.33. Figure 8.36 shows the corresponding excitation map for the same assignment of states. The

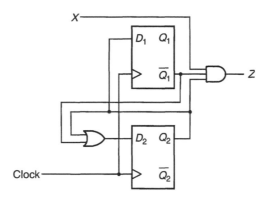

FIGURE 8.35
The *D* flip-flop based synchronous circuit to implement the reduced state table of Figure 8.32a.

Q_1Q_2	J_1K_1		J_2K_2		Z	
	$x=0$	$x=1$	$x=0$	$x=1$	$x=0$	$x=1$
00	0-	0-	1-	0-	0	0
01	0-	1-	-1	-0	0	1
11	-1	-0	-1	-0	0	0
10	--	--	--	--	0	0

FIGURE 8.36
JK excitation maps for the state table of Figure 8.32a.

map for the output Z, however, is the same as that constructed in Figure 8.34b. Since *JK* flip-flops each has two inputs, we treat the two flip-flops separately. The double entries for each of the *JK* excitation maps correspond to two simultaneously treated K-maps respectively for J and K inputs. For example, for the present state 01, the next state is 11 when $x = 1$. This implies that when $x = 1$, the first flip-flop should change from a 0 to a 1 but the second flip-flop must remain unchanged at 1. These two actions are accomplished when *JK* inputs are respectively 1- and -0. Correspondingly, 1- and -0 are entered under column for $x = 1$ in the row for which the present state is 01. The requirements as posed by Figure 8.33 are applied likewise for each of the transitions to obtain Figure 8.36. The *JK* excitation equations are obtained as follows:

$$J_1 = Q_2x$$
$$K_1 = \bar{x}$$
$$J_2 = \bar{x}$$
$$K_2 = \bar{x}$$

Thus, two *JK* flip-flops, two AND gates (one 2-input and the other a 3-input) and a NOT gate will be needed to design the sequential circuit. Although it may be too premature to conclude but this latter realization of the sequential machine using *JK* flip-flops instead of *D* flip-flops leads to a slightly more involved combinatorial module.

Example 8.8
Obtain the sequential circuits for the state diagram of Figure 8.32a respectively using *D* flip-flops and *JK* flip-flops but by making alternate binary assignments for the states: $00 \rightarrow A$, $01 \rightarrow B$, and $10 \rightarrow C$. Discuss the consequence of making this alternate state assignment.

Solution

The revised transition table is obtained as shown in Figure 8.37.

The corresponding D and JK excitation maps are obtained as shown in Figure 8.38. The excitation equations are obtained accordingly as,

$$D_1 = Q_1 + Q_2$$
$$D_2 = \overline{Q_1}\,\overline{Q_2}\,\overline{x} = \overline{Q_1 + Q_2 + x}$$

for D flip-flop based circuit; and

$$J_1 = Q_2 x$$
$$K_1 = \overline{x}$$
$$J_2 = \overline{Q_2 x} = \overline{Q_2 + x}$$
$$K_2 = 1$$

for JK flip-flop based circuit. The output equation for both options is the same and is given by,

$$Z = Q_2 x$$

The D flip-flop based sequential circuit thus requires two flip-flops, a 2-input OR gate, a 3-input NOR gate, and a 2-input AND gate. In comparison, the JK flip-flop based sequential circuit requires two flip-flops, a 2-input AND gate, a 2-input NOR gate, and a NOT gate. For the JK flip-flop based sequential circuit, however, we do not need a separate AND gate for the output.

$Q_1(t)Q_2(t)$	$Q_1(t+\Delta t)Q_2(t+\Delta t), Z(x=0)$	$Q_1(t+\Delta t)Q_2(t+\Delta t), Z(x=1)$
00	01,0	00,0
01	00,0	10,1
11	--,-	--,-
10	00,0	10,0

FIGURE 8.37
Transition table for Example 8.8.

Q_1Q_2	D_1D_2 $x=0$	$x=1$	Z $x=0$	$x=1$
00	01	00	0	0
01	00	10	0	1
11	--	--	-	-
10	00	10	0	0

Q_1Q_2	J_1K_1 $x=0$	$x=1$	J_2K_2 $x=0$	$x=1$	Z $x=0$	$x=1$
00	0-	0-	1-	0-	0	0
01	0-	1-	-1	-0	0	1
11	--	--	--	--	-	-
10	-1	-0	0-	0-	0	0

FIGURE 8.38
Excitation maps for an alternate state assignment.

PS	NS (when $x=0$), Z	NS (when $x=1$), Z
A	B,0	A,0
B	A,0	C,1
C	A,0	D,0
D	A,0	D,0

FIGURE 8.39
State diagram for Example 8.9.

$Q_1(t)Q_2(t)$	$Q_1(t+\Delta t)Q_2(t+\Delta t), Z(x=0)$	$Q_1(t+\Delta t)Q_2(t+\Delta t), Z(x=1)$
00	01,0	00,0
01	00,0	11,1
11	00,0	10,0
10	00,0	10,0

FIGURE 8.40
State table for Example 8.9.

By comparing these two sequential circuits with those obtained earlier for the other state assignment, we may conclude that the cost of logic is essentially unchanged. This is in part because the two transition tables are equally complex. In this latter state assignment as shown in Figure 8.37, one of the states (01 in this case) needs to have both of its flip-flop change values with an input ($x = 1$ in this case).

Example 8.9
Identify the sequential circuits for the state diagram shown in Figure 8.39 (same as that used earlier in Figure 8.3) respectively using D and JK flip-flops. Use the binary assignments: 00 → A, 01 → B, 11 → C, and 10 → D for its states. Realizing that the state diagram of Figure 8.32a is only a reduced form of the one in Figure 8.39, discuss the consequence of not eliminating redundancies.

Solution
The transition table is obtained as shown in Figure 8.40.

The corresponding D and JK excitation maps are obtained then as shown in Figure 8.41. Correspondingly, the excitation equations are given by

$$D_1 = Q_1 + Q_2$$
$$D_2 = \overline{Q_1}\,\overline{Q_2}\,\overline{x} + \overline{Q_1}\,Q_2 x = \overline{Q_1}\,\overline{Q_2 \oplus x}$$

for the D flip-flop based circuit; and

$$J_1 = Q_2 x$$
$$K_1 = \overline{x}$$
$$J_2 = \overline{Q_1}\,\overline{x}$$
$$K_2 = \overline{x} + Q_1 Q_2$$

for the JK flip-flop based circuit. The output equation for both option is the same and is given by,

$$Z = \overline{Q_1} Q_2 x$$

The D flip-flop based sequential circuit thus requires two flip-flops, a 2-input OR gate, a 2-input AND gate, a 2-input XNOR gate, and a 3-input AND gate. However, if we instead avoid using the XNOR gate, the alternate circuit will then require two flip-flops, two 2-input OR gates, and two 3-input AND gates. For this latter JK flip-flop based sequential circuit, however, we do not need a separate AND gate for the output. In comparison, the JK flip-flop based sequential circuit requires two flip-flops, three 2-input AND gates, one 3-input AND gate, a 2-input OR gate, and a NOT gate.

By comparing these two sequential circuits with those obtained earlier for the reduced state table of Figure 8.32a, we may safely conclude that the cost of logic is larger in case of the unreduced state table. This is in part because the two transition tables are significantly different. In this latter state assignment as shown in Figure 8.40, the next states are fully specified. If it were not the case, we would have had a larger number of don't cares present in the excitation maps that would have resulted in further circuit minimization.

The examples considered earlier should convince us that two different assignments may yield different complexity for the combinatorial module of the sequential machine. The simplest state assignment does not necessarily lead to the simplest design for the combinatorial logic circuit. We have made use of state assignments so far rather arbitrarily with no consideration of the consequences. The number of possible state assignments for any given sequential problem is impressive. For a circuit represented by n present states and consisting of p flip-flops, the number of states that may remain unused is $2^p - n$. Thus, there are $2^p! / \{n!(2^p - n)!\}$ ways of assigning the 2^p combinations of binary state assignments to n states. For each of these ways there are $n!$ permutations of assigning the n combinations to the n states. Also, for each of these assignments there are 2^p ways of interchanging 0 and 1 and there are $p!$ ways of interchanging the flip-flops. Consequently, there may be a total of $[(2^p - 1)!]/[(2^p - n)!p!]$ unique assignments. The complexity and cost of the circuit typically differ for different combinations of state assignments. The identification of the best state assignment has been the subject of considerable investigation. While the number of unique assignments for a 4-state system, for example, is 3, the number of unique assignments rises to 140 just for a 5-state system and it reaches as large as 75,675,000 for a 10-state sequential machine. Thus it becomes almost impractical for a designer to try out all these state assignments (for $n > 4$) to determine the best assignment. Most designers rely on experience and several guidelines for reasonably optimal state assignments.

Q_1Q_2	D_1D_2		Z	
	$x=0$	$x=1$	$x=0$	$x=1$
00	01	00	0	0
01	00	11	0	1
11	00	10	0	0
10	00	10	0	0

Q_1Q_2	J_1K_1		J_2K_2		Z	
	$x=0$	$x=1$	$x=0$	$x=1$	$x=0$	$x=1$
00	0-	0-	1-	0-	0	0
01	0-	1-	-1	-0	0	1
11	-1	-0	-1	-1	0	0
10	-1	-0	0-	0-	0	0

FIGURE 8.41
Excitation maps for the unreduced state table.

The optimum state assignment is one that is able to reduce the combinatorial logic of a sequential system when compared to that obtained with other assignments. Many different approaches to this state assignment problem have been developed. We can locate the best set by generating those output and excitation tables that allow the formation of large clusters of 1s and don't care values, when possible. Use of the following guidelines will probably result in simpler circuit for the combinatorial module:

1. Logically adjacent assignments should be given to those states that transitions to the same next state for a given input.
2. Two or more states that are the next states of the same present state, under logically adjacent inputs, should be given logically adjacent assignments.
3. States that have the same output for a given input should be given logically adjacent assignments.

Logically adjacent assignment implies that the states do appear next to each other on the excitation map. Rule 1 guarantees that larger number of 1s along the columns of an excitation table may be grouped. Rule 2, on the other hand, guarantees the formation of larger groups along the rows of an excitation table. Rule 3 guarantees a simpler combinatorial circuit for the output. These rules usually lead to a good but not necessarily to the optimal solution. They work a little better in the case of D flip-flops. However, it is often impossible to satisfy all three of the aforementioned rules simultaneously. In case of conflicts, Rule 1 is often preferred over Rule 2. An attempt should be made to satisfy the maximum number of suggested logical adjacencies. It must be stressed though that an ideal state assignment may not always reduce the cost, and the cost is often an insignificant component of the overall cost of a digital system.

There are other additional guidelines that are also utilized by digital designers when considering state assignments.

1. Assign the initial state typically with binary values such as either $00\ldots0$ or $11\ldots1$ depending on which can be easily entered at reset.
2. Minimize the number of state variables that change on each transition.
3. In case there are unused states, use the assignment of available don't cares in a manner so that larger group of 1s can be realized in the resulting excitation maps.
4. Maximize the number of state variables that do not change in a group of more related states.
5. Exploit symmetries, when present, in the problem specification and, thus, in the state table. If one state (or a group of states) functions similarly as the other state (or a group of some of the other states), then the state assignments of the first state (or group) may differ by only one bit from that of this other state (or group).

Since the presence of unused states typically reduces circuit cost, a designer might use more than the minimum number of flip-flops (thus making use of a larger number of don't cares) in determining the circuit. However, this must not be explored if the resulting decrease in the cost of the combinatorial component is offset by the corresponding increase in the cost of the memory component. While it is true that unused states would lead to a minimal cost, the associated sequential circuit poses certain problems as well. If the resulting circuit were to enter into one of these illegal states perhaps because of hardware failure, it may cause real frustrations since we would not know what state transition or output is going to result from such an entry. Accordingly, a designer may choose to be explicit in assigning their corresponding next states. It may be prudent to make these next states same as the initial or idle state.

8.8   Design Algorithm

In the last few sections, we have already explored the various design steps but in fragmented form to introduce the various concepts of synchronous machine. Everything put together, the comprehensive design algorithm for the synchronous sequential circuit includes the following steps:

1. Obtain a state diagram from the word description of the problem.
2. Derive the state table from the state diagram.
3. Use state reduction technique to eliminate redundant states.
4. Make state assignments.
5. Determine the type of flip-flops and then determine the corresponding excitation maps.
6. Obtain the excitation and output equations.
7. Determine and realize the logic circuit and then test it to see if it functions as expected.

The first step and, therefore, also the second step require intuition on the part of the logic designer. Accordingly, state diagram and state table obtained by one may be significantly different from those obtained by another. Step 3 helps to eliminate such discrepancy and more. In step 4, we either make state assignments arbitrarily or, when situation permits, make an almost-optimal state assignment using the concepts identified in the last section. Step 5 is often a function of what flip-flop is available to the designer. The design process may appear rather lengthy but the complexity may vary somewhat from problem to problem. The following examples illustrate application of the design algorithm.

Example 8.10
Design of a synchronous sequential circuit that recognizes the input sequence 0101. The circuit should recognize overlapped sequences such that for an input string of bits 010101, for example, the corresponding output string of bits is 000101.

Solution

Step 1. Assuming that the sequential machine begins from an initial state A, the state diagram may consist of six states, A through F, as shown in Figure 8.42a. States B, D, E, and F, respectively, represent the detection of the first (i.e., 0), second (i.e., 1), third (i.e., 0), and fourth (i.e., 1) bits of sequence 0101. The output Z goes high with the detection of the last bit of the sequence. The state C accounts for the non-occurrence of the first bit of the desired sequence. The circuit at state C thus continues to remain at state C as long as the input continues to be 1. As soon as the input is 0, however, it transitions to state B. The circuit at state F transitions to state E when the input $x = 0$, since the machine should allow for a possible overlapped sequence. With such an accommodation, the occurrence of a sequence 01010 can be interpreted as having the potential to generate another sequence of 0101 within the next input pulse. On the other hand, if the input received is a 1, it implies that the occurrence of 01011 simply does not include trailing bits that may form the leading bits of the sequence 0101.

Step 2. The state table shown in Figure 8.42b follows readily from the state diagram of Figure 8.42a.

Step 3. From visual observation of the state table of Figure 8.42b, we see that the states A and C as well as D and F are equivalent while the state E is not equal to any of the other states. In the corresponding implication table shown in Figure 8.43a, therefore, cells labeled with state E are all marked off while the one with labels A and C and the other with labels D and F are left empty. Successive passes through the implication table leaves no surviving cells except for the two that were empty. In this particular case, therefore, the implication table does not accomplish anything beyond what was already apparent from visual inspection. Typically, a designer cannot take a chance of not employing implication process since visual inspection is not always a foolproof process. The reduced state table is obtained as shown in Figure 8.43b.

Steps 4 and 5. Since the number of states is four, we need to employ only two flip-flops. For the sake of this example, let us also consider that we use JK flip-flops for the memory. The four codes 00, 01, 11, and 10 are arbitrarily assigned respectively the states $A, B, D,$ and E. Herein, we make no attempt at making an ideal state assignment. The corresponding transition table and excitation maps are obtained as shown in Figure 8.44.

Steps 6 and 7. The JK excitation and output equations are obtained next from the excitation maps shown in Figure 8.44b. They are given by

$$J_1 = Q_2 x$$
$$K_1 = \bar{x}\,\overline{Q_2} + x Q_2 = x \otimes Q_2$$
$$J_2 = \bar{x} + Q_1$$
$$K_2 = Q_1$$
$$Z = Q_1 \overline{Q_2} x.$$

The resulting sequential circuit is obtained as shown in Figure 8.45.

Example 8.11

Repeat the problem of Example 8.10 but with a Moore output, each time the circuit detects a sequence of 0101.

Solution

Since the state table, as shown in Figure 8.46a, now has a Moore output, the redundancy elimination is not any more effective as it was in the case of Mealy output. Herein, while the states A and C are equivalent, the states D and F are not equivalent. The reduced state table as shown in Figure 8.46b thus has five states instead of only four.

Implementation of this reduced state table, however, will require three flip-flops since there are now five states. Of the eight available states, only five will be used. By assigning 000, 001, 101, 100, and 110 arbitrarily respectively to states A, B, D, E, and F, we obtain a transition table as shown in Figure 8.47a. Figure 8.47b and 8.47c respectively show the excitation maps and the output map. Each of the entries next to the unused states are listed as don't cares. The map for the Moore output is such that the output is independent of the input. In the maps of Figure 8.47b and c, we list the codes 000 through 111 in Gray code order, in which case, we should be able to use them directly as K-maps. Otherwise, it would be necessary for us to construct three 4-variable K-maps for the JK excitations, and one 3-variable K-map for the Moore output.

In the maps of Figure 8.47b and 8.47c, we have used three of the variables to denote the rows. These K-maps are, therefore, a little different than the ones we encountered in Chapter 3, where we had generally used two variables to denote rows. To accommodate these 8-row K-maps, therefore, the designer needs to be cognizant that both halves (of the top four rows and of the bottom four rows) of this map may be treated separately as two different K-maps with only the first variable (i.e., Q_1) value different between them. But in addition, it is also possible to group legitimate entries between the two halves as long as the cells in question are symmetrically located from the line separating the two halves. The excitation and output equations are thus obtained as follows:

$$J_1 = Q_3 x$$
$$K_1 = \bar{x}\,\overline{Q_2}\,\overline{Q_3} + xQ_2\overline{Q_3} + x\overline{Q_2}Q_3$$
$$J_2 = xQ_1\overline{Q_2}\,\overline{Q_3}$$
$$K_2 = 1$$
$$J_3 = \bar{x}\overline{Q_2}$$
$$K_3 = xQ_1$$
$$Z = Q_2$$

This example illustrates that typically a Moore type state machine requires more states and, therefore, results in more complexity. The availability of additional don't cares does not necessarily guarantee that the complexity and cost could be reduced.

Example 8.12
Design a 3-bit synchronous sequential circuit that counts clock pulses.

Solution
A 3-bit circuit that is receiving clock pulses as its input may count up to eight pulses. Each time a clock pulse is received, the circuit should count up without the need of any control variable. Only the occurrence of the clock pulse (which now serves also as the input) is necessary for a change in state. The circuit could be reset with the eighth clock. The counter circuit can be designed such that the outputs go through the sequence $000 \rightarrow 001 \rightarrow 010 \rightarrow 011 \rightarrow 100 \rightarrow 101 \rightarrow 110 \rightarrow 111$ and repeat.

Making state assignments is an important step for at least one reason. However, we may choose to have the outputs directly from the flip-flop outputs, in which case such outputs would be classed as Moore type. The output circuits can be then eliminated completely if the state assignments are made the same as the corresponding Moore outputs of each state. The eight states are all different since they stand for completely different events. Consequently, it is safe to conclude that none of these are redundant states. This scenario for the state assignments and transitions allows the resulting circuit (with flip-flops connected to a 3-bit display device, for example) to display the clock (denoted by CK) count. We may assume further that the states change at the trailing edge of the clock input. The state diagram and the corresponding transition table for the circuit are obtained as shown in Figure 8.48.

It follows now readily from the excitation maps of Figure 8.48c,

$$J_1 = Q_2 Q_3 CK$$
$$K_1 = Q_2 Q_3 CK$$
$$J_2 = Q_3 CK$$
$$K_2 = Q_3 CK$$
$$J_3 = CK$$
$$K_3 = CK.$$

Since the clock input, CK, is already available to the edge-triggered JK flip-flops, we can set $CK = 1$ to obtain the reduced form of the JK excitations as follows:

$$J_1 = Q_2 Q_3$$
$$K_1 = Q_2 Q_3$$
$$J_2 = Q_3$$
$$K_2 = Q_3$$
$$J_3 = 1$$
$$K_3 = 1.$$

The resulting logic circuit is obtained as that shown in Figure 8.49.

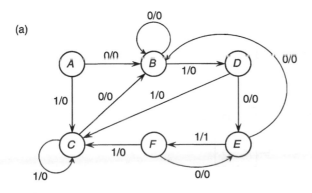

(a)

(b)

PS	NS, z (when $x=0$)	NS, z (when $x=1$)
A	B,0	C,0
B	B,0	D,0
C	B,0	C,0
D	E,0	C,0
E	B,0	F,1
F	E,0	C,0

FIGURE 8.42
A sequence detector (a) state diagram and (b) state table.

(a)

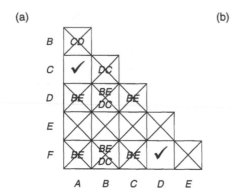

(b)

PS	NS, z (when x=0)	NS, z (when x=1)
A	B,0	A,0
B	B,0	D,0
D	E,0	A,0
E	B,0	D,1

FIGURE 8.43
Reduction of states of the state table shown in Figure 8.42b (a) implication table and (b) reduced state table.

(a)

PS	NS, z (when x=0)	NS, z (when x=1)
00	01,0	00,0
01	01,0	11,0
11	10,0	00,0
10	01,0	11,1

(b)

Q_1Q_2	J_1K_1 x=0	x=1	J_2K_2 x=0	x=1	Z X=0	x=1
00	0-	0-	1-	0-	0	0
01	0-	1-	-0	-0	0	0
11	-0	-1	-1	-1	0	0
10	-1	-0	1-	1-	0	1

FIGURE 8.44
Sequence detector design (a) transition table and (b) excitation maps.

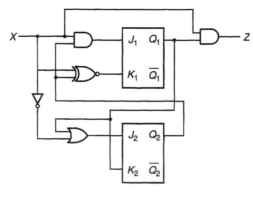

FIGURE 8.45
Logic circuit for Example 8.10.

(a)

PS	NS (when x=0)	NS (when x=1)	Z
A	B	C	0
B	B	D	0
C	B	C	0
D	E	C	0
E	B	F	0
F	E	C	1

(b)

PS	NS (when x=0)	NS (when x=1)	Z
A	B	A	0
B	B	D	0
D	E	A	0
E	B	F	0
F	E	A	1

FIGURE 8.46
State table (a) unreduced and (b) reduced.

(a)

PS	NS (when x=0)	NS (when x=1)	Z
000	001	000	0
001	001	101	0
101	100	000	0
100	001	110	0
110	100	000	1

(b)

$Q_1Q_2Q_3$	J_1K_1 x=0	J_1K_1 x=1	J_2K_2 x=0	J_2K_2 x=1	J_3K_3 x=0	J_3K_3 x=1
000	0-	0-	0-	0-	1-	0-
001	0-	1-	0-	0-	-0	-0
011	- -	- -	- -	- -	- -	- -
010	- -	- -	- -	- -	- -	- -
110	-0	-1	-1	-1	0-	0-
111	- -	- -	- -	- -	- -	- -
101	-0	-1	0-	0-	-1	-1
100	-1	-0	0-	1-	1-	0-

(c)

$Q_1Q_2Q_3$	Z
000	0
001	0
011	-
010	-
110	1
111	-
101	0
100	0

FIGURE 8.47
Sequence detector design steps for Example 8.11 (a) transition table, (b) excitation maps, and (c) output map.

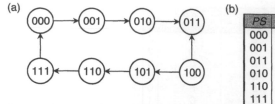

(c)

$Q_1Q_2Q_3$	J_1K_1 $\overline{CK}$	J_1K_1 CK	J_2K_2 $\overline{CK}$	J_2K_2 CK	J_3K_3 $\overline{CK}$	J_3K_3 CK
000	0-	0-	0-	0-	0-	1-
001	0-	0-	0-	1-	-0	-1
011	0-	1-	-0	-1	-0	-1
010	0-	0-	-0	-0	0-	1-
110	-0	-0	-0	-0	0-	1-
111	-0	-1	-0	-1	-0	-1
101	-0	-0	0-	1-	-0	-1
100	-0	-0	0-	0-	0-	1-

FIGURE 8.48
A 3-bit counter circuit (a) state diagram, (b) transition table, and (c) excitation maps.

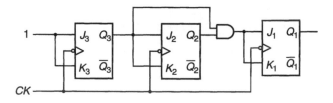

FIGURE 8.49
A 3-bit counter circuit.

Example 8.13
Analyze the circuit shown in Figure 8.50 and obtain the corresponding state diagram.

Solution
In this circuit, very much like that in Figure 8.49, the clock serves as the determining input. The clock along with past outputs of the flip-flops generates new outputs. Note that Z represents a Moore type of an output. The excitation equations corresponding to the given sequential circuit are given by,

$$J_3 = Q_2Q_1Q_0$$

$$K_3 = \overline{Q_2}Q_0 + \overline{Q_1}Q_0$$

$$J_2 = K_2 = Q_1Q_0$$

$$J_1 = \overline{Q_3}Q_0$$

$$K_1 = Q_0$$

$$J_0 = K_0 = 1$$

and

$$z = \overline{Q_3} Q_2 Q_1 \overline{Q_0}$$

Presence of four flip-flops suggests that the circuit allows for 16 states. The present states and the corresponding next states of the synchronous circuit can be obtained, as shown in Figure 8.51, by first finding flip-flop inputs for each of the present states and then by identifying the corresponding next states.

The state diagram for the circuit under consideration can now be obtained and is shown in Figure 8.52.

An observation of the state diagram shows that the circuit goes through the ten states 0000 through 1001 in succession like a modulo-10 counter. All other states make a transition to one of these ten states within two clock cycles. Provided that beginning state is one of the ten states, 0000 through 1001, the circuit of Figure 8.50 works as a modulo-10 binary counter. Such counters will be discussed more in Chapter 9.

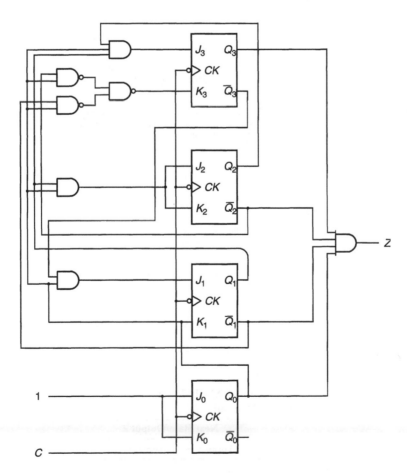

FIGURE 8.50
Logic circuit for Example 8.13.

Q_3	Q_2	Q_1	Q_0	J_3K_3	J_2K_2	J_1K_1	J_0K_0	NS
0	0	0	0	00	00	00	11	0001
0	0	0	1	01	00	11	11	0010
0	0	1	0	00	00	00	11	0011
0	0	1	1	01	11	11	11	0100
0	1	0	0	00	00	00	11	0101
0	1	0	1	01	00	11	11	0110
0	1	1	0	00	00	00	11	0111
0	1	1	1	10	11	11	11	1000
1	0	0	0	00	00	00	11	1001
1	0	0	1	01	00	01	11	0000
1	0	1	0	00	00	00	11	1011
1	0	1	1	01	11	01	11	0100
1	1	0	0	00	00	00	11	1101
1	1	0	1	01	00	01	11	0100
1	1	1	0	00	00	00	11	1111
1	1	1	1	10	11	01	11	1000

FIGURE 8.51
Excitation table for Example 8.13.

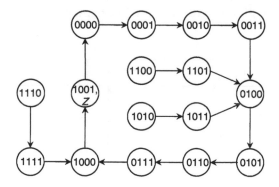

FIGURE 8.52
State diagram for Example 8.13.

TABLE 8.1

A Moore-Machine State Table

	Next State		Present
Present State	Input $x = 0$	Input $x = 1$	Output Z
A	B	A	0
B	A	C	0
C	A	D	1
D	A	D	0

TABLE 8.2

State Table of a Mealy-Machine Model
Sequential Circuit

	Next State/Output Z	
Present State	Input $x = 0$	Input $x = 1$
A	B/0	A/0
B	A/0	C/1
C	A/0	C/0

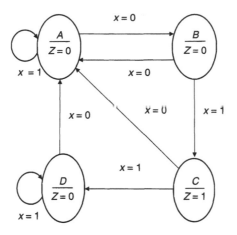

FIGURE 8.53
STG visualization of state Table 8.1

```
--  using one-hot state assignment                          -- 00
library ieee;                                               -- 01
use ieee.std_logic_1164. all;                               -- 02
                                                            -- 03
entity SREG_onehot is                                       -- 04
     port (CLK, inX: in std_logic; Z: out std_logic);       -- 05
end SREG_onehot;                                            -- 06
                                                            -- 07
architecture behavior_MoorSREG of SREG_onehot is            -- 08
-- State variables for Moor-machine SREG                    -- 09
   signal A, next_A, B, next_B, C, next_C: std_logic;       -- 10
   signal D, next_D, next_Z: std_logic;                     -- 11
   begin                                                    -- 12
      process (CLK, next_A, next_B, next_C, next_D, next_Z) -- 13
      begin                                                 -- 14
         if (CLK='1' and CLK' event) then                   -- 15
            A <= next_A; B <= next_B;                       -- 16
            C <= next_C; D <= next_D; Z <= next_Z;          -- 17
         end if;                                            -- 18
      end process;                                          -- 19
                                                            -- 20
      process (A, B, C, D, inX)                             -- 21
      begin                                                 -- 22
         if ((inX='1' and A='1') or (inX='0' and B='1') or  -- 23
            (inX='0' and C='1') or (inX='0' and D='1')) then -- 24
           next_A <='1';                                    -- 25
         else                                               -- 26
           next_A <='0';                                    -- 27
         end if;                                            -- 28
                                                            -- 29
         if (inX='0' and A='1') then next_B <='1';          -- 30
         else next_B <='0';                                 -- 31
         end if;                                            -- 32
                                                            -- 33
         if (inX='1' and (B='1') then next_C <='1';         -- 34
         else next_C <='0';                                 -- 35
         end if;                                            -- 36
                                                            -- 37
         if ((inX='1' and C='1') or (inX='1' and D='1'))    -- 38
            then next_D <='1';                              -- 39
         else next_D <='0';                                 -- 40
         end if;                                            -- 41
                                                            -- 42
         if (inX='1' and B='1') then next_Z <='1';          -- 43
         else next_Z<='0';                                  -- 44
         end if;                                            -- 45
      end process;                                          -- 46
   end behavior_MoorSREG;                                   -- 47
```

FIGURE 8.54
A behavioral VHDL description of SREG (with one-hot state assignment).

```
-- using area-optimized state assignment                    -- 01
library ieee;                                               -- 02
use ieee.std_logic_1164. all;                               -- 03
                                                            -- 04
entity SREG_area is                                         -- 05
        port (CLK, in X: in std_logic; Z: out std_logic);  -- 06
end SREG_area;                                              -- 07
                                                            -- 08
architecture behavior_Moor SREG_a of SREG_area is           -- 09
-- State variables for Moor-machine SREG                    -- 10
   type type_sreg is (A, B, C, D);                          -- 11
    signal sreg, next_sreg: type_sreg;                      -- 12
  begin                                                     -- 13
      process (CLK, next_sreg)                              -- 14
      begin                                                 -- 15
         if (CLK='1' and CLK' event) then                   -- 16
           sreg <= next_sreg;                               -- 17
          end if;                                           -- 18
      end process;                                          -- 19
                                                            -- 20
      process (sreg, inX)                                   -- 21
      begin                                                 -- 22
         Z <= '0';                                          -- 23
         next_sreg <= A;                                    -- 24
         case sreg is                                       -- 25
           when A =>                                        -- 26
                Z <= '0';                                   -- 27
                if (inX='0') then next_sreg <= B; end if;   -- 28
                if (inX='1') then next_sreg <= A; end if;   -- 29
           when B =>                                        -- 30
                Z <= '0';                                   -- 31
                if (inX='0') then next_sreg <= A; end if;   -- 32
                if (inX='1') then next_sreg <= C; end if;   -- 33
           when C =>                                        -- 34
                Z <= '1';                                   -- 35
                if (inX='0') then next_sreg <= A; end if;   -- 36
                if (inX='1') then next_sreg <= D; end if;   -- 37
           when D =>                                        -- 38
                Z <= '0';                                   -- 39
                if (inX='0') then next_sreg <= A; end if;   -- 40
                if (inX='1') then next_sreg <= D; end if;   -- 41
             when others =>                                 -- 42
           end case;                                        -- 43
         end process;                                       -- 44
      end behavior Moor SREG a;                              -- 45
```

FIGURE 8.55
A behavioral VHDL description of SREG (optimized for area).

8.9 Synchronous Sequential Circuit Implementation Using VHDL

In this section we illustrate VHDL description of sequential circuit implementation. We start with a state table (which is explained in Section 8.3 and Figure 8.8), as shown in Table 8.1. This state table describes the functions and transitions of a synchronous sequential circuit, which can be implemented with flip-flops and other basic logic-function gates. Figure 8.53 is a state transition graph (STG) that visualizes the state table. After importing the STG into Xilinx *StateCAD* software, it can generate a behavioral VHDL program based optimization selection, either for speed or for area (Figures 8.54 through 8.58).

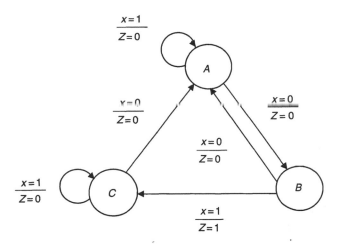

FIGURE 8.56
An STG for the Mealy-machine sequential circuit.

```
-- using one-hot state assignment                              -- 01
library ieee;                                                  -- 02
use ieee.std_logic_1164. all;                                  -- 03
                                                               -- 04
entity MealySREG3 is                                           -- 05
     port (CLK, in X: in std_logic; Z: out std_logic);         -- 06
end MealySREG3;                                                -- 07
                                                               -- 08
architecture behavior_Mealy3 of MealySREG3 is                  -- 09
-- State variables for Mealy-machine SREG                      -- 10
   signal A, next_A, B, next_B, C, next_C: std_logic;          -- 11
   signal next_Z: std_logic;                                   -- 12
  begin                                                        -- 13
     process (CLK, next_A, next_B, next_C, next_Z)             -- 14
     begin                                                     -- 15
        if (CLK='1' and CLK' event) then                       -- 16
         A <= next_A; B <= next_B;                             -- 17
         C <= next_C; Z <= next_Z;                             -- 18
        end if;                                                -- 19
     end process;                                              -- 20
                                                               -- 21
     process (A,B,C, inX)                                      -- 22
     begin                                                     -- 23
        if ((inX='1' and A='1') or (inX='0' and B='1') or      -- 24
         (inX='0' and C='1')) then next_A <='1';               -- 25
        else next_A <='0';                                     -- 26
        end if;                                                -- 27
                                                               -- 28
        if (inX='0' and A='1') then next_B <='1';              -- 29
        else next_B <='0';                                     -- 30
        end if;                                                -- 31
                                                               -- 32
        if ((inX='1' and B='1') or (inX='1' and C='1'))        -- 33
           then next_C <='1';                                  -- 34
        else next_C <='0';                                     -- 35
        end if;                                                -- 36
                                                               -- 37
        if (inX='1' and B='1') then next_Z <='1';              -- 38
        else next_Z <='0';                                     -- 39
        end if;                                                -- 40
     end process;                                              -- 41
  end behavior_Mealy3;                                         -- 42
```

FIGURE 8.57
A behavioral VHDL description of Mealy-machine sequential circuit (with one-hot state assignment).

```
--  using speed-optimized state assignment          -- 01
library ieee;                                        -- 02
use ieee.std_logic_1164. all;                        -- 03
                                                     -- 04
entity SREG_speed is                                 -- 05
      port (CLK,inX: in std_logic; Z: out std_logic);-- 06
end SREG_speed;                                      -- 07
                                                     -- 08
architecture behavior_Mealy SREG_s of SREG_speed is  -- 09
-- State variables for Mealy-machine SREG            -- 10
   type type_sreg is (A, B, C);                      -- 11
   signal sreg, next_sreg: type_sreg;                -- 12
   signal next_Z: std_logic;                         -- 13
 begin                                               -- 14
     process (CLK, next_sreg, next_Z)                -- 15
     begin                                           -- 16
        if (CLK='1' and CLK' event) then             -- 17
          sreg <= next_sreg; Z <= next_Z;            -- 18
        end if;                                      -- 19
     end process;                                    -- 20
                                                     -- 21
     process (sreg, in X)                            -- 22
     begin                                           -- 23
        next_Z <= '0'; next_sreg <= A;               -- 24
        case sreg is                                 -- 25
          when A =>                                  -- 26
             if (inX='0') thennext_sreg <= B;        -- 27
              next_Z <= '0'; end if;                 -- 28
             if (inX='1') then next_sreg <= A;       -- 29
              next_Z <= '0'; end if;                 -- 30
          when B =>                                  -- 31
             if (inX= '0') then next_sreg <= A;      -- 32
              next_Z <= '0'; end if;                 -- 33
             if (inX= '1') then next_sreg <= C;      -- 34
              next_Z <= '1'; end if;                 -- 35
          when C =>                                  -- 36
             if (inX= '0') then next_sreg <= A;      -- 37
              next_Z <= '0'; end if;                 -- 38
             if (inX= '1') then next_sreg <= C;      -- 39
                next_Z <= '0'; end if;               -- 40
          when others =>                             -- 41
        end case;                                    -- 42
     end process;                                    -- 43
  end behavior Mealy SREGs                           -- 44
```

FIGURE 8.58
A functional VHDL description of a single-stage logic unit (optimized for speed).

8.10 Summary

In this chapter, we have explored the various aspects of designing synchronous sequential circuits. The concepts of state diagram, state table, Mealy and Moore outputs, and completely specified and incompletely specified circuits were first introduced. These ideas were then used to determine equivalent states using visual inspection, partitioning, and implication methods. The impact of making a particular state assignment was investigated. Finally, the design algorithm for designing any arbitrary synchronous sequential circuit was presented and its application demonstrated. Some of the application examples covered in this chapter included counters. We shall cover more of them in the next chapter (Chapter 9) for further application in Chapter 10. In a follow-on chapter

(Chapter 11), we shall also introduce the design of asynchronous sequential circuits. The synchronous design covered in this chapter is different than that of the asynchronous type in that clock input present in the synchronous circuits makes them far more predictable.

Bibliography

Comer, D.J., *Digital Logic and State Machine Design.* 3rd edn. New York, NY. Oxford University Press, 1994.

Floyd, T., *Digital Fundamentals.* 8th edn. Englewood Cliffs, NJ. Prentice-Hall, 2003.

Johnson, E.J. and Karim, M.A., *Digital Design: A Pragmatic Approach.* Boston, MA. PWS-Kent Publishing, 1987.

Karim, M.A. and Awwal, A.A.S., *Optical Computing: An Introduction.* New York, NY. John Wiley & Sons, 1992.

Katz, R.H., *Contemporary Logic Design.* Boston, MA. Addison Wesley, 1993.

Mowle, F.J.A., *A Systematic Approach to Digital Logic Design.* Reading, MA. Addison-Wesley, 1976.

Nagle, H.T., Jr., Carroll, B.D., and Irwin, J.D., *An Introduction to Computer Logic,* Englewood Cliffs, NJ. Prentice-Hall, 1975.

Nelson, V.P.P., Nagle, H.T., and Carroll, B.D., *Digital Logic Circuit Analysis and Design.* Englewood Cliffs, NJ. Prentice-Hall, 1995.

Rhyne, V.T., *Fundamentals of Digital Systems Design.* Englewood Cliffs, NJ. Prentice-Hall, 1973.

Problems

1. Convert the following Mealy machine to its equivalent Moore machine:

PS	NS,z (when $x = 0$)	NS,z (when $x = 1$)
A	B,0	C,1
B	D,1	A,1
C	A,0	D,0
D	C,1	B,1

2. Convert the following Mealy machine to its equivalent Moore machine:

PS	NS,z (when $x = 0$)	NS,z (when $x = 1$)
A	B,0	A,1
B	C,0	D,0
C	B,1	A,0
D	A,0	D,1

3. Minimize the following state table using (a) partitioning and (b) implication method:

PS	NS,z (when $x = 0$)	NS,z (when $x = 1$)
A	D,0	B,1
B	C,0	E,1
C	E,1	F,0
D	F,1	E,0
E	B,0	C,1
F	A,0	D,1

4. Obtain the reduced state table for the sequential machine given by the following state table. Use (a) partitioning and (b) implication table.

PS	NS,z (when $x = 0$)	NS,z (when $x = 1$)
A	D,1	E,0
B	A,0	D,0
C	A,1	F,1
D	G,0	B,0
E	G,1	C,1
F	A,1	E,1
G	B,1	C,0

5. Minimize the following multi-input state table using (a) partitioning and (b) implication table. Show the reduced state tables.

PS	NS (for I_1), Z	NS (for I_1), Z	NS (for I_1), Z
A	D,1	C,0	E,1
B	D,0	E,0	C,1
C	A,0	E,0	B,1
D	A,1	B,0	E,1
E	A,1	C,0	B,1

6. Minimize the following incompletely specified state table. Discuss all valid alternatives:

PS	NS,z (when $x = 0$)	NS,z (when $x = 1$)
A	A,-	B,-
B	B,-	C,1
C	G,-	E,0
D	C,1	C,-
E	B,1	C,-
F	D,-	B,-
G	G,-	G,-
H	H,-	D,-

7. Minimize the following incompletely specified state table. Discuss all valid alternatives.

PS	NS,z (when $x = 0$)	NS,z (when $x = 1$)
A	B,0	-,0
B	A,0	E,0
C	D,0	A,-
D	-,-	C,-
E	B,1	-,1

8. Minimize the following incompletely specified state table. Discuss all valid alternatives:

PS	NS,z (when $x = 0$)	NS,z (when $x = 1$)
A	B,-	D,0
B	F,0	C,0
C	A,0	B,-
D	F,-	-,-
E	-,1	D,-
F	C,-	G,0
G	A,-	-,1

9. Design a synchronous counter using D flip-flops that counts the sequence 1, 5, 2, 3, 6, 7, and 1. Discuss what happens if the circuit were to enter any of the unused states.

10. Use JK flip-flops to design a sequence detector that produces an output when exactly two 0s are followed by two 1s.

11. Find a clocked D flip-flop realization for the following sequential circuit using each of the three unique state assignments:

PS	NS,z (when $x = 0$)	NS,z (when $x = 1$)
A	C,0	B,0
B	C,0	D,0
C	A,0	D,0
D	B,1	A,1

12. Find a clocked SR flip-flop realization for the sequential circuit of Problem 11 using each of the three unique state assignments.

13. Find a clocked JK flip-flop realization for the sequential circuit of Problem 11 using each of the three unique state assignments.

14. Find a clocked T flip-flop realization for the sequential circuit of Problem 11 using each of the three unique state assignments.

15. For the following sequential machine, use an optimal state assignment and then implement the circuit using D flip-flops:

PS	NS,z (when $x = 0$)	NS,z (when $x = 1$)
A	D,0	B,1
B	A,0	E,0
C	D,1	B,0
D	A,1	C,1
E	B,0	B,0

16. For the sequential machine given by the following excitation equations, determine its state diagram:

$$T_2 = x \otimes Q_1$$

$$z = T_1 = x\overline{Q_2}$$

17. Design a 5-bit synchronous even-parity checker working in the sequential mode that can produce an output each time there is an error in the 5-bit coded word.

18. Design a 3-bit up/down binary counter using JK flip-flops that counts up when $U = 1$ and counts down when $U = 0$.

19. Design a 3-bit up/down binary counter using D flip-flops that counts up when $U = 1$ and counts down when $U = 0$.

20. Design a sequence detector using D flip-flops that gives a high output when it has detected the sequence 10010. Allow for overlapping of sequences.

21. Design a sequence detector using D flip-flops that gives a high output when it has detected the sequence 10010 without allowing for any overlapped sequence.

22. Repeat Problem 19 but using JK flip-flops.

23. Repeat Problem 20 but using JK flip-flops.

24. Design a 4-bit up/down Gray code counter using D flip-flops. It should count up when $U = 1$ and down when $U = 0$.

25. Design a 4-bit up/down Gray code counter using JK flip-flops. It must be able to count up when $U = 1$ and down when $U = 0$.

26. Analyze the synchronous sequential circuits shown in Figure 8.P1.

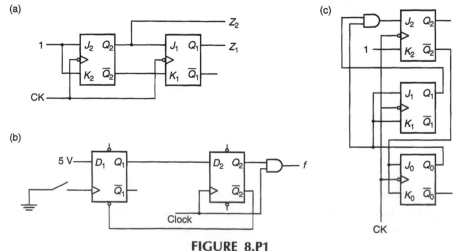

FIGURE 8.P1

9

Design of Modular Sequential Components

9.1 Introduction

Designing counters and registers will allow us to learn additional tools to then design more complex circuits found in digital computers. *Counters* are important in that they are often used to keep track of a sequence of events of interest, and *registers* are used to store and manipulate much of data that contribute to all or many of these events and functions. Most robust digital systems would in effect have two functional units: a unit wherein data is manipulated and a unit that is used for regulating the events in the first unit. Registers and associated combinational logic subunits together make the first unit, and counters are used typically for controlling the second unit.

In addition to counting, counters can track the sequence of instructions in a program, distribute the sequence of timing signals in the control and arithmetic units of processors, use frequency division to cause necessary time delays, and for a host of other similar functions. Counters can be designed to count in binary or in nonbinary sequences. They are commercially available in a large variety of medium-scale integrated devices. The basic functional characteristic of a counter is sequential, that is, for every present state there is a well-defined next state. The design of a counter, thus, involves designing combinational logic that decodes present state and enables entry into the next state of the desired sequence. All FFs in a *synchronous counter* change their states nearly synchronously with clock input whether it is a periodic clock or an aperiodic pulse.

FFs are able to store bits of information in them. When FFs are organized to store a string of multibit information, such configuration is referred to as a *register*. Registers, in turn, are classified according to the way information bits are stored and retrieved. If data are entered and removed from either end of a register, one bit at a time, it is referred to as a *serial* or *shift register*. On the other hand, a register is referred to as *parallel* when all bits of the word are entered or retrieved nearly simultaneously.

For us to grasp the functioning of any arbitrary digital system, however small or complex it may be, we need to understand how each is built and integrated from individual components, and finally, how it is controlled. The control unit of most systems can be designed using methods that were introduced and demonstrated in the preceding three chapters. In order to express and, therefore, design and orchestrate the sequences of operations of the subsystems used in manipulating data, we in effect need a powerful language tool. The tool that has been found to be useful is referred to as the *register transfer language* (RTL). RTL can be used to translate a specification then into its corresponding hardware realizations.

9.2 Synchronous Counters

Synchronous counters are distinct in that the clock pulses initiate changes in the FFs that are used in them. The simplest possible counter is often a single-bit counter that alternates between two states, 0 and 1. A JK FF whose two inputs are tied to a 1 is in effect a T FF; it can readily function as a single-bit counter alternating between two states with each clock. The output of this FF has a frequency that is one-half of that of the clock.

A two-bit binary up-counter designed using two JK FFs can provide up to four distinct states such that $Q_2 Q_1$ can be made to cycle through 00, 01, 10, 11, 00, and so on. The J and K inputs of the FFs of such a 2-bit counter are

$$J_1 = K_1 = 1$$

$$J_2 = K_2 = Q_1$$

These equations allows the outputs to be derived directly from the FFs. Here the first JK FF is set as a toggle flip-flop and its output Q_1 is fed directly to J and K input of the second JK FF. This second flip-flop works as a toggle flip-flop when $Q_1 = 1$; otherwise Q_2 remains unchanged.

Now consider synthesizing a 3-bit binary up-counter also without any separate outputs. We can go through the design steps already introduced in Chapter 8 and begin by obtaining a state diagram, then a state table from it and next make appropriate state assignments. Figure 9.1 shows the state diagram, the state table, and the excitation maps of a three-bit counter.

JK equations for a 3-bit up-counter are derived from excitation maps of Figure 9.1c

$$J_1 = K_1 = 1$$

$$J_2 = K_2 = Q_1$$

$$J_3 = K_3 = Q_1 Q_2$$

Figure 9.2 shows the resulting logic circuit.

Many a sequential design problem can be approached often without going beyond generating its state table. By examining Figure 9.1b, it is obvious that there is a certain degree of pattern in the flip-flops. With each clock, Q_1 changes and, in general a flip-flop changes its state if and when all less significant bits is a 1. This observation is consistent with the JK equations of the three counters we have considered this far. This pattern in the J and K equations leads us to conclude that these logic equations can be extended to obtain J and K equations for any arbitrary n-th bit of a multibit up-counter:

$$J_n = Q_{n-1} Q_{n-2} \cdots Q_2 Q_1 = Q_{n-1} J_{n-1}$$

$$K_n = Q_{n-1} Q_{n-2} \cdots Q_2 Q_1 = Q_{n-1} K_{n-1}$$

Inputs to the JK FFs can be introduced in two ways on the basis of how the Boolean equations for the nth term are expressed. These two schemes are shown in Figure 9.3. Figure 9.3a shows a circuit configuration where the FF outputs are all fed in parallel. The propagation delay at the input of each FF is same for all bits. However, the fan-in to the AND gate and the fan-out of each FF increase as the number of FFs. Figure 9.3b shows an

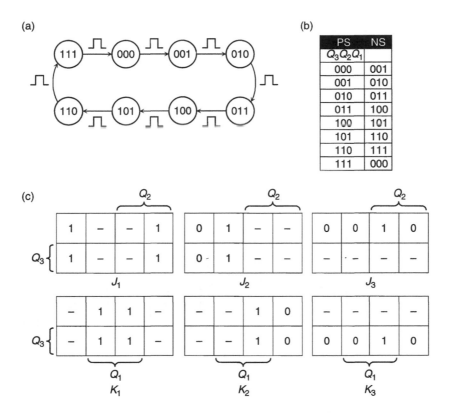

(a)

(b)

PS	NS
$Q_3Q_2Q_1$	
000	001
001	010
010	011
011	100
100	101
101	110
110	111
111	000

(c)

FIGURE 9.1
3-bit binary up-counter (a) state diagram, (b) state table, and (c) excitation maps.

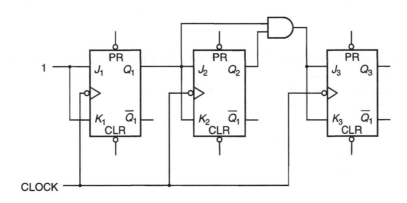

FIGURE 9.2
Logic circuit of a 3-bit up-counter.

alternative scheme that uses the second form for the J_n and K_n equations. The fan-in of the AND gates is always two; however, the propagation delay to the nth flip-flop increases with increasing n. As is obvious, the first scheme allows for faster clocking and counting but results in fan-in and fan-out problems especially when n is large.

A synchronous n-bit binary up-counter that uses JK FFs and two-input AND gates is obtained as shown in Figure 9.4. Two controls, CLEAR and COUNT, in the circuit allow for flexibility. A high on CLEAR input resets the counter; it remains reset until the CLEAR

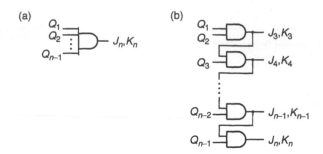

FIGURE 9.3
Configurations for J and K inputs (a) parallel and (b) serial.

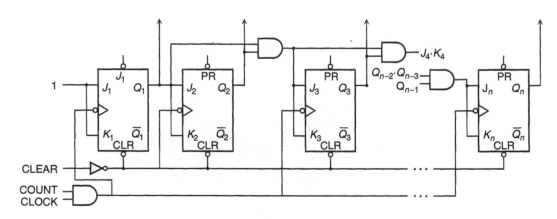

FIGURE 9.4
n-bit synchronous up-counter.

signal has been withdrawn. The COUNT signal, on the other hand, is used to disable the clock. One can make effective use of these controls to block the clock input and hold any nonzero count state over a long period or set the counter up to its maximum count state by effectively using preset (PR) inputs.

For certain applications, it may be necessary to also design a down-counter wherein a binary number is set initially into the counter and then, with each clock, the counter counts down. Down-counters can be designed in the same way as we designed the up-counters. Not surprisingly, binary down-counters, just as in the case of up-counters, also have a pattern in how their flip-flops undergo changes with clock. The J and K equations for an n-bit binary down-counter can be obtained likewise:

$$J_1 = K_1 = 1$$
$$J_2 = K_2 = \overline{Q_1}$$
$$J_3 = K_3 = \overline{Q_1}\,\overline{Q_2} = J_2\overline{Q_2} = K_2\overline{Q_2}$$
$$J_n = K_n = \overline{Q_{n-1}}\,\overline{Q_{n-2}}\cdots\overline{Q_3}\,\overline{Q_2}\,\overline{Q_1} = J_{n-1}\overline{Q_{n-1}} = K_{n-1}\overline{Q_{n-1}}$$

Two counters are quite similar as is obvious from a comparison of their respective equations. The J_n and K_n inputs are generated by two-input AND gate output, whose logic inputs in turn are $\overline{Q_{n-1}}$ and K_{n-1}. In the case of up-counters the corresponding inputs to AND logic were Q_{n-1} and J_{n-1}.

We can combine these two concepts to next design a counter that is capable of counting up as well as down on demand. A system based on such a counter can be used, for example, to keep track of total cars present in a parking garage. The counter will count up as each car enters the garage, and it will count down when a car leaves the garage. Such an up–down counter will need to be able to receive and process at least one control signal, E, to control the direction of the count. With $E = 1$, for example, the circuit can be made to count up, and when $E = 0$ the counter can count down. This counter can be synthesized following the standard steps used in the design of sequential circuits. However, we can use the equations for up- and down-counters to derive the respective J and K equations of such an n-bit, up–down counter:

$$J_1 = K_1 = 1$$

$$J_2 = K_2 = EQ_1 + \overline{E}\,\overline{Q_1}$$

$$J_3 = K_3 = EQ_1Q_2 + \overline{E}\,\overline{Q_1}\,\overline{Q_2}$$

$$\cdots\cdots \qquad \cdots\cdots \qquad \cdots\cdots \qquad \cdots\cdots$$

$$J_n = K_n = EQ_{n-1}Q_{n-2}\cdots\cdots Q_2Q_1 + \overline{E}\,\overline{Q_{n-1}}\,\overline{Q_{n-2}}\cdots\overline{Q_2}\,\overline{Q_1}$$

Here, the excitation function of the up-counter is ANDed with E, and that of the down-counter is ANDed with $\overline{E}$. These two ANDed outputs are then ORed to derive the J and K equations. With $E = 1$, these Boolean equations reduce to that of an up-counter and with $E = 0$, the two Boolean equations reduce to those of a down counter. Figure 9.5 shows, for example, a 4-bit up–down counter.

A counter can be made to also count nonbinary sequences. BCD decade counter and Gray code counter are two examples of nonbinary counters. In the former, the counter counts 0000 through 1001 and then resets back to 0000. A four-bit Gray code counter, in comparison, counts 0000, 0001, 0011,0010, 0110, 0111, 0101, 0100, 1100, 1101, 1111, 1110, 1010, 1011, 1001, and 1000 in that order and then resets to 0000 before again going through these same count sequences. Example 9.1 covers the design of a BCD decade counter.

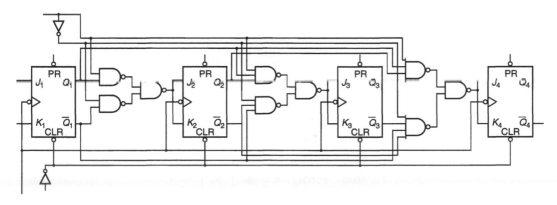

FIGURE 9.5
4-bit binary up–down counter.

Example 9.1

Derive J and K equations for a BCD up-counter.

Solution

The design steps will involve identifying its state diagram, as shown in Figure 9.6, the state and transition table, as shown in Figure 9.7, and, finally, excitation maps as shown in Figure 9.8.

The J and K equations follow directly from the K-maps of Figure 9.8:

$$J_1 = K_1 = 1$$
$$J_2 = Q_1\overline{Q_4}; \quad K_2 = Q_1$$
$$J_3 = K_3 = Q_1Q_2$$
$$J_4 = Q_1Q_2Q_3; \quad K_4 = Q_1$$

The logic circuit of a BCD down-counter can be derived using steps same as those used in Example 9.1. The J and K equations are

$$J_1 = K_1 = 1$$
$$J_2 = \overline{Q}_1(Q_3 + Q_4); \quad K_2 = \overline{Q}_1$$
$$J_3 = \overline{Q}_1Q_4; \quad K_3 = \overline{Q_1}\,\overline{Q_2}$$
$$J_4 = K_4 = \overline{Q_1}\,\overline{Q_2}\,\overline{Q_3}$$

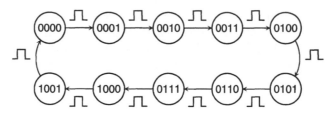

FIGURE 9.6
State diagram for Example 9.1.

PS $Q_4Q_3Q_2Q_1$	NS	J_4K_4	J_3K_3	J_2K_2	J_1K_1
0000	0001	0-	0-	0-	1-
0001	0010	0-	0-	1-	-1
0010	0011	0-	0-	-0	1-
0011	0100	0-	1-	-1	-1
0100	0101	0-	-0	0-	1-
0101	0110	0-	-0	1-	-1
0110	0111	0-	-0	-0	1-
0111	1000	1-	-1	-1	-1
1000	1001	-0	0-	0-	1-
1001	0000	-1	0-	0-	-1

FIGURE 9.7
State and transition table for Example 9.1.

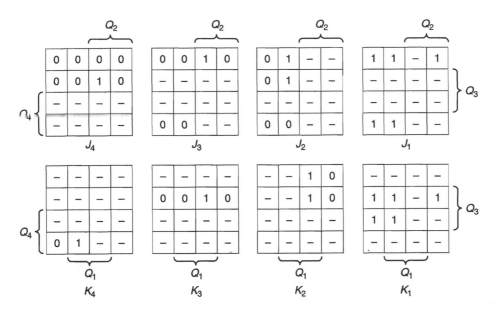

FIGURE 9.8
Excitation maps for Example 9.1.

Example 9.1 utilized the traditional design approach in realizing a BCD counter. However, it is not necessary to go through these steps if an already working circuit can be modified to suit the requirements of a BCD counter. For example, we could consider starting with a four-bit binary up-counter otherwise known as a *modulo-16* (or divide-by-16) counter. Without any modification, this counter will repeat going through 4-bit binary sequence once every 16 clocks. Note that a synchronous BCD counter will function as a four-bit, modulo-16 counter until it has reached the state $Q_4Q_3Q_2Q_1 = 1001$. With the next incoming clock, the four-bit modulo-16 counter left to itself will advance to 1010, while if it were to be adapted to become a BCD counter, it should instead start all over from 0000. To make a four-bit binary counter behave as a BCD counter, thus, it will be necessary to reset FFs when they are about ready to switch to count state 1010. The needed circuit modifications require implementing a logic circuit so that when the modulo-16 counter reaches count 1001, it also satisfies the following conditions:

Q_1, the least significant bit (LSB), resets to 0

Q_4, the most significant bit (MSB), resets to 0

Q_2 should be prevented from being set to 1

We notice that the most significant FF in the logic circuit of Figure 9.4 is set in toggle mode, which implies that Q_1 will change to a 0 by itself and requires no modification. Note the MSB, Q_4, of the four-bit counter changes only when the count reaches either 0111 or 1111. This is possible since J_4 and K_4 are both 0 during all other counts. Consequently, to accomplish the desired change, it will be necessary to supply K_4 with a 1 instead of a 0 when the count has reached 1001. This is realized by using Q_1 as K_4 input and by delinking K_4 from J_4. We need to prevent Q_2 from switching back to 1 once the count has reached 1001. This is accomplished by supplying both J_2 and K_2 with ANDed output of Q_1 and $\overline{Q}_4$. This latter modification will not contribute to any unwanted state since $\overline{Q}_4$ is already a 1 and

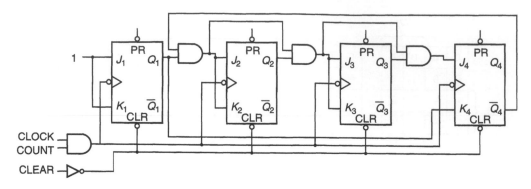

FIGURE 9.9
Synchronous BCD up-counter.

remains so until the count has reached 1000. Then when the BCD count has reached 1000, Q_2 will not need to be turned on. In summary, for this proposed modification to be effective,

> Q_1 needs to be tied to K_4
>
> K_4 and J_4 must not be linked
>
> J_2 and K_2 inputs need to be fed with $Q_1\overline{Q_4}$

Figure 9.9 shows the logic circuit of a BCD counter that has been obtained by modifying a modulo-16 counter.

Synchronous circuits such as a Gray counter that requires irregular count sequences may not be designed using a modulo-16 counter; they are designed using the standard step-by-step sequential design process. A count sequence is referred to as *irregular* when it is not magnitude ordered. Two such counters having irregular count sequences are explored in Examples 9.2 and 9.3.

Example 9.2
Derive a state table for the counter shown in Figure 9.10 where MSB is given by Q_3.

Solution
We can begin to analyze this logic circuit by starting with a present state and using our knowledge of JK FF functions to determine the corresponding next state. For each present state, the corresponding next state is determined from the J and K values that are being fed to the FFs. The resulting state table is obtained as shown in Figure 9.11. The given counter circuit goes through an irregular sequence 0, 1, 3, 4, 6, and repeat.

Example 9.3
Generate a synchronous counter that generates the count sequence 0, 2, 4, 3, 6, 7, 0, ...

Solution
Figure 9.12 shows the state table and the JK excitations of this counter. One of the ways to approach this problem would be to consider pairs of present and corresponding next states. The JK excitations necessary for causing the corresponding transitions are

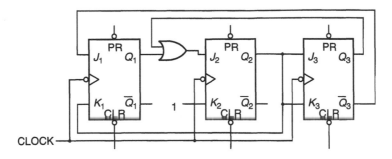

FIGURE 9.10
Logic circuit for Example 9.2.

PS	J_3K_3	J_2K_2	J_1K_1	NS
000	00	01	10	001
001	00	11	10	011
011	11	11	11	100
100	00	11	00	110
110	11	11	01	000

FIGURE 9.11
State table for Example 9.2.

PS	NS	J_3K_3	J_2K_2	J_1K_2
000	010	0-	1-	0-
001	---	--	--	--
010	100	1-	-1	0-
011	110	1-	-0	-1
100	011	-1	1-	1-
101	---	--	--	--
110	111	-0	-0	1-
111	000	-1	-1	-1

FIGURE 9.12
State table for Example 9.3.

then entered for each of the present states. Note that two states that are not included in the sequence are not considered—the corresponding next state, and excitations are treated as don't cares. These don't cares may prove to be advantageous since these may contribute to having a simpler logic circuit.

The J and K equations are obtained using K-maps.

$$J_1 = Q_3; \quad K_1 = 1$$
$$J_2 = 1; \quad K_2 = Q_1 \otimes Q_3$$
$$J_3 = Q_2; \quad K_3 = Q_1 + \overline{Q}_2$$

Figure 9.13 shows the resulting logic diagram.

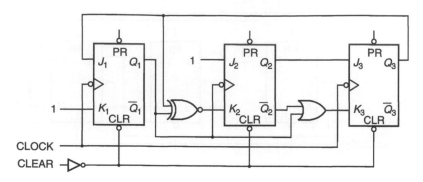

FIGURE 9.13
Logic diagram for Example 9.3.

After CLOCK	Bit pattern
0	011011001010
1	101101100101
2	110110110010
3	111011011001
4	111101101100
5	111110110110
.	.
.	.
.	.
10	111111111101
11	111111111110
12	111111111111

FIGURE 9.14
Shift-right register action.

9.3   Registers

A *shift register* consists of a group of FFs so that each FF transfers its current stored bit of information to an adjacent FF coincident with each clock. An n-bit shift register in effect can store n-bits of information, behaving in turn as a temporary memory. Upon external command it can shift data bits one position either to right or to left. The shift register happens to be one of the most extensively used sequential logic devices.

Figure 9.14 shows a 12-bit shift-right register whose shift-right serial input is tied to a 1. With each clock, the bit shifting out of the right-most FF is lost. However, with each clock, the vacated MSB takes in a 1. After 12 clock pulses, the data originally present in the register is replaced by a string of 1s. For some application, it may be useful to modify shift-right registers so that the right-most FF output is fed as an input to the left-most FF. In such a case the LSB is not lost but reappears at the MSB. After three clock pulses, for example, data bits 011011001010 stored in the register will be replaced by a new sequence: 010011011001. Such a shift-right register is known as a *circulate-right* register. In the event data bits were stored instead in a shift-left register and its MSB output was connected to the LSB FF as an input, the data would be 011001010011 after three clock pulses. Such a register is referred to as a *circulate-left* register.

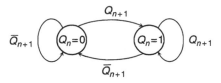

FIGURE 9.15
State diagram for the Q_n bit of a multibit, shift-right register.

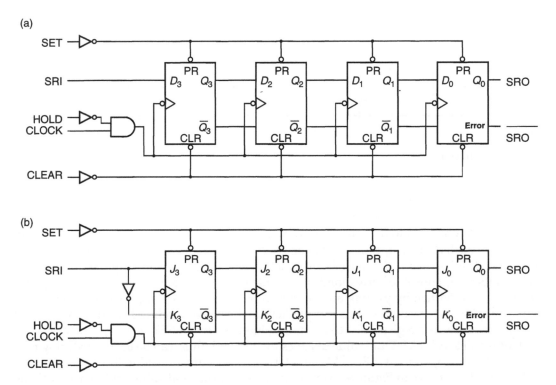

FIGURE 9.16
A four-bit shift-right register (a) using D FFs (b) using JK FFs.

Design of a n-bit multibit serial shift-right register can be approached by designing n-th FF of the register in question using procedures described earlier. Figure 9.15 shows state diagram of the n-th FF. The FF state typically reflects the current content of that bit position. If the current state of the n-th bit of this shift register is a 0 and that of the $(n+1)$-th bit is a 0, the n-th bit remains unchanged. Similarly, Q_n does not change if the present states of both Q_n and Q_{n+1} are 1. In all other cases, Q_n changes and takes the value of Q_{n+1}. Figures 9.14 and 9.15 illustrate the action of a serial shift-right register wherein the function of each FF is governed by the same next-state equation. The design of a multiple-bit shift register is thus reduced to that of a single FF. For n-bits, n such FFs are assembled and then logically combined.

The logic circuit that responds to the state diagram shown in Figure 9.15 is that of a D FF. Thus, it is customary to design a multibit, shift-right registers using edge-triggered D FFs. In case we intend to instead use JK FFs, they are converted to act as D FFs by inserting an inverter between the corresponding J and K inputs. Figure 9.16 shows these two types of

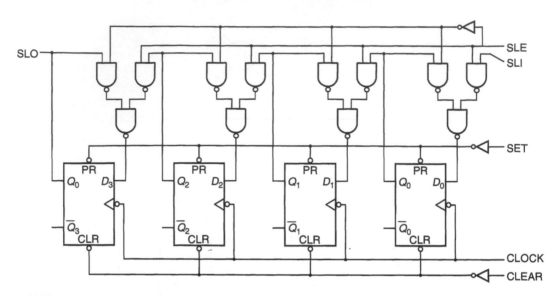

FIGURE 9.17
Controlled shift-left register.

shift-right registers. The *shift-right input* (SRI) serves as the entry point and likewise SRO serves as the exit point for *shift-right output*. In case of D FFs-based registers, the output Q of each FF is connected to the input D of the next FF. On the other hand, if JK FFs are used, Q and $\overline{Q}$ of each FF are respectively connected directly to J and K inputs of the next FF. The left-most JK FF is modified externally to function as a D FF by supplying data bit directly to J and its complement to corresponding K. Note that the remaining JK FFs in this register need not be modified by putting an inverter in between J and K inputs since each of the remaining J values, by default, is a complement of the corresponding K values. When CLR is high, it causes the register FFs to reset, and likewise when SET is high, it places a 1 at each of the FFs. A high HOLD value disables the FFs and, thus, can be used for storing bits. For serial mode operations, SET, CLR, and HOLD inputs are all held to 0.

The registers shown in Figure 9.16 are serial-in, serial-out, and shift-right registers. Same design technique can be used to obtain a shift-left registers. It is quite possible to combine shift-right and shift-left capabilities to design a bidirectional serial-in, serial-out shift register. Figure 9.17 shows a controlled shift-left register that includes a control input SLE that determines what it does on the next clock. SLI serves as the entry point for *shift-left input* and SLO is the exit point for *shift-left output*. When the *shift-left enable*, SLE, is low, the FF output is fed back as its data input. In this way, it is possible to restore information bits indefinitely. Note how the HOLD input has been eliminated in this logic circuit. The clock input excites the FFs as expected. However, in the previous case, as shown in Figure 9.16, the clock input was not allowed into the FFs. When SLE is high, the serial input sets up the right-most FF, Q_0 sets up the second FF, Q_1 sets up the third FF, and so on. With high SLE, the logic circuit functions as a shift-left register. In this case, the serial output is from the left-most FF.

The shift-left and shift-right register designs can be merged to design a bidirectional shift register as shown in Figure 9.18. This particular register and up–down counter are quite similar in their design approaches. Analogously as in the case of an up–down counter, we can use SLE and shift right enable (SRE) inputs to effect shift direction in a bidirectional register. SLE and SRE control inputs are mutually exclusive and cannot be high simultaneously except when HOLD = 1. When SRE is high, each of the FFs loads respective Q

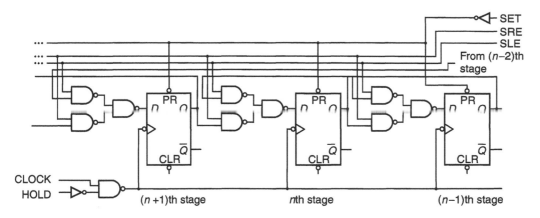

FIGURE 9.18
Bidirectional shift register.

output values of the FF on its immediate left. When SLE is high, each of the FFs loads Q output values of the FF on its immediate right. When both SLE and SRE are 0, the FFs are all reset. Note that we have eliminated the need to have a CLR control. When HOLD = 0 the register functions in its serial mode, and when HOLD = 1 the old data bits get restored. We might decide to eliminate one of these two shift controls. In that case, we may decide to keep only SRE by making sure that SLE has been replaced with the complement of SRE. The register would then function as a shift-right type when SRE = 1 and as a shift-left type when SRE = 0. This modification may cause problem for us especially if we had a need to reset the FFs at any time. Such problem, however, can be resolved by feeding the complement of CLR to each of the FF resets.

The most appropriate application of shift registers is perhaps their unique role in realizing arithmetic operations. A binary number gets multiplied by 2 when we shift the data bits by one bit to the left and divided by 2 when we shift the register content one bit to the right. As we will see later, the bits getting shifted in at one end and out at the other end are not insignificant functions; they are commonly used operations of importance to arithmetic operations.

An n-bit, serial-in shift register requires n clock pulses to load an n-bit word. A *parallel-load shift register*, in comparison, is able to load all data bits simultaneously in one clock pulse. Both serial-in and parallel-load shift registers are known for their specific applications in digital systems. A parallel-in, serial-out shift register using master-slave SR flip-flops is shown in Figure 9.19. The parallel data are loaded using a jam-entry scheme. When E is high, the data are loaded into the register in parallel. Then when E is low, the Q outputs of the FFs get shifted to the right. The combinational logic circuit that exists in between the FFs function as a gateway to either load in parallel or shift data bit to the right. In either case the HOLD control is held low. Parallel-in, serial-out shift registers allow accepting data n-bits at a time on n lines and then sending them one bit after another on one line. This very register mode is used often for data transfers in digital communications.

At the receiving end of digital systems linked by a single data line, it is often necessary to collect n-bits serially. Thereafter, these data bits are loaded on to the receiving system in parallel. Figure 9.20 shows the logic circuit of a serial-in, parallel-out register. The serial data are entered through the input S of the left-most FF while data are transferred out in parallel from the Q outputs. The register is organized in exactly the same way as that employed in Figure 9.16b; however, Q outputs are all set up for parallel unload, which may not be the case in many shift registers. Figure 9.21 shows four-bit, parallel-in, parallel-out

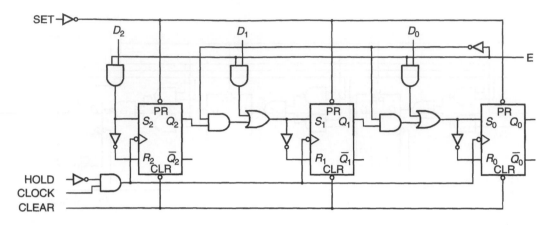

FIGURE 9.19
Three-bit, parallel-in, serial-out shift register.

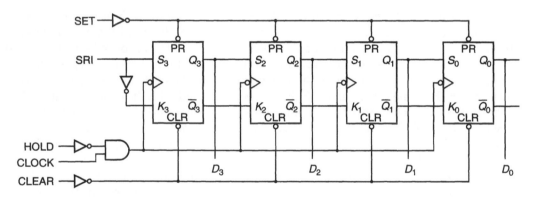

FIGURE 9.20
Four-bit, serial-in, parallel-out shift register.

register logic that allows for both parallel loading and unloading features. The CLR, SET, and HOLD are set low for normal operation. When LOAD is high, the inputs I_3, I_2, I_1, and I_0 get loaded in parallel into the register with clock. The outputs O_3, O_2, O_1, and O_0 are always available in parallel and can be accessed from the Q output of the FFs.

9.4   Shift Registers as Counters

Shift registers may often generate count sequences. Such shift registers are used in applications such as multiple address coding, parity bit generators, and random bit generators. The shift register outputs as well as their complements must be accessible for these applications. Figure 9.22 shows a circuit configuration where combinational feedback logic is driven by outputs from the register. The feedback logic in turn determines the next state of Q_n. In case of a bidirectional register, the feedback logic controls shift-left and shift-right signals that feeds either a 1 or 0 to the appropriate SLI and SRI input.

Figure 9.23 shows the state diagram of a four-bit shift register where J_3 is the input to the most significant FF. If the shift register is initially in the state $Q_3Q_2Q_1Q_0 = 1001$,

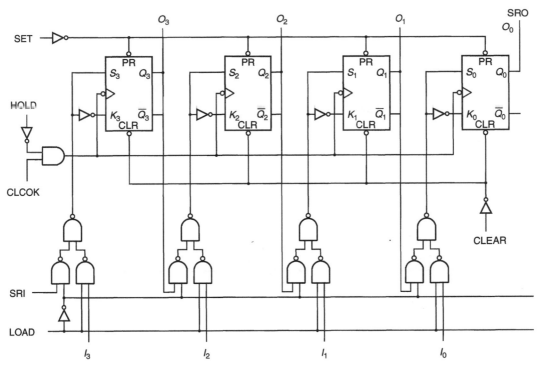

FIGURE 9.21
Four-bit, parallel-in, parallel-out shift register.

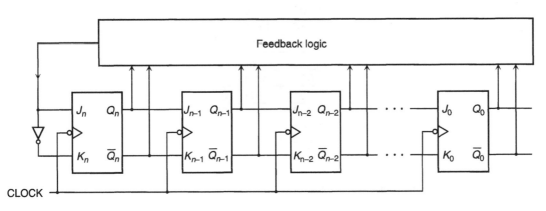

FIGURE 9.22
A feedback shift-right register configuration.

then there are two possible next states. These are 0100 when J_3 is a 0, or 1100 when J_3 is a 1. These two state values correspond to a shift-right operation. The state diagram accounts for all possible internal states of the register and all possible transitions between the states.

In order to generate an arbitrary sequence generator from the circuit of Figure 9.22, we need to assign the desired sequence of states on the universal state diagram. On the basis of the desired sequence the feedback logic is next determined so that the register will cycle through the selected sequence of states. This technique is illustrated in Example 9.4.

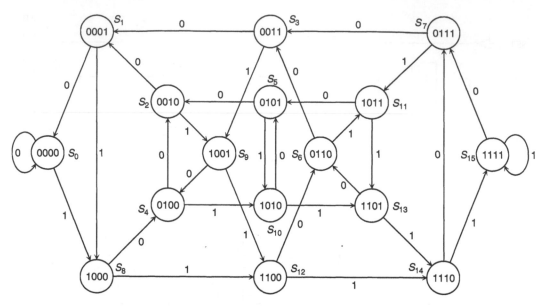

FIGURE 9.23
Universal state diagram for a four-bit feedback register.

PS	NS	J_3
0000	1000	1
1000	1100	1
1100	1110	1
1110	1111	1
1111	0111	0
0111	1011	1
1011	0101	0
0101	0010	0
0010	0001	0
0001	0000	0

FIGURE 9.24
State transition and feedback logic for Example 9.4.

Example 9.4
Design a sequence generator that contributes to the following sequence:

$$S_0 \to S_8 \to S_{12} \to S_{14} \to S_{15} \to S_7 \to S_{11} \to S_5 \to S_2 \to S_1 \to S_0$$

Solution
The state transitions and the corresponding feedback logic conditions are determined as shown in Figure 9.24.

We then follow up with an appropriate K-map as shown in Figure 9.25 for generating the feedback function J_3. The feedback function is given by,

$$J_3 = \overline{Q_1}\,\overline{Q_0} + Q_2\overline{Q_0} + \overline{Q_3}Q_2Q_1$$

The feedback logic is realized using combinational logic. It is fed to J_3 input of the shift-right register, as shown in Figure 9.26, to generate the desired sequence. Note that the states not included in the sequence may lead to the following sequences:

$$S_3 \rightarrow S_1; \quad S_4 \rightarrow S_{10} \rightarrow S_5; \quad S_9 \rightarrow S_4; \quad \text{and} \quad S_{13} \rightarrow S_6 \rightarrow S_{11}$$

If this sequence generator ever gets into an unused state due to a glitch or on power-up, it will return to the decade sequence within two clock pulses. This is evident from the state diagram of Figure 9.23.

J_3		Q_3Q_2			
		00	01	11	10
Q_1Q_0	00	1	-	1	1
	01	0	0	-	-
	11	-	1	0	0
	10	0	-	1	-

FIGURE 9.25
Excitation map for Example 9.4.

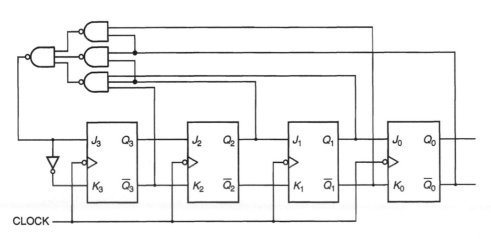

FIGURE 9.26
Logic diagram for Example 9.4.

9.5 Counter and Register Applications

The use of counters and registers in the design of digital computers is perhaps the most noteworthy. Digital computers process numbers often by repeating many different arithmetic and/or logic operations. Execution of a specific instruction usually involves moving the instruction and data between registers. The data are operated on typically by the arithmetic logic unit (ALU) as they are shuffled around between storage registers. These transfer sequences are typically controlled by sequential circuits designed for this purpose. In summary, registers provide means for storage of bits, and counters keep track of the next memory location and guarantee enough time intervals between the sequences so that complex operations may take place as desired. We will explore a few applications of counters and registers now before we consider synchronous arithmetic operations in Chapter 10.

One would prefer to accomplish operations involving data in digital computers to be executed in parallel since this lends to fast computation. Serial operations, on the other hand, slow down the process but require less complicated and, thus, less expensive sequential circuits. Consider the *add* function, for example. Parallel addition circuits already considered in Chapter 6 reveal that the techniques lend to logic circuits that were reasonably fast but they involved complex circuitry especially when number of bits is large. Each time we had to deal with an n-bit word and when n was large, the designer had to trade-off between time of execution and the number of components employed.

This same *add* operation as we shall show in this section can also be performed by loading the addend and augend into two serial shift registers and then shifting one bit at a time from each register into a single-bit FA, as shown in Figure 9.27. The carry-out of the FA is stored in a D FF which is fed back as the carry-in to the FA along with the next pair of significant bits exiting the two shift registers. The generated sum bit from the FA can be shifted then into the MSB of shift register that started off with the augend. This loading of the sum bits at the MSB should not cause any problem since with each clock one

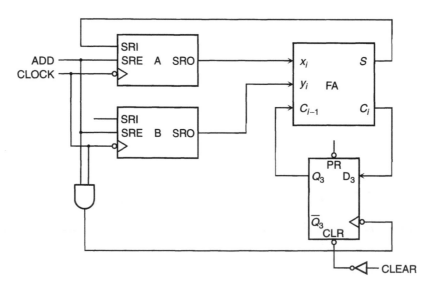

FIGURE 9.27
A sequential adder circuit.

augend bit has already been shifted out of the register leaving behind the most significant FF empty.

The shift registers A and B would initially take in the augend and addend data bits and the D FF cleared. The *add* operation can be initiated by shifting right with each clock one pair of addend and augend bits and the previously obtained carry-out into the FA circuit. In the first cycle, this carry-in is by default a 0. The sum bit so generated is serially shifted into register A. The ADD command can be used to either start or stop the operation. A high on ADD line will cause the register to perform a shift-right operation, and when ADD is low the registers will undergo a hold mode.

A control unit capable of generating a sequence of timing signals is used to control operations in digital systems. The control unit in this case will need to be able to generate a signal that remains high for the duration of a number of pulses which must at least be equal to the number of bits in the shift registers. The serial adder system also needs to be modified to incorporate the entry of an ADD control signal. Figure 9.28a shows the example of such a control circuit that generates a signal that remains high, for example, for a period of 16 clock periods. The four-bit binary counter and the SR FF are initially CLEARed. The BEGIN signal can be used to set the SR FF, which in turn can be used to get the counter to count. The FF output Q remains high for 16 pulses, as shown in the timing diagram of Figure 9.28b. When the counter has reached count state 1111, the HALT output gets activated, which in turn then resets the FF. The BEGIN signal is made to stay high for at least one clock period. It could be made to stay on for a longer period; however, if it is made to last for more than 15 clock periods, the circuit will not function as we had hoped. The HALT signal so generated

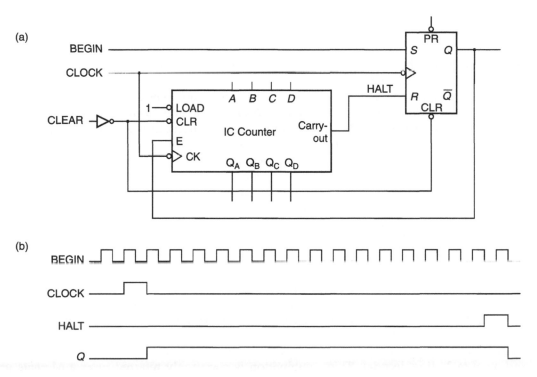

FIGURE 9.28
A timing sequencer (a) logic circuit and (b) its timing diagram.

(a)

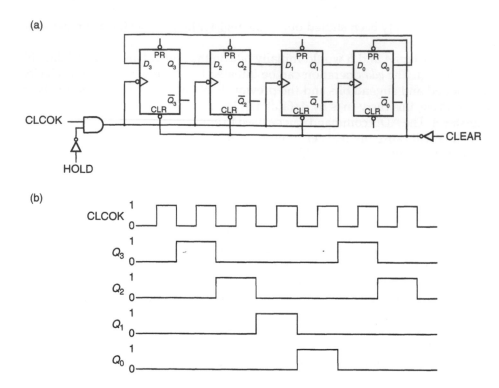

(b)

FIGURE 9.29

Example of a four-bit ring counter (a) logic circuit and (b) its timing diagram.

in Figure 9.28a can be used subsequently as an input to a follow-on logic circuit to initiate, for example, the BEGIN signal.

A single pulse can trigger the time when an operation should be executed. A shift register set up as a *ring counter*, as shown in Figure 9.29, is able to able to circulate the data bits. It has a feedback path between its serial output and serial input. Initially, CLEAR input is set which in turn presets the right-most FF and clears the remaining FFs. The starting output word of the register is then 0001, and with each successive clock pulses, the output word takes on values 1000, 0100, 0010, 0001, and repeat. n unique states are thus generated by an n-bit ring counter and as such it can be used suitably to orchestrate the timing of up to n sequential operations. By tying each of four operations' initiating lines to appropriate one bit of the ring counter, therefore, we can sequence up to n distinct operations each of which can be active for up to one-nth of the time.

A four-bit binary counter, for example, can have up to 16 distinct counting states, whereas the 4-bit ring counter just described will have only four distinct states. One may thus conclude that the ring counter makes a rather inefficient use of FFs for systems that would otherwise require a large number of timing signals. We need to realize though that in a ring counter only one FF is on during a clock and thus the ring counter output does not need to be decoded any further, thus, providing saving in additional cost. An excellent alternative to using a ring counter is to use the combination of an n-bit binary counter and an n-to-2^n line decoder. This combination is commonly referred to as a *Moebius* or *Johnson counter*. It includes a shift-right register that is connected typically in a *switch–tail* configuration.

(a)

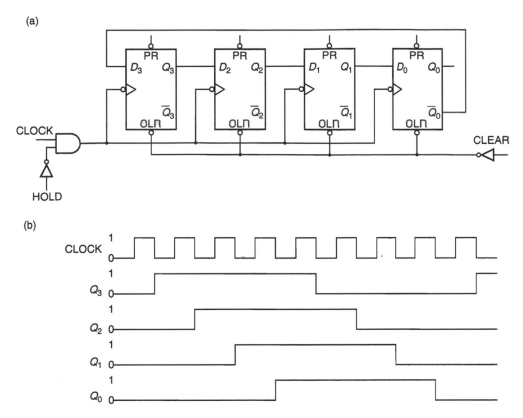

(b)

FIGURE 9.30
A four-bit Johnson counter (a) logic circuit and (b) its timing diagram.

An n-bit ring counter provides n distinguishable states. The number of states can be doubled if the shift register is connected instead in a switch-tail configuration as shown for example in Figure 9.30 in the case of a 4-bit Johnson counter. $\overline{Q_0}$ rather than Q_0 is fed back as the D_3 input. With each clock, the register shifts one bit to the right, when at the same time $\overline{Q_0}$ of the right-most FF is fed as an input to the left-most FF. The resulting counter circuit is able to generate eight counting states: 0000, 1000, 1100, 1110, 1111, 0111, 0011, and 0001 rather than just four as in the case of a standard ring counter. These eight states, however, will need to be decoded to generate eight distinct timing signals.

In many instances, sequencing of arithmetic and/or logic operations is executed only if certain condition or conditions are met. A four-state controller, for example, that can control four distinct operations is shown in Figure 9.31. This sequencer consists of a two-bit counter, a 1-of-4 MUX, and a 2-to-4 line decoder. Up to four arithmetic and/or logic operations can be executed by having the respective sequencer output to become a 0. The sequencer output D_n is a 0 only if C_n (representing the corresponding condition or conditions) is a 1. Initially, the counter is CLEARed. It STARTs at address 00 when the function F_0 is executed provided the condition C_0 has been met. While the function corresponds to a specific set of operations, the condition C_0 may be the result of one or several test results. Then when the counter reaches address 01 the function F_1 is executed, provided that the condition C_1 is met this time. The counter count allows the operations to be executed once and only at regular time intervals. Once the sequencer gets STARTed, the test conditions determine whether or not

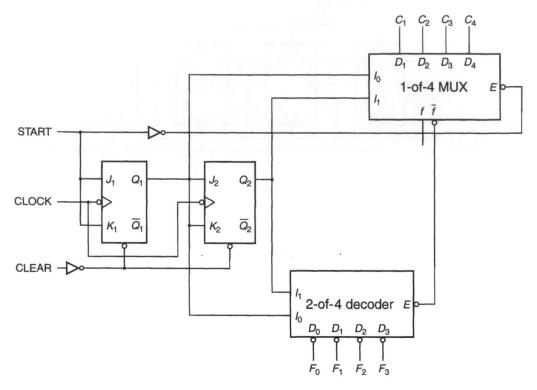

FIGURE 9.31
Two-bit operation sequencer.

the corresponding function get accessed by the decoder. If the set of conditions is not met completely, the decoder remains disabled, and as a result the corresponding function is not executed. A sequencer circuit such as that of Figure 9.31 is vitally important in executing and controlling operations in a complex digital system.

9.6 Register Transfer Language

Now that we have already introduced a number of different registers, applications point to the need for introducing a concise language for describing the flow (and processing) of information bits between registers. One of the more commonly used such language schemes is referred to as *register transfer language*, RTL. This was introduced first by I. S. Reed. A register transfer description is descriptive of two units—a data unit and a control unit. The *data unit* consists typically of registers, data paths, and logic circuit necessary to implement either one or a set of register transfers. The *control unit*, on the other hand, generates signals at appropriate time intervals to be able to effectively regulate the register transfers within the data unit. The RTL descriptor has the ability to describe the hardware functions in functional form in both data and control units.

A typical RTL operation given by $P \leftarrow Q$ indicates that the data originally stored in register P is replaced by the data contained in register Q. This RTL statement also connotes that these two registers have identical number of bits and that this operation can be completed during a single clock period and, thus, it needs to correspond to a single-state transition of a sequential machine. One clock period, referred to as the *cycle time*, may be taken as the basic unit of time. For the sake of consistency, it is customary to adopt the following guidelines for RTL:

1. Data in a register is denoted by one or more letters with the first always appearing in uppercase.

2. Data in a concatenated register (result of having joined two or more registers in a string), is represented by the respective register symbols separated by commas. It is assumed that the LSB of the first-mentioned register is one bit to the left of the MSB of the second one, and so on.

3. Data transfer between registers is considered always parallel—all of the bits will transfer at the same instant of time.

4. Data bits contained in a register are labeled right to left. A_0 and A_{n-1} respectively represent the LSB and the MSB of an n-bit register A. Note also that $A_{1,4}$ represents the bits 1 and 4 of register A, A_{1-4} represents bits 1 through 4 of register A, and A_M represents the subset of bits, M, of register A.

Table 9.1 lists examples of a variety of RTL statements representative of arithmetic, bit-by-bit logic, shift, rotate, scale, and conditional operations. To distinguish between arithmetic

TABLE 9.1

Example of RTL Operations

Type	RTL	Meaning	What Happens to Registers? ($A = 10110$, $B = 11000$, $C = 00011$)
General	$A_0 \leftarrow A_4$	Bit 4 of A to bit 0 of A	$A = 10111$
	$A_3 \leftarrow B_3$	Bit 3 of B to bit 3 of A	$A = 11110$
	$A_{1-3} \leftarrow B_{1-3}$	Bits 1-3 of B to bits 1-3 of A	$A = 11000$
	$A_{1-3} \leftarrow B_{1,3}$	Bits 1 and 3 of B to bits 1 and 3 of A	$A = 11100$
Arithmetic	$A \leftarrow 0$	Clear A	$A = 00000$
	$A \leftarrow B + C$	Sum of B and C to A	$A = 11011$
	$A \leftarrow B - C$	Difference $B - C$ to A	$A = 10111$
	$C \leftarrow C + 1$	Increment C by 1	$C = 00100$
Logic	$A \leftarrow B \wedge C$	Bit-by bit AND result of B and C to A	$A = 00000$
	$A \leftarrow B \vee C_1$	OR operation result of B with bit 1 of C to A	$A = 11111$
	$C \leftarrow \overline{C}$	Complement C	$C = 11100$
	$B \leftarrow \overline{B} + 1$	2's complement of B	$B = 01000$
	$B \leftarrow A \oplus C$	XOR of A and C to B	$B = 10101$
Serial	$B \leftarrow sr\, B$	Shift right B one bit	$B = 01100$
	$B \leftarrow sl\, B$	Shift left B one bit	$B = 10000$
	$B \leftarrow sr2\, B$	Shift right B two bits	$B = 00110$
	$B \leftarrow rr\, B$	Rotate right B one bit	$B = 01100$
	$B \leftarrow scl\, B$	Scale B one bit (shift left with sign bit unchanged)	$B = 10000$
	$B, C \leftarrow sr2\, B, C$	Shift right concatenated B and C two bits	$B, C = 0011000000$
Conditional	IF ($B_4 = 1$) $C \leftarrow 0$	If bit 4 of B is a 1, then C is cleared	$C = 00000$
	IF ($B \geq C$) $B \leftarrow 0, C_2 \leftarrow 1$	If B is greater than or equal to C, then B is cleared and C_2 is set to 1	$B = 00000$ $C = 00111$

and logic operations, addition is represented by a $+$ sign, and a logical OR and logical AND are represented respectively by $\vee$ and $\wedge$ signs. The shift, rotate, and scale operations are denoted using two lowercase letters. For shift and rotate, the first letter indicates the type of operation (r for rotate and s for shift) and the second letter indicates the particular direction (r for right and l for left). The LSB and MSB are considered adjacent in the case of rotate operations. For all shift operations, it is customary to consider that a 0 is moving in to occupy the vacated bit. scl and scr respectively represent scale-left and scale-right operations.

Examples of RTL operations such as those listed in Table 9.1 may be used and combined to write more complex functions or sequence of operations. The control conditions are included along with the operation to distinguish one set of executions from another. The execution of an operation and the transfer of data are usually regulated by one or several control conditions. For example, replacing the contents of B with that of A is expressed as follows:

$$\bar{x} \cdot \bar{y} \cdot \bar{z} : B \leftarrow A;$$

The control condition $\bar{x}\,\bar{y}\,\bar{z}$ is separated from the corresponding operation by a ":" sign, while the sign ";" is used to indicate the end of a RTL operation. This above RTL statement indicates that when $\bar{x}\,\bar{y}\,\bar{z} = 1$, the content of register A should be transferred to B. This would imply though that A still continues to hold on to its contents. When more than one operation, for example, $A \leftarrow B$ and $B \leftarrow$ sr B, are to be realized under the same control condition $P = 1$, the operation is expressed simply as:

$$P : A \leftarrow B; \quad B \leftarrow srB;$$

The operations that are performed simultaneously are executed usually by the same control condition. When P is a 1, B is transferred to A and it is also restored after being shifted right one bit.

The RTL operations describing a 2's complement logic, for example, may be described as follows:

$$Q : C \leftarrow 0; \quad S : C \leftarrow I; \quad R : C \leftarrow \bar{C} + 1;$$

where C is the register that begins by loading in the input bits from register I. These three RTL statements together describe the actions of a to be designed hardware. In Chapter 10, we shall be exploring several complex sequential circuits and how their designs are related to and derived from the corresponding RTL statements. Such one-to-one correspondence makes us appreciate the simplification that results from using RTL.

Consider the circuit shown in Figure 9.32. Here we have two registers, A and B, with four D FFs each. The FFs within the registers are not internally cascaded together, for simplicity. The FF outputs of register A are connected directly to the respective D inputs of register B. With each trailing edge of X input, the contents of register A are loaded into register B. This particular hardware operation is aptly described by the RTL statement:

$$X : B \leftarrow A;$$

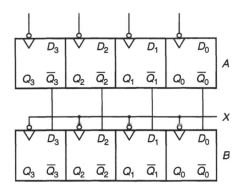

FIGURE 9.32
Copying operation.

This could also be written as follows:

$$X: B_3 \leftarrow A_3, \quad B_2 \leftarrow A_2, \quad B_1 \leftarrow A_1, \quad B_0 \leftarrow A_0;$$

These two operations are equivalent; one would rather use the first form since it is more concise.

Note that a transfer operation is in effect a copying operation wherein the contents of a source register remain unchanged. This type of RTL operation is the most common, but there are many other possibilities. Consider the four situations of Figure 9.33. In each case, the registers are of four-bit length.

In Figure 9.33a, the $\overline{Q}$ outputs of register A are connected to the respective D inputs of register B. A clock input P would cause the following transfer:

$$P: B \leftarrow \overline{A};$$

Figure 9.33b shows a register A where its MSB is tied to a 0. Each of the Q_n outputs of register A, except for the LSB, is fed to the D_{n-1} input of the same register. A clock pulse introduced at R would cause the following data transfer:

$$R: A \leftarrow sr\, A;$$

Logic circuits shown in Figure 9.33c and d respectively correspond to the following register transfers:

$$S: A \leftarrow rr\, A;$$
$$T: C \leftarrow \overline{A} \wedge B;$$

Note that in all of these RTL statements, the source and destination registers are assumed to have the same number of FFs.

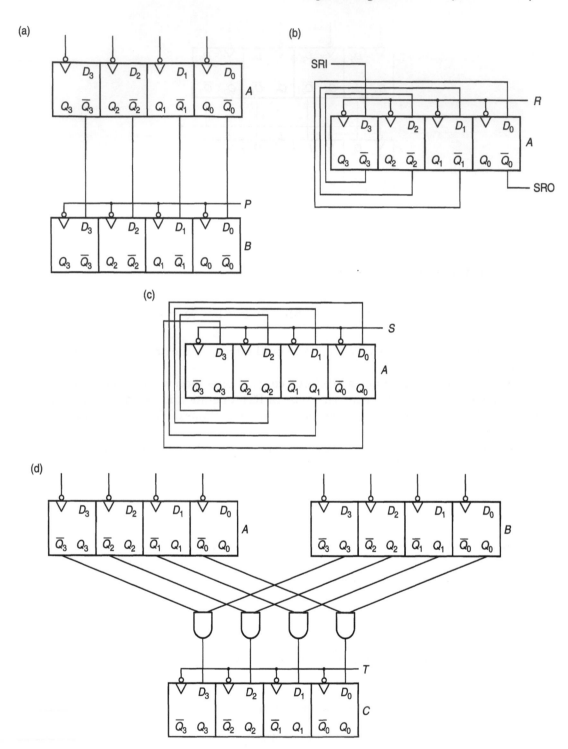

FIGURE 9.33
Logic circuits capable of realizing the following operations: (a) complement, (b) shift-right, (c) rotate-right, and (d) logical transfer.

9.7   Registers and Counters in VHDL

In this section we illustrate the design of registers and counters with VHDL descriptions. Figure 9.34 shows the schematic of a 4-bit parallel-load register, with an asynchronous reset. The load-enabled local clock signal after the AND gate is named as clk_g (for gated clock) to correspond with its use in the structural VHDL description shown in Figure 9.35. The structural VHDL clearly describes the circuit topology. A functional VHDL description for this 4-bit register is simpler, with less number of code lines and the use of *if-then-else* conditional statement, as shown in Figure 9.36. A positive edge-triggered version is shown in Figure 9.37.

Registers are one of the basic components of shift registers and counters. A shift register can be configured to perform either left- or right-shift operations, as was discussed in this chapter in Section 9.4. Figure 9.38 shows a functional VHDL description of a 4-bit left-shift register, with line 17 describing the 1-bit left-shift operation. Similarly, a register can be configured to perform counting operations. Figure 9.39 shows a schematic diagram of a 4-bit binary counter, with asynchronous reset. The structural VHDL description for this

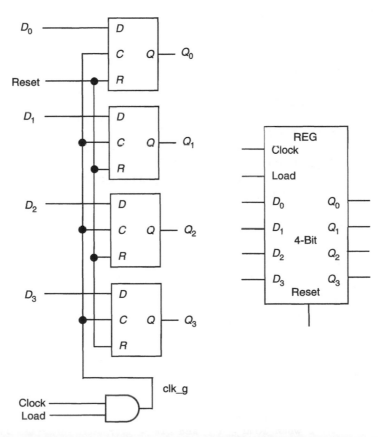

FIGURE 9.34
Schematic and block diagrams of a 4-bit parallel-load register with asynchronous reset.

```
-- A structural VHDL description of a 4-bit register      -- 01
-- with asynchronous reset and synchronous parallel load  -- 02
library ieee, lcdf_vhdl;                                  -- 03
use ieee.std_logic_1164.all, lcdf_vhdl.func_prims.all;    -- 04
entity REG_4bit is                                        -- 05
   port (Din: in std_logic_vector(0 to 3);                -- 06
         Set, Res, clk, load: in std_logic;               -- 07
         Q, Qb: out std_logic_vector (0 to 3));           -- 08
end REG_4bit;                                             -- 09
                                                          -- 10
architecture structural_REG_4bit of REG_4bit is          -- 11
                                                          -- 12
   component AND 2                                        -- 13
      port (in1, in2: in std_logic; out1:out std_logic);  -- 14
   end component;                                         -- 15
                                                          -- 16
   component D_FF                                         -- 17
      port (in1, clk, Set, Res: in std_logic;             -- 18
            Q, Qb: out std_logic);                        -- 19
   end component;                                         -- 20
                                                          -- 21
   signal clk_g: std_logic;                               -- 22
                                                          -- 23
   begin                                                  -- 24
      G1: AND 2 port map (load, clk, clk_g);              -- 25
      BIT0: D_FF port map (Din(0), clk_g,                 -- 26
                           Set, Res, Q(0), Qb(0));        -- 27
      BIT1: D_FF port map (Din(1), clk_g,                 -- 28
                           Set,Res, Q(1), Qb(1));         -- 29
      BIT2: D_FF port map (Din(2), clk_g,                 -- 30
                           Set, Res, Q(2), Qb(2));        -- 31
      BIT3: D_FF port map (Din(3), clk_g,                 -- 32
                           Set, Res, Q(3), Qb(3));        -- 33
   end structural_REG_4bit;                               -- 34
```

FIGURE 9.35
A structural VHDL description of a 4-bit parallel-load register.

```
-- A functional VHDL description of a 4-bit               -- 01
-- parallel-load register with asynchronous RESET         -- 02
library ieee;                                             -- 03
use ieee. std_logic_1164.all;                             -- 04
entity REG_4 bit is                                       -- 05
   port (D: in std_logic_vector (0 to 3);                 -- 06
         Res, load, clk: in std_logic;                    -- 07
         Q: out std_logic_vector (0 to 3));               -- 08
end REG_4bit;                                             -- 09
                                                          -- 10
architecture functional_REG_4 bit of REG_4bit is         -- 11
                                                          -- 12
   begin                                                  -- 13
      Q <= "0000" when Res = '1' else                     -- 14
           D when load ='1'  and clk = '1';               -- 17
   end functional_REG_4bit;                               -- 22
-- End of the VDHL description                            -- 23
```

FIGURE 9.36
A functional VHDL description of a 4-bit parallel-load register.

```
-- A behavioral VHDL description of a 4-bitedge-trigged    -- 01
-- parallel-load register with asynchronous RESET          -- 02
library  ieee;                                             -- 03
use ieee.std_logic_1164.all;                               -- 04
entity  4bit_REG is                                        -- 05
   port (D: in std_logic_vector (0 to 3);                  -- 06
         Res, load, clk: in std_logic;                     -- 07
         Q: out std_logic_vector (0 to 3));                -- 08
end 4 bit_REG;                                             -- 09
                                                           -- 10
architecture   functional_REG_4 bit of REG_4bit is        -- 11
                                                           -- 12
 begin                                                     -- 13
 process (Res, load, clk)                                  -- 14
 begin                                                     -- 15
     if (Res = '1') then                                   -- 16
         Q <= "0000";                                      -- 17
     else if (load = '1' and clk' event and clk = '1') then -- 18
         Q <= D;                                           -- 19
     end if;                                               -- 20
   end process;                                            -- 21
end behavioral_4bit_REG;                                   -- 22
-- End of the VHDL description                             -- 23
```

FIGURE 9.37
A behavioral VHDL description of a positive edge-trigged 4-bit parallel-load register.

```
   -- A functional VHDL description of a 4-bit           -- 01
   -- left-shift register with asynchronous RESET        -- 02
   library ieee;                                         -- 03
   use ieee.std_logic_1164.all;                          -- 04
   entity LS_REG_4bit  is                                -- 05
      port (Sin, Res, clk: in std_logic;                 -- 06
            Q: out std_logic_vector (0 to 3);            -- 07
            Sout: out std_logic);                        -- 08
   end LS_REG_4bit;                                       -- 09
                                                         -- 10
   architecture functional_LSR_4bit of LS_REG_4bit is    -- 11
                                                         -- 12
      signal hold: std_logic_vector (0 to 3);           -- 13
                                                         -- 14
    begin                                                -- 15
      hold <= "0000" when Res = '1' else                 -- 16
            Sin & hold (0 to 2) when clk= '1';           -- 17
      Sout <= hold(3);                                   -- 18
      Q <= hold;                                         -- 19
   end functional_LSR_4bit;                              -- 20
   -- End of the VHDL description                         -- 21
```

FIGURE 9.38
A functional VHDL description of a 4-bit left-shift register.

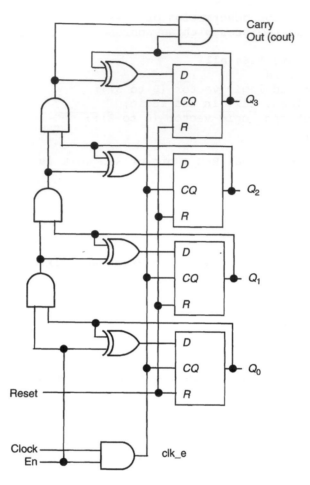

FIGURE 9.39
A 4-bit binary counter with asynchronous reset.

4-bit binary counter is shown in Figure 9.40, while a functional VHDL description is shown in Figure 9.41, in which the inputs of the registers are described in exclusive-OR function (XOR) and AND functions in lines 16 and 17. Other types of counters can be described in VHDL similarly.

9.8 Summary

In this chapter, we first introduced concepts that will be useful in designing synchronous counters; and serial, parallel, and serial-parallel registers. These concepts were used then to introduce RTL. This tool is developed enough in this chapter because of its potential to be used in designing data and control units for more complex digital systems. All these will be used effectively in Chapter 10 to design a particular class of complex systems, namely, systems that can get us to manipulate arithmetic operations.

```
-- A structural VHDL description of a 4-bit binary           -- 01
-- counter with asynchronous reset                           -- 02
library ieee, lcdf_vhdl;                                     -- 03
use ieee.std_logic_1164.all, lcdf_vhdl.func_prims.all;      -- 04
entity 4_bit_BC is                                          -- 05
    port (En, Res, clk: in std_logic;                       -- 06
          Q: out std_logic_vector (0 to 3);                 -- 07
          Cout: out std_logic);                             -- 08
end 4_bit_BC;                                               -- 09
                                                            -- 10
architecture structural_4bit_BC of 4_bit_BC is             -- 11
                                                            -- 12
   component AND 2                                          -- 13
       port (din1, din2: in std_logic; dout: out std_logic); -- 14
     end component;                                         -- 15
                                                            -- 16
     component XOR2                                         -- 17
       port (din1, din2: in std_logic; dout: out std_logic); -- 18
     end component;                                         -- 19
                                                            -- 20
     component D_FF                                         -- 21
       port (D, Res, clk: in std_logic; Q out std_logic);  -- 22
     end component;                                         -- 23
                                                            -- 24
     signal clk_e: std_logic;                               -- 25
     signal din: std_logic_vector (0 to 3);                 -- 26
     signal ct: std_logic_vector (0 to 2);                  -- 27
                                                            -- 28
     begin                                                  -- 29
        G1: AND2 port map (En, clk, clk_e);                 -- 30
        BIT0 : D_FF port map (din(0), Res, clk_e, Q(0));    -- 31
        BIT1 : D_FF port map (din(1), Res, clk_e, Q(1));    -- 32
        BIT2 : D_FF port map (din(2), Res, clk_e, Q(2));    -- 33
        BIT3 : D_FF port map (din(3), Res, clk_e, Q(3));    -- 34
        G2: AND2 port map (En, Q(0), ct(0));                -- 35
        G3: AND2 port map (ct(0), Q(1), ct(1));             -- 36
        G4: AND2 port map (ct(1), Q(2), ct(2));             -- 37
        G5: AND2 port map (ct(2), Q(3), Cout);              -- 38
     end structural 4bit; CB                                -- 39
```

FIGURE 9.40
A structural VHDL description of a 4-bit binary counter with asynchronous reset.

```
-- A functional VHDL description of a 4-bit                 -- 01
-- binary counter with asynchronous RESET                   -- 02
library ieee;                                               -- 03
use ieee.std_logic_1164. all;                               -- 04
entity BC_4bit is                                           -- 05
    port (En, Res, clk: in std_logic;                       -- 06
          Q: out std_logic_vector(0to 3);                   -- 07
          Cout: out std_logic);                             -- 08
end BC_4bit;                                                -- 09
                                                            -- 10
architecture  functional_BC_4bit of BC_4bit is             -- 11
                                                            -- 12
     signal dffin, hold, ct: std_logic_vector (0 to 3);     -- 13
                                                            -- 14
   begin                                                    -- 15
      ct <= (En & ct (0 to 2)) and hold;                    -- 16
      dffin <= (En & ct (0 to 2)) xor hold;                 -- 17
      hold <= "0000" when Res ='1' else                     -- 18
             dffin when (clk = '1' and En = '1');           -- 19
      Q <= hold;                                            -- 20
      Cout <= ct(3);                                        -- 21
   end functional_BC_4bit;                                  -- 22
   -- End of the VHDL description                           -- 23
```

FIGURE 9.41
A functional VHDL description of a 4-bit binary counter with asynchronous reset.

Bibliography

Comer, D.J., *Digital Logic and State Machine Design*. 3rd edn. New York, NY. Oxford University Press, 1994.

Floyd, T., *Digital Fundamentals*. 8th edn. Englewood Cliffs, NJ. Prentice-Hall, 2003.

Hays, J.P., *Digital System Design and Microprocessors*. New York, NY. McGraw-Hill, 1984.

Hill, F.J. and Peterson, G.R., *Digital Systems: Hardware Organization and Design*. 2nd edn. New York, NY. Wiley, 1978.

Hill, F.J. and Peterson, G.R., *Digital Logic and Microprocessors*. New York, NY. Wiley, 1984.

Johnson, E.J. and Karim, M.A., *Digital Design: A Pragmatic Approach*. Boston, MA. PWS-Kent Publishing, 1987.

Karim, M.A. and Awwal, A.A.S., *Optical Computing: An Introduction*. New York, NY. John Wiley & Sons, 1992.

Katz, R.H., *Contemporary Logic Design*. Boston, MA. Addison Wesley, 1993.

Kline, R.M., *Structured Digital Design Including MSI/LSI Components and Microprocessors*. Englewood Cliffs, NJ. Prentice-Hall, 1983.

Mowle, F.J.A., *A Systematic Approach to Digital Logic Design*. Reading, MA. Addison-Wesley, 1976.

Nagle, H.T., Jr., Carroll, B.D., and Irwin, J.D., *An Introduction to Computer Logic*, Englewood Cliffs, NJ. Prentice-Hall, 1975.

Nelson, V.P.P., Nagle, H.T., and Carroll, B.D., *Digital Logic Circuit Analysis and Design*. Englewood Cliffs, NJ. Prentice Hall. 1995.

Rhyne, V.T., *Fundamentals of Digital Systems Design*. Englewood Cliffs, NJ. Prentice Hall, 1973.

Winkel, D. and Prosser, F., *The Art of Digital Design—An Introduction to Top-Down Design*. Englewood Cliffs, NJ. Prentice-Hall, 1980.

Problems

1. Determine the state diagram of a 3-bit counter that uses D FFs and whose excitation equations are given by $D_1 = Q_3Q_1 + Q_2Q_1$; $D_2 = \overline{Q_2}(Q_1 + Q_3)$ and $D_3 = \overline{Q_3}$.

2. Design a 3-bit, Gray code counter using (a) D FFs and (b) JK FFs.

3. Design a modulo-12 counter using a modulo-3 and a modulo-4 counter. Discuss its functions and characteristics.

4. Using only D FFs, and 1-of-4 MUXs, design a 3-bit register that can hold the present data, shift right, shift left, and load new data in parallel.

5. Design a four-bit circulate-right-shift register using (a) D FFs and (b)JK FFs.

6. Verify the Boolean equations of the four-bit down counter introduced in Section 9.2.

7. Design a four-bit left-shift register using JK FFs and a minimum number of assorted logic gates.

8. Design a divide-by-2048 counter using four-bit binary counters. Each four-bit counter consists of a single FF, Q_A, followed by three cascaded FFs that form a divide-by-8 counter. It has two inputs, A and B, and four standard FF outputs, Q_A through Q_D, The two reset inputs, R_1 and R_2, clear the FFs when both are high. The counter counts in sequence when at least one of the reset inputs is low.

9. Obtain a logic circuit using as few FFs as possible to sequence 16 different operations.

10. Design a parallel Gray code-to-parallel and serial binary converter using a four-bit shift register and one FF. Explain the detailed functioning of the circuit. Note: Except for MSB, a four-bit, Gray code representing 8 through 15 are mirror images of four-bit binary equivalent representing the numbers 0 through 7 and vice versa.

11. Draw the logic diagram of a four-bit register with clocked JK FFs having control inputs for the increment, complement, and parallel transfer micro-operations. Show how the 2's complement operation can be realized using this register.

12. Design a two-bit counter that counts up when control variable C is a 1 and counts down when C is a 0. No counting occurs when the control variable D is a 0.

13. Using a four-bit binary counter with synchronous clear and asynchronous load and clock action on the leading edge, complete the necessary circuit to make a two-digit BCD counter. Make it as hardware-efficient as possible.

14. Given a four-bit adder and a large assortment of gates, counters, multiplexers, and decoders, design a logic circuit that is capable of multiplying a two-bit (plus sign) sign-magnitude quantity by 3. *Note:* You don't have to use an adder.

15. Using four-bit adder and any additional logic, design a logic circuit that can multiply a number $X_2 X_1 X_0$ (assume signed magnitude with the sign bit handled elsewhere) by 2.5. The result is to be rounded to the next highest integer if the product results in a fractional part.

16. Obtain a logic circuit using decoders, counters, MUXs, FFs, and so on that has as inputs a clock and four select lines D. The select lines determine the number by which the input clock frequency is divided. The output of this logic circuit should be a clock divided by the select line value. If F is the frequency of the input clock and D is the binary value of the select lines, the output frequency, f, is $F/(D+1)$.

17. Design a two-bit counter that counts up one count at a time when $C = 0$ and counts up two counts per clock when $C = 1$.

18. Use four-bit binary counters in parallel and assorted logic gates to realize a divide-by-39 counter.

19. Obtain the universal state diagram for a four-bit, shift-left feedback register.

20. Obtain the universal state diagram for a four-bit bidirectional feedback register.

10

Synchronous Arithmetic

10.1 Introduction

A number of important combinatorial as well as sequential subsystems design such as those of multibit adders, registers, counters, decoders, multiplexers, and comparators have already been explored. We have also introduced register transfer language (RTL) and shown how it can be used to assemble various combinatorial and sequential modules to form more complex digital systems.

One of our goals is to comprehend sufficiently how complex digital systems work in order to be able to design and assemble functional components. Digital computers, in particular, often involve multiple transfers of multibit data between registers and at appropriate times subject these data sets to multibit arithmetic and or logic operations. The register transfers are often interlinked with the control unit that coordinates data transfers as well as executes arithmetic and/or logic operations in a specific order. The RTL algorithm that describes the sequence of operations provides enough clues for the design of data as well as the corresponding control unit. This chapter introduces the RTL algorithms necessary for implementing primarily sequential arithmetic operations.

10.2 Serial Adder/Subtractor

We introduced earlier the concept of binary serial addition and in connection to that operation had discussed a sequential logic circuit as shown in Figure 9.27. In serial addition, we start with the LSB; one bit of addend and one bit of augend are shifted right into a single-bit FA. The resultant sum bit is stored then at the MSB of either the addend or augend register. Coincident with the clock, the carry-out of the current add operation is fed back as carry-in to the same FA; the other two FA inputs will include the next pair of shifted addend and augend bits. This shift and add operations cycle is repeated until all bits of the addend and augend have been processed. Realizing that this machine can be modified to also perform subtraction operation, it may be prudent to incorporate in it a 2's complement capability as well so that if and when needed it can be used to function also as a serial subtracter.

It is obvious that the serial arithmetic unit just described must be supplied with several control inputs so as to be able to control and track its various operations. For simplicity of discussion, we may limit the design to only four bits for now. As obvious from the logic circuit of Figure 9.27, the two data inputs to the system are either a four-bit addend and a four-bit augend or a four-bit minuend and a four-bit subtrahend. At least one control fed to this unit should indicate whether the intended operation is addition or subtraction. Also,

423

a system clock pulse must be supplied so that sequencing of the different operations may be synchronized and adequately controlled. Examples of such control sequencing were considered and introduced earlier in the logic circuits of Figures 9.28 and 9.31.

A preferred way to designing this serial arithmetic unit may involve splitting this unit into several subunits and then design each of these subunits separately (Figure 10.1) The list of subunits will include: four-bit registers for storing the four-bit data, a single-bit FA for addition, and a flip-flop for storing the carry-out. Recall from Chapter 6 that an X-OR gate that can function as a controllable NOT can be used to complement the subtrahend. Additionally, a combinational unit must be designed and incorporated for indicating an overflow condition. Finally for the total unit to work correctly and coherently as a whole, a control circuit needs to be designed. The serial adder/subtracter unit (Figure 10.2) thus, will consist of the following subunits:

> An augend/minuend register,
>
> An addend/subtrahend register,
>
> A single-bit FA,
>
> A carry/borrow FF,
>
> A two-input X-OR gate,
>
> An overflow logic, and
>
> A control subunit.

Figure 10.1 shows a block diagram for a serial arithmetic unit.

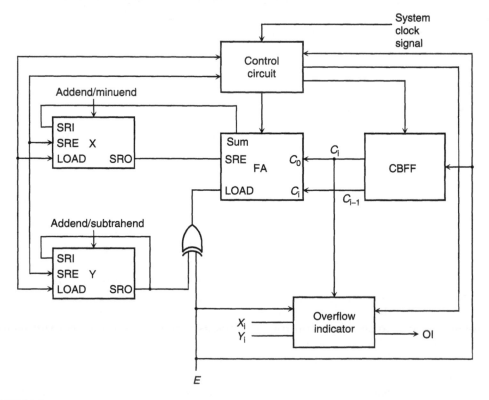

FIGURE 10.1
Block diagram of a serial adder/subtracter.

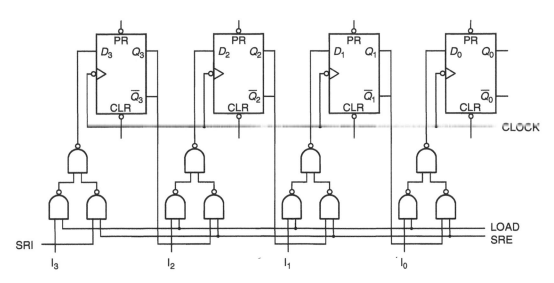

FIGURE 10.2
Shift registers capable of loading and shifting augend/minuend data bits.

Subtraction is typically realized using 2's complement arithmetic. This technique involves taking the 1's complement of each bit of register Y as it enters the FA and setting the initial content of the carry/borrow storage FF, CBFF, to a 1. This is tantamount to adding a 1 to 1's complement of the subtrahend. The control variable, E, when it is a 0 selects an addition operation and subtraction operation when it is a 1. The two values to be operated on need to be loaded in parallel into the two registers. With $E = 0$ the initial value of the CBFF gets reset to 0. The resulting operation is thus equivalent to having a simple addition operation.

The RTL fundamentals were introduced in Chapters 5 and 9. The specifications of complex digital units often take the form of one or more RTL algorithms that are implemented in its hardware. Most conveniently, algorithms are expressed in terms of a set of related RTL statements. The RTL algorithm that describes an n-bit adder/subtracter unit as well as its related operations is given by

Algorithm 1a

$S = 1: X \leftarrow$ (augend/minuend); $Y \leftarrow$ (addend/subtrahend); $\text{CBFF}_0 \leftarrow E$; $\text{OI} \leftarrow 0$; $N \leftarrow 0$;
$S = 2: X_{n-1} \leftarrow X_0 + (Y_0 \oplus E) + \text{CBFF}_0$; $Y \leftarrow \text{rr } Y$; $\text{CBFF}_0 \leftarrow$ (carry $-$ out); $X_{0\_(n-2)} \leftarrow X_{1\_(n-1)}$; $N \leftarrow N + 1$;
$S = 3: \text{IF}(N < n - 1) S \leftarrow 2 \text{ ELSE IF}(N = n - 1) S \leftarrow 4 \text{ ELSE } S \leftarrow 5$;
$S = 4: \text{IF}(X_0[Y_0 \oplus E]\overline{\text{CBFF}_0} + \overline{X_0}[Y_0 \oplus E]' \text{CBFF}_0 = 1) \text{OI} \leftarrow 1; S \leftarrow 2$;
$S = 5: \text{STOP}$;

Note that this sequential adder goes on to generate a sum even after the circuit has detected an overflow. The resultant sum obtained in case of an overflow will thus be incorrect. Appropriate control sequences that will be needed for correct operation of the circuit can be deduced from the RTL Algorithm 1a.

The addition algorithm gets started at $S = 1$. The data registers are first loaded with n-bit words and the CBFF is loaded with E. When $E = 0$ the initial carry-in is a 0, that is, CBFF ← 0. When $E = 1$, a 1 is stored at the CBFF. The addition of this 1 to the LSB completes the 2's complement operation requirement that a 1 be added to the 1's complement result that is generated when the subtrahend is serially processed using an X-OR logic. The variable N is meant to keep a count of the number of add operations performed.

A comparator can be used to determine if the current count is less than, equal to, or greater than $n - 1$. The overflow indicator, OI, is turned on when there is an overflow. An overflow condition cannot be determined until there is only one more addition step left. The RTL expression corresponding to step $S = 4$ determines whether or not there will be an overflow. There is always the possibility of an overflow if the word length of the result is constrained to be the same as that of the operands. In order for the sum to be valid the carry into and out of the sign bit must be identical. The sign bits of the two values are available respectively at X_0 and Y_0 at $S = 4$. The four cases that cause an overflow to occur include

$$X_0 = Y_0 = 1, \ E = 0, \ \text{and CBFF}_0 = 0$$
$$X_0 = Y_0 = 0, \ E = 0, \ \text{and CBFF}_0 = 1$$
$$X_0 = 1, \ Y_0 = 0, \ E = 1, \ \text{and CBFF}_0 = 0$$
$$X_0 = 0; \ Y_0 = 1, \ E = 1, \ \text{and CBFF}_0 = 1$$

When $S = 4$ has been reached, the carry-in into the sign bit will show up at CBFF_0.

Registers used for holding and shifting augend/minuend and addend/subtrahend have nearly identical characteristics. Both registers should be able to shift right and load data in parallel. The register X stores the sum bit at its MSB location while register Y should be able to simply rotate right. A four-bit register X that uses D FFs is shown in Figure 10.2 where the augend/minuend bits are loaded via input I_3 through I_0. The register Y is similar except for the external wiring. For Y to have rotate-right capability, Q_0 is connected to the shift-right input, SRI. A high LOAD will cause the data to be loaded in parallel while a 1 at the shift-right enable, SRE, line will causes the register to shift its bits to the right.

The single-bit full adder logic used in this serial adder/subtracter is no different than the one we considered in Section 6.5. There is no need to consider a CLA adder since we are adding only one bit at a time. Equations 6.4 and 6.6 may be used to now obtain Boolean expressions respectively for the sum bit, S_i, and the carry-out, C_i:

$$S_i = X_0 \oplus (Y_0 \oplus E) \oplus C_{i-1}$$
$$C_i = X_0 \oplus (Y_0 \oplus E) + C_{i-1}[X_0 \oplus (Y_0 \oplus E)]$$
$$= X_0 \oplus (Y_0 \oplus E) + C_{i-1}[\overline{X_0}\,\overline{Y_0}E + X_0 + \overline{X_0}Y_0\overline{E}X_0]$$
$$= X_0 \oplus (Y_0 \oplus E) + C_{i-1}(Y_0 \oplus E) + X_0C_{i-1}$$

Note that the CBFF inputs are controlled by SRE. When $S = 1$, E gets loaded in as the initial carry, and when $S \neq 1$, the carry-out gets shifted in. The input to the CBFF is given by,

$$D = \text{SRE} \bullet [X_0 \oplus (Y_0 \oplus E) + C_{i-1}(Y_0 \oplus E) + X_0C_{i-1}] + \overline{\text{SRE}} \bullet E$$

where SRE remains high during all steps except when $S = 1$.

The sum results in an overflow whenever for positive result the carry-out that gets shifted into the sign-bit is a 1 or when for negative result the carry-out is a 0. The overflow indicating function does not get determined until the step $S = 4$ has been reached and an *overflow control signal*, OCS, is generated. These actions help determine whether or not an overflow

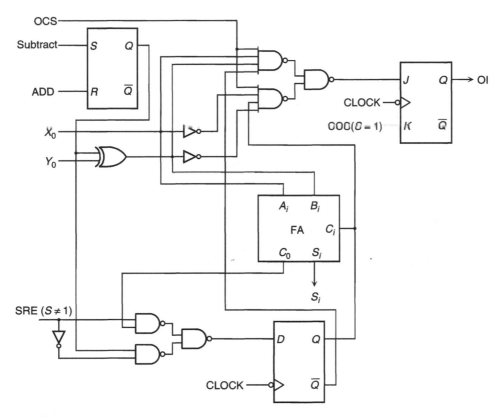

FIGURE 10.3
Logic circuit for addition/subtraction and for detecting an overflow.

has occurred. The overflow indicating function is described by,

$$OI = \text{OCS} \bullet [X_0 \oplus (Y_0 \oplus E)\overline{CBFF_0} + \overline{X_0}(Y_0 \oplus E)CBFF_0]$$

A sequential circuit capable of generating sum, carry-out, and an overflow indicating output is obtained as shown in Figure 10.3. Initially, the enable signal, E, is stored in an unclocked SR FF. Its set input is tied to SUBTRACT command and its reset input is connected to ADD command. The output of the JK FF indicates an overflow. The K input of this flip-flop gets turned on at $S = 1$ thus clearing the overflow indicator. This is realized by making sure that the control unit has generated a *clear overflow signal*, COS, when $S = 1$.

By examining the RTL statements in Algorithm 1a, we note that steps $S = 3$ and $S = 4$ do not involve any circuits actions that cannot be included in step $S = 2$. This lends to further simplification of Algorithm 1a as long as we make careful use of N. Steps $S = 2$ through $S = 5$ can be suitably combined to now yield a simpler algorithm:

Algorithm 1b

$S = 2$: $X_{n-1} \leftarrow X_0 + (Y_0 \oplus E) + CBFF_0$; $Y \leftarrow \text{rr } Y$; $CBFF_0 \leftarrow (\text{carry} - \text{out})$; $X \leftarrow \text{srX}$;
$N \leftarrow N + 1$; IF$(N = n - 1)OI \leftarrow (X_0[Y_0 \oplus E]\overline{CBFF_0} + \overline{X_0}[Y_0 \oplus E]CBFF_0)$; IF$(N = n)S \leftarrow$
3 ELSE $S \leftarrow 2$;
$S = 3$: STOP;

Since in 4-bit addition $n = 4$, a comparator will need to be employed to test if the present count is either less than, or greater than, or equal to 3. Most comparators have three output lines LT, GT, and EQ that get turned on based on whether data input A is less than, or greater than, or equal to data input B, respectively.

The significant other part of the arithmetic circuit that has not been designed yet is the control circuit. This unit must generate a clock signal, CK, as well as the control signals, SRE, OCS, and COS that we have already identified. Central to the design of this control subunit is that we first estimate the time that will be needed by the circuit to go through each of the RTL sequences. Such estimation should always account for the worst-case propagation delay for each of the gates, full adder, flip-flops, and comparator, and a safety margin to account for aging and wiring delay. Once that has been determined the control subunit design is essentially identical to designing a counter that generates needed signals at appropriate clock signals. In an earlier chapter, we have studied how a register can be used to sequence events. A multibit ring counter, for example, can be used for our current purposes. The number of counter bits would depend on the number of bits present in the data. The MSB of the counter, for example, can be used to provide LOAD and COS signals while the outputs of the next few necessary FFs can be ORed to generate the SRE signal.

10.3   Serial-Parallel Multiplication

The most straightforward multiplication technique is the one that involves signed magnitude numbers. The data in this case consists of the absolute value or magnitude and a sign bit indicating whether or not the number is negative. The sign of the product is determined by separately operating on the signs of the multiplier and the multiplicand. The sign of the product is positive if the signs of the two operands are similar; and negative, if they are otherwise. Most binary multiplication schemes use the same set of rules that are commonly used in our day-to-day decimal multiplication. Consider, for example, the multiplication of 13 and 12 in both decimal and binary forms.

$$
\begin{array}{rr}
13_{10} & 1101_2 \\
\times 12_{10} & \times 1100_2 \\
\hline
26 & 0000 \\
13 & 0000 \\
\hline
156_{10} & 1101 \\
 & 1101 \\
\hline
 & 10011100_2 \\
\end{array}
$$

As expected, $156_{10} = 10011100_2$. In both cases, the product can have up to twice as many digits (or bits) as in the multiplier. For storing multiplication result of two n-bit words, we need a register of size $2n$.

An examination of multiplication techniques reveals that it consists of a sequence of repetitive shift and add operations. Using symbolic representation of numbers, therefore,

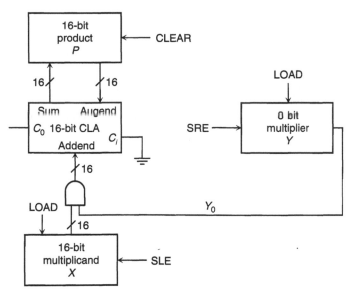

FIGURE 10.4
Multiplication block diagram for Algorithm 2a.

it should be easy to derive an appropriate multiplication algorithm. The multiplication algorithm should consist of the following steps:

Step 1. Initialize the product subtotal as 0.

Step 2. Start scanning the multiplier X from its right-most bit. If $X_n = 1$, Y is added to the product subtotal and then the partial sum is shifted right one bit. However, if $X_n = 0$, a 0 is added to the subtotal and then the sum is shifted right one bit.

Step 3. For data word of size n, Step 2 is repeated n times.

Multiplication can be approached using a number of different variations of the steps outlined above. The simplest of all is perhaps the case when the two n-bit operands are positive. The sequential multiplication logic circuit will need to include two shift registers, X and Y, for storing the operands, a $2n$-bit register P for storing the product, and a $2n$-bit adder. The register that starts out having multiplicand in it must also have $2n$ FFs since the multiplicand bits need to be shifted repetitively n times. The RTL program for realizing serial-parallel multiplication can be summed up as follows:

Algorithm 2a

$S = 1: P \leftarrow 0; N \leftarrow 1; X \leftarrow$ (multiplicand); $Y \leftarrow$ (multiplier);
$S = 2:$ IF$(Y_0 = 1)P \leftarrow X + P; N \leftarrow N + 1;$
$S = 3: Y \leftarrow$ sr $Y; X \leftarrow$ sl $X;$ IF $(N \neq n + 1)$ S $\leftarrow 2;$
$S = 4:$ STOP;

where N is the bit count. The corresponding multiplication block diagram is shown in Figure 10.4 where $n = 16$, for simplicity.

In the beginning the multiplicand is loaded into n right-most FFs of register X. The multiplier is loaded into an n-bit register Y. Since at no time would the carry-out of the adder be a 1, we make no attempt to store it. To manipulate operations of this multiplication circuit, the control unit needs to generate four control signals: CLEAR, LOAD, SRE, and SLE. The registers are set using CLEAR and LOAD inputs. Test performed at step $S = 2$ guarantees that X is added to the partial product only if the LSB of Y is a 1. This conditional add is realized often with hardware by ANDing each bit of the multiplicand, X, with Y_0.

By examining Algorithm 2a and by inspecting the block diagram shown in Figure 10.4, we note that as the multiplier bits get shifted to the right, register Y ends up having unused FFs on the left. In the very beginning only half of register P bits is used to store partial product; then the partial product starts needing to have an increasing number of FFs. By the last cycle, the final product takes up all bit positions of P. Just as register P starts making use of one more FF on its left, register Y starts emptying one more of its FFs with each clock. In fact, the total number of FFs in P and Y together that hold usable information remains invariant. This observation lends itself to deriving an improved version of RTL algorithm for multiplication.

Algorithm 2b

$S = 1$: $P \leftarrow 0$; $N \leftarrow 1$; $X \leftarrow$ (multiplicand); $Y \leftarrow$ (multiplier); $C \leftarrow 0$;
$S = 2$: IF $(Y_0 = 1)$ $C,P \leftarrow X + P$; IF $(N = n)$ $S \leftarrow 4$;
$S = 3$: $C,P,Y \leftarrow$ srC, P, Y; $N \leftarrow N + 1$; $S \leftarrow 2$;
$S = 4$: STOP;

A single FF register, C, capable of storing the carry-out is placed on the left of register P. In this revised algorithm, register Y initially stores the n-bit multiplier and eventually ends up storing the least significant n bits of the $2n$-bit product. The flip-flop C, and registers P, and Y are treated as physically concatenated such that the shift-right operation encountered in step $S = 3$ involves all three of these. The bit from C shifts right into the MSB of P, and the LSB of P occupies the MSB of Y. The block diagram corresponding to this improved algorithm is shown in Figure 10.5. The bits in register X may not be shifted left any more since the product bits are being shifted to the right. However, data bits of X are added to the more significant n bits of the product. This revised algorithm for multiplication essentially has done away with the SLE signal.

If we must consider signed complement numbers, then we need to pay attention to four different cases involving sign. One of the approaches to realizing multiplication will require that we consider the sign bits separately. We simply take the absolute values of both multiplicand and multiplier, and then employ a multiplication technique and circuit that are equipped to process positive numbers. The final product is either finally complemented or not based on the sign of the product. The corresponding RTL program for the 2's complement sign-and-magnitude, serial-parallel, n-bit multiplication is given by,

Algorithm 2c

$S = 1$: $P \leftarrow 0$; $N \leftarrow 1$; $X \leftarrow$ (multiplicand); $y \leftarrow$ (multiplier); $C \leftarrow 0$;
$S = 2$: IF $[\overline{X_{n-1}} \wedge \overline{Y_{n-1}} = 1]$ $M \leftarrow 0$;

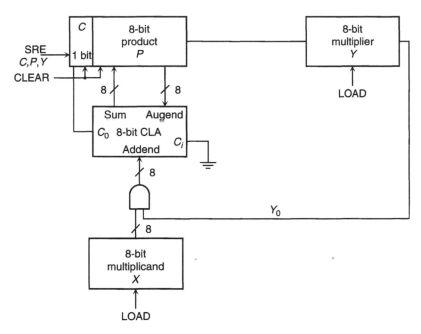

FIGURE 10.5
Multiplication block diagram corresponding to Algorithm 2b.

$$IF\ [\overline{X_{n-1}} \wedge \overline{Y_{n-1}} = 1]\ M \leftarrow 1, Y \leftarrow \overline{Y} + 1;$$
$$IF\ [\overline{X_{n-1}} \wedge Y_{n-1} = 1]\ M \leftarrow 1, X \leftarrow \overline{X} + 1;$$
$$IF\ [\overline{X_{n-1}} \wedge Y_{n-1} = 1_{n-1} = 1]\ M \leftarrow 0, X \leftarrow \overline{X} + 1, Y \leftarrow \overline{Y} + 1;$$
$$S = 3:\ C,P \leftarrow P + (X \wedge Y_0); N \leftarrow N + 1;$$
$$S = 4:\ C,P,Y \leftarrow srC, P, Y;\ IF(N \neq n + 1)\ S \leftarrow 3;$$
$$S = 5:\ IF\ (M = 1)\ P, Y \leftarrow \overline{P}, \overline{Y} + 1$$
$$S = 6:\ STOP;$$

The logical test result obtained at $S = 2$ determines the various combinations of sign bits. If the signs are equal, mode M is reset to a 0; otherwise it is set to a 1. The correct sign of the product is eventually recouped at $S = 5$.

Figure 10.6 shows the block Algorithm for the inclusion of 2's complement conversion subunits. The 2's complement operation can also be realized using an adder/subtracter. The to-be-complemented operand is loaded as B inputs while all of the A inputs are tied to a 0.

Multiplication of 2's complement numbers may result in a product that may need adjustments. As it will become obvious, it may be better to take complement of the negative operand first and then after having obtained an unsigned product, we make necessary adjustments to the sign bit to derive from it the signed product. If the multiplier is positive and the multiplicand is negative, the multiplication will not need any special consideration. However, by simply adding the multiplicand to the partial product is not a recipe for generating the correct product. Consider multiplication of decimal $6 (= 00110_2)$ and $-9 (= 10111_2)$.

The overflow that got generated indicates though that this product is negative. We can take 2's complement of this preliminary product to conclude that this generated product is indeed the negative of 00011010, which is nothing but 54_{10}. Thus the product is representative of -54_{10}.

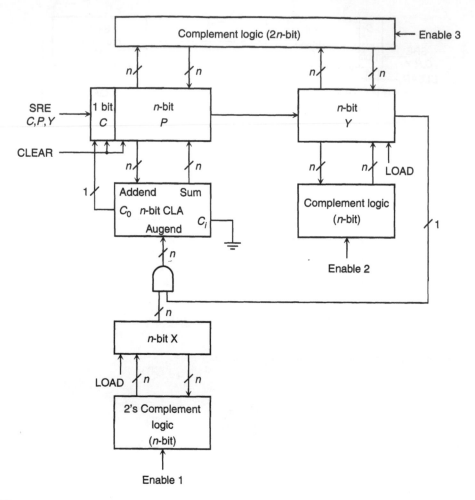

FIGURE 10.6
Multiplication block diagram corresponding to Algorithm 2c.

$$
\begin{array}{r}
10111 \\
\times 00110 \\
\hline
0000000000 \\
1111101110 \\
1111011100 \\
0000000000 \\
0000000000 \\
\hline
11111001010 \\
\end{array}
$$

We note that a string of 1s needs to be cascaded on the left of partial products. Such cascading of 1s is often realized by implementing a scale operation. The number of 1s to be put on the left will need to be such that the number of bits of the partial products is $2n$. Thus, for the case of a 2's complement negative multiplier, it is necessary that we scan the 2's complement multiplier from the right to the left. For each 1 encountered, we add to the partial product 2's complement of the multiplicand. This is repeated until all bits of the multiplier have been exhausted. The RTL program that accomplishes this type of multiplication is given as follows:

Algorithm 2d

$S = 1$: $P \leftarrow 0$; $N \leftarrow n$; $X \leftarrow$ (multiplicand); $Y \leftarrow$ (multiplier);
$S = 2$: IF $(Y_{n-1} = 0)$ $S \leftarrow 6$; IF $(Y_0 = 0)P, Y \leftarrow \text{sr}P, Y$; $S \leftarrow 2$; $N \leftarrow N - 1$;
$S = 3$: $P \leftarrow P + \overline{X} + 1$; IF $(N = 0)$ $S \leftarrow 8$;
$S = 4$: $P,Y \leftarrow \text{scr}\, P,Y$; IF $(N = 0)$ $S \leftarrow 8$;
$S = 5$: IF $(Y_0 = 1)$ $P \leftarrow \overline{X} + 1$; $N \leftarrow N - 1$; $S \leftarrow 4$;
$S = 6$: $P \leftarrow P + (X \wedge Y_0)$; $N \leftarrow N - 1$;
$S = 7$: $P,Y \leftarrow \text{scr}P,Y$; IF $(N = 0)$ $S \leftarrow 8$ ELSE $S \leftarrow 6$;
$S = 8$: STOP;

We consider next an algorithm quite similar to Algorithm 2b that can be used to realize multiplication of unsigned positive numbers. Each of three registers X, Y, and P needs to have n FFs to store the multiplicand, the multiplier, and partial product, respectively. P and Y together can house the final product. In addition, a single FF, R, is used for temporarily storing the carry-out bit. This storage FF, as we shall show later, is redundant if the shift/add operation is realized within the same clock period. The RTL algorithm for this variation of multiplication is given as follows:

Algorithm 2e

$S = 1$: $P \leftarrow 0$; $N \leftarrow 0$; $X \leftarrow$ (multiplicand); $Y \leftarrow$ (multiplier);
$S = 2$: $R,P \leftarrow P + (Y_0 \wedge X)$; $R,P \leftarrow \text{rr}\ R,P$;
$S = 3$: $R,Y \leftarrow \text{rr}\ R,Y$; $N \leftarrow N + 1$; IF $(N \leftarrow n - 1)$ $S \leftarrow 2$;
$S = 4$: STOP;

For brevity, we could use D FFs to formulate the design of registers, R, P, X, and Y. We can use the properties of MUX to load data into these registers as shown in Figure 10.7. Ten of the control lines, A through J, are used primarily for load and rotate operations. Actions initiated by the control inputs are as follows:

> A selects either P or Y,
> B selects either 0 or X,
> C selects either R or the carry-out,
> D when equal to 1 performs rotation, otherwise, selects uncirculated output,
> E selects either 0 or circulated output,
> F selects either multiplier or circulated output,
> G, H, I, J enable R, P, Y, and X, respectively, for loading input data.

Initializing the algorithm at $S = 1$ is a rather straightforward task. The add operation at $S = 2$ can be activated by manipulating A and B controls. For rotation operations (at $S = 2$

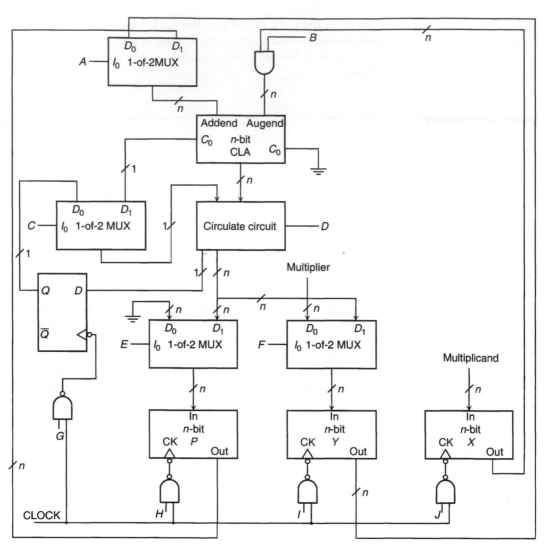

FIGURE 10.7
Multiplication logic circuit.

and $S = 3$), the content of R is loaded into the circulate unit by setting $C = 1$ while the other operand is introduced through the *carry-skip* and *carry look-ahead* (CLA) adder. For example, data bits of P can be introduced when $A = 1$ and $B = 0$.

From the RTL requirements, we can infer to have four additional subunits to control and monitor the arithmetic circuit—an event FF, a sequencer, a counter, and a sequence decoder. The event FF decides when to BEGIN and when to STOP the algorithm. The sequencer generates four state sequences, $S = 1$ through $S = 4$, and the counter counts the number of times the operation has gone through steps $S = 2$ and $S = 3$. Finally, by decoding sequencer and counter signals, the circuit gets to activate appropriate control signals at correct instants.

Since the RTL algorithm for multiplication involves four sequences, one may choose to employ a five-stage, self-starting ring counter. Such a sequencer was introduced in Chapter 9, however, it may need to be modified so that when the question "Is $N \leftarrow n - 1$?"

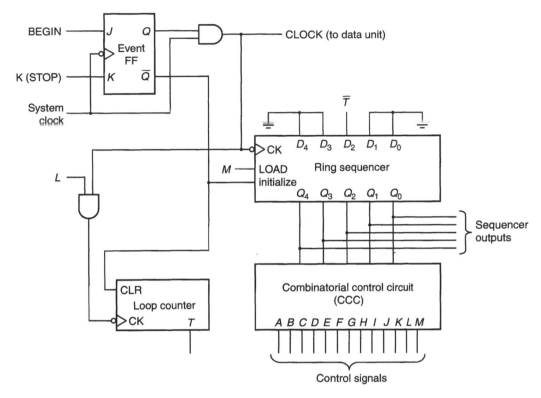

FIGURE 10.8
Control subunit for the logic circuit of Figure 10.7.

yields a negative answer, the circuit control can get switched to step $S = 2$. Figure 10.8 shows logic diagram of the control unit. The control unit would require three additional signals: K, L, and M. Here, K stops the event FF, L enables the counter to count up, and M loads the inputs, D_4 through D_0, into the sequencer.

Once the event FF gets set by BEGIN command, the sequencer is INITIALIZEd to take up count state 10000. With each clock thereafter, the sequencer goes through the states 01000, 00100, 00010, and 00001. At $S = 1$, the loop counter is CLEARed. When T is turned on, it implies that the loop count has already exceeded $n - 2$.

With a high on BEGIN line, the event FF gets to enable both the sequencer and the loop counter. The four timing cycles, corresponding to steps $S = 1$ through $S = 4$, respectively is determined by a high at Q_3, Q_2, Q_1, and Q_0. These four sequencer states are used in turn to generate 13 control inputs, A through M, to direct the algorithm. When control subunit is in state 01000, the accumulator must be cleared and multiplier and multiplicand needs to be loaded into registers Y and X respectively. This is all realized by turning on H, I, and J, provided all other control lines remain turned off.

The algorithmic function for $S = 2$ is acted upon when the counter state is 00100. This is realized by setting $A = 1$, $B = Y_0$, $D = 1$, $E = 1$, and $G = H = 1$. The content of register X is added to that of only the accumulator when the LSB of register Y is a 1. The result after it has been rotated right one bit, is stored in R and P. Then, at $S = 3$, R and Y contents need to get rotated once, and the loop count needs to increase by 1. This step is realized by setting C, D, F, G, I, and L to a 1. Additionally, a test is performed at $S = 3$ to determine the next state. If T is a 0, then the next step is given by $S = 2$; otherwise, the algorithm must stop.

State	Control signals												
	A	B	C	D	E	F	G	H	I	J	K	L	M
$Q_1(S=0)$													
$Q_2(S=1)$								1	1	1			
$Q_3(S=2)$	1	Y_0		1	1		1	1					
$Q_4(S=3)$			1	1		1	1		1		T	1	$\overline{T}$

FIGURE 10.9
Control matrix table for the logic circuits of Figures 10.7 and 10.8.

This circuit objective is met by providing T to K and $\overline{T}$ to D_2 and M. If the count is ever less than $n - 1$, 00100 (i.e., $S = 2$) is loaded in the sequencer. However, if the count equals $n - 1$, K becomes high and, consequently, the event FF is switched off.

The control conditions so identified are entered now in the *control matrix* table shown in Figure 10.9. The control equations for the design of the *combinational control circuit* (CCC) can be derived readily from Figure 10.9. The CCC equations are

$$A = E = Q_3, \quad B = B_{Y_0 \bullet Q_3}, \quad C = F = L = Q_4, \quad D = G = Q_3 + Q_4$$
$$H = Q_2 + Q_3, \quad I = Q_2 + Q_4, \quad J = Q_2, \quad K = \overline{M} = T$$

The serial-parallel multiplication schemes are sequential in nature. The algorithms explored in this section are much slower than those other schemes where all partial products are generated simultaneously and then added. In the next section, we shall deploy a fast multiplication algorithm.

10.4   Fast Multiplication

A larger fraction of multiplication time is lost by the carries that ripple through the adders during addition. There is, however, a sequential solution to this shortcoming. In one such scheme, known commonly as the *carry-save multiplication*, the carries corresponding to the addition of partial products are stored initially in a set of D FFs and used as one of the adder inputs during the next addition step. Such storing prevents the carries from rippling through different adders.

The scheme consists of two phases: *carry-save* (CS) and *carry-propagate* (CP). The CS phase involves the standard add-and-shift operation where the carry is stored temporarily in a D FF. The carry-out is prevented from moving between columns. The true product is retrieved during the final CP phase by allowing the carries to propagate and only once. The RTL algorithm that corresponds to this carry-save n-bit multiplication is described as follows:

Algorithm 3

$S = 1$: $X \leftarrow$ (multiplicand); $Y \leftarrow$ (multiplier); $CP \leftarrow 0$; $CS \leftarrow 1$; $P \leftarrow 0$; $C \leftarrow 0$; $N \leftarrow n$;
$S = 2$: $P_i \leftarrow P_i \oplus C \oplus (X \wedge Y_0)$; $C_i \leftarrow$ (carry-out); $N \leftarrow N - 1$;
$S = 3$: $P, Y \leftarrow$ sr P, Y; IF $(N \neq 0)$ $S \leftarrow 2$ ELSE $CS \leftarrow 0$, $CP \leftarrow 1$;

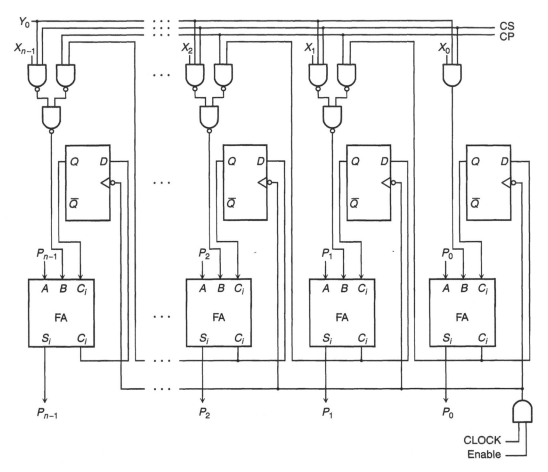

FIGURE 10.10
A n-bit carry-save adder circuit.

$S = 4 \colon P \leftarrow P + C;$
$S = 5 \colon \text{STOP};$

where n is the number of bits in the multiplier.

The RTL step at $S = 2$ involves realizing a bit-by-bit addition such that the bit-wise carry-outs are retained in register C. These registers consist of n D FFs interconnected as follows: The CS phase lasts until the process of bit-wise add-and-shift operations have been conducted n times. Finally when $S = 4$ (i.e., during CP phase), the register contributes to "add" operation, thus, allowing carry to propagate. Note that CS and CP phases are mutually exclusive. Figure 10.10 shows the logic diagram of a carry-save adder that is capable of both CS and CP operations.

ENABLE control is turned on to start the algorithm. As long as both CS and Y_0 are on, partial product is added bit-by-bit to X. The carry-outs and the sum bits are stored respectively at the respective D FFs of registers C and P. Both these registers are shift-right type and are concatenated. When Y_0 is a 0 during $S = 2$, a 0 is added to each of the P bits. However, during CP phase, each of the carry-outs propagates to the FA on the left. This last addition step is equivalent to having a ripple addition. Immediately following this CP addition, the more significant half of the product will show up in register P and the remaining half in register Y.

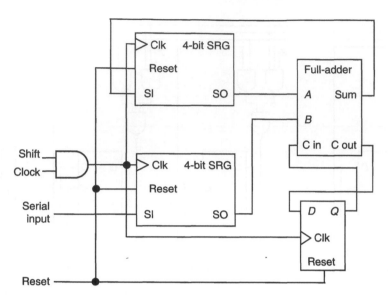

FIGURE 10.11
Block diagram of serial addition.

10.5   Implementation of Sequential Arithmetic Using VHDL

In this section, we illustrate the design of a synchronous serial addition circuit and its VHDL descriptions. Figure 10.11 shows a block diagram of the circuit. In this serial adder a single 1-bit full adder does all the addition operations. First, a reset signal clears the contents of all 8 bits of the two 4-bit registers. Next, with the Shift enabled, the 4-bits of data A are shifted into the registers one bit at a time, from Serial Input through the shift-in (SI) port, followed by the 4-bits of data B, both with the least significant bits first. After eight clock cycles, data A is stored in the top 4-bit SRG block, while data B is stored in the bottom 4-bit SRG block, with the least significant bits at the shift-out (SO) ports. The D flip-flop is holding a 0. Starting with the ninth clock cycle, the sum bits are fed back to the top 4-bit SRG block and the most recent carry-out bit is captured in the D flip-flop. At the end of the twelfth clock cycle, the resulting 4-sum bits are stored in the top SRG, while the carry-out bit (C_4) is stored in the D flip-flop.

To develop a VHDL description for this serial addition circuit, we need to define the three major components (a 4-bit SRG, a full adder, and a D-type flip-flop) first. Figure 10.12 (continue onto Figures 10.13 and 10.14) shows a hierarchical VHDL description of this serial addition circuit. A behavioral VHDL description is shown in Figure 10.15.

10.6   Summary

We considered a number of different sequential schemes for arithmetic operations. In each of these, the design schemes were translated into RTL algorithms. Most of the corresponding circuits were decomposed naturally into two parts: a data processing unit and a control unit. The RTL algorithms were shown to be the real link that bridges the design of these two subunits.

```
-- A hierarchical VHDL description of a serial adder      -- 01
library ieee;                                             -- 02
use ieee. std_logic_1164. all;                            -- 03
entity D_FF is                                            -- 04
   port (D, clk, res: in std_logic; Q: out std_logic);    -- 05
end D_FF;                                                 -- 06
                                                          -- 07
architecture functional_DFF of D_FF is                   -- 08
                                                          -- 09
 begin                                                    -- 10
   process (clk, res)                                     -- 11
     begin                                                -- 12
     if (res = '1') then Q <= '0';                        -- 13
      elsif (clk' event and clk = '1') then Q <= D;       -- 14
       end if;                                            -- 15
    end process;                                          -- 16
end functional_DFF;                                       -- 17
                                                          -- 18
                                                          -- 19
library ieee;                                             -- 20
use ieee. std_logic_1164. all;                            -- 21
entity FULL_ADDER is                                      -- 22
   port (inA, inB, inC: in std_logic;                     -- 23
   Sum, Cout:  out std_logic);                            -- 24
end FULL_ADDER;                                           -- 25
                                                          -- 26
architecture  functional_FA of FULL_ADDER is             -- 27
                                                          -- 28
  signal inA_b, inB_b, inC_b: std_logic;                 -- 29
  begin                                                   -- 29
     inA_b <=  not inA; inB_b <=not inB; inC_b <= not in C;  -- 30
     Sum < = (inA_b and inB_b and InC) or                -- 31
        (inA_b and inB and InC_b) or                      -- 32
        (inA and InB_b and inC_b) or                      -- 33
        (inA and inB and inC);                            -- 34
     Cout <= (inA and inB)or(inA and inC) or             -- 35
        (inB and inC);                                    -- 36
  end functional_FA;                                      -- 37
                                                          -- 38
library ieee;                                             -- 39
use ieee. std_logic_1164. all;                            -- 40
entity SRG4bit is                                         -- 41
   port (Sin, Clk, Res: in std_logic; Sout: out std_logic;  -- 42
   Qout: out std_logic_vector (3 downto 0));              -- 43
end SRG4bit;                                              -- 44
                                                          -- 45
architecture  structural_4SRG of SRG4 bit is             -- 46
                                                          -- 47
   component D_FF                                          -- 48
      port (D, clk, res: in std logic; Q: out std logic);  -- 49
```

FIGURE 10.12
A hierarchical VHDL description of a serial adder (part 1).

```
      end component;                                          -- 50
                                                              -- 51
        signal b3Q, b2Q, b1Q, b0Q: std_logic;                -- 52
      begin                                                   -- 53
      B3: D_FF port map (Sin, Clk, Res, b3Q);                 -- 54
      B2: D_FF port map (b3Q, Clk, Res, b2Q);                 -- 55
      B1: D_FF port map (b2Q, Clk, Res, b1Q);                 -- 56
      B0: D_FF port map (b1Q, Clk, Res, b0Q);                 -- 57
      Qout(3) <= b3Q; Qout(2) <= b2Q; Qout(1) <= b1Q;         -- 58
      Qout(0) <= b0Q; Sout <= b0Q;                            -- 59
    end structural_4SRG;                                      -- 60
                                                              -- 61
    library ieee;                                             -- 62
    use ieee. std_logic_1164. all;                            -- 63
    entity SERIAL_ADDER is                                    -- 64
      port (Sin, clk, res, shift:  in std_logic;              -- 65
            Cout: out std_logic;                              -- 66
            Sum: out std_logic_vector (3 downto 0));          -- 67
    end SERIAL_ADDER;                                         -- 68
                                                              -- 69
    architecture  hierarchical_4bitSA of SERIAL_ADDER is      -- 70
                                                              -- 71
      component AND2                                          -- 72
         port (in1, in2: in std_logic; out1: out std_logic);  -- 73
      end component;                                          -- 74
                                                              -- 75
      component D_FF                                          -- 76
         port (D, clk, res: in std_logic; Q: out std_logic);  -- 77
      end component;                                          -- 78
                                                              -- 79
      component FULL_ADDER                                    -- 80
         port (inA, inB, inC: in std_logic;                   -- 81
               Sum, Cout:out std_logic);                      -- 82
      end component;                                          -- 83
                                                              -- 84
      component SRG 4bit                                      -- 85
         port (Sin, Clk, Res: in std_logic; Sout: out std_logic;  -- 86
               Qout: out std_logic_vector (3 downto 0));      -- 87
      end component;                                          -- 88
                                                              -- 89
      signal gclk, Cdq, Cdd, Ain, bin, Stmp: std_logic;       -- 90
      signal SRG_tmpA, SRG_tmpB: std_logic_vector (3 downto 0);  -- 91
                                                              -- 92
      begin                                                   -- 93
                                                              -- 94
         G1: AND2 port map (shift, clk, gclk);                -- 95
         G2: FULL_ADDER port map (Ain, Bin, Cdq, Stmp, Cdd);  -- 96
         G3: D_FF  port map (Cdd, gclk, res, Cdq);            -- 97
         G4: SRG4bit port map in, gclk, res, Bin, SRG_tmpB);  -- 98
```

FIGURE 10.13
A hierarchical VHDL description of a serial adder (part 2).

```
         G5: SRG 4bit port map (Stmp, gclk, res, Ain, SRG_tmpA);  --  99
            Cout <= Cdq;                                          -- 100
            Sum <= SRG_tmpA;                                      -- 101
         end hierarchical_4bit SA;                                -- 102
         -- End of the VHDL description                           -- 103
```

FIGURE 10.14
A hierarchical VHDL description of a serial adder (part 3).

```
-- A behavioral VHDL description of a serial adder       -- 01
library ieee;                                            -- 02
use ieee. std_logic_1164.all;                            -- 03
entity SERIAL_ADDER is                                   -- 04
    port (En, clk: in std_logic; Cout: out std_logic;    -- 05
          Sum: out std_logic_vector (3 downto 0));       -- 06
end SERIAL_ADDER;                                        -- 07
                                                         -- 08
architecture  behavioral_SA of SERIAL_ADDER is          -- 09
                                                         -- 10
   signal A, B: std_logic_vector (3 downto 0);          -- 11
   signal shift, Ci, Co,So: std_logic;                  -- 12
   signal PS, NS: integer range 0 to 3;                 -- 13
                                                         -- 14
  begin                                                 -- 15
So <= A(0) xor B(0) xor Ci;                             -- 16
Co <= (Ci and A(0)) or (Ciand B(0))or                   -- 17
      (A(0) and B(0));                                  -- 18
Sum <= A; Cout <= Co;                                   -- 19
                                                         -- 20
  process (PS, En)                                      -- 21
      begin case PS is                                  -- 22
      when 0 =>                                         -- 23
         if (En = '1') then shift <= '1'; NS <= 1;      -- 24
         else shift <= '0'; NS <= 0; end if;            -- 25
      when 1 => shift <= '1'; NS <= 2;                  -- 26
      when 2 => shift <= '1'; NS <= 3;                  -- 27
      when 3 => shift <= '1'; NS <= 0;                  -- 28
      end case;                                         -- 29
   end process;                                         -- 30
                                                         -- 31
  process (clk)                                         -- 32
  begin                                                 -- 33
    if (clk' eventand clk = '1') then PS <= NS;         -- 34
    if (shift = '1') thenA <= So & A( 3downto 1);       -- 35
       B <= B(0) & B(3 downto 1); Ci <= Co;             -- 36
  end if;                                               -- 37
  end process;                                          -- 38
end behavioral_SA;                                      -- 39
```

FIGURE 10.15
A behavioral VHDL description of a serial adder.

Bibliography

Bywater, R.E.H., *Hardware/Software Design of Digital Systems.* Englewood Cliffs, NJ. Prentice-Hall International, 1981.

Comer, D.J., *Digital Logic and State Machine Design.* 3rd edn. New York, NY. Oxford University Press, 1994.

Floyd, T., *Digital Fundamentals.* 8th edn. Englewood Cliffs, NJ. Prentice-Hall, 2003.

Hwang, K., *Computer Arithmetic*, New York, NY. Wiley, 1979.

Johnson, E.J. and Karim, M.A., *Digital Design: A Pragmatic Approach*. Boston, MA. PWS-Kent Publishing, 1987.

Karim, M.A. and Awwal, A.A.S., *Optical Computing: An Introduction*. New York, NY. John Wiley & Sons, 1992.

Katz, R.H., *Contemporary Logic Design*. Boston, MA. Addison Wesley. 1993.

Kline, R.M., *Structured Digital Design Including MSI/LSI Components and Microprocessors*. Englewood Cliffs, NJ. Prentice-Hall, 1983.

Mowle, F.J.A., *A Systematic Approach to Digital Logic Design*. Reading, MA. Addison-Wesley, 1976.

Nelson, V.P.P., Nagle, H.T., and Carroll, B.D., *Digital Logic Circuit Analysis and Design*. Englewood Cliffs, NJ. Prentice Hall. 1995.

Winkel, D. and Prosser, F., *The Art of Digital Design—An Introduction to Top-Down Design*. Englewood Cliffs, NJ. Prentice-Hall, 1980.

Problems

1. Design a serial arithmetic unit that computes $|x| - |y|$ where x and y are two signed numbers stored respectively in registers A and B.

2. Repeat Problem 1a wherein the arithmetic unit utilizes a parallel adder and the numbers are represented in 1's complement form.

3. Design a control circuit for Algorithm 1a for which the sum would not be calculated any further once an overflow has been detected. Discuss the functions of all individual components.

4. Design a binary multiplier when the operands are all represented in 1's complement.

5. Design sequential circuit systems suitable for (a) Algorithm 2a, (b) Algorithm 2b, (c) Algorithm 2c, and (d) Algorithm 2d.

6. Obtain the logic circuits for (a) the sequencer, (b) the loop counter, and (c) the CCC of the system shown in Figure 10.8.

7. What corrections will need to be made to the logic circuits of Figures 10.7 and 10.9 if due consideration is given to the fact that a ring counter may end up in one or more unused states? Explain your designs.

8. Replace the ring counter used in the sequencer shown in Figure 10.8 with a regular binary up-counter. Discuss the impacts of this change as well as modifications necessary to have a functioning system.

11

Asynchronous Sequential Circuits

11.1 Introduction

In the preceding four chapters, we encountered circuit states only when there was a clock and a change in one or more control variables. Such circuits are somewhat more reliable in their operations and their design follows a rather straight forward approach. The control inputs are themselves pulses but are considered as levels since they are either high or low for longer than one or more clock periods. In comparison, asynchronous circuits require much closer scrutiny since there is no clock signal to regulate change of states. There are two types of asynchronous sequential circuits—pulse-mode type and fundamental-mode type.

The control inputs in *pulsed mode* sequential circuits are considered simply as pulses. This class of sequential circuits responds immediately to input changes rather than responding to the input change only in the presence of clock. This circuit characteristic is somewhat troublesome. With this in the background, we will consider only those input variables and pulses that are mutually exclusive to be able to avoid most timing problems. We need to understand that the overwhelming majority of the sequential circuits that are designed today are not necessarily the pulse-mode type. But for systems such as computer keyboards and self-serve vending machines where inputs are generally nonoverlapping and user-entered, pulse-mode design is a good and reasonable choice. Even though this type of circuit is treated separately in this chapter, the design mechanism for pulse-mode circuits is quite similar to that covered in Chapter 8 for the synchronous circuits with minor changes.

A sequential circuit is said to be *fundamental-mode* type as long as the external inputs are never allowed to change unless the system has reached already a stable state. In fundamental-mode machines circuit action occurs whenever there is a change in input. These inputs could be pulses or a clock, but for considerations of this particular type of sequential circuits, we shall be concerned only with events that occur as a result of a $0 \rightarrow 1$ or $1 \rightarrow 0$ transition in the input. The circuit state in fundamental-mode circuits is determined by both memory outputs and input variables. *SR* latches are generally used in the memory part of the fundamental-mode circuit.

In a properly designed synchronous circuit, timing problems are eliminated by waiting long enough so that the clock pulse is not introduced until after the external input change for all FF inputs have reached steady states. In many digital systems, it is quite difficult to maintain a fundamental-mode operation. Input signals are derived from different sources and are often random in nature. Consequently, special interference circuits, known as *synchronizers*, are used to guarantee continuation of normal circuit operation.

Fundamental-mode circuit design is the most difficult of all sequential circuits because of the timing problems involved. However, it lends itself as the most powerful design since the designer gets to control every aspect of the circuit action. Each of the FF types, for example, can be easily designed from scratch using fundamental mode design techniques.

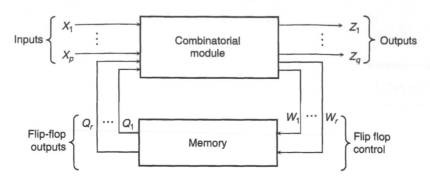

FIGURE 11.1
A pulse-mode sequential circuit model.

11.2   Pulse-Mode Circuit

A sequential circuit is referred to as pulse-mode type when it involves unclocked FFs and input signals that are pulses. Figure 11.1 shows the general pulse-mode circuit model, which is the same as that of Figure 7.60 but with clock input eliminated from it. Asynchronous nature of the pulse-mode model causes these circuits to be associated with erratic behavior if it is not constrained properly.

To avoid problems of erratic behavior, the pulse-mode operation requires that such circuits be designed with restrictions imposed on the inputs as far as their duration and interval are concerned. The inputs supplied to the circuit model shown in Figure 11.1 must satisfy the following conditions:

1. Concurrent pulses on two or more inputs are disallowed. The time interval between any two pulses must be greater than the time required for the system to reach a stable state.

2. Input pulse widths must be large enough to allow the circuit components to respond to them. For the case of level-sensitive FFs the pulse width must be short enough to have transitioned from 0 to 1 and then from 1 to 0 before the FF outputs change to their new values. For edge-sensitive FFs, the pulse width may not be as critical as long as condition 1 has been satisfied.

So long as these conditions are satisfied, the design of pulse-mode circuits is quite similar to that of synchronous circuits. Figure 11.2 shows an example of a set of three acceptable nonoverlapping pulse trains. For every n pulse trains going into a pulse-mode circuit, the circuit may be subjected to a maximum of $n + 1$ unique input conditions. One of these $n + 1$ input conditions corresponds to the time interval when all pulse trains values are 0.

In pulse-mode design, since only pulses supply information to the FFs, the input variables are used only in their uncomplemented form. The design algorithm for pulse-mode circuits include the following steps:

1. Obtain the state diagram and/or state table. There must be as many transition paths leaving each of the states as the number of inputs.

2. Obtain the reduced state table by eliminating redundancies.

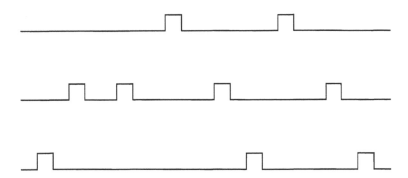

FIGURE 11.2
Nonoverlapping pulse trains suitable for being pulse-mode inputs.

3. Obtain a transition table using appropriate state assignments.
4. Derive Boolean equations from the excitation table.

Details of some of these steps may be slightly different from those discussed earlier in Section 8.8. Example 11.1 illustrates pulse-mode design considerations.

Example 11.1

Design a pulse-mode circuit that responds to two inputs X_1 and X_2 by generating an output coincident with the third consecutive X_2 pulse following at least one X_1 pulse.

Solution

An appropriate approach for obtaining a state diagram always begins from considering the sequence of inputs that results in an output and then determines what else happens to each state for all other possible input conditions. Figure 11.3 shows the state diagram for this problem. It includes all actions that result from the occurrences of pulse in any input. State B corresponds to the appearance of a pulse in X_1, while states C and D, respectively, correspond to the appearance of subsequent X_2 pulses. Once the circuit is already at state D, a subsequent X_2 pulse will cause the circuit to return to state A and, at the same time, generate an output pulse z. If an X_1 pulse appears when the circuit is in any of these states, the circuit will always return to state B.

Figure 11.4 shows the state table derived from the state diagram of Figure 11.3. Two FFs are required to cover four states as shown in the state table. Note that in the case of the synchronous circuits, the state table column headings involved all the various combination of the control variable values. In the case of the pulse mode model, however, $X_1 X_2 - 00$ implies that no pulses have occurred. The input condition $X_1 X_2 = 01$ implies that only pulse X_2 has occurred, and $X_1 X_2 = 10$ implies that only X_1 has occurred. The existence of $X_1 X_2 = 11$ would imply that both pulses appear simultaneously, and thus is unallowable in pulse-mode circuits. Pulse-mode circuits are most easily realized by using edge-triggered T FFs. The transition/output table, as shown in Figure 11.5, is obtained by arbitrarily assigning $A = 00, B = 01, C = 11$, and $D = 10$.

The T FF output and excitation maps obtained are shown in Figure 11.6. These excitation maps can be treated as reduced four-variable K-maps. Columns corresponding to both $X_1 = X_2 = 0$ and $X_1 = X_2 = 1$ are eliminated because they are of no significance to our problem. The column corresponding to X_1 is equivalent to having $X_1 = 1$

and $X_2 = 0$, and the column corresponding to X_2 is equivalent to having $X_1 = 0$ and $X_2 = 1$ in the corresponding four-variable maps.

Note that in this reduced K-map, the two columns are not adjacent and as such one cannot form any group involving entries between the two columns. From grouping of the 1s, we obtain the excitations equations for the pulse-mode circuit

$$T_2 = X_1\overline{Q_2} + X_2Q_1Q_2$$

$$T_1 = X_1Q_1 + X_2(Q_1 \oplus Q_2)$$

$$z = X_2Q_1\overline{Q_2}$$

The output z is Mealy type. An example of a timing diagram for this circuit is shown in Figure 11.7. Figure 11.8 illustrates the logic circuit for this pulse-mode circuit.

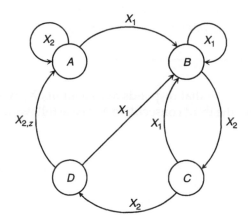

FIGURE 11.3
State diagram for Example 11.1.

	NS, z	
PS	X_2	X_1
A	A	B
B	C	B
C	D	B
D	A,z	B

FIGURE 11.4
State table for Example 11.1.

	NS, z	
Q_1Q_2	X_2	X_1
00	00	01
01	11	01
11	10	01
10	00, z	01

FIGURE 11.5
Transition table for Example 11.1

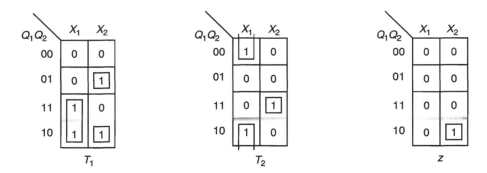

FIGURE 11.6
Excitation and output maps for Example 11.1.

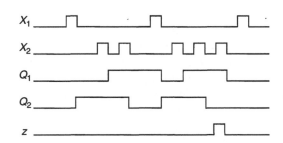

FIGURE 11.7
A timing diagram.

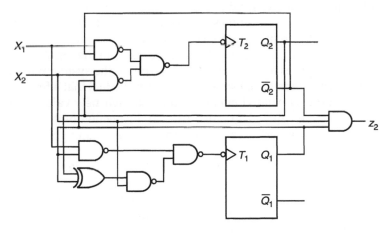

FIGURE 11.8
A pulse mode logic circuit.

11.3 Fundamental Mode Circuits

In synchronous and pulse-mode circuits, their *FF* inputs can be either stable or varying, and no action takes place until a clock or a pulse has appeared. In fundamental-mode asynchronous circuits, in comparison, the circuit action is prompted by a change in the

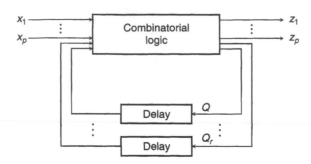

FIGURE 11.9
A model for fundamental-mode circuits.

inputs and, therefore, we shall focus on events and changes that occur as a result of either a $0 \rightarrow 1$ or a $1 \rightarrow 0$ transition in the input. As is the case in all, the circuit state is determined by a combination of both FF outputs and input variables. Figure 11.9 shows a model of the fundamental-mode circuit wherein the memory elements are shown simply as delay elements (rather than FFs as in the cases of pulse mode and synchronous circuits). The external input signals, x_1 through x_p, entering the circuit are referred to as the *external level signals*; the lines z_l through z_q available to the outside world represent the *primary outputs*; and Q_1 through Q_r represent the *secondary outputs* (state variables) that are fed back as internally generated inputs to the combinational module after they have passed through the delay elements such as the *SR* latches.

Stable as well as transitional states are encountered in fundamental-mode circuits. In pulse-mode and synchronous circuits, in comparison, there was no provision for any transitional states. For a state to be *stable* the combinational module of the circuit must generate inputs to the delay module of the circuit without causing any state to change. This is guaranteed as long the present state becomes the next state. To better appreciate issues that impact fundamental-mode circuits design, consider an asynchronous delay network such as that shown in Figure 11.10a. Whenever there is a change in the input, it results in a new state. If this new state requires a change in the FF outputs, the system will be temporarily in a *transitional* state. This transitional state accounts for the fact that a finite time, however small, is needed to set the associated latches to take on their new values. The concepts of stable and transitional states may be grasped from an inspection of the timing diagram of Figure 11.10b. The state, whether it is stable or transitional, at any point is same as the FF values, $Q_1 Q_2$.

At t_4, $Q_1 Q_2$ value is reflective of a transitional state that is still having to adjust in response to a change in x_2, $0 \rightarrow 1$, when $x_1 = 1$; the present state of $Q_1 Q_2$ soon makes the transition from 01 to the stable state 11. This time lag, albeit small, is due to the delay element. Transitional states are also encountered at t_2, t_6, t_9, and t_{11}.

The timing diagrams such as that of Figure 11.10b can never convey all possible changes in the states and may be enough for visualizing changes only for very small logic circuits. A systematic way is thus invoked as an aid to designing the fundamental-mode circuit. A tabular form is found to be the most convenient for considering all possible input combinations. First, a transition table is constructed that shows the next states of the latches as a function of the present states inputs. To describe logic circuit of Figure 11.10a, one can obtain its next-state equations:

$$Q_1(t + \Delta t) = x_1(t)x_2(t) + [x_1(t) + x_2(t)]Q_1(t)$$
$$Q_2(t + \Delta t) = x_1(t)\overline{x_2}(t) + [x_1(t) + x_2(t)]Q_2(t)$$

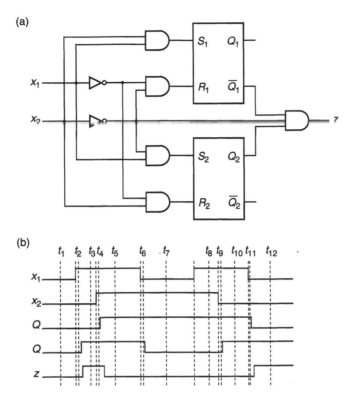

FIGURE 11.10
Asynchronous delay network: (a) logic circuit and (b) timing diagram.

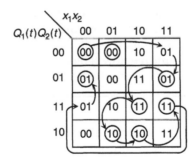

FIGURE 11.11
Flow table for the delay network of Figure 11.10a.

Figure 11.11 shows a flow table that includes a mapping of these next-state equations. Each column of this table corresponds to a specific combination of inputs, x_l and x_l, and each row corresponds to a particular stable present state. The states (shown as circled or, also denoted by using state symbol between quote and unquote signs) are representative of stable states and the uncircled or unquoted states each represent a transitional state.

We may begin, for example, from the present state $Q_1Q_2 =$ "00" when $x_1x_2 = 00$. When input x_2 changes from 0 to 1, a new stable state, "00" is entered. Since during this change, both present and next states are identical, this input change does not have to take circuit state to go through any transitional state. If, on the other hand, x_1 is changed from 0 to 1 when $x_2 = 0$, the FF state enters a new stable state "01". Since these two stable states are ‹

different, the circuit in question will temporarily be in a transitional state before stabilizing to "01". Also, since the FFs may not change instantaneously, a transitional state is entered for the circuit until the change is complete. This transitional state is typically given the same designation as that of the stable state toward which the circuit action has already begun. The fundamental-mode circuit will go through a transitional state, 01, as shown by a transition path. The transitional state is entered at the intersection of the two corresponding stable states, "00" and "01". Note that in each row the corresponding stable state is that for which the next state is also the present state. As soon as Q_1Q_2 changes from 00 to 01, a stable state is reached unless the input value of either x_1 or x_2 are changed again. The transitions corresponding to the input sequence of Figure 11.10b are illustrated by the arrows shown on the flow table of Figure 11.11. These are not the only possible transition flows of states. Different input sequences would take the circuit through different transition paths.

The basic approach of designing a fundamental-mode circuit is not too different than what has been employed already for designing pulse-mode and synchronous circuits. A flow table is drawn and completed from an understanding of the problem statement; this step is analogous to generating a state diagram in the case of synchronous and pulse-mode circuits. The flow table is reduced then to a merged flow table which is analogous to generating state table whose redundant states have been identified and eliminated. These steps are then followed by the assignment of states and derivation of excitation equations to arrive at the desired fundamental-mode logic circuit. The details of each step are slightly different for the fundamental-mode circuits.

The fundamental-mode circuit design process is begun from a primitive flow table. The *primitive flow table* is defined as a table that has exactly one stable state in each of its rows. The primitive flow table is eventually reduced to a *merged flow table* that may have fewer rows than in the primitive flow table. Some of the rows of the merged flow table may end up having more than one stable state and, thus, fewer transitional states.

Example 11.2 illustrates the various considerations necessary to arrive at a primitive flow table. We shall assume that only one input variable changes at a time and that the input changes are separated enough in time to allow the fundamental-mode circuit to reach a stable state. Every change in the input may result in a state change and only one stable state is allowed for each row of the primitive flow table.

Example 11.2
Obtain primitive flow table for a fundamental-mode circuit that responds to two inputs, x_1 and x_2 and generates an output z. The output z become high when x_2 changes from a 0 to a 1 when $x_1 = 1$; and low when x_1 changes from a 1 to a 0.

Solution
As in the case of state diagrams considered earlier, the primitive flow table will include every possible circuit action. The process is begun by considering an initial stable state (typically at the intersection of first row and first column) and considering the input transitions that make the output assume a new value. Other states are subsequently identified and entered in the table.

Each row of the primitive flow table in Figure 11.12 represents a stable state that will eventually be identified with a particular combination of latch (we shall refer to FF from here on) outputs. Assume that the circuit is in a stable state related to first row and first column for which $x_1x_2 = 00$. Note that each of the stable states symbol are shown between quotes. In order to cause an output change z, $0 \rightarrow 1$, the circuit must first enter a stable state by changing x_1 from 0 to a 1 when $x_2 = 0$. Thereafter the output z would become a 1 when x_2 is changed from a 0 to a 1. A subsequent change of x_1 from a 1 to a 0 would turn off the output. The primitive flow table shown in Figure 11.12 has four rows each corresponding to one of these four situations. Four stable states, "1," "2," "3," and "4" associated respectively with outputs of 0, 0, 1, and 0, are entered under columns $x_1 x_2 = 00, 10, 11$, and 01.

Next, we introduce appropriate transitional states in the primitive flow table. When the circuit reconfigures itself from being in one stable state to that in the next state, it has to transition through a transitional state. The transitional state should have the same designation as the destination stable state and is entered at the intersection of the same row as the source state and the same column as the destination state. Figure 11.13 shows an improved primitive flow table that includes unquoted numbers representing the transitional states. The combinatorial module of this yet to-be-designed fundamental-mode circuit will need to have x_1, x_1, and the FF outputs as its inputs. The FF inputs in turn will need to be such that no further change in the FF outputs occurs when the state has already become stable. A stable condition, by its very definition, corresponds to a situation when the next state is the present state itself.

Consider that the circuit is in its stable state "1." When the input x_1 changes from 0 to 1 while $x_2 = 0$, the circuit action transitions itself from the initial stable state "1" to a transitional location in the same row but under column $x_1 x_2 = 10$. During this transition time the combinational circuit in effect is getting FF inputs ready to cause the FF outputs to change to the binary values corresponding to row 2. This change takes a finite amount of time as determined by the gate delays. After this time delay the circuit action moves to a new stable state "2." The transitional state that is encountered when the circuit transitions from being in stable state "1" to that in "2" is indicated by entering an unquoted 2 at the intersection which indicates that the circuit in question is in transitional state but headed towards stable state "2." Similarly, while moving either from state "2" to state "3," or from state "3" to state "4," or from state "4" to state "1," the circuit goes through additional transitional states as shown in the flow table.

Next, we identify those circuit actions that occur when transitions other than the ones already considered occur. For instance, if the circuit is in stable state "1" and x_2 were to change from 0 to 1, a stable state must be entered in column 01 with an output of 0. Stable state "4" meets this particular criterion since it is already associated with an output of 0. This transition is introduced by entering a transitional state 4 at the intersection of row 1 and column $x_1 x_2 = 01$. However, if the circuit is in stable state "3" and the input x_2 were to change from 1 to 0, it will be necessary that we enter a stable state in column $x_1 x_2 = 10$ but with an output of 1. There is no such stable state defined yet in column $x_1 x_2 = 10$. Stable state "2" already present in that column is associated instead with an output of 0; so a new stable state must be introduced. This new stable state "5" is introduced by entering transitional state 5 at the intersection of row 3 and column $x_1 x_2 = 10$. Similar arguments and considerations are exhausted, which lead to a completed primitive flow table such as that in Figure 11.4.

The primitive flow table that we just completed is representative of all possible actions and combination of inputs. The don't cares represent the transitions that will not normally occur due to the restriction that two inputs may not change simultaneously.

In practice, it is virtually impossible to have two or more inputs change simultaneously. If two inputs changed but with a time difference less than the minimum response time of the circuit, the input change would then appear as simultaneous to the circuit. The possibility

x_1x_2				
00	01	11	10	z
"1"				0
			"2"	0
		"3"		1
	"4"			0

FIGURE 11.12
Primitive flow table for Example 11.2.

x_1x_2				
00	01	11	10	Z
"1"			2	0
		3	"2"	0
	4	"3"		1
1	"4"			0

FIGURE 11.13
Modified primitive flow table for Example 11.2.

x_1x_2				
00	01	11	10	Z
"1"	4	-	2	0
1	-	3	"2"	0
-	4	"3"	5	1
1	"4"	6	-	0
1	-	3	"5"	1
-	4	"6"	2	0

FIGURE 11.14
A completed primitive flow table for Example 11.2.

of having such concurrent changes can be accommodated by defining the desired circuit action when such an event would occur and by incorporating necessary additional stable states. In the example to follow, we consider possible simultaneous changes in two or more inputs. This next design problem will be worked out through to its completion with occasional pauses to look at additional design aspects.

Example 11.3

Obtain the merged flow table for a two-button electrical lock that gets activated in accordance with the following sequence:

$$A: 0111101; \quad B: 0010111; \quad z: 0000001$$

where A and B correspond to the two buttons, and z corresponds to circuit output that causes the lock to be activated. When either of these two switches is depressed it corresponds to a 1. Release of both switches clears the circuit. The lock gets opened as soon as z changes from a 0 to a 1.

Solution

Figure 11.15 shows the primitive flow table for this problem, which accounts for every possible action of the switches. We may consider that the circuit starts from stable state "1" and goes through six successive stable states, "2" through "7," until the lock is activated. This particular input sequence corresponds to the given opening sequence. When the circuit reaches stable state "7," the output becomes high. In addition, there are three states, "8," "9," and "10," that account for the consequences of entering wrong input sequence. These three stable states respectively under columns $AB = 11$, 01, and 10, account for sequences that may occur but will not activate the lock. Any time there is a mistaken input entry, the fundamental-mode sequential circuit resides in one of three states, "8," "9," and "10." Whenever both A and B are released, the circuit returns to state "1."

Next, we eliminate redundant states to affect circuit simplification. Two stable states are considered equivalent if (i) they have the same input conditions, (ii) they are associated with same output, and (iii) for each possible change in input there is a transition from these stable states to the same or equivalent other stable states. At this point, we need to consider only stable states in each column. The list of candidate state equivalencies include the following:

$$(2,4), (2,10), (3,5), (3,7), (3,8), (4,10), (5,7), (5,8), (6,9), \text{ and } (7,8)$$

whether two states are equivalent is often determined using an implication table introduced originally in Chapter 8. Figure 11.16 documents an alternate approach.

All candidate state-equivalencies are listed as column headings in Figure 11.16. The states that do not meet all three of the equivalency requirements are crossed out (with X). No state is equivalent to state "7" since only this state is associated with output z. Consequently, the candidate pairs of states $(3,7)$, $(5,7)$, and $(7,8)$ are eliminated from the list. The candidate state-equivalencies that have not been crossed out yet are now listed as row labels of this table. We insert one or more check marks in each row and only under those columns whose equivalency will need to be established first before we can determine whether or not the row-pairs are equivalent. For illustration, consider the transitional states in the rows for stable states "3" and "5." For two states to be equivalent, the corresponding transitional states for all possible input conditions must also be equivalent. The two state-pairs $(6,9)$ and $(4,10)$, therefore, must be equivalent for the state-pair $(3,5)$ to be equivalent. We note that the states 6 and 9 are equivalent only if states 7 and 8 are equivalent; however, state-pair $(7,8)$ has already been crossed out. This implies that state-pairs $(6,9)$ are not equivalent and, therefore, the column $(6,9)$ can be now slashed out. This process of elimination is repeated and leads to the elimination of two more columns, $(3,5)$ and $(5,8)$. We continue with this process of elimination until no more column headings can be eliminated. Any column heading not eliminated represents an equivalency of states. It turns out that there are no surviving columns and, therefore, no equivalencies in this primitive flow table.

Now we merge the primitive flow table to obtain a merged flow table from it. FF values will be assigned next to identify each of these rows. Since the goal is to have a uniquely specified logic circuit with a minimum number of FFs (i.e., latches), we try our utmost now to obtain a merged flow table. Up to this point, we had assigned only one stable state to

454 Digital Design: Basic Concepts and Principles

	AB				
	00	01	11	10	Z
"1"	9	8	2		0
1	9	3	"2"		0
1	9	"3"	4		0
1	9	5	"4"		0
1	9	"5"	10		0
1	"6"	7	10		0
1	9	"7"	10		1
1	9	"8"	10		0
1	"9"	8	10		0
1	9	8	"10"		0

	00	01	11	10	Z
"1"	9	8	2	—	0

FIGURE 11.15
Primitive flow table for Example 11.3.

	2,4	2,10	3,5	3,7	3,8	4,10	5,7	5,8	6,9	7,8
2,4			✓							
2,10					✓					
3,5							✓		✓	
3,8							✓			
4,10								✓		
5,8									✓	
5,9										✓

FIGURE 11.16
Implication table for Example 11.3.

each row. Recall that for fundamental-mode circuits the states were defined by the input values and the FF states. However, if we can merge more than one stable state in a row, they can still be uniquely identified, using the FF state of the row and the input values of the column.

This is accomplished using what is commonly referred as a merger diagram and is obtained as follows:

Step 1. Display all states ignoring their outputs.

Step 2. Connect all states that have the same next states where don't cares can be interpreted as either 0 or 1.

Step 3. A group of states for which every possible interconnection exists between them are referred to as completely connected and, may be merged and represented by a single merged state.

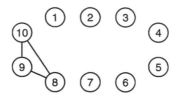

FIGURE 11.17
Merger diagram for Example 11.3.

	AB			
	00	01	11	10
	"1"	9	8	2
	1	9	3	"2"
	1	9	"3"	4
	1	9	5	"4"
	1	9	"5"	10
	1	"6"	7	10
	1	9	"7"	10
	1	"9"	"8"	"10"

FIGURE 11.18
Merged flow table for Example 11.3.

The merger diagram for the flow table in question is obtained as shown in Figure 11.17. The interconnections indicate that the states "8," "9," and "10" can have only up to three interconnections between them and no more. Given that all three are covered in the merger diagram, we may merge these three stable states to obtain a merged state from them. When this merged state is used instead of "8," "9," and "10" in the primitive flow table, it leads to a merged flow diagram as shown in Figure 11.18. When merging two or more stable states, it is important to realize that the entries in a merged row should be such that stable states always dominate the corresponding transitional states, and states of all kind should always dominate don't cares.

The example we just covered illustrates the various steps and issues we encounter in obtaining a merged flow table. The merged states are said to be *completely connected*. Note that the merged flow table has no output information associated with it. As such, we shall see later, the output equations will need to be derived separately by taking output data from the primitive flow table. Note that by reducing and or merging rows, we do not necessarily make the logic circuit operation any more accurate but only economical. We are now ready to make state assignments. This time though state assignments cannot be made arbitrarily as in clocked and pulse-mode sequential circuits. To make this consideration obvious, consider another example of obtaining a merged flow table from primitive flow table.

Example 11.4
Obtained merged flow tables for the two primitive flow tables given in Figure 11.19.

Solution
Figure 11.20 shows the corresponding merger diagrams for these primitive flow tables. In the first, pairs of flow table rows (1, 2), (1, 5), (2, 3), and (4, 6) may be merged. If

"1" and "2", for example, were to be merged, then the only other possible merger could be between "4" and "6". This would lead to having four rows in the merged table—(1, 2), (3), (5), and (4, 6). To obtain the most merged flow table, we would rather merge the pairs (1, 5), (2, 3), and (4, 6). In this latter case, in comparison, the merged flow table will have three rows. For 11.20b primitive flow table, likewise, states (1, 2, 6) and (3, 4, 5) may be merged. Note that the states (2, 3, 5, 6) may not be merged since the pairs (2, 5) and (3, 6) are not completely connected. In each of these cases, the don't cares were used in determining equivalent states. Figures 11.21a and b respectively show the resulting merged flow tables.

(a)

x_1, x_2			
00	01	11	10
"1"	5	-	2
1	-	3	"2"
-	6	"3"	2
-	6	"4"	2
1	"5"	4	2
1	"6"	4	-

(b)

x_1, x_2			
00	01	11	10
"1"	5	6	2
1	-	-	"2"
-	5	"3"	2
"4"	5	3	-
-	5	"6"	2
4	"5"	-	-

FIGURE 11.19
Primitive flow tables for Example 11.4.

(a)

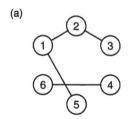

(b)

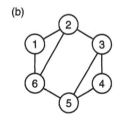

FIGURE 11.20
Merger diagrams for Example 11.4.

(a)

x_1, x_2			
00	01	11	10
"1"	"5"	4	2
1	6	"3"	"2"
1	"6"	"4"	2

(b)

x_1, x_2			
00	01	11	10
"1"	5	"6"	"2"
"4"	"5"	"3"	2

FIGURE 11.21
Merged flow tables for Example 11.4.

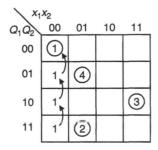
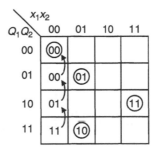

FIGURE 11.22
An example of a cycle.

11.4 Cycles, Races, and Hazards

In our earlier discussions of synchronous and pulse-mode circuits, the primary objective of choosing a state assignment is to make sure that design process lends to Boolean logic simplification. In the fundamental-mode circuits, however, the primary objective in choosing a particular state assignment is to prevent critical hazards; logic simplification is only of secondary importance. In all these other sequential circuits, there was provision for enough time for the excitations to settle down and before the next pulse arrived. In comparison, the fundamental-mode circuit that is devoid of a clock or even a pulse, we expect to encounter unwanted as well as unpredictable hazards. Thus, it is of primary importance that the state assignment results in a sequential circuit free of critical hazards. Consequently, it is important that we are familiar with the concepts of cycles and races before we try to eliminate circuit hazards.

Usually race is the problem that can plague fundamental-mode circuits and cycle is a solution to overcome the problem. Consider the primitive flow table of Figure 11.22 and that circuit state is configured currently at one of these flow table positions. By looking to the left and seeing values the FFs have at that moment and then looking at the code of position, we can determine to which row the circuit is transitioning next. If the code of current position is the code of the row it already is in, the circuit is already in a stable state and will not be transitioning to any other state at this time. If the circuit is in any one of states "2," "3," or "4" and the input variables got changed to $x_1, x_2 = 00$, then the circuit is about to transition to go to stable state "1" in row 1. Notice the arrows shown in the first column. The occurrence of two or more consecutive unstable states indicated so by such arrows is referred to as a *cycle*. What is demonstrated by the arrows here is the circuit has been forced to a cycle. If the circuit, for example, starts out from stable state "2" and the input x_2 changes from 1 to 0, the circuit is then forced to transition to the correct column ($x_1 x_2 = 00$) and examine the entry therein to see where to go next. In this particular situation, the circuit is required to first transition to row 11, then to 01, and finally to 00 before reaching the stable state "1". Similarly, from stable state "3" there could be a cycle of two successive transitions when x_1 changes from 1 to 0.

It should be noted that we have the choice to enter 00 at the intersection of $x_1 x_2 = 00$ and $Q_1, Q_2 = 10$. In this case the circuit would transition directly from the stable state "4" to the stable state "1" once x_1 changes 1 to 0. Time is the only difference between the two cases discussed. In the first case it would take three times as long to get to stable state "1." One may use cycles such as that used in the former to introduce delays in circuits if and when necessary.

A *race* is said to exist in a sequential circuit when two or more state variables must change when a fundamental-mode circuit may result in unexpected and erratic outputs. There are two types of races: critical and noncritical. Figure 11.23 shows an example a noncritical race. Here, the fundamental-mode circuit that initially is in stable state "2" can transition to stable state "1" when x_2 changes from a 1 to a 0. This state transition requires that Q_1 and Q_2 are both turned off. As pointed out earlier, it is virtually impossible for two events to occur simultaneously. Even if two FFs were literally identical, one would always turn off more quickly than the other. If Q_1 would turn off first, the circuit will transition first to row 01 and then to row 00 as Q_2 catches up with Q_1. If Q_2, on the other hand, were to turn off first, then the circuit would first transition to row 10 and then to row 00 when Q_1 catches up. The direction the circuit takes to transition to stable state "1" depends on which FF is faster. Such FFs are said to have a race condition. The particular race shown in Figure 11.23, however, is noncritical since the circuit ends up in stable state "1" irrespective of race.

Now consider the flow tables shown in Figure 11.24. This is illustrative of an example of a critical race. If the circuit were to start, for example, from stable state "3" and input x_2 were

		x_1, x_2			
		00	01	11	10
$Q_1 Q_2$	00	"1"			
	01	1		"3"	
	11	1	"2"		
	10	1			"4"

		x_1, x_2			
		00	01	11	10
$Q_1 Q_2$	00	"00"			
	01	00		"01"	
	11	00	"11"		
	10	00			"10"

FIGURE 11.23
Example of a noncritical race.

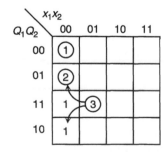
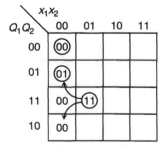

FIGURE 11.24
Example of a critical race.

to change from 1 to 0 when $x_1 = 0$, then the circuit would reach stable state "2" if Q_1 is faster and in stable state "1" if otherwise. Once the circuit transitions to stable state "2," it must be noted that this circuit gets stuck there since the next state and present state are both same. It is quite possible that we may have overlooked a race and are simply fortunate in the choice of FF used in the prototype circuit. During the testing phase the circuit might have actually worked. When several such circuits are built, however, some would have the race won by one FF, and in the other circuits another FF would win the race. Such fundamental-mode circuit conditions therefore must be avoided at all cost. By designing carefully, one would rather force the circuit to cycle through noncritical paths and thereby eliminate the race problem altogether.

Figure 11.25, for example, shows the same situation as in Figure 11.24 except that it has incorporated a cycle to eliminate critical race. As soon as x_2 would change from a 1 to a 0, the circuit state would transition from row 11 to row 10 and then to row 00. Note that the transition $11 \rightarrow 10 \rightarrow 00$ will not involve any race since no more than one flip-flop changes values in any of these two transitions $11 \rightarrow 10$ and $10 \rightarrow 00$. Whenever we have more than one FF set to change states, potential does exist to have a race when circuit is transitioning from one state to another. Race problems are usually eliminated by making state assignments in such a way as to limit transitional movements only between adjacent rows. This typically is the desired goal in fundamental-mode circuit design; however, it may not always be possible. In more complex situations, cycles are introduced to obtain race-free circuits.

We shall now explore situations that are encountered when assigning states to those included in a merged flow table. See Figure 11.26a. A letter symbol (A through D) is assigned to each row for identifying the corresponding stable state(s). Next a list of all required row-to-row transitions is prepared. For example, if the circuit were to be in stable state "2" when Y changed from a 1 to a 0, the circuit must then transition to stable state "8."

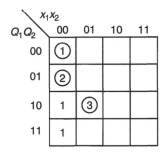

FIGURE 11.25
Elimination of race condition that exists in Figure 11.24.

(a)

	XY			
	00	01	11	10
A	"1"	5	"2"	8
B	"3"	7	"4"	6
C	3	"5"	2	"6"
D	1	"7"	4	"8"

(b)

	Q_1	
Q_2	0	1
0	A	D
1	C	B

FIGURE 11.26
(a) Merged flow table and (b) its state assignments.

	XY			
	00	01	11	10
A	"1"	9	8	2
B	1	9	3	"2"
C	1	9	"3"	4
D	1	9	5	"4"
E	1	6	"5"	10
F	1	"6"	7	10
G	1	9	"7"	10
H	1	"9"	"8"	"10"

FIGURE 11.27
Merged flow table for Example 11.2.

This would require that the circuit transitions from row A to row D. All other row-to-row transitions are determined similarly. For this merged table the required transitions include

$$A \to D; \quad B \to D; \quad C \to B; \quad D \to A$$
$$A \to C; \quad B \to C; \quad C \to A; \quad D \to B$$

Using a two-variable assignment map, A, B, C, and D are assigned $Q_1 Q_2$ codes in such a way that the transitions are limited to only neighboring rows. There are four different ways to accomplish this; only one of these assignments is shown in Figure 11.26b.

We may now return to our digital lock problem (Example 11.2) and make state assignments. The eight-row merged flow table obtained therein requires that at least three FFs be used for this circuit. The merged flow table is shown in Figure 11.27 and each of its rows is now given a letter code to identify.

To determine the most appropriate state assignments, a list of all required transitions is prepared. These include

$$B \to A; \quad G \to A; \quad B \to H; \quad E \to F$$
$$C \to A; \quad H \to A; \quad C \to D; \quad E \to H$$
$$D \to A; \quad A \to H; \quad C \to H; \quad F \to G$$
$$E \to A; \quad A \to B; \quad D \to E; \quad F \to H$$
$$F \to A; \quad B \to C; \quad D \to H; \quad G \to H$$

There are seven transitions to state A that are deemed noncritical. All the states end up reaching the same stable state "1" some time after the input XY becomes 00 since no other stable state is included in column $XY = 00$. Few of the remaining 13 transitions will need to be scrutinized since there is more than one stable state in each of the remaining columns. Every single row but the last is representative of a fraction of correct sequence that gets the digital lock to open. This may entail the transition $A \to B \to C \to D \to E \to F \to G$. It is easy to make state assignments so that the circuit state transitions only between adjacent rows. If this could be accomplished the circuit will not be subjected to any race condition. The eight states A through H are accordingly assigned binary codes as shown in Figure 11.28. The state assignments are made to guarantee that A is a neighbor of B, B is a neighbor of C, C is a neighbor of D, and D is a neighbor of E, E is a neighbor of F, and F is a neighbor of G. The unoccupied cell is assigned to H. Note that this is not the only possible state assignment that can suffice.

	Q_1Q_2			
Q_3	00	01	11	10
0	A	B	C	D
1	H	G	F	E

FIGURE 11.28
State assignment of eight states.

Figure 11.28 shows a possible assignment scheme that satisfies the adjacency require-ment for nine of thirteen state transitions. That leaves four other transitions yet to be addressed, namely,

$$B \to H; \quad C \to H; \quad D \to H; \quad F \to H$$

To guarantee race-free operation, these four transitions can be made to cycle through one or several other states. In case this is not possible, it will be necessary then and only then to add another FF. The unused states that become now available because of this additional flip-flop are used as state paths through which to cycle and avoid reaching an unwanted state because of race conditions. In the example of digital lock if we were to include an additional FF we would have twice as many cells than that present in the assignment map of Figure 11.28. As we shall see that for this current design example, we may not need more than three FFs.

The possible cycle paths that can make these four transitions to be race free are many, but we shall list only those requiring the least possible propagation delay. These are determined using Figure 11.28 assignment map:

Transition	Cycle
$B \to H$	$B \to A \to H$
	$B \to G \to H$
$C \to H$	$C \to D \to E \to H$
	$C \to F \to E \to H$
	$C \to F \to G \to H$
	$C \to B \to A \to H$
	$C \to D \to A \to H$
$D \to H$	$D \to A \to H$
	$D \to E \to H$
$F \to H$	$F \to G \to H$
	$F \to E \to H$

Before the cycle choices can be finalized, these cycles will need to be checked against the merged flow table in terms of their various consequences. We shall notice that two of the transitions, $C \to F \to E \to H$ and $C \to F \to G \to H$, are simply undesirable. These two cycles will get the circuit to end up at stable state "6" and thus must not be considered. The cycle described by $C \to B \to A \to H$ would in effect solve simultaneously two of the races, $B \to H$ and $C \to H$. There are two reasonable choices for column $AB = 01$; either $D \to C \to B \to A \to H$, or $C \to B \to A \to H$, and $D \to A \to\to H$. These two cycles are illustrated in Figure 11.29a where its rows have been appropriately rearranged to conform to a reflected binary code (thus guaranteeing each row to be a neighbor of the next). The

$Q_1Q_2Q_3$	XY				
	00	01	10	11	
000	①	9	8	2	a
001	1	⑨	⑧	⑩	h
011	1	9	⑦	10	g
010	1	9	3	②	b
110	1	9	③	4	c
111	1	⑥	7	10	f
101	1	6	⑤	10	e
100	1	9	5	④	d

$Q_1Q_2Q_3$	XY			
	00	01	10	11
000	000	001	001	010
001	000	001	001	001
011	000	001	011	001
010	000	000	110	010
110	000	010	110	100
111	000	111	011	011
101	000	111	101	001
100	000	110	101	100

FIGURE 11.29
Primitive flow table with proper state assignments and cycles.

cycle given by $D \to C \to B \to A \to H$ is capable of solving three of the races: $D \to H$, $C \to H$, and $B \to H$. Thus if this cycle were to be chosen, then we will have only $F \to H$ race to deal with. The two choices $F \to G \to H$ and $F \to E \to H$ for solving the race $F \to H$ are also identified in the primitive flow table.

Primitive flow table that has state assignment and cycles incorporated is obtained as shown in Figure 11.29b. It includes the $D \to C \to B \to A \to H$ and $F \to G \to H$ cycles. Note that in case of a noncritical race the transitional state is given the code of the steady state toward which the circuit is headed. If cycles are used for timing purposes, the transitional state is given the code of the next row in the cycle.

Next, appropriate FF control conditions are entered to obtain an excitation map from Figure 11.29b. Accordingly, inputs for the SR latches can be identified from the excitation map. Figure 11.30 shows the excitation maps for only S_1 and R_1 inputs.

The S_1 and R_1 Boolean equations are obtained readily. They are

$$S_1 = XYQ_2\overline{Q_3}$$
$$R_1 = \overline{X}\,\overline{Y} + \bar{X}Q_2\overline{Q_3} + \bar{Y}Q_3 + XQ_2Q_3$$

The Boolean equations for the remaining FF inputs are left as an exercise (see Problem 8 at the end of this chapter).

The output of the circuit becomes a 1 when the circuit has reached stable state "7". Referring to Figure 11.29, it can be seen that in state "7," $XY = 11$ and $Q_1Q_2Q_3 = 011$ and, therefore, $z = XY\overline{Q_1}Q_2Q_3$. The circuit for only Q_1 and $\overline{Q_1}$ as well as output z is shown in Figure 11.31.

Obtaining a race-free fundamental-mode circuit from a merged flow table involves the following steps:

Step 1. The state assignments are made so that as many transitions as possible are realized between adjacent rows.

$XY = 00$		Q_1Q_2			
		00	01	11	10
Q_3	0	0-	0-	01	01
	1	0-	0-	01	01

$XY = 01$		Q_1Q_2			
		00	01	11	10
Q_3	0	0-	0-	01	-0
	1	0-	0-	-0	-0

$XY = 10$		Q_1Q_2			
		00	01	11	10
Q_3	0	0-	0-	-0	-0
	1	0-	0-	01	01

$XY = 11$		Q_1Q_2			
		00	01	11	10
Q_3	0	0-	10	-0	-0
	1	0-	0-	01	-0

FIGURE 11.30
S_1 and R_1 excitation maps for the circuit described by Figure 11.29b.

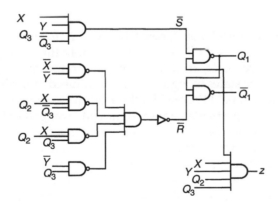

FIGURE 11.31
Q_1 excitation logic circuit for Example 11.2.

Step 2. The remaining transitions are satisfied by means of cycles if sufficient cycle paths are available. Otherwise, additional states are introduced, if needed, by including additional FFs. Cycles are employed then to satisfy all outstanding state transitions.

It is thus obvious that making state assignments is critical in the design of fundamental-mode circuits.

	x_1x_2		
00	01	11	10
"1"	2	4	"5"
3	"2"	"7"	"8"
"3"	"6"	"4"	5

FIGURE 11.32
Merged flow table for Example 11.5.

	x_1x_2		
00	01	11	10
"1"	2	4	"5"
3	"2"	"7"	"8"
3			
"3"	"6"	"4"	5

x_1x_2			
00	01	11	10
00	01	11	10
00	01	10	00
11	01	01	01
10	--	--	--
10	10	10	00

FIGURE 11.33
State assignments for Example 11.5.

Example 11.5
Show a race-free state assignment to accommodate the merged flow table of Figure 11.32.

Solution
With merged flow table rows labeled, respectively, as A, B, and C, there may be up to four transitions: $A \to B$; $A \to C$; $B \to C$; and $C \to A$. In this example no matter how the state assignments are made, at least one of these transitions contributes to a race. Since there are only three states in Figure 11.32, the unused "fourth" state can be used without adding an additional FF. Consider the arbitrary assignments: $A = 00$, $B = 01$, $C = 10$, and $D = 11$. This particular state assignment arrangement takes care of three transitions. The remaining transition, $B \to C$, requires that both FFs change their states. However, this apparent race condition can be overcome by means of the cycle $B \to D \to C$, which transitions the circuit through the unused state, D. Consequently, all of the race conditions are avoided by these state assignments and inclusion of a cycle through D. The flow table with this state assignment is obtained as shown in Figure 11.33.

Example 11.6
How will the merged flow table of Figure 11.34 be realized using as few additional FFs as possible?

Solution
The stable states in rows 1 through 4 are denoted, respectively, as A through D. The list of necessary state transitions includes

$$A \to B \quad B \to C \quad C \to B \quad D \to A$$
$$A \to C \quad B \to D \quad D \to B \quad D \to C$$

A transition diagram such as that in Figure 11.35 is obtained to see relationships between the merged states. Herein, B occurs in five state transitions. One way to approach this problem will be to introduce cycles between A and C, C and D, and between A and D. This particular cycle choices do not involve state B. The proposed choice thus will necessitate the introduction of three cycle states, A', C', and D', as shown in Figure 11.36. The need to have three cycle states requires that we introduce an additional FF as a delay element. Figure 11.37 shows the resulting primitive flow table that incorporates the new cycle states.

Figure 11.38 shows an adjacency map to determine how best to satisfy the various adjacencies, (AB), (BC), (BD), (DA'), $(A'C)$, (AD'), $(D'C)$, (AC'), and $(C'D)$. Note that this is only one example of the many ways the adjacency map can be structured. The map is completed by trial and error. Transition table may be obtained next using the state assignments shown in Figure 11.38. Figure 11.39 shows the table that includes now all possible states.

The last two examples demonstrate how we might avoid race condition either by using unused states or by introducing an additional FF. The critical race transitions were eliminated by forcing the fundamental-mode circuits to transition through one or more cycle states. There is one other way to avoid race. This last technique involves replicating states, whereby each of the states is given several binary codes so that it is a neighbor of all other states. This replication method is an otherwise expensive proposition since it requires additional FFs; however, it is much simpler than the two previous methods. In this technique, the number of FFs is increased so that the number of available states would be enough to

$x_1x_2=00$	$x_1x_2=01$	$x_1x_2=11$	$x_1x_2=10$
"1"	"2"	3	4
7	"6"	"3"	5
"7"	6	"8"	"4"
7	2	3	"5"

FIGURE 11.34
Merged flow table for Example 11.6.

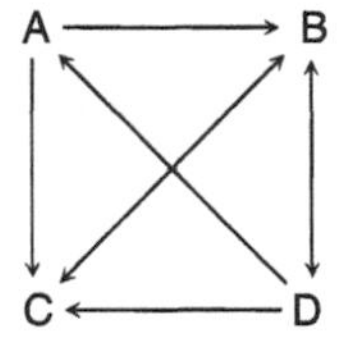

FIGURE 11.35
Transition diagram for Example 11.6.

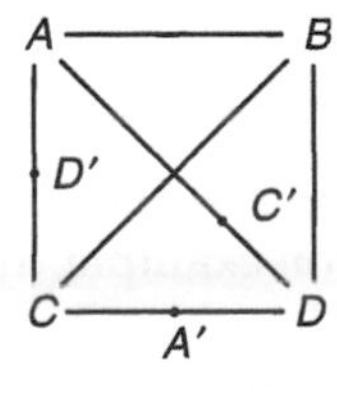

FIGURE 11.36
Transition diagram with cycle states for Example 11.6.

$x_1x_2=00$	$x_1x_2=01$	$x_1x_2=11$	$x_1x_2=10$	
"A"	"A"	B	D'	A
C	-	-	-	A'
C	"B"	"B"	D	B
"C"	B	"C"	"C"	C
-	A	-	-	C'
A'	C'	B	"D"	D
-	-	-	C	D'

FIGURE 11.37
Primitive flow table for Example 11.6.

	$Q_1Q_2=00$	$Q_1Q_2=01$	$Q_1Q_2=11$	$Q_1Q_2=10$
$Q_3=0$	B	C	D'	A
$Q_3=1$	D	A'	-	C'

FIGURE 11.38
Adjacency map for Example 11.6.

		x_1x_2 00	x_1x_2 01	x_1x_2 11	x_1x_2 10
	$Q_1Q_2Q_3$	00	01	11	10
B	000	010	000	000	001
D	001	011	101	000	001
A'	011	010	- - -	- - -	- - -
C	010	010	000	010	010
D'	110	- - -	- - -	- - -	010
-	111	- - -	- - -	- - -	- - -
C'	101	- - -	100	- - -	- - -
A	100	100	100	000	110

FIGURE 11.39
Transition table for Example 11.6.

replicate each of the original states. The number of replications and the way these states are assigned binary values are critical. The assignments should be such that each of the states (at least for one of its allowable values) becomes a neighbor of every other state.

Example 11.7

Use replication method to complete state assignment of the merged flow table given in Figure 11.34.

Solution

There can be several ways to implement replication of states. One of the assignment maps is shown in Figure 11.40 where each of the states is a neighbor of all others. This state assignment requires using three FFs. We notice that state A is a neighbor of B since $Q_1Q_2Q_3 = 000$ and $Q_1Q_2Q_3 = 010$ are adjacent; it is also a neighbor of C since $Q_1Q_2Q_3 = 001$ and $Q_1Q_2Q_3 = 011$ are adjacent; and, finally, it is also a neighbor of D since $Q_1Q_2Q_3 = 000$ and $Q_1Q_2Q_3 = 100$ are adjacent. The same arguments follow for the remaining states.

State assignments shown in Figure 11.40 can be used to derive the excitation table shown in Figure 11.41. Consider, for example, transition $A \rightarrow B$. Note State A is represented by both $Q_1Q_2Q_3 = 000$ and $Q_1Q_2Q_3 = 001$. Consequently, when the circuit is transitioning from $Q_1Q_2Q_3 = 000$ to state B, the circuit would transition through the transitional state $Q_1Q_2Q_3 = 010$. However, when transitioning from

	$Q_1Q_2=00$	$Q_1Q_2=01$	$Q_1Q_2=11$	$Q_1Q_2=10$
$Q_3=0$	A	B	B	D
$Q_3=1$	A	C	C	D

FIGURE 11.40
An assignment map for Example 11.7.

		x_1x_2	x_1x_2	x_1x_2	x_1x_2
	$Q_1Q_2Q_3$	00	01	11	10
A	000	000	000	010	001
A	001	001	001	000	011
B	010	011	010	010	110
B	110	111	110	110	100
C	011	011	010	011	011
C	111	111	110	111	111
D	100	101	000	110	100
D	101	111	001	100	101

FIGURE 11.41
Transition table for Example 11.7.

$Q_1Q_2Q_3 = 001$ to state B, the best cycle path for the circuit to follow would be $001 \rightarrow 000 \rightarrow 010$. The fundamental-mode circuit moves first internally from 001 to 000 during which time the circuit continues to be in state A and only thereafter moves on to 010. To accomplish both state transitions, 010 and 000 are entered under column 11, next to rows 000 and 001, respectively. Consider next one more state transition: $A \rightarrow C$. If the circuit were to begin from 001, it can go through transitional state 011. And if it were to start instead from 000, the fundamental-mode circuit will need to go through the cycle $000 \rightarrow 001 \rightarrow 011$. The values 001 and 011 are, thus, entered under column 10, next to rows 000 and 001, respectively. This same approach was used in case of other states to derive therefrom the excitation table shown in Figure 11.41.

The replication technique can be used to accommodate arbitrary number of states. With number of states increasing, the number of FFs that will be necessary to cover all transitions increases. The adjacency maps shown in Figure 11.42 can be used for handling respectively six and eight states.

Use of nonreplicating state assignments will result in a simpler circuit. In the non-replicating cases, there exists the possibility of encountering enough don't cares in the corresponding transition table. Such possibilities allow for simpler excitation equations.

11.5 Fundamental-Mode Outputs

Earlier discussions on fundamental-mode circuits were concerned with deriving reliable and efficient excitation equations. This far, we avoided discussing output equations. The output maps for the fundamental-mode circuits are obtained as follows: The output value for each of the merged stable states is the same as that shown in the primitive flow table. The output corresponding to the transitional states is determined by that of both source and destination stable states. If the output associated with source state is same as that associated

(a)

Q_3Q_4	$Q_1Q_2=00$	$Q_1Q_2=01$	$Q_1Q_2=11$	$Q_1Q_2=10$
00	A	A	C	D
01	B	B	C	D
11	E	F	E	E
10	E	F	F	F

(b)

$Q_3Q_4Q_5$	$Q_1Q_2=00$	$Q_1Q_2=01$	$Q_1Q_2=11$	$Q_1Q_2=10$
000	A	A	G	H
001	B	B	G	H
011	B	A	H	H
010	B	A	G	G
110	C	C	D	C
111	D	D	D	C
101	E	F	F	F
100	E	F	E	E

FIGURE 11.42
Adjacency maps using replicated states suitable for (a) six states and (b) eight states.

with destination states, then the transitional state is assigned that same output. On the other hand, if the source and destination states are associated with different output values, then output is assigned a don't-care value.

If the output values associated with both source and destination stable states are either 0 or 1, it is proper to assume transitional states to have the same output so as to avoid a momentary transitional output. If more than one stable state passes through a transitional state, then the output value must be chosen to satisfy each of the transitions to eliminate the possibility of a *glitch* (momentary 1) or *drop-out* (momentary 0). The next example illustrates these ideas.

Example 11.8
Derive the output equations for the merged flow table and the respective stable state outputs shown in Figure 11.43.

Solution
Figure 11.44 shows the resulting output map obtained as follows: The output of the transitional states are determined and then entered in the output map. Consider the transitional state at the intersection of column 01 and row 00. This is encountered when moving from either "1" or "2" to "3". The z_1z_2 values at stable states "1," "2," and "3" are, respectively, 00, 11, and 01. The value of z_1 remains 0 and unchanged in case of one transition and change from 1 to 0 in the other transition. In the first case, z_1 for the transitional state should be 0, and that for the second case should be - -. As a compromise, therefore, z_1 needs to be a 0. For nearly same kind of reasons, z_2 should be a 1. Consequently, 01 is entered at the corresponding location of the output map. Figure 11.44 output map accounts for all possible outputs. The output equations follow readily from Figure 11.44.

$$z_1 = x_1\overline{Q_2} + x_1\overline{x_2} + \overline{x_2}Q_1$$
$$z_2 = x_1Q_1 + x_2\overline{Q_1}$$

Q_1,Q_2	x_1,x_2	x_1,x_2	x_1,x_2	x_1,x_2
	00	01	11	10
00	"1"	3	"2"	4
01	1	"3"	5	"4"
11	7	8	"5"	6
10	"7"	"8"	2	"6"

Merged flow table

Q_1,Q_2	x_1,x_2	x_1,x_2	x_1,x_2	x_1,x_2
	00	01	11	10
00	00		11	
01		01		10
11			01	
10	10	00		11

Stable state output z_1,z_2

FIGURE 11.43
Merged flow and output tables for Example 11.8.

Q_1,Q_2	x_1x_2	x_1x_2	x_1x_2	x_1x_2
	00	01	11	10
00	00	01	11	10
01	00	01	01	10
11	--	0-	01	-1
10	10	00	11	11

FIGURE 11.44
Transition table for Example 11.8.

11.6 Summary

The pulse-mode circuit model and its design were introduced first. This was followed then by a discussion of the fundamental-mode sequential circuits and related concepts of races, cycles, and hazards. This latter form of asynchronous circuits, it was shown to be able to remove most of the constraints of pulse-mode circuits. Fundamental-mode circuits is the most general and most powerful of all sequential circuits. A designer would probably use clocked and pulsed sequential design almost exclusively, but in difficult timing situations the fundamental-mode sequential circuit would prove to be the most valuable.

Bibliography

Comer, D.J., *Digital Logic and State Machine Design*. 3rd edn. New York, NY. Oxford University Press, 1994.

Floyd, T., *Digital Fundamentals*. 8th edn. Englewood Cliffs, NJ. Prentice Hall, 2003.

Hill, F.J. and Peterson, G.R., *Digital Systems: Hardware Organization and Design*. 2nd edn. New York, NY. Wiley, 1978.

Johnson, E.J. and Karim, M.A., *Digital Design: A Pragmatic Approach*. Boston, MA. PWS-Kent Publishing, 1987.

Karim, M.A. and Awwal, A.A.S., *Optical Computing: An Introduction.* New York, NY. John Wiley & Sons, 1992.

Katz, R.H., *Contemporary Logic Design.* Boston, MA. Addison Wesley, 1993.

Kline, R.M., *Structured Digital Design Including MSI/LSI Components and Microprocessors.* Englewood Cliffs, NJ. Prentice-Hall, 1983.

Mowle, F.J.A., *A Systematic Approach to Digital Logic Design.* Reading, MA. Addison-Wesley, 1976.

Nagle, H.T., Jr., Carroll, B.D., and Irwin, J.D., *An Introduction to Computer Logic*, Englewood Cliffs, NJ. Prentice-Hall, 1975.

Nelson, V.P.P., Nagle, H.T., and Carroll, B.D., *Digital Logic Circuit Analysis and Design.* Englewood Cliffs, NJ. Prentice Hall. 1995.

Rhyne, V.T., *Fundamentals of Digital Systems Design.* Englewood Cliffs, NJ. Prentice-Hall, 1973.

Winkel, D. and Prosser, F., *The Art of Digital Design—An Introduction to Top-Down Design.* Englewood Cliffs, NJ. Prentice-Hall, 1980.

Problems

1. Design pulse-mode circuits using *SR* FFs that responds to two inputs, x_1 and x_2. An output is generated coincident with the last of a sequence of four input impulses if and only if the sequence contained: (a) at least two x_1 pulses, (b) at least three x_1 pulses, (c) at most two x_1 pulses, and (d) at most three x_1 pulses.

2. Design a two-input pulse-mode circuit using *JK* FFs whose output pulse, z, occurs coincident with the last pulse of a sequence of three input pulses $x_1x_2x_1$, where x_1 and x_2 are the inputs.

3. Repeat problem 2 using *T* FFs.

4. Repeat problem 2 using *D* FFs.

5. Design a pulse-mode circuit with inputs A, B, and C and output z. The output changes from 0 to 1 if and only if the sequence ABC occurs while $z = 0$. With the next B pulse, the output gets reset.

6. Design a two-input pulse-mode circuit using *JK* FFs that recognizes five successive x_1 pulses and generates an output pulse at the end of the fifth pulse. An x_2 pulse resets and initializes the counter.

7. Design 2-input pulse-mode circuit that has two pulse inputs, A and B, and one pulse output, z. The output is coincident with (a) the second of two consecutive B pulses immediately following exactly three consecutive A pulses and (b) the second consecutive B pulse immediately following three or more A pulses.

8. Derive Boolean equations for S_2, S_3, R_2, and R_3 for the problem of Figure 11.29b.

9. Make race-free state assignments for the merged table of Figure 11.P1. Determine the Boolean equations for fundamental-mode logic circuit.

State	X	
Assignments		
(Q_1Q_2)	0	1	z
	"1"	2	0
00	"3"	4	1
01	3	"2"	0
	1	"4"	1

FIGURE 11.P1

10. Obtain the merger diagram and merged flow table from the primitive flow table of Figure 11.P2.

	AB				z
	00	01	11	10	
"1"	6	5	2	0	
1	4	3	"2"	0	
8	4	"3"	7	1	
8	"4"	3	7	1	
1	6	"5"	2	0	
1	"6"	5	2	0	
8	6	5	"7"	1	
"8"	4	5	7	1	

FIGURE 11.P2

11. Determine SR and output equations for a race-free fundamental-mode circuit given by Figure 11.P3 where A and B are inputs and z is the output.

Merged flow table

State assignments $(Q_1 Q_2)$	AB						AB			
	00	01	11	10			00	01	11	10
00	"1"	"7"	8	2			1	0		
	"3"	"5"	6	4			0	1		
01	3	7	"8"	"2"					1	1
	3	7	"6"	"4"					0	1

FIGURE 11.P3

12. Determine Boolean equation for output z equation for the fundamental-mode circuit described by Figure 11.P4 where x_1 and x_2 are the inputs.

Flow table   Merged flow table

	x_1, x_2				z
	00	01	11	10	
"1"	2	-	3	0	
4	"2"	5	-	1	
1	-	6	"3"	0	
"4"	7	-	8	1	
-	2	"5"	3	1	
-	7	"6"	8	0	
1	"7"	6	-	0	
4	-	5	"8"	1	

$Q_1 Q_2$	x_1, x_2			
	00	01	11	10
00	"1"	2	"5"	3
01	1	"7"	6	"3"
11	"4"	7	"6"	8
10	4	"2"	5	"8"

FIGURE 11.P4

13. Design a fundamental-mode circuit where output, z, is a 0 when two inputs, x_1 and x_2, are equal. The output is high when $x_1 = 0$ and x_2 changes from 0 to 1, and when $x_1 = 1$ and x_2 changes from 1 to 0. No other input change may cause any change in the output.

14. Consider the fundamental-mode circuit that has two inputs, A and B, and one output, z, and satisfies the following conditions: (i). $z = 0$ when $B = 1$, (ii). z changes to 1 when $B = 0$ and A changes from 0 to 1, and (iii) z remains at 1 until B goes to 1 and forces z back to 0. Implement this asynchronous circuit using (a) two-level NAND logic and (b) two-level NOR logic.

15. For the fundamental-mode circuits shown in Figure 11.P5, how do you solve the cycle problems using as few additional states as possible?

(a)

\multicolumn{4}{c}{x_1, x_2}
00	01	11	10
"1"	3	8	"2"
6	"3"	"4"	"5"
"6"	"7"	"8"	2

(b)

\multicolumn{4}{c}{x_1, x_2}
00	01	11	10
"1"	2	8	6
5	"2"	"3"	"4"
"5"	7	3	"6"
1	"7"	"8"	4

(c)

\multicolumn{4}{c}{x_1, x_2}
00	01	11	10
"1"	2	6	10
1	"2"	"3"	"4"
1	"5"	"6"	"7"
1	"8"	"9"	"10"

FIGURE 11.P5

16. Repeat Problem 15 using replication of states.

17. For the merged table of Figure 11.P6, obtain the fundamental-mode circuit using (a) few additional states and (b) replication of states.

\multicolumn{4}{c}{x_1, x_2}
00	01	11	10
"1",0	2	3	"4",1
"5",0	6	0	7
5	"2",1	"8",1	7
1	"6",0	3	"7",0

FIGURE 11.P6

12

Introduction to Testability

12.1 Introduction

Like many other products, electronic integrated circuits are often being tested for quality assurance. These tests can be performed at different levels during manufacturing and system integration. With current semiconductor technologies, tests can be conducted at the wafer, chip, package, board and system levels, using suitable test technologies. To ensure that electronic integrated circuits and systems are testable, it is very important to make sure that circuits are testable during the design processes—hence, the terms of testability and design-for-test (DFT).

Testability is a relative measure of the effort (or cost) of testing integrated circuits. It reflects the estimated effort to *control* and *observe* internal signals via primary inputs and outputs. Integrated circuits with good testability often result in reduced test cost and improved quality. Testability analysis is often performed at different stages during design processes, making sure that defined testability objectives are satisfied.

DFT is a design practice that ensures the testability in circuit and system designs, by inserting circuit structures facilitating access to internal signals during tests. To reduce design cycles, DFT must be considered and incorporated from the very beginning of design processes while circuits and systems are being architecturally specified. DFT practiced as late-stage add-on features would lengthen design cycle times and may incur additional design iterations.

12.2 Controllability and Observability

Controllability is defined as the relative effort required to set internal signals to logic 1 (high) or logic 0 (low) via primary inputs and with respect to the effort of setting a primary input signal. Controllability is calculated for every internal signal and is characterized as 1- and 0-controllability to reflect the relative effort to set the internal signal to 1 or 0, respectively. Consider a 3-input AND gate. In order to set its output signal to 1, all its three inputs must be set to 1. Therefore, the 1-controllability of its output signal is the sum of the 1-controllability of all its inputs. On the other hand, since its output can be set to 0 by setting any of its three inputs to 0, the 0-controllability for its output is then the minimum 0-controllability of its three inputs. The two controllability numbers for other types of gates can be calculated similarly. Table 12.1 summarizes the propagation rules of 1- and 0-controllability calculations with elementary logic gates.

Calculation of 1- and 0-controllability with a circuit starts from primary inputs, by first setting all primary inputs' 1- and 0-controllability to be 1 (hence setting the reference), and

TABLE 12.1

Controllability Propagation Rules

Logic Function	Propagation/Calculation of Controllability at Output	
	1-Controllability	0-Controllability
AND	Sum of input 1-controllability	Minimum of input 0-controllability
OR	Minimum of input 1-controllability	Sum of input 0-controllability
INVERTER	Input 0-controllability	Input 1-controllability
NAND	Minimum of input 0-controllability	Sum of input 1-controllability
NOR	Sum of input 0-controllability	Minimum of input 1-controllability
BUFFER	Input 1-controllability	Input 0-controllability
XOR	$C1(a) \times C0(b) + C0(a) \times C1(b)$	$C1(a) \times C1(b) + C0(a) \times C0(b)$
XNOR	$C1(a) \times C1(b) + C0(a) \times C0(b)$	$C1(a) \times C0(b) + C0(a) \times C1(b)$

TABLE 12.2

Observability Back Propagation Rules

Logic Function	Backward Propagation/Calculation of Observability at an Input
AND/NAND	Sum of the observability of the output and the 1-controllability of other inputs
OR/NOR	Sum of the observability of the output and the 0-controllability of other inputs
INVERTER/BUFFER	Same as the observability of the output
XOR/XNOR	1-controllability plus 0-controllability of the other input

then propagating the controllability toward outputs. To improve computation efficiency, controllability is calculated by the classic breath-first search method based on circuit topology. Note the special notation used for description of controllability for XOR and XNOR devices in Table 12.1. Here, $C1(a)$, for example, refers to 1-controllability of the first input (input a). Similarly, $C0(b)$, for example, refers to 0-controllability for the second input to the device (input b).

Observability is defined as the relative effort required to observe internal signals via primary outputs with respect to the effort of observing a primary output signal. Observability is calculated for every internal signal. Consider a 3-input AND gate. The observability of one of its three inputs is the sum of the observability of its output and the 1-controllability of its other two inputs. Observability for other types of gates can be calculated similarly. Table 12.2 summarizes the back propagation rules of observability calculation with elementary logic gates.

Calculation of observability in a circuit starts from primary outputs, by first setting all primary outputs' observability to be 1 (hence setting the reference), and then propagating the observability backwards to primary inputs. To improve computation efficiency, observability is calculated by the breath-first search method. Figure 12.1 shows a circuit of a full-adder, with the observability and controllability calculations listed next to each signal. Also note that, owing to computation dependency, observability is calculated after controllability calculations are performed. When calculating the values of controllability and observability in Figure 12.1, recall the need to first set a reference (i.e., set the 1- and 0-controllability for all primary inputs to 1, while setting the observability for all primary outputs to 1 as well). Thus, notice that the 1- and 0-controllabilities for the first level AND gate are calculated to 2 and 1, respectively. Similarly, the 1- and 0-controllabilities for the first XOR gate (at left) are set to 2. Also notice that the observability for outputs of the last level of devices (the OR and second XOR gates) has been set to 1. Once the references have

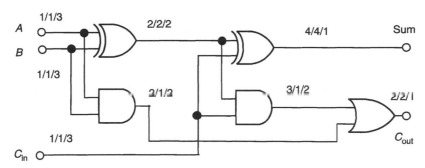

Notation:1-Controllability/0-Controllability/Observability

FIGURE 12.1
Controllability and observability measures in a full adder.

been set for controllability and observability, propagation of the remaining values can be forward and backward propagated, respectively.

12.3 Deterministic Testability versus Random Testability

Controllability and observability are calculated relative to primary input signals. Therefore, on the basis of whether inputs are to be generated randomly or deterministically, testability may have different characteristics. Random pattern testability is defined as the relative effort to achieve controllability and observability objectives with randomly generated inputs, while deterministic pattern testability is defined as the relative measures with respect to deterministically generated inputs. The controllability and observability calculations discussed in Section 12.2 are of deterministic testability measurements.

Random pattern testability, on the other hand, is based on the random occurrences of 1s and 0s at inputs that satisfy controllability and observability criteria. To illustrate random pattern testability, first consider a 3-input AND gate, that has a total of eight different inputs. Assuming that each input signal is generated randomly and independently from each other, the 1- and 0-controllability for each input is 0.5 and 0.5, respectively. The 1-controllability for the output of the AND gate is 0.125 ($0.5 \times 0.5 \times 0.5$) and 0-controllability is 0.875 ($1 - 0.125$). Table 12.3 summarizes the random controllability propagation rules with elementary logic gates.

Random pattern observability is calculated similarly as is done for calculating deterministic observability, except that internal signals' 1- and 0-controllability are now calculated on the basis of the propagation rules outlined in Table 12.3 for random pattern testability. Similar to deterministic pattern testability analysis, random controllability is calculated before observability calculations. Figure 12.2 illustrates the difference between deterministic and random testability calculations.

12.4 Test of Integrated Circuits

Test of integrated circuit (IC) is based on the fact that manufactured ICs may not function properly, owing to imperfections in the design implementation process (such as the use of wrong devices or the making of incorrect connections, etc.) as well as during

TABLE 12.3

Random Controllability Propagation Rules

Logic Function	Propagation/Calculation of Controllability at Output	
	1-Controllability	0-Controllability
AND	Sum of input 1-controllability	1 minus output 1-controllability
OR	1 minus output 0-controllability	Sum of input 0-controllability
INVERTER	Input 0-controllability	Input 1-controllability
NAND	1 minus output 0-controllability	Sum of input 1-controllability
NOR	Sum of input 0-controllability	1 minus output 1-controllability
BUFFER	Input 1-controllability	Input 0-controllability
XOR	$C1(a) \times C0(b) + C0(a) \times C1(b)$	1 minus 1-controllability
XNOR	$C0(a) \times C0(b) + C1(a) \times C1(b)$	1 minus 1-controllability

Notation: 1-/0-Controllability/Observability

(a)

```
1/1/3
1/1/3  =)D— 3/1/1
1/1/3
```

Deterministic measures

(b)

```
0.5/0.5/0.25
0.5/0.5/0.25  =)D— 0.125/0.875/1
0.5/0.5/0.25
```

Random measures

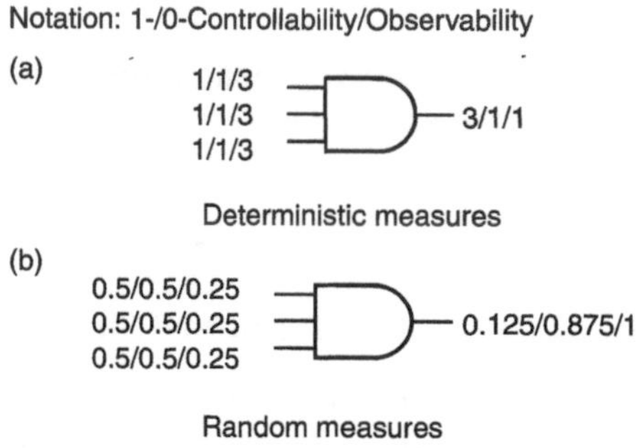

FIGURE 12.2
Comparison of deterministic and random pattern testability measures.

the manufacturing process (such as material and process imperfections, etc.). These imperfections may cause internal circuits to produce erroneous outputs. With these considerations, the objective of testing is to capture any error that may be caused by imperfections in design implementation and manufacturing processes, while minimizing cost of testing. For example, certain types of imperfections in wiring (or interconnect) may cause terminals of MOSFET devices (commonly used on ICs) used to implement logic gates, to short-circuit with voltage supply lines (Vdd - power supply voltage or Ground), resulting in logic gates being unable to function properly, including being unable to respond to changes at one or more of inputs. Design implementation errors may cause circuits to produce unexpected values at internal signals as well as primary outputs.

Test methods for ICs can be characterized as structural and/or functional, depending on how test inputs are generated. Structural tests are inputs generated on the basis of abstract structure failure models and circuit topologies to capture erroneous circuit responses due to manufacturing imperfections. Functional tests, on the other hand, are inputs generated on the basis of functional operations of the circuits. For example, an all-1 input can propagate grounded input failures of an AND gate to its output, while an input sequence of a specific counting sequence may demonstrate a counter's deviation from the normal functions owing to errors in design and/or implementation.

12.5 Fault Models

Fault models are used to mimic the erroneous signal and circuit behavior due to design and implementation errors and manufacturing imperfections. On the basis of applications

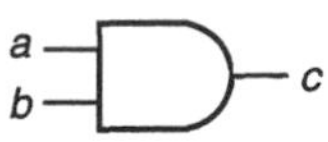

Input (*ab*)   Faults to be detected

Input (*ab*)	Faults to be detected
00	sa_1(c)
01	sa_1(c); sa_1(a)
10	sa_1(c); sa_1(b)
11	sa_0(c); sa_0(a); sa_0(b)

FIGURE 12.3
Tests of stuck-at faults with AND gate.

and levels of abstractions, many fault models have been developed with respect to ICs. The most common fault model is the well-known and widely used stuck-at fault model. With the presence of a stuck-at fault, a signal node or branch is said to be *permanently* stuck-at either logic 1 or 0. For example, the presence of a stuck-at-0 fault at the output of a 3-input NAND (due to internal implementation errors and/or physical defects) means that the output is fixed at logic 0 and will not change with any inputs.

Stuck-at faults are abstractions of internal failures of gates or circuit blocks. A signal node or fan-out branch may have stuck-at-0 stuck-at-1 faults, but not both at the same time. However, in a circuit it is often the case that multiple stuck-at faults at different sites may exist, which complicates the computation to identify most effective tests—using a minimum number of tests to detect most or all of the potential faults. Fortunately, empirical studies have shown that tests generated on the basis of the single stuck-at fault assumption are often very effective in testing multiple stuck-at faults. Therefore, many of the state-of-the-art test generation methods still use the single stuck-at fault model—test inputs are generated on the basis of the assumption that only a single stuck-at fault exists at a given time.

Consider a 2-input AND gate. There are a total of six stuck-at faults at the three terminals. Assuming that a single stuck-at fault may present at a time, three out of four possible inputs are needed to test all six stuck-at faults. Figure 12.3 illustrates the tests and faults that each test input would detect.

Bridging faults are abstraction of failures between two neighboring wires being short-circuited because of manufacturing imperfections. The results of such failures are that the otherwise independent signals become virtually "being tied together."

Design implementation errors and manufacturing imperfections may also result in higher-than-expected total circuit current. Therefore, current consumption of circuits can also be used to test ICs. Current test with CMOS technologies is often referred to as I_{DDq} test, which usually is performed during IC manufacturing with a handful of carefully selected test inputs. ICs with higher-than-expected I_{DDq} currents are often suspects of containing internal failures, which are often traced back as implementation errors and/or manufacturing imperfections. I_{DDq} faults are abstractions of internal failures that would cause larger-than-expected currents. The effectiveness of I_{DDq} tests are measured by the number of I_{DDq} faults that can be detected over the total number of I_{DDq} faults—a fraction often referred to as I_{DDq} fault coverage.

12.6   Test Sets and Test Generation

A test is an input pattern applied during testing of ICs and a test set contains multiple tests. A test set is either complete or incomplete. A test set is said to be complete if all testable faults can be detected by the test set. A test set is said to be incomplete if it only detects a subset of all testable faults.

With large circuits, tests are identified by a process known as automatic test pattern generation (ATPG) by running computer software applications. These ATPG applications take a circuit's netlist with assumed fault models as inputs, compute test patterns following predesigned algorithms, followed by running fault simulations with generated tests to identify all modeled and detectable faults. The effectiveness of test sets are measured by the number of faults that each set detects, which is often referred to as fault coverage. A test set may detect faults of multiple fault models.

12.7   Topology-Based Testability Analysis

Circuit topology defines the circuit structures connecting individual gates and blocks, often shown as circuit schematic diagrams. Circuit testability is often determined by the ways that individual gates and blocks connect to each other. On the basis of interconnections, circuit structures appear in three configurations: tree structures, fan-out structures, and reconvergent fan-out structures (Figure 12.4). Tree and fan-out structures have excellent controllability and observability, hence, excellent testability. However, reconvergent fan-out structures often cause testability concerns.

Reconvergent fan-out structures can be categorized into three classes: (A) constant functions, (B) conditional constant functions (CCFs), and (C) nonconstant functions. Class A structures are also known as redundancies in that the outputs of these functions are fixed at constant values regardless the input values to be applied. Class A structures are well known to cause testability issues because of the lack of controllability. An example of constant function circuit structures is shown in Figure 12.5, where signal b is fixed at 0 regardless the input values on signal a. A change on a produces a narrow pulse on b. The width of this pulse is determined by the delay of the inverter.

Class-B structures are nonredundant reconvergent fan-out structures that turn into constant functions under certain partially specified input conditions. Consider the simple circuit shown in Figure 12.6. Signal b fans out and reconverges at output d. Furthermore, the circuit is evidently nonredundant, since signal d is not fixed to any (1 or 0) constant value. However, further analysis indicates that when the input condition $a = 0$ and $c = 1$ is present, d is then fixed at 0 regardless of signal values on b. Here, signal b is the source while signal d is the sink of the single-stem reconvergent fan-out structure.

The probability of signal d being a constant function is determined by the ratio of the number of input conditions satisfying $a = 0$ and $c = 1$ over the number of all possible

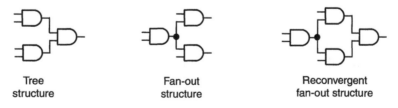

| Tree structure | Fan-out structure | Reconvergent fan-out structure |

FIGURE 12.4
Gate-level circuit configurations.

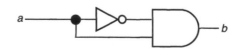

FIGURE 12.5
A simple structure of constant function.

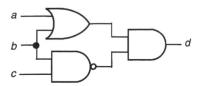

FIGURE 12.6
Example of single-stem CCF circuit structure.

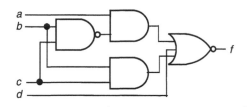

FIGURE 12.7
Example of multistem CCF circuit structures.

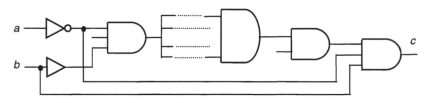

FIGURE 12.8
Blocked signal propagation by a Class-C structure.

inputs (the input space) defined by a, b, and c. In this case, the ratio is 2/8. Therefore, we name such reconvergent fan-out structures as CCFs, along with its fraction, to reflect this characteristic. Figure 12.7 shows a multistem reconvergent fan-out circuit structure, with which when $a = 1$ and $d = 0$ is present, signal f is fixed at 0 regardless of signal values on b and c.

A reconvergent fan-out circuit structure can be viewed as a fan-out cone defined by its inputs and one output, with at least one input to have fan-outs. In the example circuit shown in Figure 12.6, the fan-out cone is defined as *fanout_cone(d)* = {a,b,c}, where b is the input with reconvergent fan-outs, while inputs a and c are conditional inputs. Note that signals a, b, c, and d can be internal signals in larger circuits. Similarly, the multistem reconvergent fan-out circuit shown in Figure 12.7 can be described by *fanout_cone(f)* = {a,b,c,d}, where b and c are inputs with fan-outs, while a and d are conditional inputs.

Class-C reconvergent fan-out structures, although nonredundant, are known to often prevent signal propagations toward outputs. Current analysis methods with Class-C structures are fault-model-based and, therefore, often complex, inefficient and nonunified procedures. Consider the Class-C structure shown in Figure 12.8. A single stuck-at-1 fault at either the upper (or the lower input) of the first AND gate near the input side is prevented from being propagated to the output c. This is because the required $a = 1$ (or $b = 0$) would set output $c = 0$, effectively blocking signal propagations via other inputs of the AND gate at the output.

In practice, reconvergent fan-out circuit structures are much more complex than the simple examples shown in Figures 12.6 through 12.8. Figure 12.9 illustrates the general

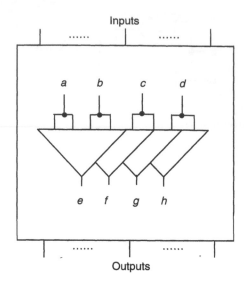

FIGURE 12.9
Illustration of reconvergent fan-outs.

form of reconvergent fan-out circuit structures, where internal signals *a*, *b*, *c*, and *d* fan out to downstream logic blocks and reconverge at signals *e*, *f*, *g*, and *h*.

12.8   Simulation-Based Testability Analysis

Instead of calculating testability measures as described in the early sections, statistical sampling techniques have also been used to collect testability characteristics. This is accomplished by simulating a number of random inputs (say 1 million with large circuits) and at the same time monitoring internal signal responses. If an internal signal does not exhibit transitions from 1 to 0 and 0 to 1 during the sampling period, it would be an indication that the signal is most likely associated with poor controllability.

During the sampling period, each internal signal maintains four counters to record the numbers of the occurrences of 1s, 0s, 1-to-0, and 0-to-1 transitions in response to the simulation. After the sampling simulations, the recorded numbers will be used to statistically profiling the testability measures. This sampling technique is also used to estimate power consumption.

12.9   Fault Analysis and Fault-Based Testability Analysis

Fault analysis is often conducted through processes known as fault simulation and fault diagnosis. Fault simulation is used to determine the effectiveness of test sets, as well as testability, during design and implementation processes, while fault diagnosis is applied to ICs that failed in various manufacturing tests to narrow down and isolate failure location(s).

Fault simulation is commonly used during ATPG for developing compact tests, as well as evaluating the effectiveness of tests. ATPG applications typically generate one test at a time, based on given criteria (such as a specific fault—a.k.a. target fault), and then use fault simulation to identify other faults that are detectable by this test. This process repeats until all modeled faults are identified by tests as detectable or undetectable.

Fault diagnosis is often a prerequisite for failure analysis whose purpose is to determine the physical defects causing the failure. Fault diagnosis is a process that analyzes failure data collected by IC testers (a.k.a. automatic test equipment). This process also generates additional tests, known as distinguishing tests, to work with the failure data and narrow down the potential failure nets in the circuit. Failure analysis will then take the potential failure nets into consideration.

Faults are classified into two classes: (i) testable and (ii) untestable. A fault is said to be testable if there exists at least one test to detect the fault. A fault is said to be untestable if it is proven that a test detecting this fault does not exist. Faults are detection equivalent if they can be detected by the exact same test(s), while these faults are also diagnostic equivalent (or indistinguishable) if the output response(s) are identical.

12.10   Testability Metrics

The effectiveness of testability analysis is determined by two measures: the first is the test coverage or fault coverage for engineering practicality; the second is the number of tests in the test set required to achieve the test coverage. Fault coverage is defined as the percentage of the number of detectable faults by a test set(s) over the total number of faults. The number of tests is also an important figure because it is directly related to test application time and, therefore, test cost.

Higher test coverage and fewer tests are often two competing objectives that IC engineers struggle with. This is because higher test coverage usually would require more tests and fewer tests usually would lower test coverage. More often than not, IC engineers have to make trade-offs between the two competing criteria. The relationship between test coverage and the number of tests achieving the test coverage follows the Pareto Law that the last 20% of test coverage would take the last 80% of the tests.

For engineering purposes, testability is often associated with fault models. This is because a test set may detect different types of faults with different coverage. During manufacturing tests, test coverage numbers are often produced at different phases of testing, such as wafer test, burn-in test, parametric test, chip test, package test and system test, and so forth.

12.11   Design-for-Testability

Design-for-testability is a design practice that ensures all modeled faults in IC designs can be tested with the lowest cost possible. It is often integrated in IC design processes such that testability is maintained at all design phases and stages, by inserting test structures—circuits that facilitate controllability and observability during many test modes. DFT practices include partial- and full-scan design, clocked scan, level-sensitive scan design (LSSD), and so forth. These scan design methods use various shift-register configurations to access internal signals in a scan-mode. During the scan-mode, external data can be shifted-in via a scan-in input, and data captured in the shift register can be shifted out via a scan-out output. The Institute of Electrical and Electronics Engineers (IEEE) 1149 boundary scan standards specify a collection of methods that facilitate core and chip level DFT.

Designs of microprocessors and application-specific ICs (ASICs) are usually very complex, often involving many millions of logic gates. With these complex designs, scan-based

DFT methods are often insufficient. Additional test structures are identified by performing testability analysis followed by test point insertion (TPI) procedure(s).

12.12 Summary

In this chapter, we briefly discussed testability measures and analysis, fault models and simulation, test generation, and fault analysis. These concepts are important and integral parts of IC design and manufacturing engineering.

Although many of the techniques used in IC design and manufacturing engineering are computerized and automated, the concepts introduced in this chapter not only shed some light on how these computer-aided applications work but also help in developing interests in electronic design automation (EDA), which has been a growing and important industry. EDA vendors offer a whole range of applications that would automatically perform scan insertion and configuration, testability analysis as well as TPI and ATPG, among many other applications.

On the other hand, skilled DFT engineers are among the most sought after in professional recruitment. This is due to the fact that the complex DFT methodologies, design and manufacturing processes require engineers to manage and often intervene in manufacturing processes. Computer-aided applications improve productivity, but cannot replace skilled engineers in solving unforeseen and ever-changing problems, which is often the case in IC design and manufacturing.

Bibliography

Abramovici, M., Breuer, M.A., and Friedman, A.D., *Digital Systems Testing and Testable Design*. Piscataway, NJ. IEEE Press, New York, revised printing, 1990.

Agrawal, V.D. and Seth, S.C., *Tutorial—Test Generation for VLSI Chips*. Washington, DC. IEEE Computer Society Press, New York, 1988.

Ed. Wang, L.-T., Wu, C.-W., and Wen, X.-Q., *VLSI Test Principles and Architectures*, Burlington, MA. Elsevier, Morgan Kaufmann Publishers, 2006.

Bushnell, M.L. and D. Agrawal, V.D., *Essentials of Electronic Testing for Digital, Memory & Mixed-Signal VLSI Circuits*, Boston, MA. Springer, 2000.

Problems

1. Calculate the deterministic controllability and observability with a 3-input XOR gate and its NAND–NOR implementation, with the assumptions that controllability at the primary inputs are 1, as well as the observability at the primary output.

2. Calculate the random testability with a 3-input XNOR gate and its NAND–NOR implementation, with the assumptions that the random controllability at the primary inputs are 0.5, while the observability at the primary output is 1.

3. Assume that 4-bit serial-load shift register contains an input stuck-at-0 fault at its second and third bits owing to manufacturing imperfection. Design an input test

sequence to demonstrate these failures at the outputs of this 4-bit serial-load shift register.

4. Consider a gate-level implementation of a full adder. Design an economical test plan to test this full adder circuits.

5. Analyze the equivalency of stuck-at faults in an AND–OR circuit implementing XOR function, and identify the minimum number of tests detecting all testable faults.

6. Analyze the equivalency of stuck-at faults in a full adder and identify the minimum number of tests detecting all testable faults.

7. Refer to Tables 12.1 and 12.2, develop general forms of controllability and observability calculations for n-input XOR and XNOR gates.

8. Similar to Table 12.2, develop a table that shows the random observability calculation rules for the basic logic functions.

9. Calculate the random testability measures with the full adder circuit shown in Figure 12.1.

10. Consider the full adder circuit shown in Figure 12.1. Generate five random inputs by flipping a coin, and then calculate random controllability and observability on the basis of the sampling with these five random inputs.

11. Consider the small circuit shown in Figure 12.4. Derive a test input to detect a stuck-at-1 fault at the output d.

12. Consider the small circuit shown in Figure 12.5. Derive a test input to detect a stuck-at-0 fault on the signal fed by input a.

13. A digital circuit consists of N number of AND, NAND, OR, NAND and INVERTER gates. Write a C/C++ procedure to calculate (deterministic) 1- and 0-controllability from primary inputs to primary outputs, as well as observability from primary outputs to primary inputs. Note that primary inputs and outputs can be considered as circuit blocks whose logic function identifications are PI and PO.

14. Similar to Problem 13, write a C/C++ procedure to calculate random controllability and observability, assuming that the random 1- and 0-controllability for primary inputs are 0.5 and 0.5, respectively, while observability for primary outputs is 1.

15. Consider the full adder circuit shown in Figure 12.1. When a test input $A = 1, B = 1, C_{in} = 1$ is applied, output Sum $= 1, C_{out} = 0$ responses are observed, indicating the circuit failed the test. Similar to the example shown in Figure 12.3, list all single stuck-at faults whose existence might cause the circuit to fail the test.

16. On the basis of Problem 15, a second input $A = 1, B = 0, C_{in} = 1$ is applied. This time the circuit responded with Sum $= 1, C_{out} = 0$. Combined with the analysis results from Problem 15, list all single stuck-at faults whose existence might cause the full adder to fail in both tests.

17. Write a paragraph with less than 50 words to summarize the testability issues caused by an internal constant-function signal. Propose a circuit structure as a test point to be inserted into a circuit to remedy the described testability issues.

Index